Social and Ecological Interactions in the Galapagos Islands

The Galapagos Islands are a "living laboratory" for the study of evolution, environmental change, and conflicts between nature and society. Free of human predators for almost all its history, these islands have developed some of the most unique life forms on the planet, adapted to their harsh surroundings and living in ecological isolation. It was not until Charles Darwin's famous visit in 1835, which helped inspire the theory of evolution that this Archipelago began to receive international recognition. The Galapagos Archipelago encompasses 11 large and 200 small islands totaling approximately 8,010 sq. km. This series will focus on the entire island archipelago, and it will emphasize the study and documentation of human-environment interactions on the four inhabited islands in the Galapagos: Isabela, Santa Cruz, San Cristobal, and Floreana. Together they constitute a well-defined "natural laboratory" for the study of human-environment interactions as they vary in fundamentally important ways.

Stephen J. Walsh • Carlos F. Mena
Jill R. Stewart • Juan Pablo Muñoz Pérez
Editors

Island Ecosystems

Challenges to Sustainability

Springer

Editors
Stephen J. Walsh
Department of Geography
University of North Carolina at Chapel Hill
Chapel Hill, NC, USA

Jill R. Stewart
Department of Environmental Sciences and Engineering
University of North Carolina at Chapel Hill
Chapel Hill, NC, USA

Carlos F. Mena
College of Biological & Environmental Studies
Universidad San Francisco de Quito
Quito, Ecuador

Juan Pablo Muñoz Pérez
College of Biological and Environmental Sciences and Galapagos Science Center
Universidad San Francisco de Quito
Quito, Ecuador

ISSN 2195-1055 ISSN 2195-1063 (electronic)
Social and Ecological Interactions in the Galapagos Islands
ISBN 978-3-031-28088-7 ISBN 978-3-031-28089-4 (eBook)
https://doi.org/10.1007/978-3-031-28089-4

Photo Credit: Community of Puerto Ayora on Santa Cruz Island, Galapagos Archipelago of Ecuador. Photo taken by Dr. Stephen J. Walsh in July, 2022.

This Springer imprint is published by the registered company Springer Nature Switzerland AG
The registered company address is: Gewerbestrasse 11, 6330 Cham, Switzerland

Foreword: World Summit on Island Sustainability

It was 4 pm on an overcast day, and I found myself sitting on the soft, cool sand of Playa Mann surrounded by dozens of snoozing Galapagos Sea Lions. I was entranced watching the dominate male patrol his shore, swimming back and forth in front of the beach. He was big and loud and taking his job very seriously. Then, two young women who had been laying on the beach, walked to the shoreline to dip their toes into the refreshing water, but he was NOT having it. He quickly swam over, waddled out of the ocean like a lightning bolt (which was incredible because of his girth) and let out a sound I can only compare to an old rusty tuba being played by a strong wind gust! Needless-to-say, the two ladies headed back to their beach towels very quickly, but the other sea lions, including females, young males, and many nursing babies on the beach, weren't even phased. Then off to my right, I caught a glimpse of a marine iguana slowly making its way out to sea to feed (oh my gosh!). Its salt crusted head swaying right to left with every step. It was then I saw the rocks in the distance were not speckled with bright colors, but it was a huge collection of Sally Lightfoot Crabs looking for their next snack. And this was within the first 30-minutes of putting my bags down after arriving on San Cristobal Island, Galapagos Archipelago of Ecuador.

The Galapagos Islands are one of the most spectacular places on our planet. Anyone who loves nature dreams of visiting them and I was grateful to be experiencing this incredible, historic, and special place myself. I was reminded that while islands make up a small percentage of the Earth's total area, they are home to a large percentage of the world's biodiversity including many threatened and endangered species. And while there is still a lot of work to be done, Galapagos has become a beacon of light for science, discovery, and successful conservation.

I am part of a family that has been forging conservation, especially in our ocean, for three generations and counting. My grandfather-in-law, Jacques Cousteau pioneered underwater exploration. You see, he fell in love with the sea and wanted to send more time under water, so he co-invented the Aqua-Lung, a device we refer to today as SCUBA. Then he wanted to share this fabulous water world with everyone he could, so he made underwater cameras, outfitted an old wooden hulled minesweeper, and set off on adventures around with world with his family in tow.

Cousteau became an international household name with his eponymous TV series, *The Under Sea World of Jacques Cousteau.* While they set out to show the beauty and wonder of the ocean, they also endeavored to share the truth about the degradation of the environment that they witnessed on their adventures. Indeed, Jacques and his son Philippe Sr. are considered two of the founders of the modern environmental movement.

When Jacques was born in 1910, there were 1.6 billion people. Today, its estimated that we have over 7.5 billion people on earth and we will most likely reach 9 or 10 billion by the middle of this century. Today, our world is facing an onslaught of severe problems: excess carbon, biodiversity loss, a warming and more acidic ocean, food shortages, sea-level rise, floods, droughts, fires… the list goes on and on. And many island nations are at the front lines of these battles. Just in my lifetime, we have lost HALF of all the biodiversity on our planet… imagine that, in just 40 years, half of the wondrous diversity of nature has disappeared. With this constant torrent of bad news, it is very easy to get upset, anxious, and even downright depressed. But there is HOPE.

Nature has an incredible ability to restore itself. And while I have witnessed this in many places on my adventures, one very unlikely place stands out. It all started in the middle of the Pacific Ocean during the Cold War when the United States detonated 23 nuclear bombs on a small chain of islands around Bikini Atoll. The largest of these bombs was a hydrogen bomb called Castel Bravo, detonated on March 1, 1954, with 1000 times more power than the bombs dropped on Hiroshima or Nagasaki. The heat from the blast registered almost as hot as the surface of our sun. So hot in fact, the sand at the blast site on Bikini Atoll turned into glass. And every living thing for miles around, above and below the surface, died instantly.

So, when I went to film the sharks of Bikini Atoll… I had no idea what to expect. But it turned out to be one of the most spectacular ecosystems I have ever seen. Slipping into the water, I was immediately surrounded by 70 grey reef sharks, massive groupers the size of German shepherds, and giant clams the size of coffee tables. The entire ecosystem wasn't just surviving, it was thriving. In just 60 years, Bikini went from nuclear waste to a pristine environment. This was because the land of the islands is still radio-active, thus no one goes there, and the area has turned into a de facto Marine Protected Area. When talking with a scientist there, they said if we compare the effects of nuclear fallout and the effects of human presence on our natural world… humans are worse. Ouch.

But in the same way humans have the power to destroy nature, we also have the power to rebuild it. We know how to do it, we have the tools, we just need the will. Because when we do, nature benefits, but so do people and as my grandfather-in-law once said, "to build environmental sustainability, we must build human sustainability."

About an hour and a half north of Cabo San Lucas at the bottom of the Baja Peninsula is a town that embodies just this. Cabo Pulmo is a small community of dusty dirt roads and small humble homes. Not unlike dozens of similar communities you might find up and down the peninsula, and yet Cabo Pulmo, as unlikely as it may seem, is a shining beacon for ocean conservation. Thirty years ago, the patriarch of this small community, a man named Juan Castro was worried. The fishing that had sustained his community for generations was declining. Every day they had to venture further and further offshore to find fish with no guarantee of success. Like any good father, Juan worried for his ability to feed his family and the future of his children.

Then, one hot summer day, a group of tourists drove up from Cabo offering him money to take them out on his boat. Much to his surprise they did not want to fish, instead they had come to scuba dive. When they returned to the surface, one of the divers tossed his mask to Juan and invited him to look underwater. This was a pivotal moment and changed Juan forever because despite growing up next to the ocean, he had never looked beneath the surface.

The beauty he witnessed gave him an idea, and after years of work, he convinced the local community to join him in a crusade to create a Marine Protected Area in their local waters. Fast forward 20 years and the 70 square kilometer no-take marine reserve in Cabo Pulmo is a paradise unlike anything in the Sea of Cortez. With upwards of a 1000% increase in living creatures in the area, even for someone like me – who has been diving all over the world, it is magical. And when you talk to Juan and his children and grandchildren, you hear a common refrain that they are proud of what they have achieved. Not because of the recovery of the reef (though that matters too), they are most proud because in the words of one of his sons Mario… they have something that they can pass on to their children that will only increase in value.

It's these stories of success that give me hope. As you will learn in the chapters to follow, scientists, researchers, and local people have made fascinating discoveries and found viable solutions for many of the problems we face. Humans have a steadfast drive to survive and when we come together and work WITH nature, incredible things can happen.

While nature has islands, knowledge does not. And knowledge is best shared.

Ashlan Cousteau
Journalist, Adventurer & Ocean Advocate
EarthEcho International,
Washington, DC, USA

Preface: Galapagos Science Center and Island Sustainability

Overview Statement

On June 26, 2022, a group of scientists from around the globe convened at the Charles Darwin Convention Center in the Galapagos capital of Puerto Baquerizo Moreno. They arrived on San Cristobal Island to participate in the *2022 Galapagos World Summit on Island Sustainability*, hosted for over four days by the USFQ-UNC Chapel Hill, Galapagos Science Center.

The event was significant if for nothing more than it represented the first full-scale global science conference hosted on San Cristobal since the archipelago closed to visitors at the onset of the COVID-19 pandemic in February 2019.

But the timing could not have been more appropriate. By June 2022, Galapagos and other island systems around the world were seeing tourism economies that had lain dormant during the pandemic begin to revive, reintroducing familiar stresses to fragile archipelagos.

The topic of the summit was island sustainability – a comprehensive examination of the threats to island ecosystems and strategies and solutions for their sustainable management and growth. Launched with an inspirational welcome from EarthEcho International's Ashlan Cousteau, the event brought together speakers with unique perspectives from locations such as French Polynesia, the Caribbean, the Hawaiian Islands, Guam, Australia, Chile, New Zealand, and the USA to discuss common challenges and to learn from one another.

While the Galapagos Islands are well-known for the endemic fauna and flora that inspired Darwin's *On the Origin of Species*, the focus of the conference was much broader than the preservation and study of unique species and habitats. The summit examined the topic of island sustainability comprehensively, exploring the unique role archipelagos play as geographically isolated systems to understand interactions between humans and the natural world and the larger forces facing the planet itself. With the global pandemic subsiding – and faced with the mounting threats of global climate change and human impacts on the planet – the discussion could not have been more-timely, nor held in a more appropriate place.

The work of the scientists who contributed to the 2022 Galapagos World Summit is contained in the pages of this publication, collected under appropriately broad sections that address the general foci of island sustainability. These range from the unique challenges to sustainability in island ecosystems to the critical role of social sub-systems in island communities and the attributes of terrestrial and marine sub-systems. It embraces the interdisciplinary nature of sustainability research and concludes with thoughts on the future of Earth's archipelagos.

The magnet that drew these scholars to this remote corner of the world was the Galapagos Science Center – the only university-owned research center in Galapagos. A partnership of the University of North Carolina at Chapel Hill and Universidad San Francisco de Quito, this modern three-story laboratory facility is located at the foot of the volcanic hill that rises behind Playa Mann on San Cristobal Island. Opened in 2011, the 20,000 square foot facility houses four specialty labs outfitted separately for the study of terrestrial ecology, marine ecology, microbiology and genetics, and GIS and data science. It maintains a full-time staff of 15 employees under the leadership of co-directors from USFQ and UNC-Chapel Hill.

The vision for the Galapagos Science Center was the product of the close relationship and collaborative work of Dr. Steve Walsh at UNC-Chapel Hill and Dr. Carlos Mena at USFQ – and relationships of all kinds have proven the key to the GSC's success.

USFQ and UNC-Chapel Hill combined resources to construct and launch the center, which has grown steadily since its founding into a highly successful platform for scientific work in Galapagos. Unlike one-off research engagements where scientists visit, collect their samples and data, and depart, the GSC has embedded itself in the community and established a close *working relationship with the Galapagos National Park and Marine Reserve.*

Together, the Park and the GSC have developed a streamlined permitting process for scientists working through the Galapagos Science Center. The GSC has sought to align its work with the Park's needs and priorities and, where mutually beneficial, dedicated its own resources to support the Park's work. A joint initiative led through UNC and USFQ genetics and microbiology faculty and developed in consultation with the Ecuadorian authorities has established a GSC biobank and launched a DNA sequencing and storage project in the islands.

The GSC has also deeply embedded itself in the community it serves, offering career and educational opportunities to *local citizens, hiring locally, and contributing actively to public life in the islands.* Each year, a symposium is held at which GSC scientists present their work to the Galapagos National Park, their peers, and members of the local community. When the Covid-19 pandemic reached Galapagos, the islands locked down and tourism that supports their economy suddenly came to a crashing halt. As a key employer in the San Cristobal Island community, the GSC kept its staff on the job and launched the GSC REACCT program, reallocating a share of its budget to fund community projects that both created local jobs and advanced resource conservation and island sustainability goals. Information about some of the GSC's citizen science and sustainability projects, such as the Galapagos Barcode Project, can be found in this volume.

The GSC's attention to local and Park relationships is complemented by the global research partnerships that have been a part of the center since its inception. The International Galapagos Science Consortium was created early on to offer enhanced GSC access and benefits to key global partners and to provide a platform for strategic engagement with faculty and students from around the globe.

The 2022 Galapagos World Summit on Island Sustainability leveraged all of these relationships – local, park, and global – that have made possible the success of the Galapagos Science Center. In the pages that follow, we offer the fruits of that convocation of scholars. We hope that, in the wake of the COVID-19 pandemic, their work may guide others toward a healthy balance between the human and natural world in our planet's precious archipelagos.

Donald Hobart

Office of the Vice Chancellor for Research
University of North Carolina at Chapel Hill
Chapel Hill, NC, USA

UNC-USGQ Galapagos Advisory Board
UNC-USFQ Galapagos Science Center
San Cristobal Island, Ecuador

Contents

Editors and Contributors

About the Editors

Stephen J. Walsh Co-Guest Editor, is the Founding Emeritus Director of the UNC Center for Galapagos Studies and the UNC-USFQ Galapagos Science Center. He is also an Emeritus Distinguished Professor of Geography and is a Research Professor in the UNC Department of Geography. His research focuses on understanding the multi-scale and multi-thematic drivers of land use/land cover change in vulnerable and fragile ecosystems through remote sensing, spatial analysis, and spatial simulation model of "what if" scenarios of change.

Carlos F. Mena Co-Guest Editor, Co-Director of the Galapagos Science Center, Director of the Institute of Geography, and Professor in the College of Biological and Environmental Sciences, Universidad San Francisco de Quito, Ecuador. His research assesses the importance of land use/land cover change on social and terrestrial systems in the Galapagos Islands as well as in the Ecuadorian Amazon.

Jill R. Stewart Co-Guest Editor, is the Philip C. Singer Distinguished Professor, Department of Environmental Sciences and Engineering, University of North Carolina at Chapel Hill, USA. She develops novel techniques to detect and track pathogens in water. She is particularly interested in linkages between ecosystems and human health and well-being.

Juan Pablo Muñoz Pérez Co-Guest Editor, has more than 20 years of experience in diving and surfing in the Galápagos and 10 years of conducting research. Juan Pablo's research focuses on animal movement and ecology, especially sea turtles and other marine animals such as whales and dolphins. Since 2014 he has been a pioneer in investigating plastic pollution in Galápagos with the primary objective of using science to find solutions. Currently, he is completing his PhD at the University of the Sunshine Coast.

Contributors

Nikhil K. Advani World Wildlife Fund, Washington, DC, USA

Christopher M. Aiken CQ University, Coastal Marine Ecosystems Research Centre, Gladstone, Rockhampton, QLD, Australia

Daniela Alarcón-Ruales School of Science, Technology & Engineering, University of the Sunshine Coast, Sunshine Coast, QLD, Australia

Juan Manuel Alava-Jurado Marine Biology and Aquaculture, James Cook University, Douglas, QLD, Australia

Javier A. Arce-Nazario Department of Geography, University of North Carolina at Chapel Hill, Chapel Hill, NC, USA

Francisco Arreguín-Sánchez Instituto Politécnico Nacional, Centro Interdisciplinario de Ciencias Marinas, La Paz, México

Sol Aspinosa Galapagos Conservation Trust, London, UK

M. Isidora Ávila-Thieme Instituto Milenio de Socio-Ecología Costera (SECOS); Departamento de Ecología, Facultad de Ciencias Biológicas, Pontificia Universidad Católica de Chile; Advanced Conservation Strategies, Midway, UT, USA

Amelia A. Bain Research Fellow, School of Geosciences & School of Architecture and Landscape Architecture, University of Edinburgh, Edinburgh, UK

Claire Barrington Department of Health Behavior, University of North Carolina at Chapel Hill, Chapel Hill, NC, USA

Julia G. Barzyk Army Research Office, Research Triangle Park, Triangle Park, NC, USA

Margaret E. Bentley Department of Nutrition, University of North Carolina at Chapel Hill, Chapel Hill, NC, USA

Richard E. Bilsborrow Carolina Population Center, University of North Carolina at Chapel Hill, Chapel Hill, NC, USA

Camille Bonneaud Center for Ecology and Conservation, University of Exeter, Exeter, UK

Margarita Brandt College of Biological & Environmental Sciences, Universidad San Francisco de Quito, Quito, Ecuador

Laura Brewington Arizona State University, Tempe, AZ, USA

East-West Center, Honolulu, HI, USA

Jeff Burgett US Fish and Wildlife Service, Washington, DC, USA

Tom Chaigneau Environment and Sustainability Institute, University of Exeter, Exeter, UK

Stephanie Crane De Narvaez Research Fellow, School of Architecture and Landscape Architecture, University of Edinburgh, Edinburgh, UK

Mimi Chapman School of Social Work & Global Social Development Innovations Center, University of North Carolina at Chapel Hill, Chapel Hill, NC, USA

Jaime A. Chaves Department of Biology, San Francisco State University, San Francisco, CA, USA
Department of Ornithology & Mammalogy, California Academy of Sciences, San Francisco, CA, USA

Gina Chowa School of Social Work & Global Social Development Innovations, University of North Carolina at Chapel Hill, Chapel Hill, NC, USA

María Emilia Menoscal Coello Universidad San Francisco de Quito, Pichincha, Ecuador

Ashlan Cousteau EarthEcho International, Washington, DC, USA

Marilyn Cruz Agency of Regulation and Control of the Biosecurity and Quarantine for Galapagos, Quito, Ecuador

Sebastián Cruz Tierramar Research Group, Galápagos, Quito, Ecuador

Neil Davies Gump South Pacific Research Station, University of California, Moorea, French Polynesia

Judith Denkinger College of Biological & Environmental Sciences, Universidad San Francisco de Quito, Quito, Ecuador

Diane Deresienski School of Aquatic, Wildlife, and Zoological Medicine, North Carolina State University, Raleigh, NC, USA

Santiago Díaz-Pazmiño Animal Behavior Lab, University of Sydney, Sydney, NSW, Australia

Eric Dorfman Marine Earth and Atmospheric Sciences, North Carolina State University, Raleigh, NC, USA

John Dumbacher Department of Ornithology and Mammalogy, California Academy of Sciences, San Francisco, CA, USA

Bradley Eichelberger CNMI Division of Fish and Wildlife, USA

Ana Eguiguren Department of Biology, Dalhousie University, Halifax, Canada

Eduardo Espinoza Dirección del Parque Nacional Galápagos, Quito, Ecuador

Elizabeth Frankenberg Department of Sociology & Public Policy, University of North Carolina at Chapel Hill, Chapel Hill, NC, USA

Tamara S. Galloway University of Exeter, Exeter, UK

Maria Soledad Garcia Ferrari Senior Lecturer & Director of the Center for Latin America Studies, University of Edinburgh, Edinburgh, UK

Stefan Gelcich Instituto Milenio de Socio-Ecología Costera (SECOS), Center for Applied Ecology and Sustainability (CAPES), Pontificia Universidad Católica de Chile and CQ University, Coastal Marine Ecosystems Research Centre, Gladstone, Callemondah, QLD, Australia

Alexandre Génin Departamento de Ecología, Facultad de Ciencias Biológicas, Pontificia Universidad Católica de Chile and Copernicus Institute of Sustainable Development, Utrecht University, Utrecht, The Netherlands

Matt Gibson Department of Biology, Indiana University, Bloomington, IN, USA

Madeline Giefer Department of Geosciences, Austin Peay State University, Clarksville, TN, USA

Ben L. Gilby School of Science, Technology and Engineering, University of the Sunshine Coast, Sunshine Coast, QLD, Australia

Jonathan Guillén Naturalist Guide, Galápagos, Quito, Ecuador

Heather D. Hanna Department of Earth, Marine and Environmental Sciences, University of North Carolina at Chapel Hill, Chapel Hill, NC, USA

Salome Herrera College of Biological & Environmental Sciences, Universidad San Francisco de Quito, Quito, Ecuador

Maximilian Hirschfeld James Cook University, Douglas, QLD, Australia

Donald Hobart Office of the Vice Chancellor for Research, University of North Carolina at Chapel Hill, Chapel Hill, NC, USA

UNC-USGQ Galapagos Advisory Board, UNC-USFQ Galapagos Science Center, San Cristobal Island, Ecuador

Bonnie J. Holmes School of Science, Technology & Engineering, University of the Sunshine Coast, Sippy Downs, QLD, Australia

Jess Howard Galapagos Conservation Trust, London, UK

Jacques Idechong Palau Community College, Ngerulmud, Republic of Palau

Jen S. Jones Galapagos Conservation Trust, London, UK

Heather Kerkering Pacific Islands Climate Adaptation Science Center, Pacific Island, USA

Michael J. Kingsford Department of Marine Biology and Aquaculture & Center of Excellence for Coral Reefs Studies, James Cook University, Douglas, QLD, Australia

Noah Kittner Department of Environmental Sciences and Engineering, University of North Carolina at Chapel Hill, Chapel Hill, NC, USA

Francisco J. Laso Environmental Studies, Western Washington University, Bellingham, WA, USA

Gregory A. Lewbart School of Aquatic, Wildlife, and Zoological Medicine, North Carolina State University, Raleigh, NC, USA

Xiao-Ming Liu Department of Earth, Marine, and Environmental Sciences, University of North Carolina at Chapel Hill, Chapel Hill, NC, USA

Kenneth J. Lohmann Department of Biology, University of North Carolina at Chapel Hill, Chapel Hill, NC, USA

Juan Pablo Luna School of Government, Pontificia Universidad Catolica de Chile, Santiago, Chile

Christy Martin Hawai'i Coordinating Group on Alien Pest Species, Honolulu, HI, USA

Carlos F. Mena College of Biological & Environmental Studies, Universidad San Francisco de Quito, Quito, Ecuador

Jason P. van de Merwe Coastal and Marine Research Centre, Griffith University, Gold Coast, QLD, Australia
School of Environment and Science, Griffith University, Gold Coast, QLD, Australia

Wendy Miles US Fish and Wildlife Service, Washington, DC, USA

Leonie Moyle Department of Biology, Indiana University, Bloomington, IN, USA

Juan Pablo Muñoz Pérez College of Biological and Environmental Sciences and Galapagos Science Center, Universidad San Francisco de Quito, Quito, Ecuador
School of Science, Technology & Engineering, University of the Sunshine Coast, Sippy Downs, Australia

Sergio A. Navarrete Department of Ecology & Millennium Nucleus for Ecology and Conservation of Temperate Mesophotic Reefs Ecosystems (NUTME), Pontificia Universidad Católica de Chile and Center COPAS-COASTAL, Universidad de Concepción, Concepción, Chile

Khristopher M. Nicholas Department of Nutrition, University of North Carolina at Chapel Hill, Chapel Hill, NC, USA

Lucía Norris Crespo Galapagos Conservation Trust, London, UK

Jaime Eduardo Ocampo Trujillo School of Public Health, Universidad San Francisco de Quito, Pichincha, Ecuador

Valeria Ochoa College of Biological & Environmental Studies and USFQ-UNC Galapagos Science Center, Universidad San Francisco de Quito, Quito, Ecuador

Diego A. Ortiz Universidad San Francisco de Quito USFQ, Galapagos Science Center, Quito, Ecuador

Diego Páez-Rosas College of Biological & Environmental Sciences, Universidad San Francisco de Quito, Quito, Ecuador

Elliott Parsons University of Hawai'i at Mānoa, Honolulu, HI, USA

Diana A. Pazmino College of Biological & Environmental Sciences, Universidad San Francisco de Quito, Quito, Ecuador

Carolina Proaño Tierramar Research Group, Galápagos & College of Biological and Environmental Sciences, Universidad San Francisco de Quito, Quito, Ecuador

Diego Quiroga Ferri College of Social Sciences and Humanities, Universidad San Francisco de Quito, Quito, Ecuador

Nicole Read Duke University, Durham, NC, USA

Bryan Reatini Department of Biology, University of North Carolina, Chapel Hill, NC, USA

Gunter Reck College of Biological and Environmental Sciences, Universidad San Francisco de Quito, Quito, Ecuador

Marjorie Riofrío-Lazo College of Biological & Environmental Sciences, Universidad San Francisco de Quito, Quito, Ecuador

Cynthia Fraga Rizzo School of Social Work & Global Social Development Innovations Center, University of North Carolina at Chapel Hill, Chapel Hill, NC, USA

Andy Russell Center for Ecology and Conservation, University of Exeter, Exeter, UK

Caitlin E. Smith School of Science, Technology & Engineering, University of the Sunshine Coast, Sunshine Coast, QLD, Australia

Jill R. Stewart Department of Environmental Sciences and Engineering, University of North Carolina at Chapel Hill, Chapel Hill, NC, USA

Cecep Sumantri SurveyMETER, Yogyakarta, Indonesia

Javan Sutton North Carolina Museum of Natural Sciences, Raleigh, NC, USA

Enrique Teran School of Medicine, Universidad San Francisco de Quito, Quito, Ecuador

Duncan Thomas Department of Economics & Global Health Institute, Duke University, Durham, NC, USA

Amanda L. Thompson Department of Anthropology & Center for Galapagos Studies, University of North Carolina at Chapel Hill, Chapel Hill, NC, USA

Stella de la Torre College of Biological & Environmental Studies and USFQ-UNC Galapagos Science Center, Universidad San Francisco de Quito, Quito, Ecuador

María de Lourdes Torres College of Biological & Environmental Studies and USFQ-UNC Galapagos Science Center, Universidad San Francisco de Quito, Quito, Ecuador

Kathy A. Townsend School of Science, Technology and Engineering, University of the Sunshine Coast, Sunshine Coast, QLD, Australia

Diego Urquía College of Biological & Environmental Studies and USFQ-UNC Galapagos Science Center, Universidad San Francisco de Quito, Quito, Ecuador

Daniel Valencia Department of Ecology, Pontificia Universidad Católica de Chile, Santiago, Chile

Carlos Valle College of Biological & Environmental Studies and USFQ-UNC Galapagos Science Center, Universidad San Francisco de Quito, Quito, Ecuador

Evelyn Vega Barrera Criminal Procedural Law, Universidad Complutense de Madrid, Madrid, Spain ESCOEX (Social and Legal Sciences), Universidad del Atlántico Medio, Gran Canaria, Spain

Alberto Velez Agency of Regulation and Control of the Biosecurity and Quarantine for Galapagos, Santa Cruz, Ecuador

Isabel Villarruel-Oviedo College of Biological & Environmental Studies and USFQ-UNC Galapagos Science Center, Universidad San Francisco de Quito, Quito, Ecuador

Todd Vision Galapagos Science Center, USFQ-UNC, San Cristóbal, Galápagos, Ecuador

Department of Biology, University of North Carolina, Chapel Hill, NC, USA

Stephen J. Walsh Department of Geography, University of North Carolina at Chapel Hill, Chapel Hill, NC, USA

David Wiedenfeld American Bird Conservancy, Washington, DC, USA

Manuel J. Zetina-Rejón Instituto Politécnico Nacional, Centro Interdisciplinario de Ciencias Marinas, La Paz, México

Leo Zurita-Arthos College of Biological & Environmental Sciences, Universidad San Francisco de Quito, Quito, Ecuador

Part I
UNC & USFQ Galapagos Science Center, 10th Year Anniversary

Chapter 1
Connected Places and Social-Ecological Forces that Impact Small Island States and Their Sustainability: An Essay

Stephen J. Walsh and Carlos F. Mena

Introductions

We are motivated by the general concern for the fragility and uniqueness of island ecosystems and the tensions between human-environment interactions, often manifested through economic development, disturbance regimes, and challenges to resource conservation (Baldacchino 2018; Walsh and Mena 2016). Risks to islands, and particularly, Small Island States, are often associated with the intensity and type of environmental and socio-economic processes that impact local island conditions (Gardner and Grenier 2011). For instance, island ecosystems are influenced by the type of economic development associated with residential and touristic processes, such as, population migration, urbanization, deforestation, and agricultural extensification (Brewington 2013). Climate change and the related factors of sea-level rise and changes in the intensity and frequency of ENSO events are confounding processes that further shape island ecosystems (Uyarra et al. 2005). Environmental change, including, coastal erosion, invasive species, and land use dynamics further mediate islands, thereby, necessitating the development of mitigating strategies. Such strategies often involve resource conservation and protected area designation to ensure the wise stewardship of marine and terrestrial sites, with the goal to minimize the social-ecological challenges to the sustainability of island ecosystems (Ernoul and Wardell-Johnson 2013; Ghosh et al. 2001; Walsh et al. 2018).

S. J. Walsh (✉)
Department of Geography, University of North Carolina at Chapel Hill, Chapel Hill, NC, USA
e-mail: swalsh@email.unc.edu

C. F. Mena
College of Biological & Environmental Studies, Universidad San Francisco de Quito, Quito, Ecuador

S. J. Walsh et al. (eds.), *Island Ecosystems*, Social and Ecological Interactions in the Galapagos Islands, https://doi.org/10.1007/978-3-031-28089-4_1

Focal areas of island sustainability discourses generally include the importance of self-reliance of islanders and their social-ecological systems, protection and development of human and cultural resources, and the promotion of partnerships between government and non-government organizations to ensure sustainable human-environment interactions as an approach to reduce uncertainty (Pazmino et al. 2018). Critiques of island sustainability often address issues that focus solely on the future, such as, climate change, that may neglect persistent social challenges, including, household livelihood alternatives, healthcare, education, water quality, and food security on islands (Mai and Smith 2015).

As described by Mena et al. (2020), *Sustainability* may be defined as a socio-ecological process used to maintain critical multi-scale functions and corresponding ecosystem services through time, often in response to disturbances imposed by human and/or natural processes. *Sustainable Systems* may require adaptation to accommodate change without permanently transitioning the system into an alternate state, thereby, requiring interventions to restore sites to a previous condition. *Sustainable Development* is a process to improve the quality of life of island people, while maintaining the ability of social–ecological systems to continue to provide valuable ecological services that social systems require, and physical systems depend. In the Galapagos Islands of Ecuador, for instance, the maintenance of amenity resources to support tourism and the quality of life of residents are explicitly linked to ecosystem goods and services, particularly, the accessibility to high-quality natural environments and the terrestrial and marine visitation sites that showcase iconic species and iconic environments (Dong et al. 2009).

Islands – Space and Place

Islands, circumscribed often by irregular coastlines and shaped by complex historical and multi-scale geographies, are strongly influenced by ocean conditions, but so too by the mutual relationships among space, place, society, globalization, and the social-ecological forces of change (Dodds and Royle 2003; Ratter 2018). The general context of islands is that they can be geographically remote or positioned close to the mainland, connected directly or indirectly to the continent and to other islands, singularly located or arrayed in archipelagos, irregular in shape and size, locationally distributed according to the combination of geological, biological, and anthropogenic forces, and varied in their ecological, economic, social, cultural, and topographic settings (Johannes de Haan et al. 2019).

Island vulnerability often is influenced by a collection of factors, for instance, their geographic remoteness, land cover/land use change patterns, intensity and type of human uses, orientation to the ocean, biophysical and social factors of formation and change, terrain configurations, settlement patterns, social and ecological connectivity, disturbance regimes, and the human dimension linked to consumptive behavior, economic development, and conservation management (Royle 2001). Islands are further shaped by their endogenous dynamics including, soil and freshwater conditions to support residential and transitory populations as well as subsistence and/or commercial agriculture, shoreline conditions to support trade and ocean transport, and settlement patterns and colonial legacies that impact populations as well as tourism (Baldacchino 2018; Stamoulis et al. 2018). Exogenous factors also impact islands, for instance, through situational factors of wind, currents, climate, and migratory pathways of marine species as well as through distant political forces associated with social and environmental policies, economic development programs, and resource conservation initiatives (Esther et al. 2012). From a biological perspective, isolation affects island biota in complex ways, for instance, through the amounts of surrounding landmasses that determine the number of arriving propagules and the over-water distance that may act as dispersal biases that favor arriving species that are good swimmers, flyers, or floaters as well as seeds that can maintain their integrity for relatively long periods of time and under changing environmental conditions (Weigelt et al. 2013). Also, when compared to mainland sites, oceanic islands are known for their high percentage of endemic species, but only moderate levels of species richness, both highly relevant to the global prioritization of tourists as well as conservation efforts.

Often the focus on islands is on their complex connectivity among nearby and distant places and the spatiality of islands and the role they play in shaping the evolutionary characteristics of habitats as well as land cover/land use change patterns on islands, mediated by local inhabitants and visitors, including, people, plants, and animals (Pazmino et al. 2018). While accounting for only about 2-percent of the world's land area and formed and continuously shaped by geologic, geomorphic, and biological forces, islands are connected to mainland locations and other islands in strategic ways (Royle 2001). Often borne from the fire of volcanic activity, islands are shaped by indigenous populations, colonial legacies, and migration streams whose contexts are further influenced by social, cultural, economic, geopolitical, and biophysical domains. Further, islands are often influenced through insularity, that is, generation of a local sense of place and an island identity in which inhabitants and their associated political structures become agents of change and conveyors of development (Royle 2001). The complex and precarious nature of small islands are often defined through their limited resources, multi-layered social systems, and the close social interactions among population groups, social externalities, environmental change, and global networks (Ratter 2018). Small islands are also often portrayed as vulnerable to external pressures, ranging from colonization, geopolitical forces, sea-level rise, natural hazards, and resource scarcities, as well as internal pressures, including under-employment, depopulation or over-population, diversification of household livelihoods, and loss of historical identities (Baldacchino 2018).

Islands are also the home to vulnerable iconic species and unique habitats that are the focus of global tourism activities and conservation efforts. Microcosms of larger mainland systems, islands exhibit smaller sizes, crisp boundaries, restricted access, and often historical isolation that make them more manageable to study and more effective to measure the factors that threaten their social-ecological sustainability. Tourism has become one of the most important economic sectors in recent decades, and island tourism is one of the main components of world tourism. Tourism has become the most important and dynamic driving force of economic change on islands, especially for islands with high amenity resources. Tourism can reduce the seemingly remoteness of islands as islands connect to the world through constructs that shape place and alter the use and imagination of space (Machado and Almeida 2014). Human factors associated with residential populations, migrations, and tourism frequently drive land cover/land use change through direct and indirect ways, even more strongly than ecology and geography. Deforestation, agricultural extensification, and urbanization shape natural and human circumstances that act upon islands by episodic and continuous forces of change. The Human Impact Index, a measure of current threats to islands and mainland sites through development, is significantly higher for islands. Further, land cover/land use change on islands for the year 2021 indicates significantly higher levels as compared to mainland locations.

Tourism is directly influenced by location, which plays a central role in the development of mass tourism on islands. Tourism has led to the restructuring of island economies, for instance, through clustering of accommodations, impact of national and international airports, food imports from the mainland or from other islands, port and harbor development, and community infrastructure development that has tended to reframe economic activities from agriculture, agroforestry, and/or fisheries to the marketing of alternate, exotic products, including, amenity resources and visitation sites that are consumed by resident and tourists. Often higher revenues are generated through the direct employment of residents and/or a migrant work force incentivized to locate on islands by the service industry and its emphasis on island tourism. Placed-based and boat-based tourism are often elements of island tourism, but their impacts on the environment are considerably different as a consequence of their respective demands on the community infrastructure. Island branding often seeks to establish a special niche in tourism, products marketed locally or globally (e.g., coffee), special services rendered (e.g., banking), and even the island's identity, as more international visitors seek to indulge in the island experience, interact with iconic wildlife species, and marvel at special landscapes and/or island cultures. Island challenges to tourism include the possible saturation of an island as a tourist destination, degradation of alternate sources of island revenues, surrendering to tourism as the only island product, and the challenges of securing sufficient and dependable sources of freshwater and approaches for handling associated waste and sanitation demands (Fiorini et al. 2017). Sustainability in the context of tourism is the development and maintenance of islands in a manner that ensures the region remains viable over an indefinite period, does not degrade the environment, and does not prohibit the successful development of alternate activities conducted on islands. Island resilience links to island sustainability in which islanders and government organizations engage local people in the development and management of tourism, strengthens the local capacity and social capital on islands through participatory engagement and the development of innovative mechanism to achieve positive and long-lasting results from tourism that extends throughout the community.

As seen on several occasions around the globe, *nature-based tourism* can rejuvenate a socio-economic system through a destination's social, physical, and natural capital. *Social capital* is the source of cultural and touristic identify of a place, *physical capital* includes the infrastructure, transport, and services offered, and *natural capital* is the source of environmental services provided, such as, access to iconic species and special landscapes and seascapes (Juan et al. 2008). Islands, however, that depend on a specialized environmental niche are vulnerable to environmental change, heighted by social-ecological stressors, such as, climate change and in-migration of population for jobs in the tourism industry (Gil 2003). *Vulnerability* refers to characteristics of a place, person, or group and the situations that influence their capacity to anticipate, cope, resist, and recover from the impacts of disturbances and/or shocks to human and ecological systems. Generally speaking, islands, particularly small islands, are more affected by external pressures and more immediately than larger islands, island archipelagos, and mainland areas. *Coastal Vulnerability Index* that assesses vulnerability by indicators of exposure and coping capacity shows that all Small Island Developing States (SIDS) face a moderate or high risk from factors, such as, sea-level rise, El Nino events, alien species, flooding and erosion, population migration, and environmental degradation. Development challenges applied to small islands include diseconomies of scale, fluctuating commodity prices, difficulties in accessing global markets, lack of adequate infrastructure, and, often, inadequate availability of local labor and high transit costs. In the Galapagos Islands, for instance, the economic benefits of tourism are unequally distributed at the local, national, and international levels (González et al. 2008; Taylor et al. 2008). The inequality in the tourism sector in the Galapagos is supported by tension among local people and government institutions that impact household livelihoods, lack of trust in the permit system imposed by government, and a sense that the process is not fair to all stakeholders.

The geography of islands, particularly, small islands, is closely related to space, place, and time and the on-going context of globalization, and the social-ecological drivers of change. Small islands are bounded spaces that are limited in size, land area, resources, economic and population potential, and political power. Small islands also face specific problems associated with isolation, which impacts the relative accessibility to services and markets, however, exotic branding of islands can reduce the anxiety of marketing single products to the region or world. By far, marketing of special and iconic places through tourism has elevated islands to new financial levels, but not without the expressed concern of over-population, resource exploitation, and diminishing returns on investment over time. Loss of island identity and its "islandness," a term used to embody the essence of island living and the attributes that make an island what it fundamentally represents, hangs in the balance of complex human-environment interactions and understanding and mitigating the threats to island sustainability (Baldacchino 2018; Ratter 2018; Royle 2001).

Island Sustainability

The challenges associated with sustainable management of island ecosystems, the impact of local, regional, and global context linked to endogenous and exogenous forces of change, and complex and multi-scale human-environment interactions necessitates an integrated approach to island studies and the ability to simulate "what if" scenarios applied to islands (Pizzitutti et al. 2017). Island biocomplexity is a term used to describe the study of complex adaptive systems that combine a new island ecology that incorporates human induced change on the environment. Island biocomplexity encompasses the complex interactions within and among ecological systems, physical systems on which they depend, and human systems with which they interact (Walsh et al. 2019). Island ecosystems are complex adaptive systems because their macroscopic properties emerge from the interactions among the individual components of the ecosystem (Pizzitutti et al. 2014). Global changes, including the forces associated with tourism, migration, and land cover/land use change, exert exogenous pressures on island ecosystems, but their systems have their own spatially contingent endogenous dynamics (Miller et al. 2018). Development of quantitative and spatial simulation models can help to understand the forces and elements influencing the specific patterns of tourism growth (Miller et al. 2010). Simulation models suggest that the type of tourism determines the pattern of tourism growth and the characteristics of the post-stagnation phase in the absence of policy changes. The simulation models are designed to consider social and ecological threats to diverse and fragile island settings, informed through multi-scale and multi-thematic geospatial data. Often, the models are designed to improve the decision-support systems to examine various elements of tourism management (Pizzitutti et al. 2014; Walsh and Mena 2016). Participatory approaches integrate the views of multiple stakeholders and to build an understandable, graphical representation of the impacts of tourism and resident populations on complex human-environment interactions. Dynamic systems models rely upon data to specify rules to model behavior, complex relationships, and rates of exchange that are derived through statistical functions specified in theory or practice (Thanh and Carl 2015). Figure 1.1 is a conceptual representation of the concepts and perspectives that shape island sustainability and influence the development of dynamic simulation models to understand the implication of policy, disturbances, development, and social-ecological pattern-process relationships on island futures (Zhang and Walsh 2018).

Concluding Thoughts

People are an important driver of change on islands, for instance, through urbanization, tourism, fisheries, agriculture, and the influx of people and products from mainland settings. The introduction of invasive species, depletion of natural resources, and excessive reliance on terrestrial and marine assets may jeopardize the

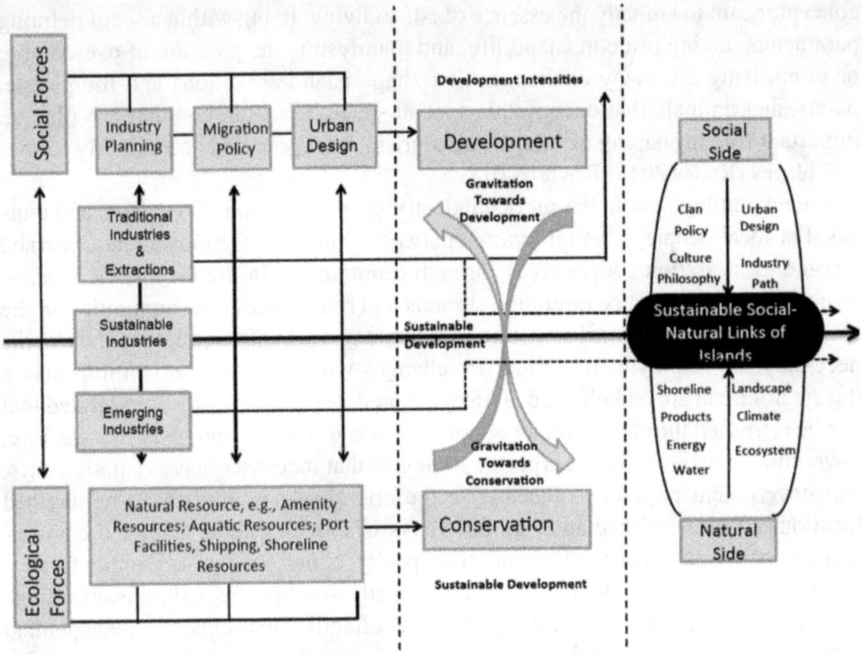

Fig. 1.1 Conceptual representation of an economic development and a resource conservation gradient, shaped by social and ecological forces of change and the characteristic elements of each (Zhang and Walsh 2018)

long-term provisioning of ecosystem goods and services which local residents, tourists, and human-environment interactions depend. Small islands, in particular, and their social-ecological systems, are threatened by forces of globalization, natural hazards, and the worsening impacts of climate change. Shaped by geographic isolation and unique environments, cultural diversity and biological endemism of islands are at risk from the very forces that encourage tourism and the associated development brought to their shores (Shi et al. 2004).

As in colonial periods, islands continue to be exploited through resource extraction and production, now linked to the niche tourism marketing of iconic species, amenity resources, and island identities. Tourism development makes islands attractive to migration from within and outside the region. Often placed-based vs. boat-bases tourism impacts the environment through excessive use, but also moves revenues to off-island locations and entrepreneurs. As outposts of globalization, islands are constantly in motion from social and ecological perspectives, changing the discourse about islands and their ability to seize opportunities and to protect themselves from threats to their sustainability. Often with limited political power, islanders may find it difficult to address the complexities of contemporary life and to prepare for alternative futures among the intrinsic challenges to islands. Embracing the island experience is often craved by populations that are economically enabled to relocate to islands and to realize the benefits of "islandness," a concept meant to embody the essence of island living, living within a set of defining parameters, taking pride in island life, and manifesting the pleasure of overcoming or minimizing the many challenges that shape island conditions and the people, plants, and animals that occupy this special space. Location continues to play an important role in shaping development, tourism, transport, and accessibility to special places (Ratter 2018; Royle 2001).

Direct employment in the service industry spurs migration of residents and tourists. For local people, island migration networks influence the movement of earned income to migration source areas through remittances. In the Galapagos Islands, social networks influence migration streams and remittances to communities on the mainland. The type of tourism determines the magnitude of migration and the skills necessary for employment. As tourism changes with time and tourism impacts on the environment are embellished, a lifecycle model of tourism may be exhibited that can be refreshed through a greater reliance on eco-tourism, improved infrastructure, government incentives, and corporate strategies that renew services, visitation sites, and government policies to incentivize the arrival of new tourists to reimagined locations (Butler 1980; Juan et al. 2008). As island sites show degraded environmental quality due to excessive touristic capacity, deficient socio-economic factors, and the appearance of other competitors, islands may require a significant refresh and reconceptualization, including, a more effective participatory management scheme that integrates residents into the decision-making process as a vital stakeholder group in defining future trajectories and charting a new course into the future.

References

Baldacchino G (2018) The Routledge international handbook of Island studies. Routledge Publishing, London, UK

Brewington L (2013) The double bind of tourism in Galapagos society. In: Walsh SJ, Mena CF (eds) Science and conservation in the galapagos Islands: frameworks and perspectives. Springer Science þ Business Media, NY, pp 105–125. Social and Ecological Interactions in the Galapagos Islands (Galapagos Book Series) New York

Butler R (1980) The concept of a tourist area cycle of evolution: implication for the management of resources. Can Geogr 24:5–12

de Haan JF, Quiroga D, Walsh SJ, Bettencourt L (2019) Scales and transformative change – transitions in the galapagos. In: Kvan T, Karakiewicz JA (eds) Urban galapagos – transitions to sustainability in complex adaptive systems. Social and Ecological Interactions in the Galapagos Islands. Springer, pp 43–58

Dodds K, Royle SA (2003) The historical geography of islands introduction: rethinking islands. J Hist Geogr 29(4):487–498

Dong LS, Hong SR, Jun G (2009) Research on tourism environmental carrying capacity of Chong Ming Island. In: International conference on environmental science and information application technology, pp 177–181

Ernoul L, Wardell-Johnson A (2013) Governance in integrated coastal zone management: a social network analysis of cross-scale collaboration. Environ Conserv. https://doi.org/10.1017/S0376892913000106

Esther JM, Tatsuro M, Kazuhiko A (2012) Sustainability of fishers' communities in tropical Island fisheries from the perspectives of resource use and management: a comparative study of Pohnpei (Micronesia), mafia (Tanzania), and Guimaras (Philippines). Jpn Soc Fish Sci 78:947–964

Fiorini L, Zullo F, Romano B (2017) Urban development of the coastal system of the Italian largest islands: sicily and Sardinia. Ocean Coast Manag 143:184–194

Gardner MR, Grenier C (2011) Linking livelihoods and conservation: challenges facing the Galapagos Islands. In: Baldacchino G, Niles D (eds) Island futures: conservation and development across the Asia-Pacific region. Springer. https://doi.org/10.1007/978-4-431-53989-6_6

Ghosh T, Bhandari G, Hazra S (2001) Assessment of land use/land cover dynamics and shoreline changes of Sagar Island through remote sensing. Asian Confer Remote Sensing Singapore 2:848–852

Gil SM (2003) Tourism development in the Canary Islands. Ann Tour Res 30(3):744–747

González JA, Montes C, Rodríguez J, Tapia W (2008) Rethinking the Galapagos Islands as a complex social-ecological system: implications for conservation and management. Ecol Soc 13(2) [online] http://www.ecologyandsociety.org/vol13/iss2/art13/

Juan ROR, Eduardo PL, Vanessa YE (2008) The sustainability of Island destinations: tourism area life cycle and teleological perspectives, the case of Tenerife. Tour Manag 29:53–65

Machado LP, Almeida A (2014) The on-going process of reinventing classic tourism destinations-dthe case of nordic tourists in Madeira Island. Scand J Hosp Tour 13(1):24–43

Mai T, Smith C (2015) Addressing the threats to tourism sustainability using systems thinking: a case study of cat Ba Island. Vietnam J Sustain Tour 23(10):1504–1528

Mena CF, Quiroga D, Walsh SJ (2020) Threats to sustainability in the Galapagos Islands: a social - ecological perspective. In: Sarmiento FO, Frolich LM (eds) International handbook of geography and sustainability. Edward Elgar Publishing, UK, pp 342–358

Miller BW, Breckheimer I, McCleary AL, Guzman-Ramirez L, Caplow SC, Walsh SJ (2010) Using stylized agent-based models for population-environment research: a case from the Galapagos Islands. Popul Environ 31(6):401–426

Miller ML, Lieske SN, Walsh SJ, Carter RW (2018) Understanding the interaction between a protected destination system and conservation tourism through remote sensing. In: Walsh SJ, Liang S (eds) Comprehensive remote sensing, applications for societal benefits. Elsevier Publisher, London, pp 123–143

Pazmino A, Serrao-Neumann S, Choy DL (2018) Towards comprehensive policy integration for the sustainability of small islands: a landscape-scale planning approach for the Galapagos Islands. Sustain For 10:1228. https://doi.org/10.3390/su10041228

Pizzitutti F, Mena CF, Walsh SJ (2014) Modeling tourism in the Galapagos Islands: an agent-based model approach. J Artif Soc Soc Simulat 17(1):14. http://jasss.soc.surrey.ac.uk/17/1/14.html

Pizzitutti F, Walsh SJ, Rindfuss RR, Reck G, Quiroga D, Tippett R, Mena CF (2017) Scenario planning for tourism management: a participatory and system dynamics model applied to the Galapagos Islands of Ecuador. J Sustain Tour 25(8):1117–1137

Ratter BMW (2018) Geography of small Islands. Springer. https://doi.org/10.1007/978-3-319-63869-0_6

Royle SA (2001) A geography of islands: small Island insularity. In: Routledge studies in human geography. Routledge Publishing, London

Shi C, Hutchinson SM, Xu S (2004) Evaluation of coastal zone sustainability: an integrated approach applied in Shanghai municipality and Chong Ming Island. J Environ Manag 71:335–344

Stamoulis KA, Delevaux JMS, Williams ID, Poti M, Lecky J, Costa B, Kendall MS, Pittman SJ, Donovan MK, Wedding LM, Friedlander AM (2018) Seascape models reveal places to focus coastal fisheries management. Ecol Appl 28(4):910–925

Taylor JE, Hardner J, Stewart M (2008) Ecotourism and economic growth in the Galapagos: an Island economy-wide analysis. Environ Dev Econ 14:139–162

Thanh M, Carl S (2015) Addressing the threats to tourism sustainability using systems thinking: a case study of cat Ba Island, Vietnam. J Sustain Tour 23(10):1504–1528
Uyarra MC, Cote IM, Gill JA, Tinch RRT, Viner D, Watkinson AR (2005) Island-specific preferences of tourists for environmental features: implications of climate change for tourism-dependent states. Environ Conserv 32(1):11–19
Walsh SJ, Mena CF (2016) Interactions of social, terrestrial, and marine sub-systems in the Galapagos Islands. Proc Natl Acad Sci 113(51):14536–14543
Walsh SJ, Page PH, Brewington L, Bradley JR, Mena CF (2018) Beach vulnerability in the Galapagos Islands: fusion of worldview 2 imagery, 3-D laser scanner data & unmanned aerial systems. In: Walsh SJ, Liang S (eds) Comprehensive remote sensing, applications for societal benefits. Elsevier Publisher, London, pp 159–176
Walsh SJ, Engie K, Page PH, Frizzelle BG (2019) Demographics of change: modelling the transition of fishers to tourism in the Galapagos Islands. In: Kvan T, Karakiewicz JA (eds) Urban Galapagos – Transitions to sustainability in complex adaptive systems. Social and ecological interactions in the Galapagos Islands. Springer, pp 61–83
Weigelt P, Jetz W, Kreft H (2013) Bioclimatic and physical characterization of the world's islands. Proc Natl Acad Sci. https://doi.org/10.1073/pnas.1306309110
Zhang H, Walsh SJ (2018) Comparison of the Zhoushan Islands, China and the Galapagos Islands, Ecuador: Island sustainability and forces of change. In: Walsh SJ, Liang S (eds) Comprehensive remote sensing, applications for societal benefits. Elsevier Publisher, London, pp 306–329

Stephen J. Walsh, Co-Guest Editor, is the Founding Emeritus Director of the UNC Center for Galapagos Studies and the UNC-USFQ Galapagos Science Center. He is also an Emeritus Distinguished Professor of Geography and is a Reserch Professor in the UNC Department of Geography. His research focuses on understanding the multi-scale and multi-thematic drivers of land use/land cover change in vulnerable and fragile ecosystems through remote sensing, spatial analysis, and spatial simulation model of "what if" scenarios of change.

Carlos F. Mena, Co-Guest Editor, Co-Director of the Galapagos Science Center, Director of the Institute of Geography, and Professor in the College of Biological & Environmental Sciences, Universidad San Francisco de Quito, Ecuador. His research assesses the importance of land use/ land cover change on social and terrestrial systems in the Galapagos Islands as well as in the Ecuadorian Amazon.

Part II
Communique of the World Summit on Island Sustainability

Chapter 2
Communique – Goals and Objectives of the World Summit on Island Sustainability

Stephen J. Walsh and Carlos F. Mena

Introduction

On June 26–30, 2022, the Galapagos Science Center and the broader UNC & USFQ Galapagos Initiative celebrated its 10-Year Anniversary. As the crowning event of the anniversary celebration, the *World Summit on Island Sustainability* was held on San Cristobal Island, Galapagos Archipelago of Ecuador. The intent of the World Summit was to bring together leading experts on island ecosystems and, particularly, on island sustainability from across the globe that represented a diversity of

10-Year Anniversary Celebration of the Galapagos Science Center, San Cristobal Island, Galapagos Archipelago of Ecuador; University of North Carolina at Chapel Hill & Universidad San Francisco de Quito – Strategic Partners

S. J. Walsh (✉)
Department of Geography, University of North Carolina at Chapel Hill,
Chapel Hill, NC, USA
e-mail: swalsh@email.unc.edu

C. F. Mena
College of Biological & Environmental Studies, Universidad San Francisco de Quito,
Quito, Ecuador

S. J. Walsh et al. (eds.), *Island Ecosystems*, Social and Ecological Interactions in the Galapagos Islands, https://doi.org/10.1007/978-3-031-28089-4_2

Fig. 2.1 Participants at the World Summit on Island Sustainability, June 26–30, 2022

perspectives, approaches, and stakeholder groups. The World Summit was exclusive and global and featured an "expert convening" of scholars and practitioners that addressed the social, terrestrial, and marine sub-systems of the Galapagos Islands and other similarly challenged island ecosystems from around the globe to assess island vulnerability, resilience, and sustainability (See Fig. 2.1 – group photo of World Summit participants).

The content of the World Summit will be distributed through a book, part of the Galapagos Book Series (Stephen J. Walsh & Carlos Mena, Editors), published by Springer Nature and edited by Stephen J. Walsh, University of North Carolina at Chapel Hill, Carlos Mena, Universidad San Francisco de Quito, Jill Stewart, University of North Carolina at Chapel Hill, and Juan Pablo Muñoz Pérez, University of the Sunshine Coast. The book, scheduled to be published by mid-year 2023, will include 34-chapters generated by outstanding scientists working in the Galapagos Islands and around the globe. The book will be broad in its context, innovative in its perspectives, and accessible to the broader island community including scientists, managers, residents, tourists, and government and non-government organizations.

While the most obvious goal of organizing the World Summit on Island Sustainability was to celebrate the tenth year anniversary of the GSC and the UNC-USFQ Galapagos Initiative, other goals were vigorously addressed. For instance, we seek to elevate and highlight the Galapagos in the island conservation discourse, seeking to interact with other island networks in more obvious and conspicuous ways to benefit the Galapagos Islands, the UNC-USFQ Galapagos Initiative, and the world. We highlight multiple visions of a sustainable future for the Galapagos Islands, thereby, engaging the Ecuadorian Ministry of Environment and the Ministry of Tourism as well as the Government Council of Galapagos and the Galapagos National Park as well as local Galapagos authorities, including government and non-government organizations and local citizen groups. Borrowing from Green Growth Programs from around the globe and the Global Island Partnership, we examine existing global programs that emphasize island sustainability and their incorporation into life, policies, and circumstances in the Galapagos Islands. We also enhance our connections with the institutional members of our *International Galapagos Science Consortium* and expand the Consortium through recruitment of other member institutions and reaching out to institutional collaborators. We also work to benefit islands and their local communities by working with citizen groups as well as important NGOs who seek to improve the natural and social conditions in the Galapagos and diminish the impact of the human dimension on the future of Galapagos' ecosystems.

The Galapagos Initiative – An Overview

The Galapagos Islands are facing increasing danger. Globalization, climate change, and international tourism conspire to threaten island sustainability through local and global forces of change. The critical loss of endemic species, wildlife trafficking, and the unrelenting pressure of human impacts has placed these "enchanted islands" at risk. Further, these islands act as an early warning system to recognize the threats, for instance, from ENSO events on marine conditions, including marine productivity. Saving the Galapagos Islands requires a dedicated and innovative strategy that is transformative, interdisciplinary, and sustained. Together with the USFQ, UNC leverages its long-term commitment to the Galapagos Islands by scaling key projects that extend across the sciences to save these islands and, in so doing, create a template for saving similarly challenged island settings around the globe.

In 2011, UNC, and our strategic partner, the Universidad San Francisco de Quito, dedicated the *Galapagos Science Center* (see Figs. 2.2 & 2.3) *on San Cristobal Island*, a facility of 20,000 square feet that contains four laboratories – Microbiology & Genetics, Terrestrial Ecology, Marine Ecology, and Spatial Analysis & Modeling. The GSC is staffed by 15 professionals supporting over 125 collaborative scientists from UNC, USFQ, Consortium members, and other global institutions. The Galapagos Science Center is the only institution of its kind in the Galapagos, and it represents a unique facility and a special opportunity to make the world a better place by working to understand island ecosystems and the social-ecological-physical threats to sustainability.

The Galapagos Science Center has as its goals conducting outstanding and integrative research and education that contributes to protecting Earth's unique and special places – animals and plants (from the cell to the landscape level) that make their homes there and the ecosystems in which they thrive. This work specifically involves the human dimension and the inter-connections among the social, terrestrial, and marine sub-systems of the Galapagos Islands. Work in the Galapagos is approached through a rich and varied interdisciplinary and integrative perspective aimed at identifying the proper balance between the natural environment and the people who live in and visit these special places. The location for this work is the Galapagos Archipelago of Ecuador – home to some of the world's most iconic and unique

Fig. 2.2 The Galapagos Science Center (foreground, right) on San Cristobal Island, Galapagos Islands, Ecuador, and the harbor of Puerto Baquerizo Moreno

Fig. 2.3 UNC & USFQ Galapagos Science Center (GSC)

species and one of the world's few isolated and largely undeveloped natural habitats. The problems the Galapagos Science Center seeks to solve are complex, so they cannot be solved by a single scientist working in a narrow field. They require scientists from different disciplines, working together in teams to fuse studies of the environment with studies of human and animal populations, their health and well-being, and their direct and indirect consequences to understand island ecosystems and the threats to their sustainability.

UNC and USFQ are uniquely suited for this role. One of their greatest strengths as institutions is the interdisciplinary, problem-solving approach they bring to research in these fields. Together, they are among the world's foremost leaders in interdisciplinary research that engages natural, social, spatial, and computational sciences. This is the way UNC & USFQ are wired to work – their DNA. The Galapagos Science Center represents a major commitment by both universities to put its research strength in these fields into action – to preserve the Galapagos and the beloved species that live there and, at the same time, enable people of the Galapagos and people of the world to experience and sustain this unique ecosystem. The Galapagos Science Center is the only institutions of its kind in the Galapagos. The Galapagos Initiative represents a unique opportunity to make the world a better place and to enrich our understanding of human-natural systems in a world-renowned National Park, Marine Reserve & World Heritage Site, and to extend that understanding to address similar challenges to island ecosystems from around the globe.

Our vision of globalized learning is to transform research, education, and outreach by creating links to international partners whose ideas and energies flow between UNC, USFQ, Galapagos Islands, global island ecosystems, and world institutions to innovate, enlighten, and transform. The Galapagos Initiative draws from the varied talents and experiences of our faculty, students, staff, alumni, and special friends, and applies the lessons learned and new knowledge generated to the challenges facing our planet.

The Galapagos Islands are a highly charismatic place of world-renowned stature and reputation. Iconic species and iconic landscapes, and their accessibility, have justified their placement on most everyone's "bucket list" for a future visit and, possibly, for financial and emotional support. With a UNC-USFQ building on one of the most famous and iconic places in the world, the Galapagos Science Center, attracts the "best and brightest" scholars by providing our scientists an amazing environmental setting as well as a cutting-edge infrastructure that supports innovative and transformative science and education. While we have come a long way in a relatively short period of time since the dedication of the Galapagos Science Center in 2011, our full potential remains relatively untapped. In the next 5- years, we will take the necessary steps to accelerate and enhance our capacities in very meaningful and substantial ways to build upon our many successes that benefit UNC, USFQ, Consortium members, Galapagos Islands, and the planet.

Highlights of the World Summit on Island Sustainability

The World Summit attracted 150 scientists to the Galapagos Islands. They discussed projects conducted in the Galapagos Islands, Hawaii, Guam, French Polynesia, Chile, Australia, Caribbean, and well beyond by faculty, students, and staff from a host of affiliated institutions. Featured Speakers included Nikhil Advani, World Wildlife Fund; Laura Brewington, East-West Center, Honolulu, Hawaii & Arizona State University; Ashlan Cousteau, EarthEcho International; Jenny Daltry, Re:wild; Neil Davies, Gump South Pacific Research Station; Monty Halls, Galapagos Conservation Trust; Mike Kingsford, James Cook University; Juan Pablo Luna & Sergio Navarrete, Pontifica Universidad Catolica de Chile; and Austin Shelton, University of Guam. Other notable speakers at the World Summit included GSC Advisory Board Members from UNC – Corbin Jones, Biology; Adrian Marchetti, Earth, Marine & Environmental Sciences; Diego Riveros-Iregui, Geography; Amanda Thompson, Anthropology & GSC Co-Director; Gina Chowa, Social Work; Jill Stewart, Environmental Sciences & Engineering; Steve Walsh, Geography; and Don Hobart, Associate Vice Chancellor for Research. From USFQ, the following added additional richness to the program as speakers and members of the GSC Advisory Board – Carlos Mena, Life & Environmental Sciences & GSC Co-Director; Jaime Ocampo, Public Health; María de Lourdes Torres, Microbiology; Carlos Valle, Biology; Gunter Reck, Resource Management and Tourism; and Stella de la Torre, Life and Environmental Sciences. In addition to the invited featured speakers and representatives of the GSC Advisory Board, members of the UNC & USFQ Leadership Team also attended the World Summit. The UNC Leadership Team consisted of Kevin Guskiewicz, Chancellor; Chris Clemens, Provost; Barbara Stephenson, Vice Provost for Global Affairs & Chief Global Officer; Penny Gordon-Larsen, Interim Vice Chancellor for Research; Mike Barker, Vice Chancellor for Information Technology Services; and Terry Rhodes, Dean, College of Arts & Sciences. The USFQ Leadership Team consisted of Diego Quioga, Rector; Andrea Encalada, Vice-Rector; Cesar Zambrano, Dean of Research; and Carlos Valle, Dean of Biological & Environmental Sciences. Also, important guests at the World Summit included Fausto Arcos, Associate Vice President & General Manager, Galapagos Celebrity Cruises; Thomas Krise, President, University of Guam; Ross Young, Deputy Vice Chancellor for Research & Innovation, University of the Sunshine Coast; Jon Horowitz, Associate Vice Chancellor for Research, North Carolina State University; Eric Dorfman, Director, North Carolina Museum of Natural Sciences; and Norman Wray, Coordinator of the Galapagos Hub for Sustainability, Innovation, and Resilience.

The World Summit's program consisted of organizational topics selected to showcase the innovative work on island ecosystems that is being conducted by a diverse group of outstanding scientists. Below we highlight several of the presentations to convey a sense of the science that was presented though oral and poster presentations as well as a panel discussion on "Island Sustainability: Paths Forward" and breakout sessions around vital topics in the Galapagos and other island ecosystems around the globe – "Sustainable Agriculture and Food Security," "Sustainable Fisheries," "Sustainable Tourism," and "One Health: People, Organisms, and Environment."

The World Summit was launched at the Welcoming Reception held on Sunday, June 26th that included a pre-recorded welcome to all participants by Sarah Darwin, the Great, Great Grand Daughter of Charles Darwin, followed by remarks by the UNC and USFQ Leadership Teams and comments by Vice President and General Manager of Galapagos Celebrity Cruises, the special host of the Welcoming Reception. As the focal point of the opening event, Ashlan Cousteau of EarthEcho International delivered a passionate "call-to-arms" to protect the oceans and their island ecosystems by engaging a diverse set of global partners in this noble endeavor. To that end, a new member of the *International Galapagos Science Consortium* was welcomed, North Carolina State University (NCSU), through a signing ceremony that involved the leadership of NCSU, UNC and USFQ. The Welcoming Reception was concluded with tours of the Galapagos Science Center, the 20,000 square foot facility that was dedicated in 2011 and whose construction and operational support is equally shared by UNC and USFQ.

Monday, June 27th was the first full day of scientific presentations organized around the topics of "Challenges to Island Sustainability – Paths Forward," "Island Ecosystems – Marine Subsystems," and "Island Ecosystems – Terrestrial Subsystems." The invited, featured speakers led the way through fascinating presentations on "Galapagos, the Next Generation," "Invasive Species and Climate Change – Lessons from Pacific Islands," and "Harnessing Wind in our Sails to Sustainable 2030," followed by a presentation on "What We Learned from COVID-19 in the Galapagos." This opening session was followed by presentations on "Globalization and the Challenging Political Economy of Governing (and Researching) Islands in Contemporary Times," "Island Digital Ecosystem Avatars Consortium Infrastructure for Democratic Ecological Action," "Restoring Resilient Island Ecosystems – Lessons Learned from the Caribbean," and "On-the-Ground Solutions to Help People and Wildlife in a Changing Climate." The afternoon sessions focused on marine and terrestrial subsystems through presentations on "Connecting Marine Protected Areas in the Eastern Tropical Pacific: The Science and the First Steps" and the "Sustainability Targets & Economic Benefits from Marine Protection in Islands: A Review for Galapagos and other Archipelagoes in the Pacific Ocean" as well as a presentation on "Darwin and the Microbes: Investigating the Linkages between Ocean Physics, Carbon Cycling, and Galapagos Microbiome" and "Spatial Patterns of Upwelling that Influence Fish Assemblages and Environmental Records in the Otoliths of Fishes." The final session of the day featured presentations on the "Environmental Sculpting of Genomes: Lessons from the Grasses of the Galapagos," and "Unraveling the Interactions between Endemic and Invasive Plant Species in the Galapagos," followed by papers on "Galapagos Land Snails & Environmental Sustainability," and "Galapagos Petrels Conservation and a Sustainable Future in the Archipelago."

Tuesday, June 28th began with a session on "Island Ecosystems – Social Subsystems." Four papers were presented that examined "Human Challenges in

Small Island States: Human Health and Nutrition in the Galapagos," "Human Dimensions in the Galapagos Islands," "Water Quality on the Galapagos Islands: Linking Human and Environmental Health," and "Infectious Diseases in the Galapagos: From a Paradise to an Isolation." "Island Ecosystems – Marine Subsystems" was the topic of the final session of the morning and day. Presentations described "Species-Specific Thermal Sensitivity in the Galapagos Marine Ecosystem: Predicting Warming Induced Changes in Composition and Function," "Trophic Web Structure and Ecosystem Attributes of the Galapagos Marine Shelf," "Leaf to Reef: Generating Critical Knowledge to Support Resilience-Based Management of the Great Barrier Reef," and "Our Transforming Coastal Marine Ecosystems: The Urgent Need for an Effective and Science-Based Conservation Network along the Southeastern Pacific." No afternoon sessions were scheduled, opting, rather to move interactions to the field and marine settings through a bus trip to *La Loberia*, a nearby coastal setting on San Cristobal Island that is known for its dynamic environments as well as the presence of marine mammals and seabirds, and a boat trip to Cerro Brujo to highlight marine productivity, species assemblages, and coastal settings.

Wednesday, June 29th started with a session on "Island Ecosystems – Social Subsystems" that featured papers on "Improvements in the Galapagos Health System – Telemedicine, Research, and Medical Assistants," "Mapping of Galapagos Environmental and Social Values," and "SARS-CoV-2 Variants in Galapagos and Ecuador During the Pandemic." The following session returned to "Island Ecosystems – Terrestrial Subsystems" and featured presentations on the "Galapagos Barcode Project: Where Science and Local Communities Meet," "Mapping the Shrinking Scalesia Forest and Blackberry Invasion in Galapagos," "Microclimate as a Strong Predictor of the Native and Invasive Plant-Associated Soil Microbiota on San Cristobal Island, Galapagos Archipelago," "Chemical and Mineralogical Composition of Soils on San Cristobal Island, Galapagos Archipelago," and "Island Conservation and Sustainability." With the social, terrestrial, and marine subsystems examined through multiple sessions, the topics for the afternoon sessions transitioned to "Interdisciplinary Science: Conservation and Sustainability." Papers were presented on "Ten Years of Wildlife Health & Conservation in the Galapagos, 2013–2022," "Galapagos Marine Apex Predators: Does Geographic Isolation Shield Them against Global Pollution," "Science to Solution in Galapagos: How Research, Community, NGOs, and Policy Actors are Coming Together to Improve Ocean Protection, Tackle Plastic Pollution, and Inspire Climate Action," and "Cyber Infrastructure in Island Settings: Challenges and Opportunities." The afternoon also featured a poster session that included 15 presentations on a wide range of topics – "Whales and Dolphins of the Galapagos Archipelago: A Multidisciplinary Approach to Understand the Most Common Cetacean Species in the Ecuadorian Whale Sanctuary," "Baseline Analysis of Plastic Pollution Issues within the Galapagos Archipelago, "State of Water Quality in the Galapagos Islands: Challenges, Opportunities, and Household Preferences for Improved Water Services," "Guam Green Growth Initiative," "Development of Biomaterials through Valorising Abundant Shellfish, Marine, and Agricultural Waste in the Galapagos," "Drivers of

Marine Protist Diversity and Connectivity in the Galapagos Archipelago," "Population Monitoring and Ecological Studies of the Galapagos Pinnipeds and their Management and Conservation, "Mapping Maternal Lineages of Sharks Using Environmental DNA," "Characterizing Thermal Tolerances and Sensitivities in Galapagos Cnidarians," "Restoring Darwin's Reefs: Pioneer Coral Gardening Initiatives in the Galapagos Islands," "Trophic Interactions Mediated by Ecological Parameters in Galapagos Upwelling System, "Effect of Temperature and Ecological Interactions Using Galapagos as a Model System," "Differences in Soil Microbiomes and Chemical Characterization in Four Islands of the Galapagos Archipelago," and "Exploring Fungal Pathogens to Control the Invasive Raspberry from San Cristobal Island of Galapagos Archipelago."

Thursday, June 30th, the final day of World Summit included two sessions on "Interdisciplinary Science: Conservation and Sustainability" that featured presentations on "The Galapagos Islands and the Fight Against Climate Change: Interdisciplinary Ocean and Atmospheric Research," "Establishing Comparable Health Baselines for Marine Turtle Populations," "The Role for Scientific Collections and Public Museums in Island Conservation," and "The Museum Effect: Platforms for Advocacy, Sustainability, and Conservation in Constrained Environments." The second session included presentations on "Enhancing Sustainability of Wildlife Tourism in Galapagos," "Integrating Science and Sustainability in Galapagos Cruise Tourism," and "Illegal Trade and Galapagos Futures." Shifting gears from paper and poster presentations, the final two sessions included special opportunities for additional interactions among participants through conversations organized through a Panel Discussion, "Island Sustainability – Paths Forward," and a Breakout Session on "Sustainable Island Ecosystems, Galapagos and Beyond" that featured small group discussions to address identified topics – "Sustainable Agriculture and Food Security," "Sustainable Fisheries," "Sustainable Tourism," and "One Health: People, Organisms, and Environment." Session chairs coordinated the conversations and noted comments for presentation to the reassembled participants through a Plenary conversation on each topic and then all topics in a more integrative and holistic manner. A Concluding Reception closed the World Summit, after thanking the participants for their insights, enthusiasm, and intentions to participate in the forthcoming book, part of the current 10-volume Galapagos Book Series, published by Springer Nature.

Conclusions

The *World Summit on Island Sustainability*, held on San Cristobal Island in the Galapagos Archipelago of Ecuador, was a fitting location to celebrate the crowning event of the 10-year anniversary celebration of the UNC and USFQ Galapagos Science Center. The diversity and depth of topics that were examined offered applicability to the Galapagos Islands, but also to island ecosystems around the world that are similarly stressed by the expanding human dimension as well as other forces of change, including climate change, land use/land cover dynamics, population migration, globalization, and other complex and integrated social-ecological-physical factors that shape and reshape islands. The World Summit specified presentation titles, speakers and co-authors, and affiliations to further appreciate the Summit's content, stature of participating speakers, and the intention to address the multi-scale and multi-dimensional forces of change on islands through exogenous and endogenous dynamics that alter island conditions, status, and future trajectories. Motivated to understand the challenges that island ecosystems face, including local people and visitors to these special places, the World Summit in 2022 and the Springer Nature book that will follow in 2023 contribute to the global focus and discourse on islands that have engendered the important work to educate, manage, conserve, and steward islands for future generations. Understanding island vulnerabilities, resilience, and sustainability are vital to informed policy, stakeholder engagement, participatory management, and island futures that are responsive to threats to their sustainability and to their very existence. In subtle and obvious ways, the *World Summit on Island Sustainability* contributed to an improved understanding of the complex interactions among social, terrestrial, and marine subsystems and their integration on islands, and the stated and unstated pledge among Summit participants to push the envelope of science through innovative methods and practices, create new knowledge on islands that better guides human-environment interactions, and educate and engage local people and citizens of the world to ensure the sustainability of islands benefits local places and global settings now and into the future.

In closing, there are many messages that can be distilled from the presentations, but chief among them includes the importance of Marine Protected Areas and their expansion through government and non-government engagement as well as the realization of economic benefits to islands through the protection of marine resources. Also, the essential engagement of "citizen scientists" to conduct studies that advantage science and local people. The recruitment, training, and engagement of local Galapagos citizens in the Barcode Project helped expand the collection and enhancement of our understanding of species found throughout the Galapagos Islands. In addition, the collaboration of UNC and USFQ scientists with the Galapagos National Park allowed for monitoring of the Galapagos Marine Reserve to assess base-line conditions and deviations from normal, particularly, during EL Nino events. Lastly, the assessment of the expanding human dimension in the Galapagos, and on islands around the globe, confirms the vital work of social scientists in addressing, for instance, human health, nutrition, water quality, and family well-being.

A persistent theme throughout the presentations was the analysis of complex processes and mechanisms that shape the social, terrestrial, and marine subsystems. Whether the topic was human health and well-being, invasive species, or marine productivity, scientists created innovative approaches to data collection and analysis that yields new insights into system behaviors and dynamics. Collectively, the topics discussed at the World Summit confirm our programmatic emphasis on integrative and interdisciplinary science and global visions of island sustainability. The assessment of the threats to island ecosystems, critical linkages among education, research, and community outreach, and the importance of fully understanding the direct and indirect effects of human-environment interactions are examined to ensure the sustainability of fragile and sensitive islands for future generations. Together, we can make a meaningful difference in the sustainability of island ecosystems, as the alternative of island degradation is unacceptable!

Stephen J. Walsh, Co-Guest Editor, is the Founding Emeritus Director of the UNC Center for Galapagos Studies and the UNC-USFQ Galapagos Science Center. He is also an Emeritus Distinguished Professor of Geography and is a Reserch Professor in the UNC Department of Geography. His research focuses on understanding the multi-scale and multi-thematic drivers of land use/land cover change in vulnerable and fragile ecosystems through remote sensing, spatial analysis, and spatial simulation model of "what if" scenarios of change.

Carlos F. Mena, Co-Guest Editor, Co-Director of the Galapagos Science Center, Director of the Institute of Geography, and Professor in the College of Biological & Environmental Sciences, Universidad San Francisco de Quito, Ecuador. His research assesses the importance of land use/ land cover change on social and terrestrial systems in the Galapagos Islands as well as in the Ecuadorian Amazon.

Part III
Island Ecosystems: Challenges to Sustainability

Chapter 3
Globalization & The Challenging Political Economy of Governing (and Researching) Islands in Contemporary Times

Juan Pablo Luna

Introduction: Islands and Sustainability

Island territories are geopolitically diverse. Some powerful countries in the international arena are islands (e.g., Australia, New Zeeland, or the United Kingdom). Many other islands are tiny and powerless. While some islands are independent nation-states, others make part as provinces of a larger country. Furthermore, those provinces can be governed as any other province or through a special regime (e.g., if they are a national park or under a specific protected area regulation). Given such diversity, it is necessary to specify the unit of analysis we have in mind when analyzing the nature of sustainability challenges in islands.

In what follows, the case of islands as "small states," (Maass 2009) or in Keohane's terminology, as "Lilliputians" in world politics (Keohane 1969) is considered. In other words, the analysis is centered on the challenges small and relatively powerless islands face in the contemporary world. Those islands are "price-takers" in geopolitical terms, and also the ones currently facing deeper sustainability challenges. Given their status in world politics, those islands' larger sustainability challenges cannot be addressed autonomously by islands and their local authorities. Although the implications of the arguments that follow might apply differently to independent island territories and to islands that are part of a larger nation-state, explicit distinctions are not made between those two alternatives.

The chapter will first discuss the notion of governance and its relation to sustainability. Second, three external conditions that impact small islands will be discussed: democratic recession, declining state capacity, and a shifting international order. Subsequently, I address some specific characteristics of islands jurisdictions, and their relation to the sustainability challenges they face. Finally, I analyze the

J. P. Luna (✉)
School of Government, Pontificia Universidad Catolica de Chile, Santiago, Chile
e-mail: jlunaf@uc.cl

S. J. Walsh et al. (eds.), *Island Ecosystems*, Social and Ecological Interactions in the Galapagos Islands, https://doi.org/10.1007/978-3-031-28089-4_3

governance frameworks often applied in small island jurisdictions and discuss their limitations in the current world. The chapter closes with a minimalist proposal to enhance sustainability while governing and researching small islands.

What Do We Mean by Governance for Sustainability?

In society, effective democratic governance schemes can be thought of as institutional designs that distribute scarce resources while achieving two interrelated goals: regulations and institutional mechanisms become socially legitimate and collectively binding (Fukuyama 2013). The fact that these are collectively binding schemes implies that they are institutional schemes that generate legitimate decisions for critical actors and communities at large. The fact that decisions are binding, means that they can be effectively implemented and enforced throughout the territory (Mann 2008). Therefore, when combined with normatively desired outcomes (e.g., enhancing sustainability) good governance induces a (Pareto-optimal) equilibrium (i.e., it produces public outcomes that enhance collective welfare.) When governance schemes generate welfare-enhancing results, they should also induce self-reinforcing equilibrium over time.

How does this notion of governance translate to the problem of sustainability in contemporary societies and islands? The challenge of sustainability governance resides in finding socially legitimate (and, therefore, collectively binding) institutional designs that contribute incrementally and permanently to development and sustainability (Baldacchino 2006). In other words, sustainable governance in islands implies designing public policies that promote and encourage social practices oriented towards sustainability while discouraging and eventually punishing unsustainable practices in island territories. In the contemporary world, a fundamental component of the legitimacy and capacity to effectively implement these public policies is that decisions must comply with democratic practices and hopefully be decided in the context of open and participatory processes.

There are two critical determinants of contemporary public policy decision-making processes. On the one hand, democratic participation must be achieved in a context where the number of social actors who need to make their voices heard in the decision-making process has multiplied. In short, legitimacy must be achieved through the broadest possible participation of groups and interests in the community. However, the breadth of participation makes decision-making more costly and difficult (Sartori 1988). On the other hand, the scale of sustainability governance schemes varies. Suppose some sustainability challenges can be tackled and successfully addressed locally. However, most other challenges need to involve the participation of agents nested in a multi-level game, which takes place across distinct local, regional, and global arenas (Ostrom and Janssen 2004).

While we usually focus on large-scale and encompassing public policy schemes and ambitious cooperation agreements for the sustainable management of islands, it is also necessary (and more productive today) to think "small" and "local." Indeed,

I will argue that much of what could be currently achieved in terms of improving governance for sustainability is through local, pragmatic interventions. Those interventions often can emerge in interaction with scientific research currently pursued in island territories. In other words, researchers, local leaders, and community activists should also actively seek to contribute to enhancing governance for sustainability through their respective local practices.

The latter is particularly significant for island territories, where minimal interventions and public policy shifts (e.g., incorporating an additional frequency of connecting flights or introducing new regulations for water treatment or fishing) generate massive knock-on effects (see cf. Ratter 2018; Médail 2017). In other words, those who ignore governance challenges in their decision-making risk creating massive externalities while unleashing acute social conflicts (Celata and Sanna 2012). Besides its technical and scientific pillars in enhancing sustainability, each institutional mechanism, regulatory policy, or new intervention considered for island communities will turn out socially contested (de jure or de facto) if governance issues are ignored. In that case, sustainability remains unreachable.

Context: Democratic Governance Today

This section discusses three threats to democratic governance for sustainability in the contemporary world. Those threats are democratic recession at the country level, nation-states' declining power, and the international system's reconfiguration.

Liberal Democracy in Nation-States

Beyond their particular territorial characteristics, island governance cannot be considered in isolation. Even less so when globalization and technological convergence have radically diminished the disconnection between island territories and the mainland. It is necessary, therefore, to briefly discuss the prevailing trends in democratic governance in the contemporary world.

The concepts of "democratic backsliding," "democratic recession," and "authocratization" have become more and more frequent in specialized literature (Diamond 2015; Waldner and Lust 2018; Haggard and Kaufman 2021). Figure 3.1 presents the electoral democracy index of the Varieties of Democracy project. The index shows how after a period of boom and consolidation of liberal democracy in the different regions of the world, a period of democratic backsliding has begun in the last decade.

In line with this, Adam Przeworksi, one of the most connoted living theorists of democracy, has recently described the expansion of "democratic crises" on a global scale and argues that neither the causes nor effective solutions to those crises are fully identified (Przeworski 2019). These crises have, in turn, facilitated the

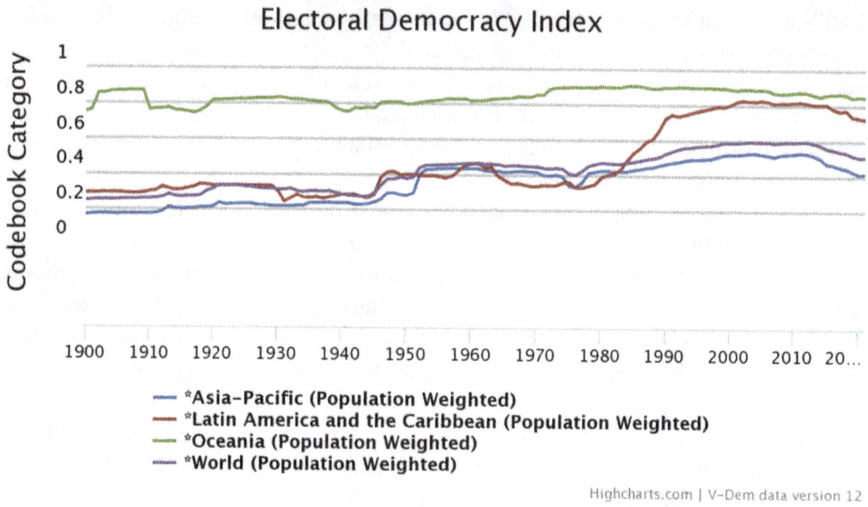

Fig. 3.1 Electoral Democracy Around the World. (Source: Own construction on the basis of V-DEM project (https://www.v-dem.net))

irruption of populist leaderships (left and right) and the actions of such leaderships, once elected, have generally contributed to deepening the erosion of democratic institutionality.

In this context, current societies are increasingly polarized. Unlike in previous periods when polarization also surged (e.g., the 1930s), today, social conflicts have radically escaped the traditional mediation structures of liberal democracy and the nation-state: political parties (Carothers and O'Donohue 2019). In this sense, we are witnessing a surge of affective polarization (instead of the ideological polarization of the past). This type of polarization, facilitated by the increasing circulation of information and the growing succession of corruption scandals that foster citizen indignation, has compressed governments' time to govern (Luna 2021). In several countries, affective polarization does not crystallize into stable alignments but generates permanent cycling. While contributing to producing party system turmoil, electoral movements favor emerging leaders. However, those leaders tend to alternate frequently. They often lose their popularity and government capacity a short while after being elected into office (e.g., in recent times, France, Chile, Brazil, Costa Rica).

In Latin America (considering 17 countries), for example, the first three presidencies after the transition to democracy enjoyed, on average, 35 months without a 10% drop in popularity (Luna 2021). The last three presidencies, again on average, managed to maintain their adhesion once appointed for seven months. In this sense, more and more frequently, currents of opinion consolidate the irruption of leadership in an election. However, those leaders rapidly fall in adhesion once elected, quickly becoming less capable of governing.

Alternatively, successful populist leaderships can cleave society between two poles, thus constituting a new polarizing cleavage that stabilizes electoral competition (Roberts 2021). At the same time, those leaderships have induced institutional polarization, thus threatening democratic institutions worldwide (e.g., in recent times, the USA, Turkey, Poland, Argentina, Bolivia, El Salvador). Moreover, in cases such as Venezuela or Nicaragua, this type of polarization brought democracy to an end (Munck 2022).

Contemporary societies also register changes in the patterns of political participation and the configuration of interest groups, which have become more fragmented, both in functional and territorial terms. For example, activism around single-issue movements has multiplied in recent years (Luna 2021). That has resulted, for example, in an increase in environmental activism. Figure 3.2 illustrates that recent trend. However, such activism has become atomized, making it difficult for traditional political systems to represent salient grievances in society. Atomization is also territorial, which has decreased the vertical and horizontal integration of political activism. These trends are associated, in aggregate terms, with an increase in highly disruptive and violent protests (Beissinger 2022). While highly disruptive, those protest events turn discontent difficult to channel back into the institutional realm (i.e., governance structures). Moreover, they tend to make existing institutional mechanisms less legitimate and stable. Figure 3.3 shows, for example, the evolution of protest and violent protest in recent years worldwide.

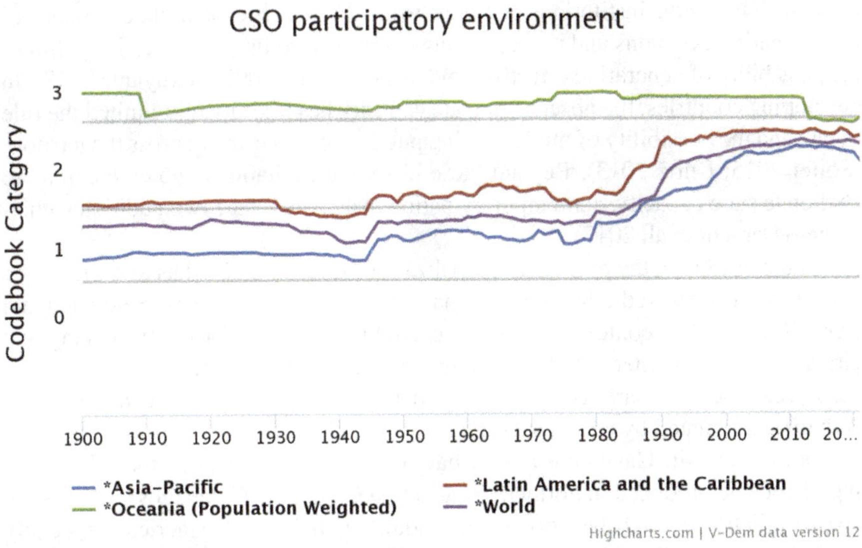

Fig. 3.2 Civil Society Organizations' Participation in Environmental Participatory Frameworks Around the World. (Source: Own construction on the basis of V-DEM project (https://www.v-dem.net))

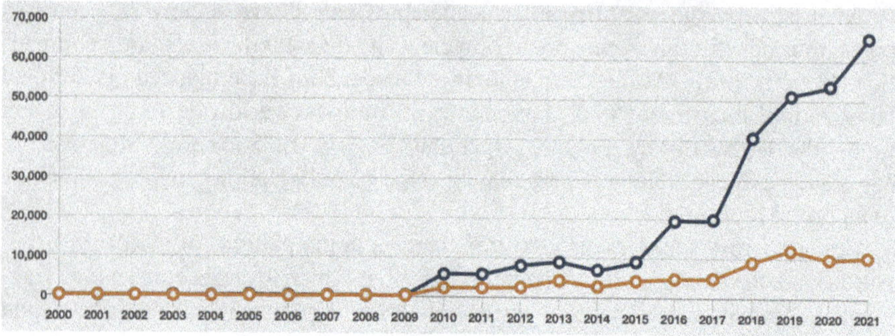

Fig. 3.3 Incidence of Protest (Blue) and Riot (Orange) Activities Around the World. (Source: Own construction on the basis of ACLED Project (https://acleddata.com))

In sum, recent social transformations and the erosion of current political systems have contributed to deep crises in contemporary liberal democracies. Countries are thus facing difficulties in structuring a legitimate order capable of generating effective liberal-democratic governance at the national level.

State Capacity in Contemporary Nation-States

Even if democratic institutions function well, the absence of a state capable of implementing decisions and monopolizing coercion throughout the territory dilutes the possibility of generating effective governance (Mann 2008; Fukuyama 2013). In developing countries, the absence of state capacity has traditionally limited the rule of law and the possibility of implementing public policies evenly across the territory (Soifer 2015; Kurtz 2013). Beyond these historical limitations, which continue to challenge state capacity, contemporary nation-states must face two additional challenges (Dargent et al. 2017).

On the one hand, the process of globalization since the 1970s has also meant the expansion of organized crime and international criminal syndicates (Feldmann and Luna 2022). In this context, islands are desirable territories for the trafficking and circulation of prohibited goods (including trafficking in protected species) and for the development of various activities used for laundering the proceeds of illegal activities dominated by organized crime.

For example, the Galapagos Islands have become a logistical hub for the trafficking of cocaine produced in northern Peru and in Colombia. Galapagos connects the coast of Ecuador (and the port of Guayaquil) with Central America (especially Costa Rica), providing a convenient transit route for speedboats operated by organizations trafficking drugs from South America to Central America, Mexico, and the

United States.[1] In turn, islands with significant tourist activity provide an ideal context for money laundering. Islands provide a convenient infrastructure for money laundering across different economic activities, from real estate developments and the provision of different tourist services to casinos or money exchange houses.

On the other hand, with the emergence and development of Big Tech, nation-states have also seen their ability to establish a monopoly on coercion weakened (Sheng 2022; Staab 2021). This monopoly was primarily based on the unilateral capacity of the state to generate data to guide its actions and govern society (Scott 2008). The technological disruption generated by the information revolution has contributed to decentralizing the production and circulation of critical information for governance (e.g., census records or technologies for environmental monitoring). In this sense, state sovereignty is no longer constrained by the loss of control over critical information flows, which have begun to be dominated by private companies and interest groups with the capacity to generate alternative narratives to the official ones, with which they dispute the government agenda. Although this information revolution has strong democratizing potential, in theory, it has nonetheless weakened nation-states' capacity to implement binding decisions across the territory.

In sum, contemporary nation-states are relationally weaker than in the past, as they are challenged by criminal and corporate actors that limit their coercive power and legitimacy. Contemporary states have also lost their historical monopoly on strategic data for sovereignty and the ability to govern society successfully. For the first time in several centuries, politico-institutional power has thus escaped its (main) locus.

International Cooperation and Governance Frameworks

Since the end of World War II, two international cooperation and governance frameworks have been especially promising in terms of providing a valuable infrastructure for advancing sustainability issues. On the one hand, under US hegemony and with the consolidation of the United Nations, there was an international cooperation framework capable of advancing reasonably binding agreements for member countries (Foot et al. 2003; Norrlof 2010). On the other hand, in Europe, the European Union process had advanced significantly towards consolidating a multi-level governance system, which provided a promising framework for anchoring coordinated sustainability policies among many nation-states and local territories (Marks and Hooghe 2001).

---

[1] See e.g.,: https://www.businessinsider.com/cocaine-drug-smuggling-increasing-around-galapagos-islands-in-pacific-2018-11; https://es.insightcrime.org/investigaciones/lagunas-legales-facilitan-comercio-aletas-tiburon-ecuador/; https://es.insightcrime.org/noticias/guerra-narcotrafico-ecuador-alimentada-colombia-mexico/; https://es.insightcrime.org/noticias/analisis/ecuador-autopista-de-la-cocaina-hacia-estados-unidos-y-europa/; https://es.insightcrime.org/noticias/noticias-del-dia/guayaquil-ecuador-puerta-salida-cocaina-europa/

Both structures of international cooperation are now under tension. On the one hand, the rise of China (and Russia) as powers competing with the United States for hegemony in the international system has diluted the influence of the former and its allies (Mastanduno 2019; Mansbach and Ferguson 2021; Rapanyane 2021). The US role has also weakened under the Trump Presidency (Falahi and Heidari 2019). On the other hand, the process of European integration is in question (Van Apeldoorn 2009; Tuori 2012). The EURO crisis, the failure of the ratification process of the European Constitution, Brexit, and the rise of populist leaderships with nationalist agendas in several EU countries have complicated the progress of integration and the scope of supra-national regulation efforts emaneting from Brussels (Tuori 2016).

In addition to both structural factors, there is a conjunctural one. In the post-COVID-19 context, and within the framework of a deep economic and social crisis in large part of the planet, it is highly probable that economic necessity and short-term horizons will dominate over the medium and long-term planning required by the sustainability agenda and democracy (Rapeli and Saikkonen 2020). In this context, and given the growing weakness of international cooperation frameworks, it will also be more likely to observe situations of free-riding by countries (and islands) seeking to improve their short-term situation in a context of scarcity.

Islands as Hotspots for Researching Governance for Sustainability

Although it may seem counterintuitive, islands are places where the problems associated with governance and sustainability are most radically present today. If isolated in the past, globalization and interconnection have contributed to placing islands at center stage. Moreover, the growing weight of environmental and sustainability issues also places islands at the center since environmental challenges are more present in island territories. Those challenges have additionally increased due to interconnection and the growing strategic value of islands for various legal and illegal industries.

How the alluded factors transform islands varies across specific cases. The reconfiguration of islands'economic, social, and political dynamics will depend on each case's specific characteristics and competitive advantages concerning its insertion into the global economy. History and other structural characteristics (i.e., demographic, geopolitical, etc.) will also matter greatly. Nevertheless, it is possible to point to some stylized trends.

First, islands have increasingly become a hotspot for international tourism (which produces various backward and forward linkages in the local economy) (Parra-López and Martínez-González 2018). Those backward and forward linkages are likely to be massive. Tourism also puts significant stress on local ecosystems charging capacity.

However, at the same time, islands remain pivotal in traditional extractive industries with growing international demand, such as fishing. Due to their characteristics, islands have always been territories suitable for establishing and developing illegal industries such as the trafficking of species and drug trafficking. Island characteristics also make them an ideal territory for laundering the proceeds of illegal activities.

Secondly, as a result of this emerging economic dynamic and the technological shock of increased connectivity with the mainland facilitated by new information technologies, islands are also suffering a shock at the social level. Connectivity implies a process of convergence with multiple possible impacts in cultural terms and even in terms of the population dynamics of the islands (e.g., an accelerated demographic transition). At the same time, the diversification of economic activity tends to exacerbate a process of social dualization between social sectors capable of successfully inserting themselves into the new industries and social sectors that choose to remain in traditional activities or fail to incorporate themselves into the new industries.

Finally, the emerging economic and social dynamics are also associated with a reconfiguration of political conflict. Depending on the initial configuration present on each island, ethnic grievances, between sectors working in emerging vs. traditional industries, between generations (i.e., young people adopting consumption patterns and cultural references of globalization vs. older generations defending traditional values), and between islanders and newcomers will reconfigure past patterns of political conflict in a highly contextual way (Vierros et al. 2020).

Change in all three arenas (economic, social, political) directly impacts sustainability (Petzold and Magnan 2019). New industries and the intensification of the extraction and production process of goods and services translate into stronger dilemmas between short-term economic growth and development and sustainability. Population dynamics also stress ecological services and island territories' charging capacity. The new dynamics of social conflict interact in a complex way with the politicization of sustainability dilemmas arising from economic transformation and social change (Connell 2018). This complexity also calls into question the management and governance schemes traditionally deployed in island territories (Mawyer and Jacka 2018).

Island Governance in a "Less Governable" World

Based on a literature review of island governance frameworks, it is possible to identify three general governance schemes. They have virtues and disadvantages with respect to two dilemmas: the tension between maximizing the legitimacy of decisions or their contributions to sustainability and the tension between maximizing sustainability and the risk of breaking or radicalizing island communities.

The first model is bottom-up and returns sovereignty to island communities over their practices and decisions regarding managing essential resources for sustainability (water, endemic species, etc.). This model promotes legitimacy, especially in communities with little conflict where a significant proportion of the population validates decisions. The model can also incorporate democratic innovations and participatory practices that provide greater legitimacy to decision-making. However, this model has a high probability of institutionalizing and promoting ancestral practices that, although highly valued by the communities, are generally suboptimal (or openly detrimental) regarding sustainability. In communities strained by emerging conflicts, they risk escalating conflict and deepening discontent.

The second model is top-down and is more frequent, especially in the case of islands that are part of continental nation-states. In this case, the legitimacy of regulatory frameworks for resource management tends to be increasingly problematic, particularly when social conflict has increased. In this scheme, the lack of legitimacy may translate into less enforcement capacity on the part of national authorities with respect to the implementation of regulation in island territories. In this sense, the need to implement (at a distance) sustainability regulations that local communities may resist generates principal-agent dilemmas that reduce the leverage of regulation in real terms. Regulations can be designed, but they will be challenging to implement in island territories.

Third, especially in the case of the Western Pacific, attempts have been made to move towards multi-level governance schemes associated with the development of an agreed framework between islands and countries with island territories based on multilateral international agreements. This model usually focuses on resource management (global public goods) or specific agendas (fisheries, combating organized crime and terrorism). Although promising from the point of view of seeking coordinated solutions that apply to larger ocean areas, this type of framework is likely to be seen as threatening the sovereignty of local communities. It may also thus reduce the legitimacy of the agreed regulation. On the other hand, the ongoing transformation of the international order (i.e., the loss of hegemony of the US and its allies in the international order and the emergence of China and Russia as relevant powers with alternative agendas) makes this model less feasible. For example, in a context in which the climate crisis has crystallized in policy circles as a significant and pressing global challenge, it has not yet been possible, after five failed attempts, to make progress towards the signing of a global treaty for the protection of the oceans.[2]

What Can Be Done?

Despite the urgency and relevance of the climate crisis, governance for sustainability today faces enormous challenges. First, the liberal democratic order is weakened and eventually in recession. Second, challenges to state capacity, and in particular,

[2]https://www.bbc.com/news/science-environment-62680423

the capacity of states to enforce regulation in practice, have multiplied. Third, the international system is also weakened in its ability to generate intergovernmental cooperation or implement effective multi-level governance schemes. Finally, in the wake of the COVID-19 pandemic and its effects on economic dynamics and welfare, we now face an incentive structure that makes it more likely that individuals and politicians will prefer to trade short-term benefits for medium- and long-term sustainability. This incentive structure also promotes the proliferation of informality and illegality, evading the regulation.

In this general context, islands, due to their characteristics, concentrate, on a relatively small territorial scale, the most profound dilemmas facing humanity today. At the same time, also because of their characteristics, islands provide an attractive field for developing applied research and relevant interventions regarding novel formulas for governance and the promotion of sustainability. Island's structural diversity (i.e., there is a wide variance in demographic structure, geographic characteristics, and economic activity, ethnic and social diversity, relationship with the mainland, etc.) turn them into ideal sites for researching governance for sustainability. Such diversity provides the chance to conduct research through the controlled comparison of different case studies. For this reason, islands constitute strategic territories in terms of investigating the effects and possibilities of formulas capable of mitigating social conflict and promoting sustainability through processes of democratic innovation.

How should we approach research on island governance and sustainability challenges? In my opinion, as a function of the enormous emerging challenges, the search for new institutional mechanisms could only be pursued locally and at a very incremental pace. In the wake of the crisis of different grand theories and their associated political projects, in the 1970s, Albert Hirschman (2013) coined the term "possibilism." Hirschman's possibilism framework was predicated on the notion that the search for development should be orienteded towards a search for solutions to local problems. Such search should be incremental and tentative and should be disciplined by highlighting how contextual factors limit the portability of the institutional mechanisms developed in one locality to other contexts.

Although minimalist, tentative, and incremental, Hirschman's approach is also realistic (especially if we bring the current state of global affairs into account). This minimalist framework provides opportunities around which policymakers, researchers, and local communities can cooperate in co-creating solutions to further islands'sustainability. Those solutions should also actively integrate governance mechanisms that are instrumental at mitigating conflicts and endowing emerging regulatory mechanisms with a modicum of legitimacy. Without the latter, regulations turn out to be paper tigers. Finally, those local solutions and governance mechanisms cannot automatically be expected to work well in contexts other than those in which they originated. Nevertheless, parallel advances across islands can jointly contribute to enhancing sustainability on a global scale. Moreover, comparative analyses of successes and failures can eventually illuminate (more portable or adaptable) possible ways forward.

References

Baldacchino G (2006) Innovative development strategies from non-sovereign Island jurisdictions? A global review of economic policy and governance practices. World Dev 34(5):852–867
Beissinger MR (2022) The revolutionary city: urbanization and the global transformation of rebellion. Princeton University Press
Carothers T, O'Donohue A (eds) (2019) Democracies divided: the global challenge of political polarization. Brookings Institution Press
Celata F, Sanna VS (2012) The post-political ecology of protected areas: nature, social justice and political conflicts in the Galápagos Islands. Local Environ 17(9):977–990
Connell J (2018) Islands: balancing development and sustainability? Environ Conserv 45(2):111–124
Dargent E, Feldmann AE, Luna JP (2017) Greater state capacity, lesser stateness: lessons from the Peruvian commodity boom. Polit Soc 45(1):3–34
Diamond L (2015) Facing up to the democratic recession. J Democr 26(1):141–155
Falahi, Sara, Nosrat Heidari. An examination of Donald Trump's foreign policy with an emphasis on the hegemony of the liberal world order. (2019): 76–100.
Feldmann AE, Luna JP (2022) Criminal governance and the crisis of contemporary Latin American states. Annu Rev Sociol 48
Foot R, Neil MacFarlane S, Mastanduno M (eds) (2003) US hegemony and international organizations: the United States and multilateral institutions. OUP Oxford
Fukuyama F (2013) What is governance? Governance 26(3):347–368
Haggard S, Kaufman R (2021) The anatomy of democratic backsliding. J Democr 32(4):27–41
Hirschman AO (2013) The essential Hirschman. Princeton University Press
Hooghe L, Marks G, Marks GW (2001) Multi-level governance and European integration. Rowman & Littlefield.
Keohane RO (1969) Lilliputians' dilemmas: small states in Internatinal politics. Int Organ 23(2):291–310
Kurtz MJ (2013) Latin American state building in comparative perspective: social foundations of institutional order. Cambridge University Press
Luna JP (2021) La chusma inconsciente: La crisis de un país atendido por sus propios dueños. Editorial Catalonia
Maass M (2009) The elusive definition of the small state. International Politics 46(1):65–83
Mann M (2008) Infrastructural power revisited. Stud Comp Int Dev 43(3):355–365
Mansbach RW, Ferguson YH (2021) The return of geopolitics and declining US hegemony. Cham, Populism and Globalization. Palgrave Macmillan, pp 89–140
Mastanduno M (2019) Partner politics: Russia, China, and the challenge of extending US hegemony after the cold war. Secur Stud 28(3):479–504
Mawyer A, Jacka JK (2018) Sovereignty, conservation and Island ecological futures. Environ Conserv 45(3):238–251
Médail F (2017) The specific vulnerability of plant biodiversity and vegetation on Mediterranean islands in the face of global change. Reg Environ Chang 17(6):1775–1790
Munck GL (2022) and Juan Pablo Luna. Cambridge University Press, Latin American politics and society: a comparative and historical analysis
Norrlof C (2010) America's global advantage: US hegemony and international cooperation. Cambridge University Press
Ostrom E, Janssen MA (2004) Multi-level governance and resilience of social-ecological systems. In: Globalisation, poverty and conflict. Springer, Dordrecht, pp 239–259
Parra-López E, Martínez-González JA (2018) Tourism research on Island destinations: a review. Tour Rev
Petzold J, Magnan AK (2019) Climate change: thinking small islands beyond Small Island developing states (SIDS). Clim Chang 152(1):145–165
Przeworski A (2019) Crises of democracy. Cambridge University Press

Rapanyane MB (2021) The new world [dis] order in the complexity of multi-polarity: United States of America's hegemonic decline and the configuration of new power patterns. J Public Aff 21(1):e2114
Rapeli L, Saikkonen I (2020) How will the COVID-19 pandemic affect democracy? Democratic Theory 7(2):25–32
Ratter BMW (2018) Island vulnerability and resilience. In: Geography of Small Islands. Springer, Cham, pp 173–199
Roberts KM (2021) Populism and polarization in comparative perspective: constitutive, spatial and institutional dimensions. Gov Oppos:1–23
Sartori G (1988) Teoría de la democracia, vol 2. Alianza, Madrid
Scott JC (2008) Seeing like a state. In: Seeing like a state. yale university Press
Sheng L (2022) Big tech and the nation-state. In: Big tech firms and international relations. Springer, Singapore, pp 1–41
Soifer HD (2015) State building in latin america. Cambridge University Press
Staab P (2021) Crises of sovereignty. Practicing Sovereignty: Digital Involvement in Times of Crises 54:107
Tuori, Kaarlo. "The European financial crisis: constitutional aspects and implications." (2012).
Tuori K (2016) The Eurozone crisis as a constitutional crisis 1. In: Polity and crisis. Routledge, pp 3–40
Van Apeldoorn B (2009) The contradictions of 'embedded neoliberalism'and Europe's multi-level legitimacy crisis: the European project and its limits. In: Contradictions and limits of neoliberal European governance. Palgrave Macmillan, London, pp 21–43
Vierros MK et al (2020) Considering indigenous peoples and local communities in governance of the global ocean commons. Mar Policy 119:104039
Waldner D, Lust E (2018) Unwelcome change: coming to terms with democratic backsliding. Annu Rev Polit Sci 21(1):93–113

Juan Pablo Luna is the author of Segmented Representation, Political Party Strategies in Unequal Democracies (OUP, 2014), and has co-authored Latin American Party Systems (CUP, 2010) and Latin American Politics and Society, and A Comparative Historical Perspective (CUP, 2022). He co-edited The Resilience of the Latin American Right (JHUP, 2014) and Diminished Parties (CUP 2022). His work focuses on political representation, state capacity, organized crime, and the political economy of democratic governance. The author acknowledges research support from the Center for Applied Ecology and Sustainability (CAPES, Proyecto Basal CONICYT-PIA) and from the Millenium Institute VIODEMOS.

Chapter 4
Changing Land Use in Island Countries: A Meta Perspective on Effects of Demographic Processes and Tourism

Richard E. Bilsborrow

Introduction

This is the second of two book chapters providing a meta-analysis of the demographic-socio-economic aspects of LCLUC (Land Cover/Land Use Change) in island states and countries, focusing on developing regions of the world. This chapter covers three broad topics in terms of their linkages with land use: population growth and migration, urbanization, and tourism, recognizing that there are inter-linkages across them as well as with the topics covered in the earlier chapter – agriculture, forest cover, economics, and modelling of linkages with land use. These broader linkages are also important for understanding how land use on islands is likely to continue in the future, as populations grow and move, economies grow, global warming continues, technology evolves, and country and global policies change.

Methodology for and Scope of Literature Review

The methodology used for identifying and screening the large and growing literature pertaining to the four topics here is described in more detail in the prior meta-study cited and can only be summarized here. The process began in 2017 with a computer search based on key words crossed with land use/land cover, leading to

R. E. Bilsborrow (✉)
Department of Geography and Carolina Population Center, University of North Carolina at Chapel Hill, Chapel Hill, NC, USA
e-mail: richard_bilsborrow@unc.edu

S. J. Walsh et al. (eds.), *Island Ecosystems*, Social and Ecological Interactions in the Galapagos Islands, https://doi.org/10.1007/978-3-031-28089-4_4

large numbers of initial references, which were classified by the apparent main topic, then pared down based on the titles to tier 2 (potentially relevant) titles for which abstracts were obtained, and then finally to tier 1 items (apparently relevant).[1] The time reference was 1990 to 2017.[2] The goal was to identify studies which, while possibly also including valuable descriptive material, had some analytical, modeling or statistical analysis of data linking the topic to LCLUC. For each study reviewed in this chapter, I will note the author, publication date, island (and Archipelago or country) studied, time reference of the data and analysis, model structure, linkages to LCLUC, statistical methods and parameters estimated (if any), data used, main findings and conclusions, and (if relevant) potential utility for investigating or modeling similar relationships in other island contexts and/or for simulation/forecasting models. For each topic below, the discussion will mostly draw on tier 1 items. Given the passage of five years since the original computer search, a supplemental search was carried out in mid-2022 to update the references. The geographic distribution of the research across islands covered in this meta-assessment is indicated in the global color map below (Fig. 4.1).

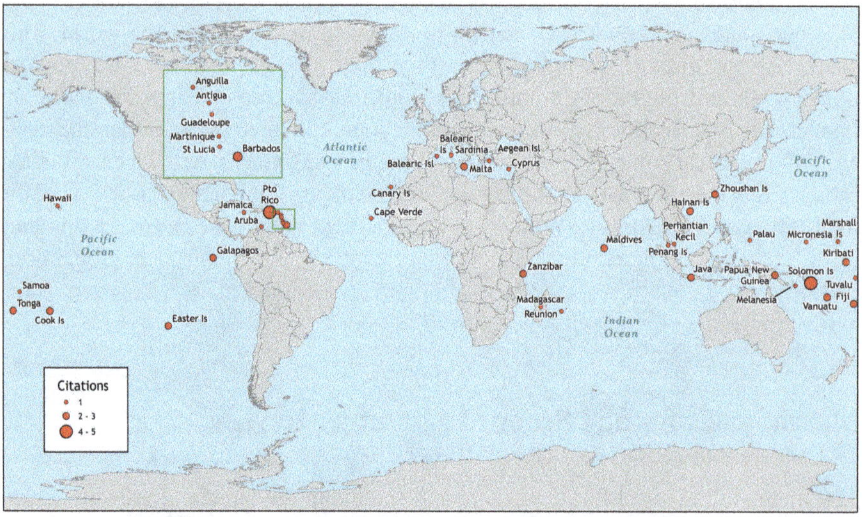

Fig. 4.1 Map showing number of studies reviewed here per country/study site

[1] I am grateful for excellent bibliographic assistance from Sommer Barnes and Lori Delaney of the Carolina Population Center library at UNC; to Brian Frizzelle of CPC for the informative color map; and to doctoral students in Geography at UNC Sara Schmidt & Francisco Laso for help in organizing and obtaining documents. Philip McDaniel of the UNC Davis graduate library helped in updating the references to 2022.

[2] For example, for tourism and land use, the 1990 to 2017 computer search led to 1740 journal references, of which 309 appeared potentially relevant based on their title, which was reduced to 99 for which abstracts were sought and read, and finally to 29 for which full articles were obtained. Similarly, the computerized update found over 500 references but only 23 were reviewed in full.

Topic 1. Population Aspects of Land Use (LU) in Islands, Including Migration

This covers key dimensions of population, including population size, growth, and distribution, as well as changes resulting from internal or international migration, all to the degree they may relate to LCLUC in island states. Little attention is given to studies focusing on the two demographic factors which together account for population growth, viz., fertility and mortality, nor to studies on only population trends/forecasts. Demographic processes are, of course, linked to most other topics, much covered in the prior meta-study chapter (viz, on changes in forest cover, agricultural land area, and economic models of land use change) (Bilsborrow 2020). Theoretical aspects of the relationships between population and agriculture/forest cover including reviews of the literature are found in Bilsborrow (2021), and references therein.

Both internal and international migration are covered under this topic when central aspects of the paper and when there appear implications for LCLUC. For example, internal rural-urban migration is of special relevance here when contributing to urban population growth and urban area expansion, and international migration for its possible effects on population size and distribution, with emigration reducing population pressures on space and immigration increasing them. International labor migration often also leads to large remittances to origin island households, which may stimulate consumption, investment, and economic growth (and the expansion of urban and agricultural land areas) and may also reduce poverty though it can increase inequality and local prices. In addition, return migration of international migrants may affect LU if the migrants invest in housing, farming, or farmland, set up or expand a business, and bring technological innovations.

Topic 2. Urbanization

This is key for understanding LCLUC in many islands, as urban areas grow with population growth in the usually limited areas of islands. This may also occur due to rural-urban migration, effects of an expansion of tourism (leading to more infrastructure/facilities needed for tourists and more labor to come from other sectors of the economy), effects of the flow of remittances from international migrants living abroad on consumption and investment in urban areas, or other factors stimulating economic growth on the island. Urban areas of islands often have substantial economic sectors and land areas tied up in services and manufacturing, whether for domestic needs, export, or tourism, and all have water, electricity, transportation, housing, construction, etc., infrastructure and land use needs, with implications for the urban land use area and its expansion over time into agricultural and forested areas.

Topic 3. Tourism

The basic data on tourism in island states – numbers of tourists, their origins, expenditures in countries, etc., is found in yearbooks of the UN World Tourist Organization and the World Bank. Statistics on international tourism should be better for island states than others since people can only arrive by air and sea, so arrivals can be easily tracked, though identifying which are tourists still requires excluding residents returning and people arriving for business or other non-tourism reasons. In recent decades, economic growth in most of the world and vast improvements in communications and transport have led to big increases in global tourism. The size and growth of tourism can have important effects on land use on an island, requiring and stimulating infrastructure expansion – much in coastal and urban areas. This may be at the expense of forest lands and add to pressures to expand agricultural lands to meet growing demands for food from more tourists. Different types of tourism will have different effects, varying from nature or ecological tourism to beach/surfing/snorkeling tourism, fishing tourism, luxury hotel tourism, etc. Many Island States are small and highly dependent on tourism, and if not, hope to increase it. Tourism often provides significant monetary resources and jobs in the island state, increasing incomes of those involved with tourists who provide lodging (hotels, etc.), food (whether produced locally or imported), and recreation (beaches and coral reefs, protected areas, and other tourist locations). This can have important implications for urban areas for space for lodging and restaurants, and for recreation and agricultural land use, and therefore for employment in the sectors catering to tourists, which has multiplier effects on the larger economy and labor force. However, there may also be negative implications for the environment, including damage to protected areas, coastal areas or other sites frequented by tourists. This review will include studies on the determinants of tourism and linkages with other sectors, including multiplier effects on economic growth. As in the prior meta-study on islands, this one will also be organized by geographic area – Asia-Pacific, Caribbean, Africa (Atlantic and Indian Ocean regions), and Other (mostly Mediterranean).

Topic 1. Population Aspects of Land Use (LU) in Islands, Including Migration

Theoretical underpinnings for assessing the impacts of population on land use, including agriculture and forests, are covered in Bilsborrow (2021), with applications to islands in Bilsborrow (2020). The focus here is therefore on empirical relationships between population growth and land use on island states, and secondly on the effects of migration, internal and international. This will thus exclude studies focusing on the direct impacts of population growth or migration on forests, agricultural lands, biodiversity, coral reefs, water shortages, or urban land area (latter covered in following section).

Population Growth Effects

As with the earlier collapse of the classic Maya populations in Central America, the disappearance of the human population on Easter Island in the Pacific Ocean has attracted much research interest. Two recent examples include, first, Reuveny and Decker (2000), who postulate that its collapse was a function of high fertility and limited resources, such that cycles of technological changes generated fluctuating periods of population growth that put ever more pressure on resources, resulting eventually in a Malthusian collapse. A subsequent modeling study (Cole and Flenley 2008) confirms the possibility of such a "far from equilibrium" situation. An earlier study by Rapaport (1990) reviews the carrying capacity of Tuvalu, Kiribati, the Marshall Islands, and the Maldives and the effects of high population growth, finding it leading to high urbanization, pollution and epidemics, emigration, and foreign aid dependence, while in contrast lower population growth results in less urbanization and out-migration in the Cook Islands and Micronesia. Jones (1993) notes that Java (Indonesia) already had a population of 180 million in 1990, projected to grow by 50% by 2020 even with ongoing fertility decline, and was already clearing land moving up mountain slopes to expand cultivation, with the underlying problem being population growth and its needs for food and employment. Similar studies for other large islands of Indonesia, Madagascar, and the Philippines are described in Bilsborrow (2020). Ningal et al. (2008) linked the fast rise in population in Papua New Guinea from 2.3 to 5.2 million in 1975–2000 to a 60% increase in the agricultural land area, associated with 85% of the population depending on subsistence agriculture, reducing the primary forest by 60% as well. They conclude that this trend will continue unless improved farming systems are implemented. Fazey et al. (2011) note that the accumulated responses of people to increasing population pressures can be coped with by people for a while but eventually leads to "maladaptive outcomes." They develop a conceptual model for the Solomon Islands with feedbacks, showing the two drivers of population growth and peoples' aspirations for a better life led to mal-adaptations. They recommend that both excessive population growth and unrealistic aspirations should be addressed, the latter to focus on well-being rather than higher incomes per se, along with orienting development projects toward those with less impact on the environment. Also, on the Solomon Islands, Reenberg et al. (2008) use satellite imagery and air photos to show that agricultural land use intensity did not change much despite the increase in population. They interview 48 households, finding traditional agriculture supplemented increasingly by other activities (shopkeeping, other businesses, government employment). Eriksson et al. (2020) study the popular hypothesis that livelihood diversification is a good strategy for farm households to cope with increasing populations on limited land. They analyze survey data from 235 households in the Solomon Islands and eight types of diversification projects, finding they led to *higher* rates of food insecurity. Pilbeam et al. (2019) call for greater efforts to integrate traditional knowledge and modern Western scientific knowledge to support conservation in Palau, showing how this is challenging.

There are only a few good relevant studies outside the Asia-Pacific region, including Tole (2001) on Jamaica, Hugh et al. (2018) on Zanzibar, and Conrad and Cassar (2014) on Malta. Tole found at the parish level that forest loss was more correlated with population size than any other variable studied. She recommends the government change its policies of promoting export agriculture to instead enhancing smallholder productivity, but neglected to say anything about reducing population growth. Looking at coastal development scenarios, Huge et al. used a Delphi survey of local and international "experts" to seek a balance between tourism and local fisheries' land use, recognizing that increasing demographic pressures will exist. On Malta, the authors seek how to decouple economic growth from environmental degradation – how to achieve this in 4 sectors, energy, water, land, and waste – noting that no decoupling was found between land development and either economic or population growth, both being major sources of pressure.

Migration Effects

Connell and Brown (1995; see also Skeldon, 1991) review the situation up to that time in 22 island states of the South Pacific. All of the local indigenous populations engaged in international migration except the Melanesians who (along with Micronesians) engaged in internal migration. Earlier migration tended to be circular but has been becoming more permanent. The major goal is to improve the standard of living of the migrant in the destination as well as in the origin household via remittances. Besides often being the largest source of origin household incomes, remittances also provide insurance and social ties, while reinforcing the social hierarchy by going mainly to senior household heads. While their major use is for daily consumption needs, some go to churches, for celebrations, school fees, and modern houses. Women migrants are more likely to send remittances than men. They recommend governments develop policies to encourage more investment to help economic growth. Sofer (1993) describes internal labor migration's contribution to development in Fiji, noting migration is from the lower income to the higher income provinces, much circular. This mobility provides cheap labor to the higher income areas, while savings from wages taken back to origin areas improves their living standards, in this core-periphery model.

For Caribbean islands, Walters (2016) states that migration is a central feature, though its consequences for land use have been rarely studied. He focuses on St. Lucia from the 1950s to the early 2000s. An initial wave of emigrants to the UK and North America led to reduced land in farming and some upland reforestation in the 1950s to 1970s. The growth of tourism and construction then led to further rural-urban migration, followed by a wave of immigrants returning from the earlier flow, to live in the growing urban areas. This led to a residential home construction boom, transforming agricultural and forest lands near the coast to peri-urban suburbs, and to further afforestation in the highlands. Except for the wave of return migrants, this is similar to what has been observed in Puerto Rico, where tourism and urban economic growth led to massive rural-urban migration, declining lands in agriculture, and afforestation (see also Del Mar Lopez et al. 2001; Helmer 2004; Bilsborrow 2020). Finally, Condon and Ogden (1996) study emigration, circulation and return migration from the French islands of Guadeloupe and Martinique to France for both employment and education. They note how emigration reduced population growth directly and then return migrants came back with lower fertility norms, lowering overall fertility by the 1980s.

The only study on Africa here is on Cape Verde by Brandao and Zoomers (2010). Traditionally, people out-migrated from Cape Verde, with about half of its one million people living abroad. Economic growth started to lead people to return, encouraged by government policies. Return migrants brought money and human and social capital, including guest workers, young workers, and graduates from abroad. Despite the huge return migration, the development impact was limited, so they recommend that policies switch from a focus on encouraging return to promoting better integration and investment.

The last item is on the determinants of return migration to the Greek islands of the East Aegean Sea. Robolis and Xideas (1996) investigate the sensitivity to economic factors, and the consequences of their investments for economic growth.

Topic 2. Urbanization

Here we look directly at the expansion of urban land areas and its causes. First, Seto et al. (2011) conducted their own meta-analysis of 326 studies of urban land conversions at the global level in 1970–2000, noting that "across all regions… urban land expansion rates were higher than or equal to urban population growth rates". Zhang (1994) synthesized the Alonso (1964) location model, the two-sector neoclassical growth model, and endogenous population theory to show the interdependences between economic growth, residential structures, and population growth over space and time. With reasonable assumptions for an island, cellular automata models (e.g., White and Engelen 1994) could also be used to explore past and future patterns of urban land expansion.

In the Asia and Pacific region, Jones and Lea (2007) explore future urban destinations for island countries, noting past improvements from successful urban reform efforts in Kiribati and Samoa. These required political will, external technical assistance, and having a baseline of modest economic growth and interest in environmental planning. Tan et al. (2010) study urban expansion on the island of Penang, Malaysia in 1999–2007 based on spatial imagery, macroeconomic data, and climate data, and use multiple regression to find urban expansion and higher density linked to large declines in barren and forest areas, a smaller decline in grasslands, and cycles of out-migration and return migration. Pan et al. (2016) study urban expansion on the Zhoushan Islands, China, linked to building a bridge to the mainland, the overall economic boom in China, and a big increase in tourism. Hotspots were coastal areas of ports, harbors, and tourist facilities. But the land expansion was inefficient, suggesting the need for better land use planning, as well as ongoing reclamation from the ocean and ecological protections. See also Shen et al. (2017) and Xu et al., (2002) on Hainan Island, China. Finally, Pedersen Zari et al. (2020) describe devising urban ecosystem-based adaptation (EbA) projects for Vanuatu, to plan for adaptation to rapid urbanization, climate change, and ecosystem degradation. Besides pulling together multidisciplinary data sets and experts, they include a participatory approach which leads to more "soft" cost-effective solutions instead of "hard" engineered solutions. EbA begins with a comprehensive overall qualitative assessment of the causes and impacts of degradation of ecosystems, climate change, and social systems, considers future scenarios and possible policies, and then prioritizes solutions.

In the Caribbean, several studies have linked urban expansion to the loss of agricultural lands, especially on Puerto Rico (e.g., del Mar Lopez et al. 2001, Martinuzzi et al. 2007, Parés-Ramos et al., 2008). Del Mar Lopez et al. (2001) note that after four centuries of dependence on agriculture, in the twentieth century industry boomed leading to abandonment of agriculture and migration to cities, with new urban lands increasingly coming from prime agricultural lands rather than non-agricultural lands. Martinuzzi et al. (2007) describe the history of ineffective land use planning and urban sprawl and note how integrating satellite imagery and population census data should be done for more effective planning.

The Africa region provides two especially stimulating studies here. Kukkonen and Käyhkö (2014) use satellite imagery and aerial photos to investigate relationships between urban sprawl and forest clearing on Zanzibar Island, Tanzania. They note a rapid increase in deforestation rates from 1975–1996 to 1996–2009 and recommend examining the three main land tenure regimes separately to better understand it. In community forests, increased population and shifting cultivation occurred, while agroforest areas were most impacted by urban sprawl. In the third area, government forests, forest cover actually increased due to tree planting programs. Overall, causes were population growth ("the biggest hindrance to economic development", p. 195), in-migration, urbanization, and tourism, which combined led to increasing demand for agricultural and forestry products. The second study is by Lestrelin et al. (2017) on the small 2500 km2 island of Reunion, a hotspot of biodiversity but undergoing rising population sizes and density and urban sprawl. A spatially explicit model was developed and simulated over time using a modeling language (Ocelet) developed by the French government Agricultural Research Center. CIRAD, facilitating interacting ecological and social processes dynamically. The model facilitated bringing together stakeholders and planners in a participatory approach, to first understand the model structure, then develop and test alternative future scenarios, provoking discussions, and finally revising and rerunning the model, involving multiple meetings over 3-years. Most local participants found it very useful – and a new way to go beyond the usual biophysical-economic models to integrate social and political factors.

Finally, a study on Sardinia, Italy (Manca and Clarke 2012), use the SLEUTH (acronym for slope, land cover, excluded regions, urban land cover, transportation, and hill shade) urban planning model. It includes an urban growth cellular automata sub-model (where cells transform over time from one form or use to another based on neighboring cell attributes). It was used to project the urban expansion of an interior municipality Mancomer based on its characteristics (of cells) from 1998 to 2006, which it did with 98% accuracy. The model takes into account population growth, parks and archeological sites, road networks, water resources, etc. It could be useful in testing scenarios of future expansion for urban areas on other islands once sufficient socio-economic and satellite imagery baseline information is brought together.

Topic 3. Tourism

Given the focus of this book on land use, and space constraints, some topics of special interest regarding tourism will not be considered, including negative impacts of tourists on local cultures or the environment; increasing water pollution, scarcity; benefit-cost studies; how to make islands more attractive to tourists; optimal construction of infrastructure for tourists (hotels, etc.); surfing tourism; attitudes of locals to tourists or of tourists to their experience; historical studies, etc. Due to other articles in this volume on the Galapagos and Hawaii, they are not covered here much, nor are the effects of the pandemic (70–90% reductions in tourism globally in 2020–21).

Some Theoretical Considerations

Since tourism is a major economic sector (in terms of share of Gross Domestic Product and employment) in many if not most of the small island states which are the subjects of this chapter, then increasing tourism stimulates the economy, so most governments seek such an expansion. Narayan et al. (2010) provides an excellent review of the global literature, then a statistical analysis of data over time for four Pacific Islands (Solomons, Fiji, Tonga, Papua-New Guinea). They find a 1% increase in tourism earnings linked to a .5 to .9% growth in GDP, depending on the relative size of the tourist sector and the size of leakages resulting from any increased imports of food. In practice, some countries experience a "tourist lifecycle" in which expansion is eventually followed by stagnation, then excess capacity (Hernandez and Leon 2006). Evidently this depends on the size and carrying capacity of the island and the type of tourism *(op. cit.)* as well as its size. In a complex economic model with various assumptions, Stauvermann and Kumar (2017) show that government investments in education (human capital) can contribute to lowering population growth and increasing economic growth more with rising tourism than without it. In a related, non-mathematical study, Mihalic (2020) studies the issue of over-tourism, which takes into account not only economic and environmental but also socio-cultural sustainability (negative impacts on environmental/cultural attractions, drawing on the influential *Our Common Future* report (WCED 1987).

This involves a broader dimension of sustainability, which she says is still generally lacking despite promotions of "responsible tourism." She calls for greater regulations and more government leadership in island states to achieve this.

McElroy et al. (2010) review the literature on economic aspects of tourism in small islands using a "tourist penetration index", with different socio-economic and demographic profiles and postulating low, medium, and high levels of tourism for 36 small islands, finding that tourism is generally an effective engine of island development. Similarly, Marsiglio (2015) describes how tourism contributes to development, which can be green with environmentally conscious tourists. Scheyvens and Momsen (2008) consider whether tourism contributes to poverty reduction in small island states. They say that this is more likely if social sustainability is taken into account along with the usual economic and environmental sustainability and suggest ways in which national planners could encourage the private (tourist) sector to support poverty reduction. Finally, structural models and agent-based models have been applied to simulate linkages between tourism, infrastructure, employment, food demand and supply, urban and rural land use, etc., for the Galapagos Islands of Ecuador (Miller et al. 2010; Pizzitutti et al. 2020), which could be improved and adapted for other islands. (See also White and Engelen 1994, on agent-based models.)

Asia-Pacific Region

This appears to be region where tourism has been most studied in the journals. Telfer and Wall (1996) investigate a successful effort on the island of Lombok, Indonesia, to expand the production and distribution of quality fish and local vegetables to meet the increased demands of tourists for food, thanks to an enterprising Chef of the Sheraton Hotel. On Perhentian Kecil Island off the east coast of Malaysia, Hamzah and Hampton (2013) discuss how the resilience of the local population accustomed to backpackers and independent tourists helped fight off state government attempts to replace that tourism with high-end tourist resorts. Two interesting studies on islands of China are also noteworthy. Wang and Liu (2013) note Hainan is the most important domestic tourism site in China and its growth has led to rapid economic growth, confirmed by remote sensing and socio-economic data for 1991–2007, with forests and farmland declining and replaced by tropical fruit orchards and more urban land. Xie et al. (2021) report on the effects of construction of the sea-cross bridge and tunnel from Shanghai mainland to Zhujiajian Is., but increasing its ecological vulnerability, with 90% of the potential land for built-up areas already used up and areas of severe vulnerability expected to grow by 78% in 2015–35. Finally, Diedrich and Aswani (2016) interviewed people in the Solomon Is., finding most have negative feelings about tourism – too much foreign influence and diminishing social capital – yet most still support it.

In an economic study on changes over time in the Galapagos Is., Taylor (2009) found increased tourism since 1999 resulting in restructuring towards larger hotels and cruise ships and overall 78% growth in GDP in 6 years. Nevertheless, rapid population growth from migrants coming from mainland Ecuador nearly erased the benefits in terms of growth in *per capita* income, raising questions about the compatibility of ecotourism and conservation in this unique ecological setting. Kumar et al. (2016) study the bi-directional relationships between tourism and growth in GDP per capita in 2009–2014 in the Cook Islands, recommending policies focusing on inclusive and income-generating tourism projects to achieve positive feedbacks. Finally, for the Cook Islands plus Fiji, Tonga, Samoa & Vanuatu, Kumar et al. (2022) study tourism competitiveness over 2002–2019, finding islands more alike face more competition from each other.

Caribbean

Surprisingly, only four studies made the cut here. Maul (1996) investigated linkages between tourism and sustainable development in 36 Caribbean and other island states with under one million in population and under 5000 km2 in area, using the tourism penetration index. He found rising tourism associated with rising income, in-migration, rising literacy and life expectancy, and falling unemployment, fertility and infant mortality, with the benefits greater in islands still linked to colonial powers than in sovereign islands. Modeste (1995) reviewed evidence on links between tourism and economic expansion in three Caribbean islands (Barbados, Antigua & Barbuda and Anguilla), concluding that tourism growth was overall good for the economy except for agriculture. Similarly, Steenge and van de Steeg (2010) develop an input-output model from data for Aruba in 1999 to explore tourism multipliers. And for Barbados, Bunce (2008) discusses the transformation of upland rural areas via “leisuring” into tourist areas, by external elites insensitive to local interests, calling for more research into the issue.

Africa

In designing a strategic plan for sustainable tourism for the Gran Canaria Islands, Garcia-Falcon & Medina-Muñoz (1999) recommends a methodology based on socio-cultural aspects, the environment, and the economy. López-Guzmán et al. (2013) examine demand-side impacts for Cape Verde. On the other side of the continent, for the Maldives, Buckley et al. (2017) use stakeholder interviews to study competition for control over the best surfing sites in previous decades, involving battles over property rights for guest houses in 88 islands, observing increasing crowding.

Other

In the Balearic Islands, Polo and Valle (2008) use a general equilibrium economic model with data from an input-output table to explore the effects of a 10% fall in tourism, with first, a reduction in non-resident consumption followed by effects on other sectors. Kammas and Salehi-Esfahani (1992) study the development of tourism in Cyprus, which was successful initially due to little leakage of tourist demand for imports (as in other countries) since the domestic economy could provide the necessary food and labor (indeed, 20% of the labor force). However, they conclude that these economic benefits need to be balanced against environmental costs. Loumou et al. (2000) examine how tourism led to ecosystem conservation on Lesbos Is., Greece.

Conclusions

Figure 4.1 above shows the distribution of studies covered here. Perhaps because it is the nature of the beast, viz., most of the islands covered being, after centuries of colonialism, now small independent states in tropical areas, this recent topic meta-analysis finds more good studies on tourism than on population/migration or urbanization, and given the location of these states, more in the Asia-Pacific region. The concentration on tourism may also be linked to the ease of access to data from the WTO, in contrast to many small countries having limited socio-economic-demographic data (e.g., from censuses and surveys) due to limited human resources/staff, which needs to be addressed to improve research.

The lack of quality analytical studies on population/migration impacts and on urbanization processes, with implications for policies on islands, is a bit disappointing. In principle, it should be easier to implement specialized household surveys in small islands than in larger continental countries; given the growing wealth of satellite imagery, there is a bright future for further research on islands, especially if linked with improved administrative data and household survey data.

Finally, the three topics here should be considered in combination with the topics of forestry, agriculture, economics and land use models in the previous meta-analysis for islands (Bilsborrow 2020), given the many interdependent linkages among the topics.

References

Alonso W (1964) Location and land use. Harvard University Press, Cambridge

Bilsborrow RE (2020) Economic and related aspects of land use on islands: A meta perspective. In: Walsh SJ, Riveros-Iregui D, Acre-Nazario J, Page PH (eds) Land cover and land use change on islands: social & ecological threats to sustainability. Springer, pp 11–62

Bilsborrow RE (2021) Population and agricultural change. Ch. 17. In: Hunter L, Gray C, Veron J (eds) Handbook on population & environment. Springer, pp 374–420

Brandao F, Zoomers A (2010) Sodade di Nhos Terra: The development potential of return migration to Cape Verde. Int Dev Plan Rev 32(34):267–289

Bunce M (2008) The 'leisuring' of rural landscapes in Barbados. New spatialities and the implications for sustainability in small Island states. Geoforum 39(2):969–979

Buckley R, Guitart D, Shakeela A (2017) Contested surf tourism resources in the Maldives. Ann Tour Res 64:185–199

Cole A, Flenley J (2008) Modelling human population change on Easter Island far-from-equilibrium. Quat Int 184(1):150–165

Condon SA, Ogden PE (1996) Questions of emigration, circulation and return: Mobility between the French Caribbean and France. In J Popul Geogr 2(1):35–50

Connell J, Brown RP (1995) Migration and remittances in the South Pacific: Towards new perspectives. Asian Pac Migr J 4(1):1–33

Conrad E, Cassar LF (2014) Decoupling economic growth and environmental degradation: Reviewing progress to date in the small Island state of Malta. Sustainability 6(10):6729–6750

Del Mar Lopez T, Aide TM, Thomlinson JR (2001) Urban expansion and the loss of prime agricultural lands in Puerto Rico. Ambio 30(1):49–54

Diedrich A, Aswani S (2016) Exploring the potential impacts of tourism development on social and ecological change in the Solomon Islands. Ambio 45(7):808–818

Eriksson H, Sulu R, Blythe JL, van der Ploeg J, Andrew N (2020) Intangible links between household livelihoods and food security in Solomon Islands: Implications for rural development. Ecol Soc 25(4):18ff

Fazey I, Pettorelli N, Kenter J, Wagatora D, Schuett D (2011) Maladaptive trajectories of change in Makira, Solomon Islands. Glob Environ Chang 21(4):1275–1289

García-Falcón JM, Medina-Muñoz D (1999) Sustainable tourism development in islands: A case study of Gran Canaria. Bus Strategy Environ 8(6):336–357

Hamzah A, Hampton M (2013) Resilience and non-linear change in Island tourism. Tour Geogr 15(1):43–67

Helmer EH (2004) Forest conservation and land development in Puerto Rico. Landsc Ecol 19(1):29–40

Hernandez J, Leon C (2006) The interactions between natural and physical capitals in the tourist lifecycle model. Ecol Econ 62:184–193

Hugh J, Van Puyvelde K, Munga C, Dahdouh-Guebas F, Koedam N (2018) Exploring coastal development scenarios for Zanzibar: a local microcosm-inspired Delphi survey. Ocean Coast Manag 115:83–92

Jones GW (1993) Population, environment, and sustainable development in Indonesia. Majalah Demografi Indones 20(40):1–20

Jones P, Lea JP (2007) What has happened to urban reform in the Island Pacific? Some lessons from Kiribati and Samoa. Pac Affairs 80(3):473–494

Kammas M, Salehi-Esfahani H (1992) Tourism and export-led growth: The case of Cyprus, 1976-1988. J Dev Areas 26(4):489ff

Kukkonen M, Käyhkö N (2014) Spatio-temporal analysis of forest changes in contrasting land use regimes of Zanzibar, Tanzania. App Geogr 55:193–202

Kumar R, Stauvermann P, Patel A, Kumar N, Prasad S (2016) Exploring the nexus between tourism and output in Cook Islands: An ARDL bounds approach. Soc Indic Res 128(3):1085–1101

Kumar NN, Patel A, Singh R (2022) Modelling tourism competitiveness in small Pacific Island countries. Tour Econ 28:692–713

Lestrelin G, Augusseau X, David D, Bourgoin J, Lagabrielle E, Lo Seen D, Degenne P (2017) Collaborative landscape research in Reunion Island: Using spatial modelling and simulation to support territorial foresight and urban planning. App Geogr 78:68–77

López-Guzmán T, Borges O, Hernández-Merino M, Cerezo JM (2013) Tourism in Cape Verde: An analysis from the perspective of demand. Tour Econ 19(3):675–688

Loumou A, Giourga C, Dimitrakopoulos P, Koukoulas S (2000) PROFILE: Tourism contribution to agro-ecosystems conservation: The case of Lesbos Island, Greece. Environ Manag 26(4):363–370

Manca G, Clarke KC (2012) Waiting to know the future: A SLEUTH model forecast of urban growth with real data. Cartographica Intern J Geogr Inform Geovisualiz 47(4):1–9

Marsiglio S (2015) Economic growth and environment: Tourism as a trigger for green growth. Tour Econ 21(1):183–204

Martinuzzi S, Gould W, Ramos Gonzalez O (2007) Land development, land use, and urban sprawl in Puerto Rico integrating remote sensing and population census data. Landsc Urban Plan 79:288–297

Maul GA (1996) Ocean science in relation to sustainable economic development and coastal area management of small islands. Coast Estuar Sci 51:1–17

McElroy J, Hamma P (2010) SITEs revisited: Socioeconomic and demographic contours of small Island tourist economies. Asia Pac Viewp 51(1):36–46

Miha T (2020) Conceptualizing overtourism: A sustainability approach. Ann Tour Res 84:103025

Miller BW, Breckheimer I, McCleary AL, Guzman-Ramirez L, Caplow SC, Jones-Smith JC, Walsh SJ (2010) Using stylized agent-based models for population-environment research: A case study from the Galapagos Islands. Popul Environ 31(6):401–426

Modeste N (1995) The impact of growth in the tourism sector on economic development: The experience of selected Caribbean countries. Economia Internazionale 48:375–385

Narayan P, Narayan S, Prasad A, Prasad B (2010) Tourism and economic growth: A panel data analysis for Pacific-Island countries. Tour Econ 16(1):169–184

Ningal T, Hartemink AE, Bregt A (2008) Land use change and population growth in the Morobe Province of Papua New Guinea between 1975 and 2000. J Environ Manage 87(1):117–124

Pan Y, Zhai M, Lin L, Lin Y, Cai J, Deng JS, Wang K (2016) Characterizing the spatiotemporal evolutions and impact of rapid urbanization on Island sustainable development. Habitat Int 53:215–227

Parés-Ramos IK, Gould WA, Aide TM (2008) Agricultural abandonment, suburban growth, and forest expansion in Puerto Rico between 1991 and 2000. Ecol Soc 13(2)

Pedersen Zari M, Blaschke P, Jackson B, Komugabe-Dixson A, Livesey C, Loubser D, Martinez-Almoyna Gual C, Maxwell D, Rastandeh A, Renwick J, Weaver S, Archie K (2020) Devising urban ecosystem-based adaptation (EbA) projects with developing nations: a case study of Port Vila, Vanuatu. Ocean Coast Manag 184:105037

Pilbeam V, van Kerkhoff L, Weir R (2019) Conservation decision-making in Palau: An example of the parallel working of scientific and traditional ecological knowledge. Environ Manag 64:564–579

Pizzitutti F, Brewington L, Walsh SJ (2020) Human and natural environments, Island of Santa Cruz, Galapagos: A model-based approach to link land cover/land use changes to direct and indirect socio-economic drivers of change. In: Walsh SJ, Riveros-Iregui D, Acre-Nazario J, Page PH (eds) Land cover and land use change on islands: Social & ecological threats to sustainability. Springer, pp 183–204

Polo C, Valle E (2008) A general equilibrium assessment of the impact of a fall in tourism under alternative closure rules: The case of the Balearic Islands. Intern Region Sci Rev 31(1):3–34

Rapaport M (1990) Population pressure on coral atolls: Trends and approaching limits. Atoll Res Bull 340:1–33

Reenberg A, Birch-Thomsen T, Mertz O, Fog B, Christiansen S (2008) Adaptation of human coping strategies in a Small Island society in the SW Pacific—50 years of change in the coupled human–environment system on Bellona, Solomon Islands. Hum Ecol 36:807–819

Reuveny R, Decker CS (2000) Easter Island: Historical anecdote or warning for the future? Ecol Econ 35(2):271–287

Robolis S, Xideas E (1996) The economic determinants of Greek return migration to the islands of the East Aegean. Intern Migrat 34(2):297–319

Scheyvens R, Momsen J (2008) Tourism and poverty reduction: Issues for small Island states. Tour Geogr 10(1):22–41

Seto K, Fragkias M, Guneralp B, Reilly M (2011) A meta-analysis of global urban land expansion. PloS One 6(8):e23777

Shen G, Abdoul N, Zhu Y, Wang Z, Gong J (2017) Remote sensing of urban growth and landscape pattern changes in response to the expansion of Chongming Island in Shanghai, China. Geocarto Int 32(5):488–502

Skeldon R (1991) The relationship between migration and development in the ESCAP region. Popul Res Leads 38:1–22

Sofer M (1993) Uneven regional development and internal labor migration in Fiji. World Dev 21(2):301ff

Stauvermann PJ, Kumar R (2017) Productivity growth and income in the tourism sector: Role of tourism demand and human capital investment. Tour Manag 61:426–433

Steenge A, van de Steeg A (2010) Tourism multipliers for a small Caribbean Island state: The case of Aruba. Econ Syst Res 22(4):359–384

Tan T, Chooi K, San Lim H, M, MatJafri, & K. Abdullah. (2010) Landsat data to evaluate urban expansion and determine land use/land cover changes in Penang Island, Malaysia. Environ Earth Sci 60(7):1509–1521

Taylor JE (2009) Ecotourism and economic growth in the Galapagos: An Island economy-wide analysis. Environ Dev Econ 14(2):1–16

Telfer D, Wall G (1996) Linkages between tourism and food production. Ann Tour Res 23(3):635–653

Tole L (2001) Jamaica's disappearing forests: physical and human aspects. Environ Manag 28(4):455–467

Walters B (2016) Migration, land use and forest change in St. Lucia, West Indies. Land Use Policy 51:290–300

Wang JY, Liu YS (2013) Tourism-led land-use changes and their environmental effects in the Southern Coastal Region of Hainan Island, China. J Coast Res 29(5):1118–1125

Wang S, Ouyang Z, Zhang C, Xu W, Xiao Y (2012) The dynamics of spatial and temporal changes to forested land and key factors driving change on Hainan Island. Shengtai Xuebao/Acta Ecologica Sinica 32(23):7364–7374

White R, Engelen G (1994) Cellular dynamics and GIS: Modelling spatial complexity. Geogr Syst 1(3):237–253

World Commission on Environment and Development (WCED) (1987) Our Common Future. Report of the Bruntland Commission

Xie Z, Li X, Chi Y, Jiang D, Zhang Y, Chen M (2021) Ecosystem service value decreases more rapidly under the dual pressures of land use change and ecological vulnerability: A case study in Zjiajian Island, China. Ocean Coast Manag 201:105493

Xu X, Zeng L, Zhuang D (2002) Analysis on land-use change and socio-economic driving factors in Hainan Island during 50 years from 1950 to 1999. Chin Geogr Sci 12(3):193–198

Zhang W (1994) Capital, population, and urban patterns. Region Sci Urban Econ 24(2):273–286

Richard E. Bilsborrow is currently Research Professor in the Department of Geography and a Fellow in the Carolina Popualtion Center, University of North Carolina at Chapel Hill, USA. Bilsborrow (PhD, Economics, Michigan; Postdoctoral Certificate, Demography, Princeton) has dedicated his professional life to seeking to better understand how to achieve economic development and reduce poverty in developing countries. In the middle of his career his focus turned to migration, and he became a pioneer in the new subfield of population and the environment. He has developed, implemented and analyzed household surveys on livelihoods, migration and the environment in many countries.

Chapter 5
Pacific Island Perspectives on Invasive Species and Climate Change

Laura Brewington, Bradley Eichelberger, Nicole Read, Elliott Parsons, Heather Kerkering, Christy Martin, Wendy Miles, Jacques Idechong, and Jeff Burgett

Introduction

Ecosystem health and sustainability worldwide are threatened by the independent and interacting impacts of invasive species and climate change. A primary driver of global biodiversity loss is the intentional or unintended introduction, through trade, tourism, and human migration, of non-native species that can become invasive in their new environments (Levine and D'Antonio 2003; Lodge et al. 2006; Pejchar and Mooney 2009; Mace et al. 2012). Invasive species also directly or indirectly degrade ecosystem services, affect food crops, damage infrastructure and industry,

L. Brewington (✉)
Arizona State University, Phoenix, AZ, USA

East-West Center, Honolulu, HI, USA
e-mail: brewingl@eastwestcenter.org

B. Eichelberger
CNMI Division of Fish and Wildlife, Saipan, MP, USA

N. Read
Duke University, Durham, NC, USA

E. Parsons
University of Hawai'i at Mānoa, Honolulu, HI, USA

H. Kerkering
Pacific Islands Climate Adaptation Science Center, Honolulu, HI, USA

C. Martin
Hawai'i Coordinating Group on Alien Pest Species, Honolulu, HI, USA

W. Miles · J. Burgett
US Fish and Wildlife Service, Tishomingo, OK, USA

J. Idechong
Palau Community College, Koror, PW, Palau

S. J. Walsh et al. (eds.), *Island Ecosystems*, Social and Ecological Interactions in the Galapagos Islands, https://doi.org/10.1007/978-3-031-28089-4_5

and harm public health (RRISC 2022). Meanwhile, climate change presents short-, medium-, and long-term risks through impacts to ecosystems and the communities that depend on them. Climate change-related challenges include sea level rise, warming air and ocean temperatures, changing rainfall patterns and drought, wildfires, and increased frequency and intensity of storms and cyclones (IPBES 2019; IPCC 2022), all of which render certain ecosystems more vulnerable to new invasive species arrivals, and facilitate range expansions of existing invasive species or the transport of species to new regions where they may become invasive (Hellmann et al. 2008; Pyke et al. 2008; Ad Hoc WG 2014; Seebens et al. 2018; IPBES 2019). On islands, the stakes may be even higher (Fernández-Palacios et al. 2021).

In recent years, several regional networks were created in the United States to better understand the interacting effects of invasive species and climate change and provide natural and cultural resource managers with tools to respond to them. Known collectively as the Regional Invasive Species and Climate Change (RISCC) management network, RISCC teams now cover more than one-half of the US, including the Northeast RISCC,[1] the Northwest RISCC,[2] the North Central RISCC,[3] the Southeast RISCC,[4] and the Pacific RISCC.[5] Each regional network consists of invasion researchers, climate scientists, resource managers, policymakers, and interested members of the broader public.

As boundary organizations working at the interface between science production and user groups (Guston 2001), RISCC teams co-produce actionable research and usable products, and facilitate multi-directional dialog between researchers, resource managers, and community members. Although management issues and information needs vary considerably between regions, all RISCCs share the following objectives: (1) synthesize and translate relevant science; (2) share the needs and knowledge of managers; (3) build stronger researcher-manager communities; and (4) conduct priority research (Northeast 2020).

This chapter first considers the state of knowledge about the compounding impacts of climate change and invasive species, with an emphasis on sustainability in the Pacific Islands region. We then describe the evolution of the Pacific RISCC that represents the US Territories of Guam and American Sāmoa, the Commonwealth of the Northern Mariana Islands (CNMI), the Republic of Palau, the Federated States of Micronesia, and the Republic of the Marshall Islands (collectively referred to as the US-Affiliated Pacific Islands, USAPI), and the US State of Hawaiʻi (Fig. 5.1). Finally, we present two regional case studies as examples of how Pacific RISCC research is promoting Pacific Island resilience and sustainable outcomes through climate-literate resource management.

---

[1] https://www.risccnetwork.org

[2] https://nwriscc.org

[3] https://www.risccnetwork.org/north-central

[4] https://secasc.ncsu.edu/home/partners/academic-partners/southeast-regional-invasive-species-and-climate-change-management-network-se-riscc/

[5] https://www.pacificriscc.org

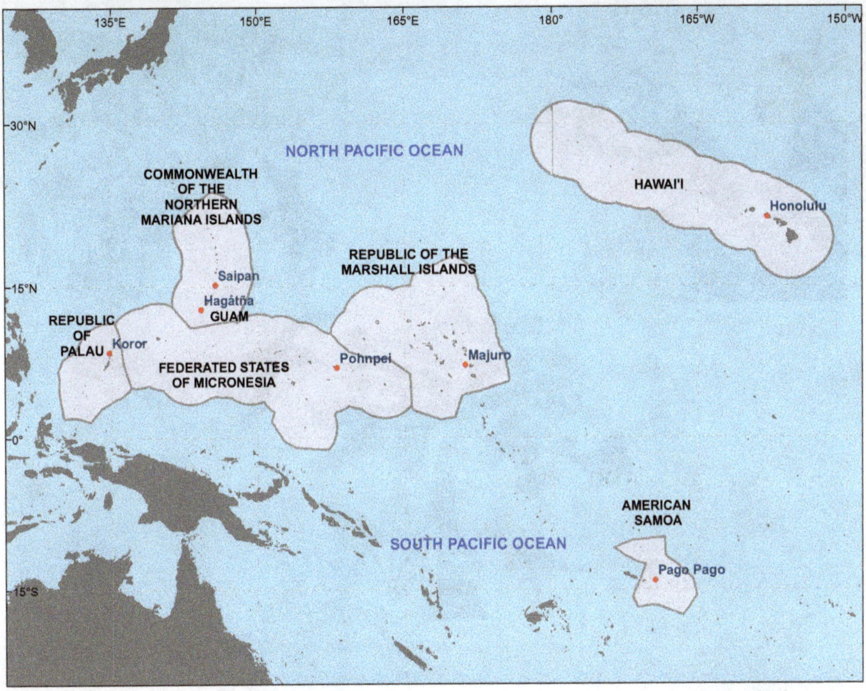

Fig. 5.1 Locations of the US-Affiliated Pacific Islands within the North and South Pacific Ocean

Invasive Species and Climate Change: Threats to Island Sustainability

Invasive species and climate change pose serious and imminent threats to the Pacific Islands region (Fig. 5.2). Invasive species are the most significant driver of biodiversity loss in the region and contribute directly to losses in ecosystem function and resilience (Loope and Mueller-Dombois 1989; D'Antonio and Vitousek 1992; Meyer 1996; Wilcove et al. 1998; Steadman 2006). Invasive rats (primarily *Rattus rattus, R. Norvegicus, and R. exulans*) have resulted in myriad predation, competition, and extinction outcomes throughout the region (Harper and Bunbury 2015), whereas the brown tree snake (*Boiga irregularis*) invasion is responsible for the extinction of 11 of Guam's native bird species and has contributed to the losses of several native bat and lizard species (Rodda and Savidge 2007). Invasive plants and animals also displace and destroy native forests along with the ecosystem and cultural services they provide (Le Roux et al. 2008; Cordell et al. 2009; Holmes et al. 2019). One example of this deep cultural connection can be seen in Hawaiian traditions that view humans as the youngest kin of the plants and animals of native ecosystems (Kealiikanakaoleohaililani and Giardina 2016), a familial relationship that is disrupted by non-native species invasions.

Fig. 5.2 This figure represents some of the interacting impacts of climate change and invasive species on Pacific Island ecosystems. The Hawaiian honeycreepers serve as an illustrative example: many of these endemic forest bird species are at risk of extinction in this century due to avian malaria and avian pox, two diseases transmitted by invasive mosquitos. These birds are already restricted to the islands' high elevation forests and climate change is exacerbating the problem as temperatures increase and mosquitos move up in elevation to the birds' last disease-free habitats. Novel solutions are urgently needed to protect them from these twin threats (Atkinson et al. 2014; Fortini et al. 2020; Paxton et al. 2022; Rivera et al. 2021)

Cascading impacts across ecosystems also occur – invasive rats on islands have been linked to declines in nearshore water quality, reef health, and fish populations (Graham et al. 2018), and invasive trees have altered soils, faunal populations, watershed function, and food webs, resulting in ecosystem state changes (Vitousek et al. 1987; McCauley et al. 2012; Young et al. 2016; Jorgensen et al. 2021). Further, invasive species threaten food security in Pacific Islands in two direct ways: decreasing production in both subsistence and economic sectors and reducing native biodiversity and ecosystem resilience (Vargas et al. 2016; Aristizábal 2018; Shiels and Kalodimos 2019; Kappes et al. 2021).

Tropical islands worldwide are also already experiencing the negative impacts of climate change (Mycoo et al. 2022). Recent research has shown that Pacific Island ecosystems and communities are particularly vulnerable to climate-induced threats: sea level rise and flooding are detrimental to freshwater resources (Keener et al. 2018; Leta et al. 2018; Frauendorf et al. 2019; Clilverd et al. 2019); cyclones and coastal inundation damage natural or green/blue infrastructure (Kane and Fletcher 2020; Reguero et al. 2021; Ward et al. 2016; Buffington et al. 2021); declines to coral and reef structure, along with ocean temperature increases and acidification, cause marine and fisheries losses (Lehodey et al. 2013; McManus et al. 2021); and damage to land and water systems jeopardizes agricultural and subsistence activities, resulting in food security concerns (Kurashima et al. 2019). These impacts are projected to become "catastrophic" for Pacific Islands should global heating exceed the Paris Agreement target of 1.5 degrees Celsius (Mycoo et al. 2022; Lesa 2022).

The Pacific RISCC Management Network

Despite having a broad understanding of the impacts of climate change and invasive species in Pacific Islands, the interactions between them remain poorly studied, leaving resource managers ill-prepared to respond to these compound threats, especially when they are confounded by additional anthropogenic pressures. A growing interest in the region about how to manage invasive species under climate change led to the creation of the Pacific RISCC management network in 2020. By investigating the interacting drivers and effects of climate change and invasive species, the Pacific RISCC provides information and tools that managers need to prioritize strategies with sustainable co-benefits for ecosystems and communities in the Pacific Islands. Our efforts and the expansion of the network over time can be summarized in three main phases:

During Phase I (2020–2021), a Core Team was established made up of 14 members representing agencies throughout Hawai'i and the USAPI.[6] Subsequently, the Core Team developed a network listserv and website, initiated a webinar series, and conducted a survey[7] of resource manager needs with respect to the intersection of climate change and invasive species. The survey (n = 59) established a baseline of managers' level of concern about the influence of climate change on invasive species management, assessed their access to and understanding of the available climate information, and identified barriers to successful incorporation of science-based climate change knowledge into management practices (Table 5.1).

[6] The Pacific RISCC Core Team is currently comprised of representatives from American Sāmoa Community College (https://www.amsamoa.edu); CNMI Division of Fish and Wildlife (http://www.dfwcnmi.com); Hawai'i Coordinating Group on Alien Pest Species (https://www.cgaps.org), Guam Department of Agriculture (https://doag.guam.gov), Hawai'i Invasive Species Council (https://dlnr.hawaii.gov/hisc/), Micronesia Regional Invasive Species Council, Pacific Islands Climate Adaptation Science Center (https://pi-casc.soest.hawaii.edu), Pacific Regional Integrated Sciences and Assessments program (https://www.pacificrisa.org), Palau Community College (https://pcc.palau.edu); University of Guam (https://www.uog.edu), University of Hawai'i at Mānoa (https://manoa.hawaii.edu), and US Fish and Wildlife Service (https://www.fws.gov/office/pacific-islands-fish-and-wildlife).

[7] The survey described in this report was organized and implemented by the East-West Center and was not conducted on behalf of the US Federal Government.

Table 5.1 Key findings from the Pacific RISCC manager survey

Over 90% of survey respondents said that considering climate change was important, or very important, to them in accomplishing their management goals (Brewington et al. 2021). Their access to climate information, however, was largely limited to informal conversations, whereas technical reports and peer-reviewed literature sources were utilized much less often. This reflected the lower levels of satisfaction respondents had regarding the number of climate information products available to meet their needs, with over 40% being only somewhat or not at all satisfied, and a substantial 28% indicated that they simply did not know about available products.	When asked about their decision-making needs, respondents identified priorities for greater knowledge, products, and services from the climate information community (Fig. 5.3). Three-quarters of respondents said that more climate projections over a range of scales (e.g., rainfall projections at the refuge or island scale) would be useful. Nearly 70% identified collaborative research projects and better communication between managers and the climate science community as opportunities to support decision making. Over half said they need a better understanding of uncertainty around future projected conditions, and how to manage under uncertainty.

In response to needs identified in the survey, during Phase II (2021–2022) the Core Team launched a research synthesis effort, expanded the webinar series themes to be more regionally and thematically representative, and hired a dedicated research specialist to lead science and communications efforts across the network. The Pacific RISCC prioritized manager-relevant research needs, creating literature summaries organized by thematic categories that were identified in the survey, such as range-shifting species, native community resilience, changes in extreme events, and new introduction pathways (Fig. 5.4), and identifying existing climate monitoring and prediction tools to support adaptive management priorities. During Phase II, the Pacific RISCC also contributed to the expansion of the national RISCC network by sharing information and lessons learned with new teams in the continental US.

Assessing the state of knowledge within the literature about climate change-invasive species interactions was a key focal area during Phase II. For example, the Fourth National Climate Assessment chapter on Hawaiʻi and the Pacific Islands (Keener et al. 2018) considered how it is anticipated that climate change will promote the spread of invasive species and reduce native ecosystem resilience against future invasions, but the assessment relied heavily on global research with limited Pacific-focused studies. Nevertheless, a growing body of regional literature is contributing essential information for managers concerned about changing species ranges, interactions and behaviors, and introductory pathways under climate change, as well as the multiple impacts on sustainability, including agriculture, livelihoods, water resources, cultural practices, and health. Recent work has focused on shifting invasive species ranges, including those of mosquitos (example in Fig. 5.1, caption; Fortini et al. 2020), slugs (Sommer and Cowie 2020), ants (Lee et al. 2021), and plants (Taylor and Kumar 2014), as well as identifying potential future refugia for native insects and plant species (Pouteau and Birnbaum 2016; da Silva et al. 2021).

In Hawaiʻi, where mean annual rainfall has been declining since the early 1900s amid steadily increasing temperatures (Keener et al. 2018), researchers are concerned about the impacts of invasive grasses on fire regimes and species behavior

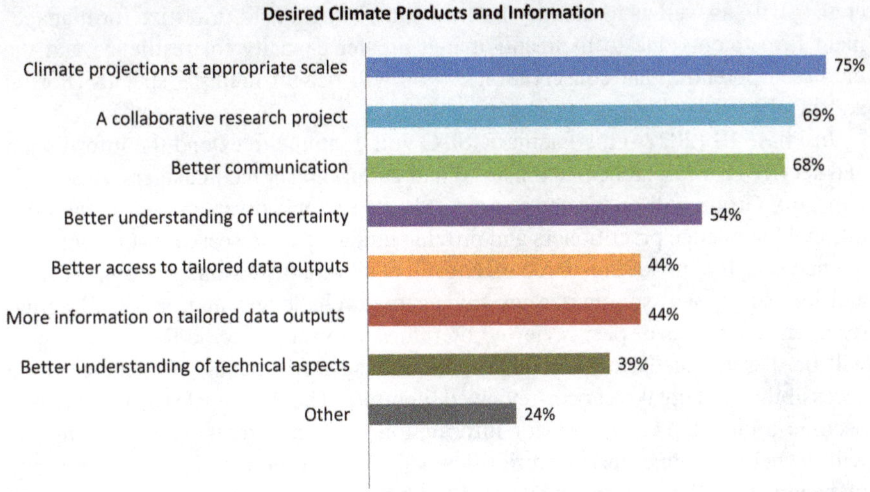

Fig. 5.3 Types of products and information that resource managers wish to see from the climate science community. (From Brewington et al. 2021)

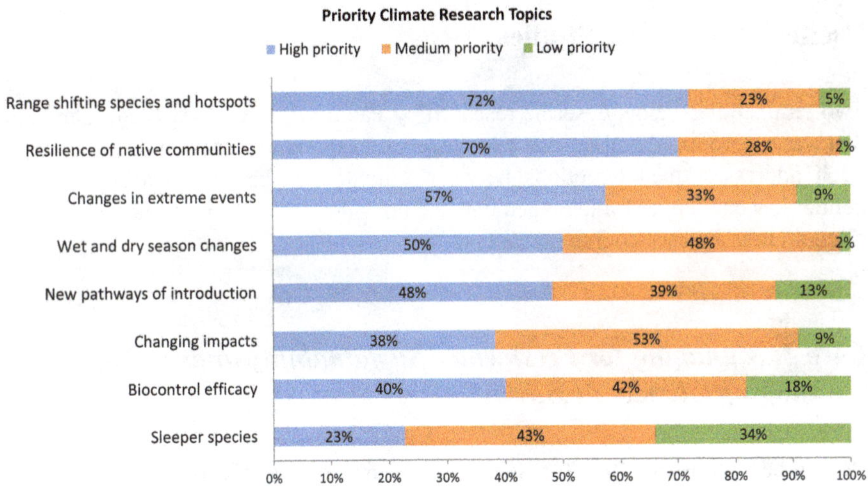

Fig. 5.4 Priority research topics for resource managers at the intersection of climate change and invasive species. (From Brewington et al. 2021)

(Ellsworth et al. 2014; Trauernicht 2019). Elsewhere in the Pacific, mangrove systems, which form a key barrier for coastal communities to buffer against climate impacts like storms and king tides, are at risk from both sea level rise and displacement by invasive plants, so more accurate predictions of how mangroves will respond to these twin threats are needed (Gilman et al. 2008). Finally, novel modeling approaches are being developed to predict suitable habitat for invasive marine and terrestrial plant species under a changing climate (Vorsino et al. 2014; Veazey et al. 2019), as well as to identify biomes and ecoregions to prioritize for management investments due to their anticipated greater capacity for resilience and the increased potential that conservation efforts will benefit multiple species (Fortini and Jacobi 2018).

In Phase III (2022–) the Pacific RISCC will continue to extend the information and services for researchers and natural and cultural resource managers. A Science Advisory Group will be created to represent the region's diverse scientific and traditional knowledge practitioners and provide guidance for research that responds to the needs of the broader resource management community. Manager-relevant tools and fact sheets will summarize top management challenges and options. Regular research summaries of peer reviewed literature relevant to Pacific RISCC concerns will be disseminated with key take-home messages for managers to increase the accessibility and uptake of peer-reviewed literature. The Pacific RISCC website will become a one-stop repository for information, data, and tools. Annual symposia will be held to refine network priorities and facilitate new regional research and management collaborations, while also promoting exchanges with researchers and managers from other RISCC teams in the continental US.

Pacific RISCC Case Studies

Two case studies below describe research by Pacific RISCC Core Team members that was co-produced with resource managers to respond to these compound threats. Each addresses the interactions between climate change and invasive species, defines key challenges for management, and considers how this information is being or could be used by resource managers.

Case 1: Managing for Freshwater Sustainability under a Changing Climate in Hawai'i

Researchers within the RISCC network in Hawai'i have been working with resource managers, urban planners, agricultural producers, landowners, and state and county water utilities to predict the sustainability of groundwater resources under future climate and land management conditions (Brewington et al. 2017; Brewington et al. 2019; Bremer et al. 2021). Groundwater, which supplies 99% of the state's drinking water, is directly impacted by land management, temperature, and rainfall, all of which have been changing in recent decades. The detrimental effects of invasive ungulates, trees, and grasses, especially at the upper elevations of watersheds where groundwater recharge and cloud-water interception are the highest, are particularly concerning to the islands' water managers. Furthermore, both global and regional data and models suggest that Hawai'i's future climate may be significantly different from the past, making prior assumptions about future water resource availability invalid.

On the Island of Maui, researchers conducted a participatory, stakeholder-driven process to model groundwater recharge under a set of future land and water management scenarios and climate projections (Brewington et al. 2019). The work aimed to 1) integrate future climate information and land and water management decisions on Maui with a water budget model for evaluating freshwater sustainability; and 2) identify tradeoffs and optimized solutions for managing Maui's land and water ecosystems, urban development, and agriculture under a changing climate. A hydrological model was created that encompassed a diverse range of urban and natural ecosystems to aid in resource management and planning for the future, and two downscaled, end-of-century rainfall projections for the island were selected to represent futures in which the island becomes much drier on average, versus a future in which it becomes much wetter. To ensure that the groundwater modeling analyses addressed the needs of Maui's decision makers, the research team also solicited input from over 100 stakeholders to generate four feasible future land cover scenarios (Brewington et al. 2017; Fig. 5.5). The scenarios reflected realistic management decisions that affect the island's water resources, such as the presence or absence of native or non-native forest species, adding or removing fences to keep feral ungulates out of priority watersheds, changing agricultural regimes and ranching practices, and different levels of urban development.

A key finding of this study was that climate change, regardless of whether Maui becomes more wet or dry, is projected to have an enormous impact on groundwater recharge on the island. Under the wet climate projection, the upper elevations where recharge is currently the highest would receive even more recharge, but that effect diminishes nearer to the coast (Fig. 5.6a). Under the dry climate projection, sharp declines in recharge are projected in Maui's upper elevation watersheds (Fig. 5.6b). Areas that already have low recharge—including many coastal ecosystems and some of the most heavily populated areas or places zoned for development—are expected to receive even less recharge in the future. Such areas must be carefully monitored and managed to ensure continued freshwater ecosystem services and traditional Hawaiian water uses, and alternative irrigation sources for golf courses and other urban landscaping should be explored. Equally critical, however, was the finding that land management decisions, particularly the protection of native forest ecosystems, could also mitigate the impacts of climate change on groundwater recharge in some of Maui's aquifer systems. Even under the dry future climate scenario, recharge increased where native forest protection or restoration was prioritized over invasive forest or grassland cover (Fig. 5.6c, d). The results from this work are helping resource managers on Maui prioritize conservation efforts in high-recharge areas like the upper elevation native forests, with clear ecological, cultural, and hydrological co-benefits for sustainability.

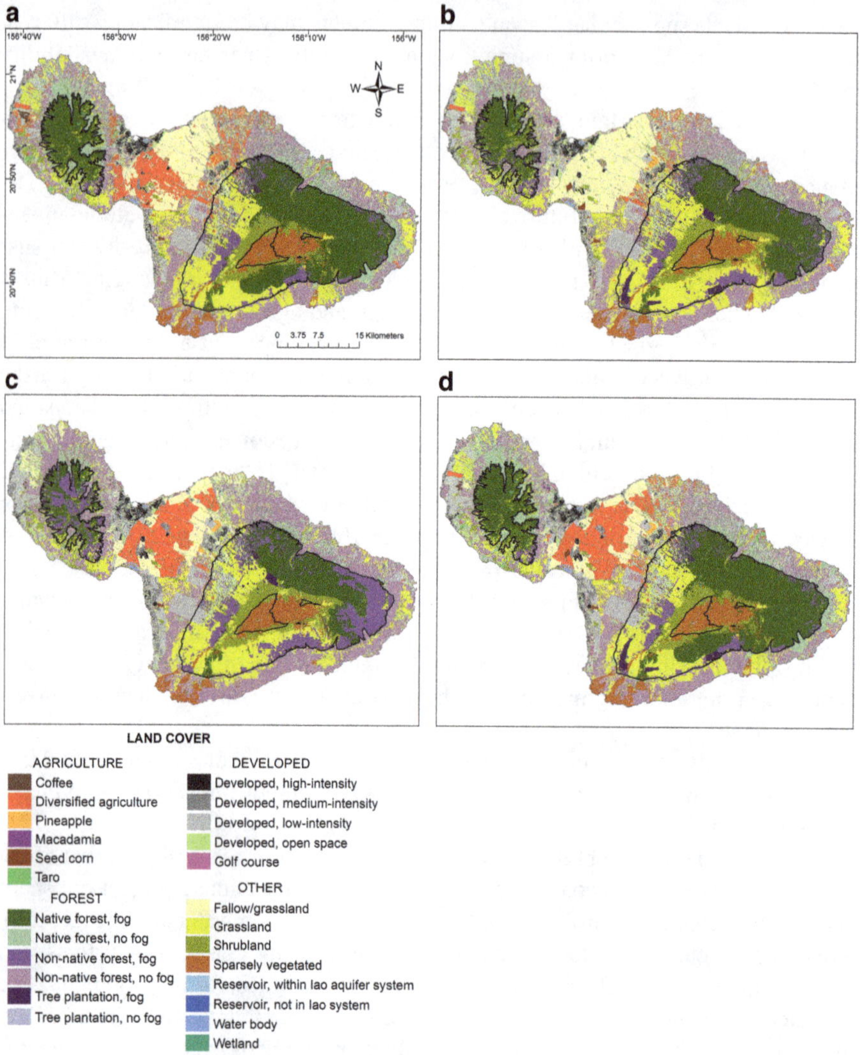

Fig. 5.5 Four future land cover maps representing different management scenarios for the Island of Maui: (**a**) Future 1: characterized by high native forest conservation, low urban development, and local agriculture production; (**b**) Future 2: a business-as-usual management trajectory with modest investments in forest conservation, urban build out of currently approved projects, and no changes to agricultural and ranching lands; (**c**) Future 3: very high urban development, low forest conservation, and biofuel production for local energy; and (**d**) Future 4: high development combined with high forest conservation and biofuel production. The dark black lines indicate the location of the cloud interception zone between 610 m and 2100 m above sea level. (From Brewington et al. 2019)

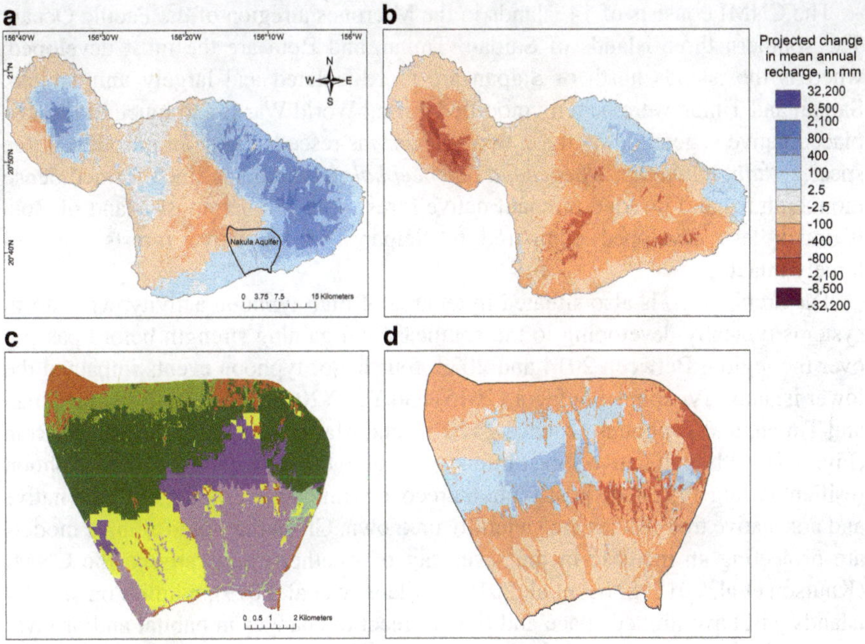

Fig. 5.6 (**a**) Island-wide change in mean annual recharge under the wet climate projection and 2017 land cover, showing recharge increases in the upper elevations of West Maui and East Maui leeward areas; (**b**) change in recharge under the dry climate projection and 2017 land cover, showing decreases in all of West Maui and most of East Maui; (**c**) Future 1 land cover under the high conservation management scenario in the Nakula aquifer system (see Fig. 5.4 land cover legend); (**d**) change in recharge in Nakula under the dry climate projection and land cover scenario Future 1, showing increases where native forest is protected/restored, and decreases under non-native forest cover. (From Brewington et al. 2019)

Case 2: Monitoring Typhoon and Land Cover Impacts on Endemic Birds in CNMI

In the Commonwealth of the Northern Mariana Islands (CNMI), Pacific RISCC biologists and resource managers are investigating the combined effects of invasive species and climate change on endemic bird populations and their supporting forest habitat. Most of the forest in the CNMI has been modified through World War II occupation, human development, and frequent typhoon events, which have resulted in the dominance of non-native, invasive tree species throughout several of the islands (Dendy et al. 2020). The close proximity and frequent trade and transportation between the CNMI and Guam also create pathways for the introduction of the brown tree snake that has decimated bird populations on Guam. Several endemic bird species are only found on specific islands within the CNMI island chain, so effective conservation and management of these species is critical to prevent extinctions.

The CNMI consists of 14 islands in the Micronesia region of the Pacific Ocean. The southern three islands of Saipan, Tinian, and Rota are the most developed, whereas the islands north of Saipan are more isolated and largely uninhabited. Saipan and Tinian were heavily modified during World War II and much of the displaced native vegetation on these two islands was reseeded with an introduced tree species, "*tångantångan*" (*Leucaena leucocephala*), which has since formed dense monocultures and invaded remnant native forest. The southernmost island of Rota is much less developed compared to Saipan and its native forests are still largely intact.

The archipelago is also situated in an area of high typhoon activity, with storm systems typically developing to the southeast and gaining strength before passing over the region. Between 2014 and 2022, four major typhoon events impacted the lower islands: Typhoons Souldeor (2015) and Yutu (2018) made landfall on Saipan and Tinian, and Typhoons Vongfong (2014) and Mangkhut (2018) devastated Rota (Fig. 5.7). Although native forest tree species are generally believed to be typhoon resilient (Craig 1992), the post-typhoon recovery time and responses of both native and non-native tree species are generally unknown. Given that some climate models are projecting an increase in the severerity of weather events around the CNMI (Knutson et al. 2015; Sobel et al. 2016; Widlansky et al. 2019), wildlife on smaller islands will have limited space and time to react to changes in habitat and/or invasive species expansions. The varying degrees of habitat modification, invasive species encroachment, and climate impacts in the CNMI therefore create a unique opportunity to study the sustainability of ecosystems and endemic species in small island systems.

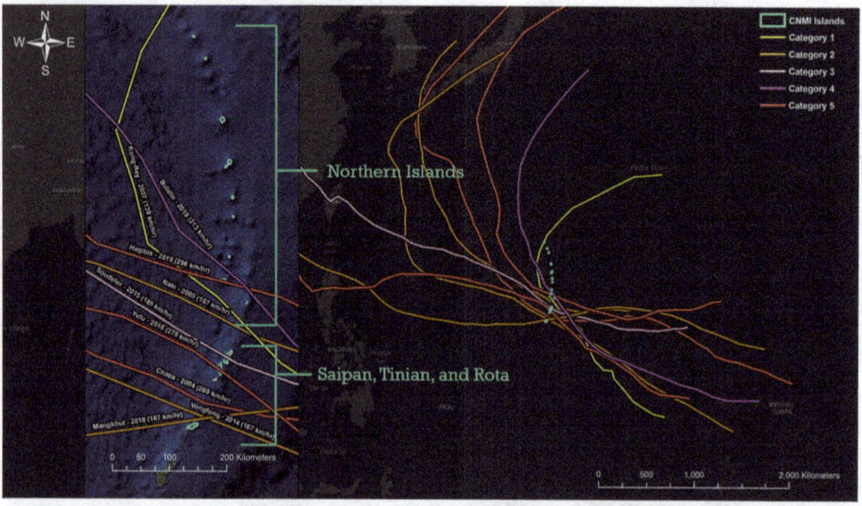

Fig. 5.7 Typhoon event categories and pathways for the Northern Mariana Islands from 2000 to 2019

Using 32-day composite enhanced vegetation indices (EVI) derived from Landsat-8 satellite data during the dry (December to May) and wet (June to November) seasons from 2014 to 2021 as a metric for vegetation growth status and density, CNMI biologists compared annual seasonal EVI averages to seasonal averages across the entire 2014–2021 timeframe to approximate how long it takes vegetation to return to an eight-year average baseline condition (Eichelberger et al. unpublished data). Vegetation on both Saipan and Tinian exhibited a full year (dry and wet season) recovery period following Typhoon Soudelor to return to 2014–2021 baseline averages and one full dry season to return to baseline following Typhoon Yutu. During a drought year (2020) following the typhoon events on Saipan and Tinian, EVI fell below the baseline average (Fig. 5.8a, b). Vegetation on Rota only needed one dry season to recover to the 2014–2021 baseline following Typhoon Vongfong and no recovery time in response to Typhoon Mangkhut (Fig. 5.8c). Although the strength and impact zones varied across typhoon events for each island, Saipan and Tinian exhibited similar vegetation recovery times in response to typhoons and droughts, whereas Rota exhibited shorter recovery times.

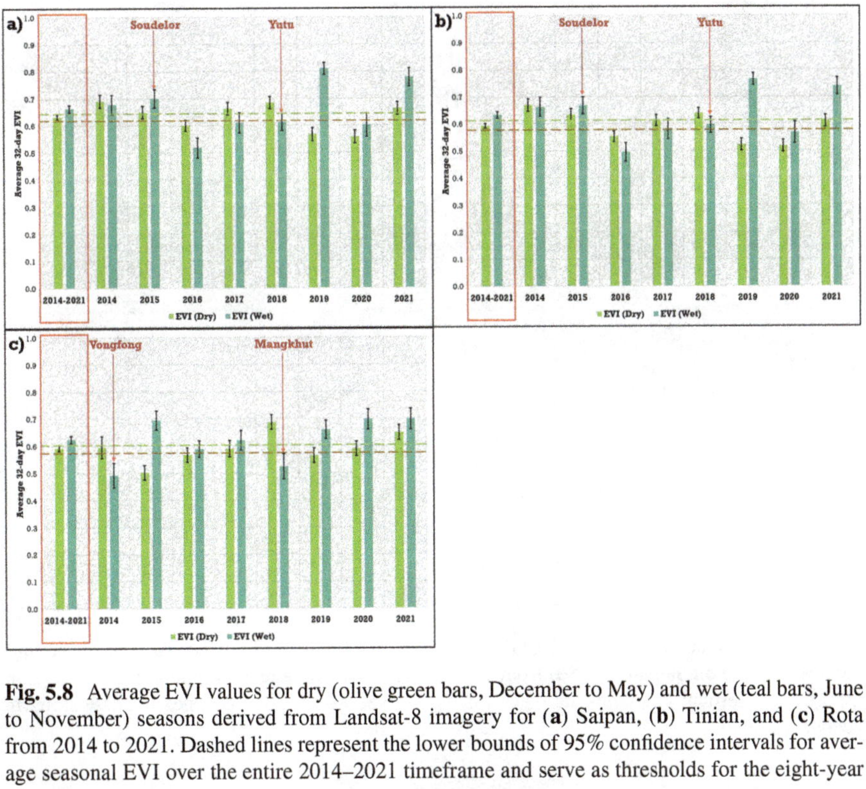

Fig. 5.8 Average EVI values for dry (olive green bars, December to May) and wet (teal bars, June to November) seasons derived from Landsat-8 imagery for (**a**) Saipan, (**b**) Tinian, and (**c**) Rota from 2014 to 2021. Dashed lines represent the lower bounds of 95% confidence intervals for average seasonal EVI over the entire 2014–2021 timeframe and serve as thresholds for the eight-year baseline

To examine the impacts of recent typhoons on wildlife abundance and recovery, CNMI biologists analyzed bird species abundance collected from auditory surveys at 50 stations sampled quarterly every year from 1999 to 2021. Bird species were subdivided into frugivores/nectivores (FN) and carnivores/omnivores/insectivores (COI) to further examine typhoon impacts on food availability for different dietary guilds. On Saipan, there was variability in wildlife responses but generally decreases in detected abundance for both FN and COI species following each successive typhoon event (Fig. 5.9), implying that the cumulative impacts of multiple typhoons further impede population recovery of several species. These results, combined with the results from the EVI analysis, are guiding researchers to re-examine EVI responses across native- and *tångantångan*-dominated forests, the latter of which offer little resources to FN species. This information will help wildlife managers understand if reduced food availability could be contributing to decreases in detected bird abundance and develop approaches to ensure food security for wildlife. Researchers will also resample pre-typhoon (2017 and 2018) vegetation plots on Saipan to quantify changes in vegetation structure and composition and investigate whether typhoons are providing pathways for non-native species invasions into native forests.

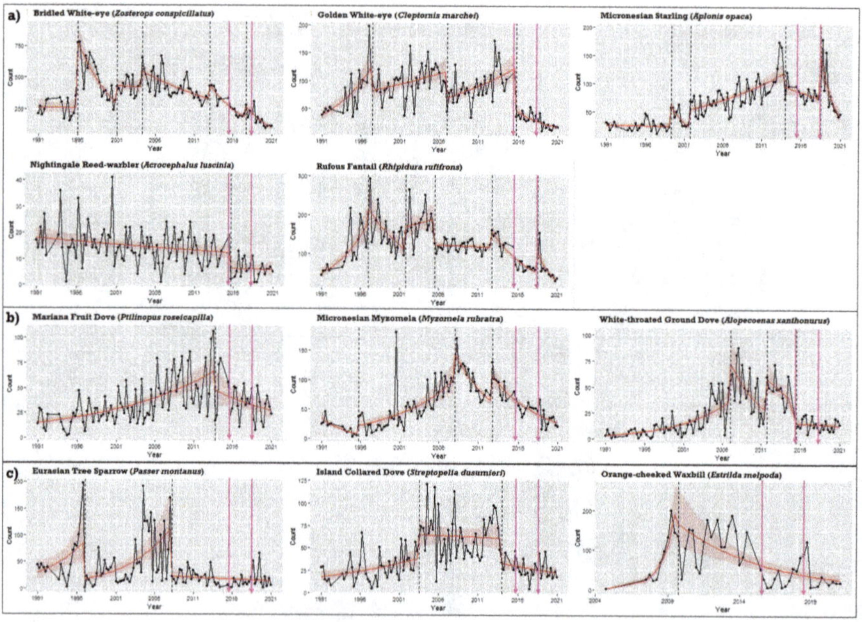

Fig. 5.9 Species abundance for (**a**) COI forest birds, (**b**) FN forest birds and (**c**) non-native bird species sampled on Saipan ($n = 50$) from 1999–2021 fitted with a negative binomial changepoint model. Pink vertical lines indicate dates for typhoon events and dashed lines indicate shifts in abundance trends

This research suggests that both wildlife populations and vegetation in the CNMI have experienced negative responses and potential recovery periods following typhoon events; however, the cumulative impacts of high-intensity and successive typhoon events may place wildlife in a vulnerable position. This is particularly worrisome for species conservation as humanitarian relief efforts following typhoon events typically increase cargo traffic from Guam and provide new pathways for the introduction of brown tree snakes that could decimate the already stressed wildlife populations. While future analyses may provide more detailed answers, information on the interactions between native and invasive species, and the potential impacts of increasing extreme weather events under climate change, is critical to sustaining endemic species populations and decreasing the likelihood of extinctions.

Conclusions

There are compounding impacts from invasive species and climate change in the Pacific Islands region. Climate events like storms and floods introduce invasive species into new areas, while both native and invasive species ranges are shifting as long-term conditions change. The resilience of native ecosystems and protected areas can be reduced by invasive species, making them more vulnerable to the impacts of climate change, and vice-versa. These complex interactions, intertwined with other anthropogenic disturbances and pressures, can have cascading negative effects on many aspects of island sustainability, including livelihoods, food and water resources, cultural practices, biodiversity, and human and ecosystem health.

In response to a growing need from natural and cultural resource managers for more information about the interactions between invasive species and climate change, the Pacific RISCC management network was formed in 2020 and joined four regionally focused teams addressing similar concerns in the continental United States. As a novel manager-researcher collaborative, the Pacific RISCC serves diverse science generation and communication needs in Hawaiʻi and the USAPI. On the Island of Maui, for example, water resource managers can prioritize conservation efforts and protect freshwater sustainability under an uncertain future climate using results from research co-developed with them at relevant spatial and temporal scales. Pacific RISCC researchers in the CNMI are studying vegetation changes and recovery times from typhoon events between native and invasive-dominated forests, which will help them understand if post-typhoon food security for wildlife is affecting the viability of endemic bird populations. As the Pacific RISCC network continues to expand and evolve, the group endeavors to become the trusted source for regional information, data, and tools at the nexus of climate change and invasive species, while accelerating sustainability solutions in a region that is uniquely vulnerable to these compounding threats.

References

Ad Hoc Working Group on Invasive Species and Climate Change [Ad Hoc WG] (2014) Bioinvasions in a changing world: A resource on invasive species-climate change interactions for conservation and natural resource management. Available online: https://bugwoodcloud.org/mura/naisn/assets/File/BioinvasionsDec2014.pdf

Aristizábal F (2018) Challenges faced by coffee growers establishing an integrated pest management for coffee berry borer in Hawaii. Agricult Res Technol Open Access J 14(3):555919. https://doi.org/10.19080/ARTOAJ.2018.14.555919

Atkinson CT, Utzurrum RB, LaPointe DA, Camp RJ, Crampton LH, Foster JT, Giambelluca TW (2014) Changing climate and the altitudinal range of avian malaria in the Hawaiian islands – An ongoing conservation crisis on the Island of Kaua'i. Glob Chang Biol 20:2426–2436. https://doi.org/10.1111/gcb.12535

Bremer LL, Elshall AS, Wada CA, Brewington L, Delevaux JMS, El-Kadi AI, Voss CI, Burnett KM (2021) Effects of land cover and watershed protection futures on sustainable groundwater management in a heavily utilized aquifer in Hawai'i (USA). Hydrogeol J 29:1749–1765. https://doi.org/10.1007/s10040-021-02310-6

Brewington L, Keener V, Finucane M, Eaton P (2017) Participatory scenario planning for climate change adaptation using remote sensing and GIS. In: Walsh SJ (ed) Remote sensing for societal benefits. Elsevier, pp 236–252. https://doi.org/10.1016/B978-0-12-409548-9.10434-8

Brewington L, Keener V, Mair A (2019) Simulating land cover change impacts on groundwater recharge under selected climate projections, Maui, Hawai'i. Remote Sens 11(24):3048. https://doi.org/10.3390/rs11243048

Brewington L, Burgett J, Martin C, Kerkering H, Arnott C (2021) When invasive species and climate change intersect: Survey of Hawai'i natural resource managers. The Pacific Regional Invasive Species and Climate Change Management Network. Available at: https://pi-casc.soest.hawaii.edu/wp-content/uploads/2022/01/Pacific_RISCC_2021_Survey_Report_FINAL_web.pdf

Buffington KJ, MacKenzie RA, Carr JA, Apwong M, Krauss KW, Thorne KM (2021) Mangrove Species' Response to Sea-Level Rise Across Pohnpei, Federated States of Micronesia. US Geological Survey Open-File (Report 2021–1002). https://doi.org/10.3133/ofr20211002

Clilverd HM, Tsang Y-P, Infante DM, Lynch AJ, Strauch AM (2019) Long-term streamflow trends in Hawai'i and implications for native stream fauna. Hydrol Process 2019(33):699–719. https://doi.org/10.1002/hyp.13356

Cordell S, Ostertag R, Rowe B, Sweinhart L, Vasquez-Radonic L, Michaud J, Cole TC, Schulten JR (2009) Evaluating barriers to native seedling establishment in an invaded Hawaiian lowland wet forest. Biol Conserv 142:2997–3004. https://doi.org/10.1016/j.biocon.2009.07.033

Craig RJ (1992) Ecological characteristics of a native limestone forest on Saipan, Mariana Islands. Micronesica 25:85–97

D'Antonio CM, Vitousek PM (1992) Biological invasions by exotic grasses, the grass/fire cycle, and global change. Annu Rev Ecol Syst 23:63–87. https://doi.org/10.1146/annurev.es.23.110192.000431

da Silva CRB, Beaman JE, Dorey JB, Barker SJ, Congedi NC, Elmer MC et al (2021) Climate change and invasive species: A physiological performance comparison of invasive and endemic bees in Fiji. J Exp Biol 224(1):jeb230326. https://doi.org/10.1242/jeb.230326

Dendy J, Kuegler O, Lehman A, Marquez WR (2020) Forest status across Micronesia from an assessment of Micronesia Challenge terrestrial measures and forest inventory and analysis data. Micronesica 2:1–15

Ellsworth LM, Litton CM, Dale AP, Miura T (2014) Invasive grasses change landscape structure and fire behaviour in Hawaii. Appl Veg Sci 17:680–689. https://doi.org/10.1111/avsc.12110

Fernández-Palacios JM, Kreft H, Irl SDH, Norder S, Ah-Peng C, Borges PAV et al (2021) Scientists' warning – The outstanding biodiversity of islands is in peril. Global Ecol Conserv 31:e01847. https://doi.org/10.1016/j.gecco.2021.e01847

Fortini LB, Jacobi JD (2018) Identifying opportunities for long-lasting habitat conservation and restoration in Hawaii's shifting climate. Reg Environ Chang 18(8):2391–2402. https://doi.org/10.1007/s10113-018-1342-6

Fortini LB, Kaiser LR, LaPointe DA (2020) Fostering real-time climate adaptation: Analyzing past, current, and forecast temperature to understand the dynamic risk to Hawaiian honeycreepers from avian malaria. Global Ecol Conserv 23:e01069. https://doi.org/10.1016/j.gecco.2020.e01069

Frauendorf TC, MacKenzie RA, Tingley RW, Frazier AG, Riney MH, El-Sabaawi RW (2019) Evaluating ecosystem effects of climate change on tropical Island streams using high spatial and temporal resolution sampling regimes. Glob Chang Biol 25:1344–1357. https://doi.org/10.1111/gcb.14584

Gilman EL, Ellison J, Duke NC, Field C (2008) Threats to mangroves from climate change and adaptation options: A review. Aquat Bot 89(2):237–250. https://doi.org/10.1016/j.aquabot.2007.12.009

Graham NAJ, Wilson SK, Carr P, Hoey AS, Jennings S, MacNiel MA (2018) Seabirds enhance coral reef productivity and functioning in the absence of invasive rats. Nature 559:250–253. https://doi.org/10.1038/s41586-018-0202-3

Guston DH (2001) Boundary organizations in environmental policy and science: An introduction. Sci Technol Hum Values 26(4):399–408. https://doi.org/10.1177/016224390102600401

Harper GA, Bunbury N (2015) Invasive rats on tropical islands: Their population biology and impacts on native species. Global Eco Conserv 3:607–627. https://doi.org/10.1016/j.gecco.2015.02.010

Hellmann JJ, Byers JE, Bierwagen BG, Dukes JS (2008) Five potential consequences of climate change for invasive species. Conserv Biol 22(3):534–543. https://doi.org/10.1111/j.1523-1739.2008.00951.x

Holmes ND, Spatz DR, Oppel S, Tershy B, Croll DA, Keitt B et al (2019) Globally important islands where eradicating invasive mammals will benefit highly threatened vertebrates. PLoS One 14(3):e0212128. https://doi.org/10.1371/journal.pone.0212128

Intergovernmental Panel on Climate Change [IPCC] (2022) Climate change 2022: Impacts, adaptation, and vulnerability. In: Pörtner H-O, Roberts DC, Tignor M, Poloczanska ES, Mintenbeck K, Alegría A et al (eds) Contribution of working group II to the sixth assessment report of the intergovernmental panel on climate change. Cambridge University Press. In Press

Intergovernmental Science-Policy Platform on Biodiversity and Ecosystem Services [IPBES] (2019) In: Brondizio ES, Settele J, Díaz S, Ngo HT (eds) Global Assessment Report on Biodiversity and Ecosystem Services. IPBES Secretariat

Jorgensen N, Leary J, Renz M, Mahnken B (2021) Characterizing the suitable habitat of Miconia calvescens in the east Maui watershed. Manag Biol Invas 12(2):313–330. https://doi.org/10.3391/mbi.2021.12.2.07

Kane HH, Fletcher CH (2020) Rethinking reef Island stability in relation to anthropogenic sea level rise. Earth's Futures 8:e2020EF001525. https://doi.org/10.1029/2020EF001525

Kappes PJ, Benkwitt CE, Spatz DR, Wolf CA, Will DJ, Holmes ND (2021) Do invasive mammal eradications from islands support climate adaptation and mitigation? Climate 9:172. https://doi.org/10.3390/cli9120172

Kealiikanakaoleohaililani K, Giardina CP (2016) Embracing the sacred: An indigenous framework for tomorrow's sustainability science. Sustain Sci 11:57–67. https://doi.org/10.1007/s11625-015-0343-3

Keener VW, Helweg DA, Asam S, Balwani S, Burkett M, Fletcher C et al (2018) Ch. 27: Hawai'i and U.S. affiliated Pacific Islands. In: Reidmiller DR, Avery CW, Easterling DR, Kunkel KE, Lewis KLM, Maycock TK, Stewart BC (eds) The fourth National Climate Assessment Volume II: Impacts, risks, and adaptation in the United States. US Global Change Research Program, pp 1242–1308

Knutson TR, Sirutis JJ, Zhao M, Tuleya RE, Bender M, Vecchi GA et al (2015) Global projections of intense tropical cyclone activity for the late twenty-first century from dynamical downscaling of CMIP5/RCP4.5 scenarios. J Clim 28:7203–7224. https://doi.org/10.1175/JCLI-D-15-0129.1

Kurashima N, Fortini LB, Ticktin T (2019) The potential of indigenous agricultural food production under climate change. Nature Sustain 2:191–199. https://doi.org/10.1038/s41893-019-0226-1

Le Roux JJ, Wieczorek AM, Meyer J-Y (2008) Genetic diversity and structure of the invasive tree Miconia calvescens in Pacific islands. Divers Distrib 14:935–948. https://doi.org/10.1111/j.1472-4642.2008.00504.x

Lee CM, Lee D-S, Kwon T-S, Athar M, Park Y-S (2021) Predicting the global distribution of Solenopsis geminata (Hymenoptera: Formicidae) under climate change using the MaxEnt model. Insects 12:229. https://doi.org/10.3390/insects12030229

Lehodey P, Senina I, Calmettes B, Hampton J, Nicol S (2013) Modelling the impact of climate change on Pacific Skipjack Tuna population and fisheries. Clim Chang 119:95–109. https://doi.org/10.1007/s10584-012-0595-1

Lesa S (2022) Pacific countries lament threat of catastrophic weather events, Amplify Call for Urgent Actions. Secretariat of the Pacific Environment Programme. Available online: https://www.sprep.org/news/pacific-countries-lament-threat-of-catastrophic-weather-events-amplify-call-for-urgent-actions

Leta OT, El-Kadi AI, Dulai H (2018) Impact of climate change on daily streamflow and its extreme values in Pacific Island watersheds. Sustain For 10(6):2057. https://doi.org/10.3390/su10062057

Levine JM, D'Antonio CM (2003) Forecasting biological invasions with increasing international trade. Conserv Biol 17(1):322–326. https://doi.org/10.1046/j.1523-1739.2003.02038.x

Lodge DM, Williams S, MacIsaac HJ, Hayes KR, Leung B, Reichard S et al (2006) Biological invasions: Recommendations for U.S. policy and management. Ecol Appl 16(6):2035–2054. https://doi.org/10.1890/1051-0761

Loope LL, Mueller-Dombois D (1989) Characteristics of invaded islands, with special reference to Hawaii. In: Drake JA, Mooney HA, di Castri F, Groves RH, Kruger FJ, Rejmanek M, Williamson M (eds) Biological invasions: A global perspective. John Wiley & Sons, pp 257–280

Mace GM, Norris K, Fitter AH (2012) Biodiversity and ecosystem services: A multilayered relationship. Trends Ecol Evol 27:19–26. https://doi.org/10.1016/j.tree.2011.08.006

McCauley DJ, DeSalles PA, Young HS, Dunbar RB, Dirzo R, Mills MM, Micheli F (2012) From wing to wing: The persistence of long ecological interaction chains in less-disturbed ecosystems. Sci Rep 2:409. https://doi.org/10.1038/srep00409

McManus LC, Forrest DL, Tekwa EW, Schindler DE, Colton MA, Webster MM et al (2021) Evolution and connectivity influence the persistence and recovery of coral reefs under climate change in the Caribbean, Southwest Pacific, and coral triangle. Glob Chang Biol 27(18):4307–4321. https://doi.org/10.1111/gcb.15725

Meyer J-Y (1996) Status of Miconia calvescens (Melastomataceae), a dominant invasive tree in the Society Islands (French Polynesia). Pac Sci 50:66–76

Mycoo M, Wairiu M, Campbell D, Duvat V, Golbuu Y, Maharaj S et al (2022) Small Islands. In: Pörtner H-O, Roberts DC, Tignor M, Poloczanska ES, Mintenbeck K, Alegría A et al (eds) Climate change 2022: Impacts, adaptation, and vulnerability. Contribution of working group II to the sixth assessment report of the intergovernmental panel on climate change. Cambridge University Press, In Press

Northeast RISCC (2020) Guiding principles. Available online: https://static1.squarespace.com/static/5e0e3042e0bd7760c6f2ca0c/t/5fa1aa266a00107ddd1da4b1/1604430388153/RISSC_GP_D1.pdf

Paxton E., Laut M, Enomoto BM (2022) Hawaiian Forest bird conservation strategies for minimizing the risk of extinction: Biological and biocultural considerations. Hawaiʻi cooperative studies unit technical report HCSU-103. University of Hawaiʻi at Hilo. Available online: http://hdl.handle.net/10790/5386

Pejchar L, Mooney HA (2009) Invasive species, ecosystem services and human Well-being. Trends Ecol Evol 24(9):497–504. https://doi.org/10.1016/j.tree.2009.03.016

Pouteau R, Birnbaum P (2016) Island biodiversity hotspots are getting hotter: Vulnerability of tree species to climate change in New Caledonia. Biol Conserv 201:111–119. https://doi.org/10.1016/j.biocon.2016.06.031

Pyke C, Thomas R, Porter R, Hellmann J, Dukes J, Lodge D, Chavarria G (2008) Current practices and future opportunities for policy on climate change and invasive species. Conserv Biol 22:585–592. https://doi.org/10.1111/j.1523-1739.2008.00956.x

Reduce Risks from Invasive Species Coalition [RRISC] (2022) How are invasive species harmful? Available online: https://www.rrisc.org/why-are-invasives-bad/

Reguero BG, Storlazzi CD, Gibbs AE, Shope JB, Cole AD, Cumming KA et al (2021) The value of US coral reefs for flood risk reduction. Nature Sustain 4:688–698. https://doi.org/10.1038/s41893-021-00706-6

Rivera SN, Fortini LB, Plentovich S, Price MR (2021) Perceived barriers to the use of assisted colonization for climate sensitive species in the Hawaiian islands. Environ Manag 68:329–339. https://doi.org/10.1007/s00267-021-01491-w

Rodda GH, Savidge JA (2007) Biology and impacts of Pacific Island invasive species. 2. Boiga irregularis, the Brown Tree Snake (Reptilia: Colubridae). Pac Sci 61(3):307–324. https://doi.org/10.2984/1534-6188(2007)61[307:BAIOPI]2.0.CO;2

Seebens H, Blackburn TM, Dyer EE, Essl F (2018) Global rise in emerging alien species results from increased accessibility of new source pools. Proc Natl Acad Sci 115(10):E2264–E2273. https://doi.org/10.1073/pnas.1719429115

Shiels AB, Kalodimos NP (2019) Biology and impacts of Pacific Island invasive species. 15. Psittacula krameri, the rose-ringed parakeet (Psittaciformes: Psittacidae). Pac Sci 73(4):421. https://doi.org/10.2984/73.4.1

Sobel AH, Camargo SJ, Hall TM, Lee CY, Tippett MK, Wang AA (2016) Human influence on tropical cyclone intensity. Science 353:242–246. https://doi.org/10.1126/science.aaf6574

Sommer RM, Cowie RH (2020) Invasive traits of veronicellid slugs in the Hawaiian islands and temperature response suggesting possible range shifts under a changing climate. J Molluscan Stud 86(2):147–155. https://doi.org/10.1093/mollus/eyz042

Steadman DW (2006) Extinction and biogeography of tropical Pacific birds. University of Chicago Press

Taylor S, Kumar L (2014) Climate change and weed impacts on small Island ecosystems: Lantana camara L. (Magnoliopsida: Verbenaceae) distribution in Fiji. Pac Sci 68(1):117–133. https://doi.org/10.2984/68.1.11

Trauernicht C (2019) Vegetation—Rainfall interactions reveal how climate variability and climate change alter spatial patterns of wildland fire probability on big Island, Hawai'i. Sci Total Environ 650(Part 1):459–469. https://doi.org/10.1016/j.scitotenv.2018.08.347

Vargas RI, Piñero JC, Leblanc L, Manoukis NC, Mau RFL (2016) Area-wide management of fruit flies (Diptera: Tephritidae) in Hawai'i. In: Ekesi S, Mohamed S, De Meyer M (eds) Fruit fly research and development in Africa—towards a sustainable management strategy to improve horticulture. Springer, pp 673–693. https://doi.org/10.1007/978-3-319-43226-7_29

Veazey L, Williams O, Wade R, Toonin R, Spalding HL (2019) Present-day distribution and potential spread of the invasive green alga Avrainvillea amadelpha around the main Hawaiian islands. Front Mar Sci 26. https://doi.org/10.3389/fmars.2019.00402

Vitousek PM, Walker LR, Whiteaker LD, Mueller-Dombois D, Matson PA (1987) Biological invasion by Myrica faya alters ecosystem development in Hawaii. Science 238(4828):802–804. https://doi.org/10.1126/science.238.4828.802

Vorsino AE, Fortini LB, Amidon FA, Miller SE, Jacobi JD, Price JJ et al (2014) Modeling Hawaiian ecosystem degradation due to invasive plants under current and future climates. PLoS One 9(5):e95427. https://doi.org/10.1371/journal.pone.0095427

Ward RD, Friess DA, Day RH, MacKenzie RA (2016) Impacts of climate change on mangrove ecosystems: A region by region overview. Ecosyst Health Sustain 2(4):e01211. https://doi.org/10.1002/ehs2.1211

Widlansky MJ, Annamalai H, Gingerich SB, Storlazzi CD, Marra JJ, Hodges KI et al (2019) Tropical cyclone projections: Changing climate threats for Pacific Island defense installations. Weather Clim Soc 11:3–15. https://doi.org/10.1175/WCAS-D-17-0112.1

Wilcove DS, Rothstein D, Dubow J, Phillips A, Losos E (1998) Quantifying threats to imperiled species in the United States: Assessing the relative importance of habitat destruction, alien species, pollution, overexploitation, and disease. Bioscience 48(8):607–615. https://doi.org/10.2307/1313420

Young HS, Miller-ter Kuile A, McCauley DJ, D.J., Dirzo, R. (2016) Cascading community and ecosystem consequences of introduced coconut palms (Cocos nucifera) in tropical islands. Can J Zool 95(3):139–148. https://doi.org/10.1139/cjz-2016-0107

Laura Brewington co-directs the Pacific Regional Integrated Sciences and Assessments center, a climate adaptation research partnership in the Pacific Islands between Arizona State University, the East-West Center, and the University of Hawai'i. She is a Core Team member of the Pacific Invasive Species and Climate Change management network and leads island-based research projects at the intersection of climate, environmental sustainability, and health.

Bradley Eichelberger is Chief Geographic Information Officer at the Commonwealth of the Northern Mariana Islands' Division of Fish and Wildlife. His work is focused on the spatial ecology of threatened and endangered species and remote sensing applications.

Nicole Read is an Intern with the National Regional Invasive Species and Climate Change management network and an undergraduate student at Duke University.

Elliott Parsons is the Specialist for the Pacific Regional Invasive Species and Climate Change management network and has 10+ years in conservation, management, and research in Hawai'i.

Heather Kerkering is Federal Acting Deputy Director and Science Coordinator for the US Geological Survey Pacific Islands Climate Adaptation Science Center.

Christy Martin is Program Manager for the Coordinating Group on Alien Pest Species, a project of the Pacific Cooperative Studies Unit at the University of Hawai'i at Mānoa.

Wendy Miles is Science Collaboration Coordinator for the US Fish and Wildlife Service's Science Applications program in the Pacific Islands.

Jacques Idechong is a researcher for the Palau Community College Cooperative Research and Extension Program, currently focused on developing Palau's research capacity to help address invasive pests.

Jeff Burgett is Science Applications Conservation Coordinator for the US Fish and Wildlife Service. He works on a range of projects including climate change, invasive species and diseases, and coral reef restoration.

Chapter 6
On-the-Ground Solutions to Help People and Wildlife in a Changing Climate

Nikhil K. Advani

Introduction

Climate change is one of the most pressing issues of our time, and the evidence base on observed impacts to date, and projected impacts, is growing. The IPCC sixth assessment report (IPCC 2022) details how climate change, including more frequent and intense extreme events, have caused widespread adverse impacts to people and nature. Increasingly since the fifth assessment report, the extent and magnitude of climate impacts are larger, and these observed impacts have been increasingly attributed to human-induce climate change. The impacts to island ecosystems and low-lying nations is particularly severe, including sea level rise, heavy precipitation events, and ocean warming, resulting in destruction of property, declining fish yields, and coral bleaching, among many others. Adaptation, across all regions, is generating multiple benefits helping to reduce vulnerability of people and nature. Decision support tools and other climate services are increasingly being used, and pilot projects are being implemented across many different sectors. Among larger adaptation projects worldwide, approximately 20% focus on the agricultural sector, 20% focus on ecosystems, and 20% focus on water or infrastructure (UNEP 2021). But many gaps remain, and financing and implementation fall far short of what is needed (UNEP 2021).

Over the past few years, World Wildlife Fund has launched initiatives to better understand how climate change is impacting people and wildlife, and use the research to inform on-the-ground interventions which help people and wildlife adapt to a changing climate.

N. K. Advani (✉)
World Wildlife Fund, Washington, DC, USA
e-mail: Nikhil.Advani@wwfus.org

S. J. Walsh et al. (eds.), *Island Ecosystems*, Social and Ecological Interactions in the Galapagos Islands, https://doi.org/10.1007/978-3-031-28089-4_6

Helping wildlife Adapt to a Changing Climate

Through WWF's Wildlife Adaptation Innovation Fund (worldwildlife.org/WAIF), an annual call for proposals has been conducted since 2016, for innovative project ideas that help to reduce species vulnerability to climate change. The fund supports project ideas from around the world that reduce the vulnerability of wildlife to changes in weather and climate. Successes and lessons learned from these pilot projects can be replicated and scaled to help wildlife endure under conditions of rapid global change. Project criteria include:

- Must address climate vulnerability of one or more target species through interventions that directly support those species
- Must be implemented in one year or less with plans to monitor results in following years
- Focus on on-the-ground project implementation rather than research

To date, 15 projects have been funded, 4 of which focus on island or marine species.

Project spotlight 1: Constructing Artificial Nests for Shy Albatross (*Thalassarche cauta*) in Tasmania – in collaboration with the Tasmanian and Australian Governments, CSIRO, and the Tasmanian Albatross Fund.The shy albatross is a threatened and endemic Tasmanian species, which faces a variety of threats across its range. Their different life history stages make them particularly sensitive to the changes in weather and climate occurring in both their marine foraging habitats and their terrestrial breeding environments. Higher air temperatures during the chick-rearing period are associated with fewer eggs successfully producing chicks at the end of the breeding season, and their nests are susceptible to extreme rainfall events and wind (Thomson et al. 2015).

This project tested two different kinds of artificial nests, as a way to boost the reproductive success of shy albatross. Specially built mudbrick and aerated concrete nests were constructed and airlifted to Bass Strait's Albatross Island. A total of 123 artificial nests were constructed and installed on Albatross Island in July 2017. The albatross readily adopted their new nests, even personalizing them with mud and vegetation.

Follow-up monitoring throughout the breeding season confirmed high rates of uptake, with eggs laid in 90% of the artificial nests. By the end of the season, breeding success (that is, the proportion of eggs laid that produce chicks that survive to fledging) in the artificial nests was more than twice as high as in the naturally built nests in the study (Fig. 6.1).

Project spotlight 2: Improved nesting, management and monitoring for African penguins (*Spheniscus demersus*) in South Africa – in collaboration with the Southern African Foundation for the Conservation of Coastal Birds (SANCCOB) and South African National Parks (SANParks).

African penguins breed, or have bred, at 32 island and mainland colonies between central Namibia and South Africa's Eastern Cape province, and the global population has declined by almost 65% since 1989 (Sherley et al. 2020). This population

Fig. 6.1 Shy albatross chicks on artificial nests (Matthew Newton/WWF-Aus)

decline has been attributed to a number of threats, including egg and guano collection, changes in abundance and distribution of prey, pollution, habitat destruction, predation, competition with fisheries, and climate change.

Fishing and climate change have rapidly reduced fish abundance, leading to insufficient food for penguin colonies (Sherley et al. 2017). There has also been an increase in the frequency of extreme weather events, and storm surges and heavy rainfall events have resulted in destruction of penguin habitat. During heat waves, chick-rearing penguins often abandon their nests to cool themselves in the water, leaving their eggs and chicks exposed and vulnerable to hyperthermia and/or dehydration.

This project tested three types of artificial nest boxes to assist breeding by providing protection to chicks from predation, heat waves and winter storms. The nest boxes were constructed from cement, ceramic and fiberglass, and included new and old designs. The project monitored breeding success at each nest in conjunction with data from a newly installed weather station and temperature sensors installed within artificial nests and adjacent areas where penguins breed in the open or under bushes. Ceramic nests had the highest usage levels, and also the highest hatching success. Overall breeding success (chicks fledged as a percentage of eggs laid) however was similar between ceramic and fiberglass nests. Older cement nests were less frequently used and also had the lowest breeding success.

The data is now being used to assess the most favorable nest types and placements to improve breeding success. Gaining an understanding of how extreme

Fig. 6.2 (a) (left): Ceramic (left) vs. fiberglass (right) artificial nests (Nikhil Advani). (b) (right): African penguins nesting in the ground (Nikhil Advani)

weather events play out locally and their effects in different parts of the colony will also feed into an early warning system being developed to alert managing authorities to extreme weather events for coordinated rescue of eggs and chicks in danger. On the ground Penguin rangers also monitor breeding of the birds and assist with habitat maintenance and rescue of eggs for hand-rearing, when chicks or eggs are abandoned during extreme weather. Over the course of one breeding season, 60 chicks and 112 eggs were rescued in this way from the colony (Fig. 6.2).

Helping Communities Adapt to a Changing Climate

Indigenous, local, and traditional knowledge systems and practices, including indigenous peoples' holistic view of community and environment, are a major resource for adapting to climate change, but these have not been used consistently in existing adaptation efforts (IPCC 2014). Additionally, communities lack the resources to help them improve, implement and scale these efforts, and ensure that protecting nature is at the heart of the project outcomes.

WWF Climate Crowd (wwfclimatecrowd.org) is a bottom-up community-driven approach. Working with communities and local NGOs in over 30 countries, we collect data on climate impacts to communities, analyze the data, present the data back to the communities, and work with them to develop, fund and implement on-the-ground solutions that help people and nature adapt to a changing climate. The Climate Crowd model therefore provides a rapid way to gather data, pilot projects, and mobilize financial resources for the most vulnerable communities, through a participatory method.

On-the-ground projects focus on increasing water security (e.g., rainwater harvesting, fog catchers), climate-smart agriculture (e.g., conservation agriculture techniques, improved irrigation systems, planned grazing, agroforestry), alternative livelihoods (e.g. seaweed farming, beekeeping), reforestation (e.g., wildlife corridors, watershed restoration, mangrove restoration, agroforestry), educating communities and school children on climate impacts and solutions, and installing weather stations in data scarce regions.

Project Goals

- Give a voice to the most vulnerable communities and learn from local/traditional/indigenous knowledge.
- Scale the amount and diversity of data collected from the field, data disseminated, and create evidence-based recommendations for better adaptation strategies.
- Pilot projects that increase the resilience of communities and nature, and scale successful pilots through Climate Crowd as well as through governments and other large institutions such as the Green Climate Fund.
- Raise awareness through compelling stories from the front lines of climate change.

Project Spotlight 1: Madagascar

Climate Crowd interviews were conducted with communities on the west coast of Madagascar in 2017. Key informants reported a decline in rainfall, changes in the timing of seasons, drought, heatwaves/hotter days, and changes in wind patterns. The impacts of these changes in weather and climate included reduced abundance of fish, crop failure, and reduced availability of freshwater. These all have a significant impact on the main livelihoods in the village. Communities responded by fishing in new areas, employing new fishing techniques, pursuing alternative livelihoods, gathering non-timber forest products, constructing wells for freshwater access, and migrating to other areas (Climate Crowd 2017) (Fig. 6.3).

In response to the identified climate impacts, a Climate Crowd project, completed in 2021, implemented seaweed farming and beekeeping to help households diversify their income and lessen the burden on marine ecosystems (Climate Crowd 2021). Project activities included identification of sites, training of local communities,

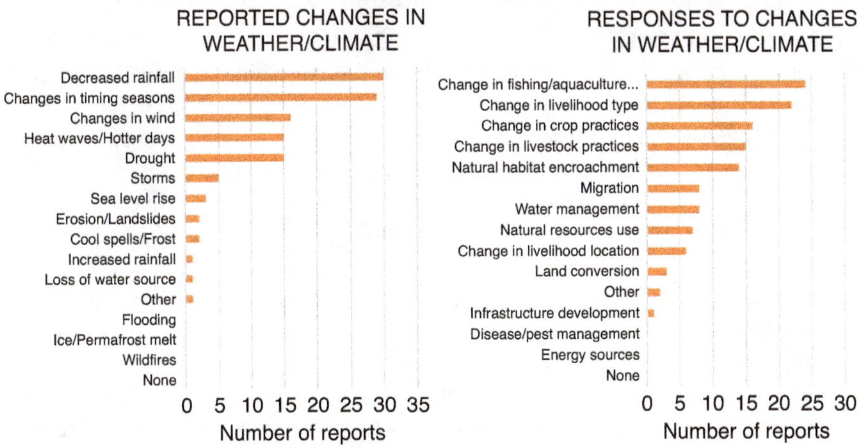

Fig. 6.3 Summary data from Climate Crowd interviews conducted in Madagascar (Climate Crowd 2017)

Fig. 6.4 Seaweed farming in Madagascar (WWF Madagascar)

purchase and installation of materials for beekeeping and seaweed production, and providing market access for the sale of products to the private sector. 32 households are involved in beekeeping, and 42 households in seaweed farming. The project outcomes include increased household income, reduced pressure on fisheries, and increased involvement of women in income generating activities (Fig. 6.4).

Project Spotlight 2: Fiji

Climate Crowd interviews were conducted with communities across Fiji in 2022. Key informants reported changes in the timing of seasons, variations in rainfall with both prolonged wet and dry seasons, drought, increased flooding and storm surge, more intense and frequent heatwaves, and sea level rise and coastal erosion. The impacts of these changes in weather and climate included a decline in crop yields, a decline in fish abundance, reduced water availability, and property damage, amongst other impacts. Communities have responded by switching to alternative crops, conservation agriculture, changing farming location, purchasing goods, spending more time fishing, changing fishing location, travelling longer distances to collect water, drilling wells, and relocating their homes further inland (Fig. 6.5).

Implementation of project activities is yet to commence, however, based on the data collected and analyzed, projects are likely to focus on increasing water security, alternative livelihoods, and climate-smart agriculture. Some interventions, such as rainwater harvesting, are already widely implemented on many of the islands (Fig. 6.6).

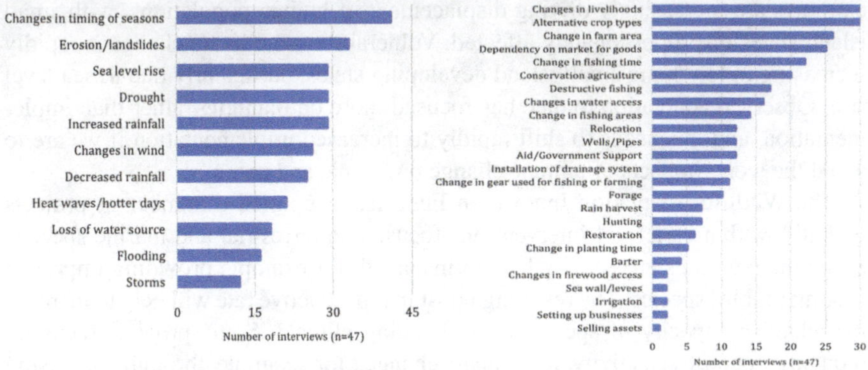

Fig. 6.5 Summary data from Climate Crowd interviews conducted in Fiji (2022)

Fig. 6.6 Coastal erosion in Fiji (Nikhil Advani)

Discussion & Conclusions

Climate change is one of the most pressing issues of our time, and the impacts are already evident in many ecosystems and human systems worldwide. Small Islands have seen changes in ecosystem structure and species range shifts, with high confidence in their attribution to climate change (IPCC 2022). For humans, small islands are experiencing impacts to water scarcity and food production, health and wellbeing, and impacts on cities, settlements and infrastructure. Climate and weather extremes are increasingly driving displacement of human populations, with small island states disproportionately affected. Vulnerability to climate change is rapidly increasing in low lying small island developing states, particularly due to sea level rise. Observed adaptation to date has focused more on planning rather than implementation, and this needs to shift rapidly to increased implementation if we are to avoid the worst impacts of climate change on people and nature.

The Wildlife Adaptation Innovation Fund has supported a number of projects globally, with a variety of interventions focused on terrestrial and marine species. Some interventions are heavily human managed, for example providing improved nesting for bird species. The resulting boost in reproductive rate will help to increase the adaptive capacity of species to a changing climate. Some projects focus on reducing species sensitivity to climate change, for example through increasing water access, or mitigating the impact of increased temperatures. Some projects have a longer term goal of restoring and preserving ecosystems, with a view to facilitating range shifts to higher elevations for example. These projects are all relatively small and innovative, with a goal of scaling or replicating successful interventions.

An often unrecognized threat to biodiversity comes from human responses to climate change. That is, the ways in which people are being impacted by changes in weather and climate, and how they cope with that, can often have negative impacts on biodiversity. Examples include increasingly unsustainable use of natural resources, or shifting livelihood activities to new areas, often areas that are richer in biodiversity. Climate Crowd works with communities all over the world, to better understand the climate impacts they are experiencing, and works with them to design, fund and implement projects that will reduce community vulnerability to climate change, in ways that are nature-based, benefit nature, or do no harm to nature. The Climate Crowd model ensures that community voices are at the heart of the project, and this in turn ensures that interventions have strong community buy-in, thus increasing the long term sustainability and success of the projects. To date, projects have largely focused on increasing water security, climate-smart agriculture, and alternative livelihoods.

If we are to avoid the worst impacts of climate change, it is imperative that governments, the private sector, and individual citizens take steps to rapidly reduce our emissions of greenhouse gases. However, climate impacts are already being felt all over the world, and these will increase in frequency and severity. No matter the emissions reductions, earth will continue to warm for decades due to the concentration of greenhouse gases already in the atmosphere. It is therefore essential that we help people and nature all over the world adapt to these changes, in ways that are nature-friendly.

References

Climate Crowd (2017) Madagascar summary report. World Wildlife Fund, Washington, DC

Climate Crowd (2021) Climate Crowd on the ground: A compilation of climate Crowd projects. World Wildlife Fund, Washington, DC

IPCC (2014) Summary for policymakers. In: Climate change. impacts, adaptation, and vulnerability. Part A: global and sectoral aspects. Contribution of working group ii to the fifth assessment report of the intergovernmental panel on climate change. Cambridge University Press, Cambridge, United Kingdom and New York, NY, USA, pp 1–32. Intergovernmental Panel on Climate Change WGII AR5

IPCC (2022) Summary for Policymakers [H.-O. Pörtner, D.C. Roberts, E.S. Poloczanska, K. Mintenbeck, M. Tignor, A. Alegría, M. Craig, S. Langsdorf, S. Löschke, V. Möller, A. Okem (eds.)]. In: Climate Change (2022): Impacts, Adaptation and Vulnerability. Contribution of Working Group II to the Sixth Assessment Report of the Intergovernmental Panel on Climate Change [H.-O. Pörtner, D.C. Roberts, M. Tignor, E.S. Poloczanska, K. Mintenbeck, A. Alegría, M. Craig, S. Langsdorf, S. Löschke, V. Möller, A. Okem, B. Rama (eds.)]. Cambridge University Press, Cambridge, UK and New York, NY, USA, pp. 3–33, doi:10.1017/9781009325844.001

Sherley RB, Ludynia K, Dyer BM, Lamont T, Makhado AB, Roux JP, Scales KL, Underhill LG, Votier SC (2017) Metapopulation tracking juvenile penguins reveals an ecosystem-wide ecological trap. Curr Biol 27(4):563–568., ISSN 0960-9822. https://doi.org/10.1016/j.cub.2016.12.054

Sherley RB, Crawford RJM, de Blocq AD et al (2020) The conservation status and population decline of the African penguin deconstructed in space and time. Ecol Evol 10:8506–8516. https://doi.org/10.1002/ece3.6554

Thomson RB, Alderman RL, Tuck GN, Hobday AJ (2015) Effects of climate change and fisheries bycatch on shy albatross (Thalassarche cauta) in Southern Australia. PLoS One 10(6):e0127006. https://doi.org/10.1371/journal.pone.0127006

United Nations Environment Programme (2021) Adaptation gap report: The gathering storm – Adapting to climate change in a post-pandemic world. Nairobi

Nikhil K. Advani was born and brought up in Kenya, and he moved to the US to pursue his degrees at the University of Texas at Austin. He's now been at World Wildlife Fund for almost 10 years, where his role is at the intersection of communities, wildlife, and the varied threats they face, from climate change to the global COVID-19 pandemic.

Part IV
Island Ecosystems: Social Sub-systems

Chapter 7
Climate and Health Challenges in Small Island States: Identifying Vulnerability in Water and Food Resources in the Galapagos Islands, Ecuador

Amanda L. Thompson, Jill R. Stewart, Margaret Bentley, Jaime Eduardo Ocampo Trujillo, Enrique Teran, and Valeria Ochoa

Introduction

Climate change has multi-faceted impacts on human health and development that are already being felt through rising temperatures and more frequent extreme weather events (Watts et al. 2021). Along with the direct impacts of climate change on morbidity and mortality through injury and heat-related illness, the indirect impacts through compromised food systems, water contamination, ecosystem disruption, and social instability are likely to be considerable, persistent stressors on human health and well-being. Changes in air and water quality and exposure to new or more frequent environmental contaminants and vector-borne pathogens may increase the likelihood of respiratory, gastrointestinal and zoonotic illnesses (McMichael et al. 2006). At the same time, increased heat and periods of drought may shift food consumption from locally grown crops to prepared, packaged foods and reduce the amount of physical activity individuals engage in for work or leisure, increasing the risk of developing noncommunicable diseases (NCDs) (Frumkin and Haines 2019). Community health may also suffer as the result of increased crime, violence and instability and decreased social cohesion, leading to problems with mental health (McIver et al. 2016).

These health and well-being impacts of climate change are likely to be even stronger in the Small Island States (SIS), a United Nations-defined group of 58

A. L. Thompson (✉) · J. R. Stewart · M. Bentley
University of North Carolina at Chapel Hill, Chapel Hill, NC, USA
e-mail: althomps@email.unc.edu

J. E. Ocampo Trujillo
Universidad San Francisco de Quito, Quito, Ecuador

E. Teran
School of Medicine, Universidad San Francisco de Quito, Quito, Ecuador

V. Ochoa
College of Biological & Environmental Studies and USFQ-UNC Galapagos Science Center, Universidad San Francisco de Quito, Quito, Ecuador

S. J. Walsh et al. (eds.), *Island Ecosystems*, Social and Ecological Interactions in the Galapagos Islands, https://doi.org/10.1007/978-3-031-28089-4_7

countries mostly located in the Caribbean and Pacific (Weir and Pittock 2017). These islands are already vulnerable due their geographic isolation, limited human and natural resources, high costs, and distance from global markets (Hickey and Unwin 2020). The small size of many SIS limit the amounts of food that can be grown locally. At the same time, changes in ocean temperature and over-fishing may limit the availability of marine resources, contributing to food insecurity for island residents (Hanna and McIver 2014). Due to their low-lying geography, fresh water sources may also be limited, and water quality may be comprised by extreme weather events or insufficient sanitation infrastructure (Akpinar-Elci and Sealy 2014). At the same time, the relatively small populations and limited human resources contribute to poor health care infrastructure, exacerbating the treatment of both acute and chronic health conditions (McIver et al. 2016). Together these environmental and social stressors contribute to increasing risk of mental health disorders, such as depression and post-traumatic stress disorders.

SIS that rely upon tourism are particularly vulnerable, since climate change and other exogenous shocks, like pandemics or natural disasters, can prevent or significantly reduce travel with dramatic consequences for islands' economies. For example, in contrast to the 1.7% contraction of gross domestic product (GDP) seen in developing economies in 2020, contractions as high as 30% were seen in Pacific Ocean SIDS (Small Island Developing States) (World Bank 2022), where normally 12% to 87% of islands' GDP come from tourism (Scheyvens et al. 2021). Along with the direct impacts on employment and household income, tourism and fluctuations in the number of visitors can also indirectly impact available water, food, and health care infrastructure. The growth of tourism, for example, can change the food environment by driving higher costs and increasing the importation of "Western" diets (Burke 2021). Similarly, limited water resources may be preferentially distributed to hotels and resorts (Mateus and Quiroga 2022). Thus, the economic impacts of tourism also shape island infrastructure and household food security, often exacerbating the already existing social and economic inequalities contributing to population health in SIS.

While not an independent state, the Galapagos Islands, a province of Ecuador, suffer from many of the same vulnerabilities as these Pacific SIS. The islands, which have approximately 30,000 inhabitants over the main islands, are 1000 kilometers from the mainland, leading to issues with small populations, geographic isolation, and distance from markets. As a national park, 97% of the islands' land is off limits for farming and the marine reserve limits the areas available for fishing. Only one of the three main inhabited islands, San Cristobal, has a surface freshwater source, leaving the majority of residents dependent on wells or desalinated water sources. Like islands such as the Maldives of Fiji, the Galapagos are dependent on tourism, with over 80% of the economy reliant on the tourism industry (Burbano et al. 2022). The islands have seen population growth of over 300% over the past several decades partly to meet the demands of tourism. Prior to the Covid pandemic, over 250,000 tourists visited the islands annually (Walsh and Mena 2016), a figure that is higher than many other Pacific Island destinations and, relative to population size, is considerably higher than even Fiji. Thus, the islands serve as a model of the challenges faced by SIS more broadly and, in some ways, provide a more extreme example. In this chapter, we describe the health situation in the Galapagos, discuss the health impacts of existing water and food insecurity, and identify the households that are most vulnerable to the effects of climate-related impacts on food and water resources.

Dual Burden of Disease in Pacific SIS and Galapagos

Pacific SIS, like many low- and middle-income countries, suffer from the dual burden of disease, the co-occurrence of conditions associated with undernutrition and poverty, such as child stunting and infectious illness from pathogenic environmental exposures, and those associated with economic development and changing lifestyles, such as overweight and noncommunicable diseases (NCDs). An estimated 27% of the population in the Pacific Island region is undernourished; at the same time, NCDs account for 75% of deaths in the region (Hanna and McIver 2014). Similar to several Pacific SIS (Fig. 7.1), the population of the Galapagos also suffers from this dual burden of disease. The percentage of child stunting, a reflection of chronic undernutrition and frequent infection, remains higher than some other Pacific SIS as does the proportion of residents lacking access to improved water sources (ENSANUT-ECU 2012, INEC-CREG 2010). At the same time rates of adult obesity and NCDs are also high, with 30% of adults considered obese and over 70% of adults considered overweight or obese (ENSANUT-ECU 2012). More positively, the rate of infant mortality is considerably lower than many Pacific SIS, likely reflecting the availability of better health care services (Jahnke et al. 2022).

Overall, these patterns reflect the impact of economic development and tourism. The dual burden has been documented at the population level with some individuals

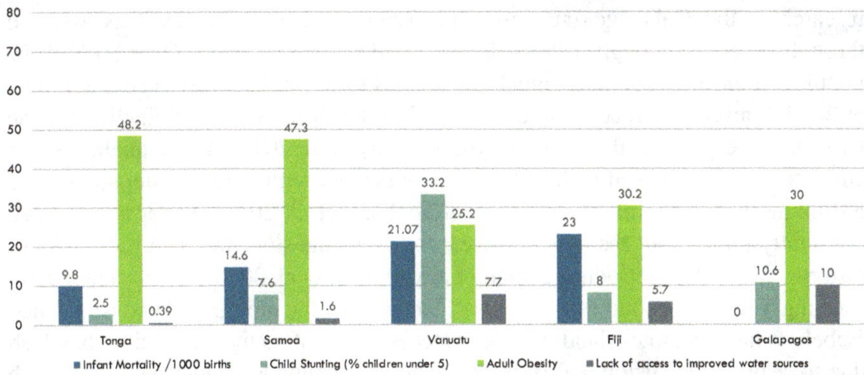

Fig. 7.1 Health Challenges in Pacific Small Island States and Galapagos. Data for SIS come from the Global Health Observatory, World Health Organization, 2022; Data for Galapagos come from ENSANUT-ECU (Freire et al. 2015)

having one condition and some the other, at the household level when mothers, for example, have iron deficiency and their children are overweight, or even within individuals. These patterns reflect differences in access to new foods and occupations, intra-household food allocation, or dietary and disease environments across the life course that are seen with economic development (Tzioumis and Adair 2014). In tourism-focused SIS, along with the shifts in "traditional" diets, which tend to be higher in fresh fruits and vegetables and lean protein, to "Western" diets characterized by higher fat and energy-dense foods seen with economic development more generally (Popkin et al. 2012), the food environment is also influenced by the direct and indirect demands of tourism. Directly, tourism may change the types of foods available to reflect visitor preferences, limit access to higher quality foods that preferentially go to restaurants and hotels or increase residents' food insecurity by driving higher prices. Indirectly, tourism may also shift traditional diets through changing livelihoods as individuals leave farming for work in the tourism sector (Hickey and Unwin 2020; Scheyvens et al. 2021). These changes in household economics and diet often occur before broad-scale improvements in water and sanitation infrastructure, leading to issues with water availability and quality. Tourism may then further exacerbate these issues with water quality and availability by placing additional demands on existing systems (Vásquez et al. 2021; Mateus and Quiroga 2022). Together, these food and water environments shape the health and well-being of island residents. In the following sections, we will discuss the health impacts of the water and food insecurity in Galapagos.

Water Insecurity and Health

Water security, defined as reliable access to sufficient quantities of acceptable quality water (Connor 2015), is included in the United Nations Sustainable Development Goals, partly due to its role as an important determinant of health. Our and others' research in the Galapagos suggests that water insecurity – at least as assessed through access to enough water— is not a widespread problem. Over 80% of the population on the three main inhabited islands have public water access and over 90% of households receive treated, piped water from the municipality in San Cristobal, the provincial capital (Guyot-Tephany et al. 2013). Water quality is less consistent. San Cristobal is the only inhabited island with a fresh water source and a drinking water treatment plant (Guyot-Tephany et al. 2013). Water is not piped constantly throughout the day, however, and residents store water in roof tanks or cisterns, which may introduce contamination (Grube et al. 2020; Houck et al. 2020). Lacking fresh water sources, residents in Santa Cruz, the most populated island, and Isabela, a more remote island, rely on water from aquafers that are mainly brackish due to seawater intrusion and over-exploitation (Mateus and Quiroga 2022). Both islands have municipal desalination plants for water treatment, but these plants cannot keep up with the demands for water and household water tends to be brackish and/or contaminated (Mateus and Quiroga 2022; Badhwa et al. 2022). Cisterns and roof tanks are also employed for water storage since water is rationed to a few hours per day (Vásquez et al. 2021). On all three islands, residents tend to rely on bottled water for their main source of drinking water, leading to significant household expenditures on water (Reyes et al. 2015; Houck et al. 2020; Thompson et al. 2022) and potentially exacerbating insecurity in lower income households (Stoler et al. 2020).

Numerous studies have documented that inadequate water and sanitation services contribute to the global burden of infectious diseases, such as diarrheal and respiratory infections, particularly in low- and middle-income countries (LMICs) (Prüss-Ustün et al. 2019). The lack of sufficient high-quality water is associated with poorer health for Galapagos residents. Physicians on Isabela and San Cristobal Islands report that many of the conditions they treat, such as diarrhea, gastroenteritis, and urinary tract infections, are associated with exposure to contaminated water (Walsh et al. 2010; Jahnke et al. 2022). Research by Houck and colleagues documented higher measured levels of *E coli* in household water were associated with a greater likelihood of experiences urinary infections in mothers and children and GI infections in children (Houck et al. 2020). Repeated incidences of infection due to repeated and/or chronic exposure to pathogens in water can have longer term health impacts. Preliminary analysis from *The Healthy Families Study,* a household-based survey of water, food and health in San Cristobal conducted in 2018–2019, documents that the presence of *E coli* in household water is associated with greater risk of both stunting, low height-for-age associated with chronic undernutrition and/or infection, and overweight and obesity. This finding that water quality is associated with the dual burden of disease may stem from the physiological consequences of repeated exposure to contaminated water, which may contribute to bacterial overgrowth in the small intestine and subclinical environmental enteropathy. Environmental enteropathy is associated with increased infection and stunting in children (Donowitz & Petri 2015; Brown et al. 2013), but when coupled with high fat diets may also induce inflammation and increase the risk of overweight and NCDs (Kau et al. 2011).

Along with these direct effects of exposure to contaminants in water, perceived water insecurity is also associated with health outcomes. Work in numerous settings (Brewis et al. 2020) documents that the perception that water resources are costly and of poor quality adds to the burden of disease, through both physiological and behavioral pathways. Physiologically, water insecurity may influence physical and mental health through several direct and indirect pathways (Rosinger and Young 2020). Behaviorally, household water insecurity may lead to the consumption of more foods away from home and greater intake of sugar sweetened beverages (Miller et al. 2021; Rosinger and Young 2020). In the Healthy Families Study, greater household water insecurity, measured through a validated scale (Young et al. 2019), was associated with an increased risk of stunting (Fig. 7.2).

Our results that both overall water insecurity and concerns about having sufficient water were associated with a greater risk of stunting in children and adults provides preliminary evidence that long-term exposure to water insecurity may be associated with health outcomes that persist across generations. Water insecurity was associated

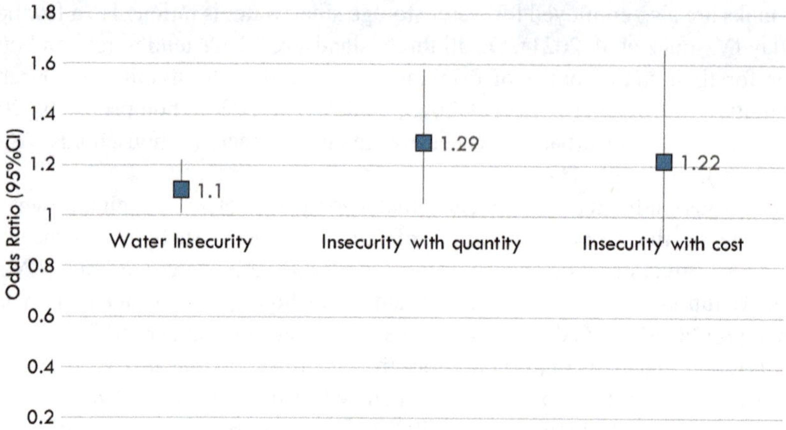

Fig. 7.2 Water insecurity and risk of stunting. Results from logistic regression models controlling for age, sex, socioeconomic index, and clustering by household (n = 389 individuals aged 2wk-59 years)

with poorer mental health in adults. Participants living in households with even mild water insecurity were more likely to report symptoms of depression, anxiety, or high stress (18%) than those with secure water (11%; Thompson et al. 2022).

Food Insecurity and Health

Like water insecurity, food insecurity, the limited or uncertain availability of nutritionally adequate, safe, and culturally appropriate foods, or the uncertain ability to acquire such foods in socially acceptable ways (UN), is a problem for many Galapagos residents. Food availability is a critical issue and contributes to episodic food insecurity. Nearly all the land in the islands is designated as a national park limiting the amount of land that can be used for farming and animal husbandry. Even this available land is underutilized as residents have shifted from working in agriculture to working in the more lucrative tourism industry (Sampedro et al. 2018), further exacerbating the lack of locally grown and produced foods. Consequently, residents rely on food shipped or, increasingly, flown in from the mainland. The distance of the islands from the mainland, over 1000 km, means that produce often arrives spoiled and the higher cost of foods sent by air means that residents tend to rely on processed, packaged foods, which are viewed as better quality and less expensive. While few, if any, households on the islands suffer from severe food insecurity, the limited availability, and high costs of food mean that over 40% of households experience at less mild food insecurity (Thompson et al. 2022). Residents describe their food insecurity as stemming from the boat schedules and inventories, which determine the quantity and variety of foods available for purchase on the island (Pera et al. 2019). Further, exacerbating this insecurity is the unequal distribution of available foods. Our work in San Cristobal suggests that when supplies are limited, resources may be concentrated in areas frequented by tourists and be less available to island residents (Thompson et al. 2022).

Certainly, these constraints influence the residents' diet quality. In our analysis of the nationally representative ENSANUT data, for example, we found that overall dietary adequacy was lower in food insecure households (Watson 2019). Similarly, research from the 2011 Mothers Study (Pera et al. 2019) documented that diet diversity, a marker of diet quality that quantifies the number of food groups consumed each day based on FAO recommendations, was lower in households with higher insecurity. However, this study also documented that the impacts of food insecurity may differ within households, since mothers tended to have poorer diet diversity than their children across all levels of food security. Qualitative research in the households demonstrated that mothers tended to preferentially feed their children available fruits and vegetables. Preliminary analysis from our larger Healthy Families Study finds similar patterns. We find that mothers' diet diversity remains relatively constant across levels of food insecurity, but that children have higher diet quality in households with mild food insecurity. These patterns are even more pronounced when looking at consumption of foods from "healthy" food groups, defined from the Global Dietary Recommendations (Herforth et al. 2020). Mothers' consumption of healthy food groups decreases across mild and moderate food insecurity, while children's consumption is higher in households with mild insecurity, suggesting that mothers may alter their diets to buffer their children when resources are low. The lower diet diversity and healthy food consumption seen in children in moderately food insecure household suggests that this buffering may not be possible in more extreme situations.

Food insecurity, with its associated impacts on diet quality and diversity, is an important determinant of the dual burden of disease in many settings. In LMIC settings, household food insecurity has been linked to inadequate food intake, poor diet quality, anemia, higher infectious disease burden, and, in children, growth faltering (Weigel & Armijos 2015; Schmeer and Piperata 2017). However, household food insecurity is also increasingly linked to a greater risk of overweight/obesity and NCDs in LMIC (Jones et al. 2016; Jones et al. 2018; Pérez-Escamilla et al. 2014). In Galapagos, we find little evidence that food insecurity is associated with acute undernutrition; indeed, in our Healthy Families Study <3% of children and only 1% of adults were underweight and there was no significant association between food insecurity and undernutrition, anemia or infection. This lack of acute malnutrition likely reflects the relatively higher socioeconomic status of the island residents, compared to other sites on the mainland, and the lack of abject poverty (Granda et al. 2012).

Conversely, we do find that food insecurity is associated with a greater likelihood of being overweight or having risk factors for NCDs, including hypertension and elevated glucose. As shown in Fig. 7.3, scores for both overall food insecurity and insecurity in both food quantity and quality are higher among those with overweight or NCDs. This association of household food insecurity with overnutrition and

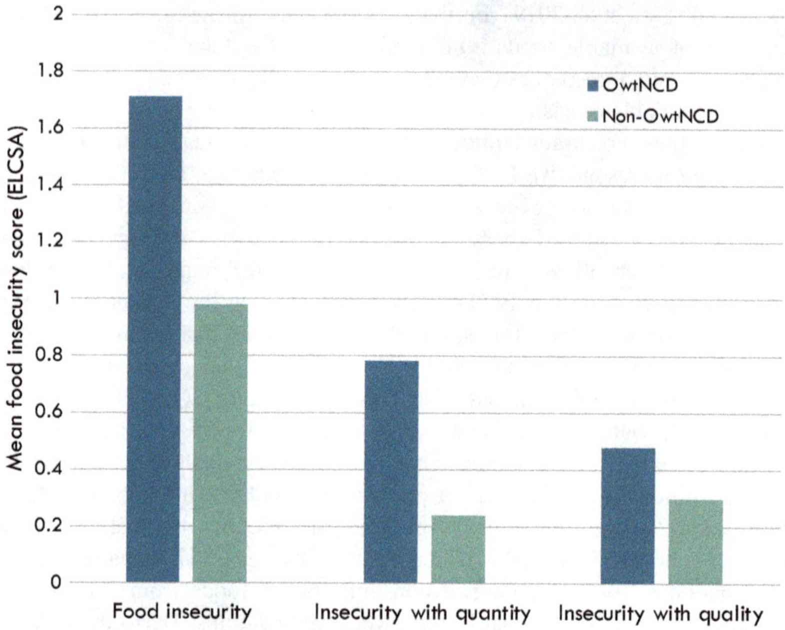

Fig. 7.3 Food insecurity scores are higher in those with overweight and NCDs. Food insecurity scores come from the Latin American and Caribbean Food Insecurity Scale (ELCSA; Pérez-Escamilla et al. 2014)

NCDs may be due to that fact that the type of food insecurity seen on the Galapagos, like that of other tourism-based economies (Himmelgreen et al. 2012, Ruiz et al. 2014), is generally mild-to-moderate and episodic. This type of food insecurity is thought to be particularly risky for the development of overweight and NCDs, since it promotes poorer diet quality with reduced intake of fruits, vegetables, and dairy (Weigel et al. 2016; Pérez-Escamilla et al. 2014) and over consumption of easily stored processed foods (Freire et al. 2018). The fluctuating and unpredictable availability of food can also contribute to stress, with concomitant increases in risk factors for overweight and NCDS such as cortisol, visceral adiposity, inflammation, and insulin dysregulation (Pérez-Escamilla et al. 2014).

Identifying Vulnerable Households

As work in other LMIC and SIS has shown (Jones et al. 2018; Stoler et al. 2020; Brewis et al. 2020), water and food insecurity are not evenly distributed throughout the population. Identifying the households that are most vulnerable is critical for intervention. In many LMIC and SIS settings, both water and food insecurity and the burden of disease are higher among low-income, indigenous and otherwise vulnerable segments of the population (e.g. women, children and the elderly) (Lee et al. 2010; Schnitter et al. 2018).

Similarly, our work in Galapagos has shown that water and food insecurity are more common in rural areas and among households that are indigenous Ecuadorian, female-headed, or lower income (Thompson et al. 2022; Nicholas et al. 2020; Watson 2019). In our secondary analysis of ENSANUT data, diet quality, one component of food insecurity, was lower in women and girls compared to men and in indigenous residents compared to *mestizo* residents (Watson 2019). Both dietary adequacy and quality were lower in rural households, highlighting the dependence of residents on market foods which tend to be concentrated in the urban areas. These urban/rural differences are also seen in water insecurity, but research by Nicholas et al. (2020) suggests that the impact of geography may also differ by income. Urban households had significantly better water security than rural households, but this gap narrowed at higher income levels, with the wealthiest rural households being protected from water insecurity. Interestingly, social connections, or lack of them, may also serve to buffer against or, conversely, worsen food and/or water security. Residents who were born elsewhere and migrated to the islands had poorer water access (Nicholas et al. 2020), but better diet quality (Watson 2019). Prior research in Latin America has shown that social networks may help lower-income families' buffer against the impacts of water insecurity (Wutich et al. 2018), by offering opportunities for water sharing among other strategies. However, these relationships may also introduce considerable stress into social relationships and concerns over reciprocity (Wutich and Ragsdale 2008; Brewis et al. 2021) and may be less effective than other strategies employed by higher resourced households (Stoler et al. 2020).

Given the social determinants underlying water and food insecurity, it is not surprising then that water and food insecurity are also likely to co-occur (Brewis et al. 2020). In both our analysis of ENSANUT and the Healthy Families Study, we found that 20 to 35% of households experienced both water and food insecurity, respectively (Thompson et al. 2020; Thompson et al. 2022). As shown in Fig. 7.4, household water and food insecurity are associated with household income. The probability of a household experiencing both water and food insecurity decreases as household economic quintile increases. Conversely, the probability of households having no limitations with food or water is highest for those in the two highest income brackets. Interestingly, the association between income and water insecurity alone and between water and food insecurity alone is less pronounced. These patterns suggest that, while households may be able to cope with one limitation, the presence of both water and food limitations stretches households' resources, leaving those with fewer resources less able to respond to these dual pressures. Importantly, having both water and food insecurity simultaneously was also associated with a greater likelihood of a household experiencing the dual burden (Thompson et al. 2020). These findings suggest that households that are the most vulnerable to insecurity due to economic and social constraints are also those most likely to suffer from ill health. Managing conditions associated with both undernutrition and infection and overweight and NCDs simultaneously then may act as another significant stressor to households' income and resources.

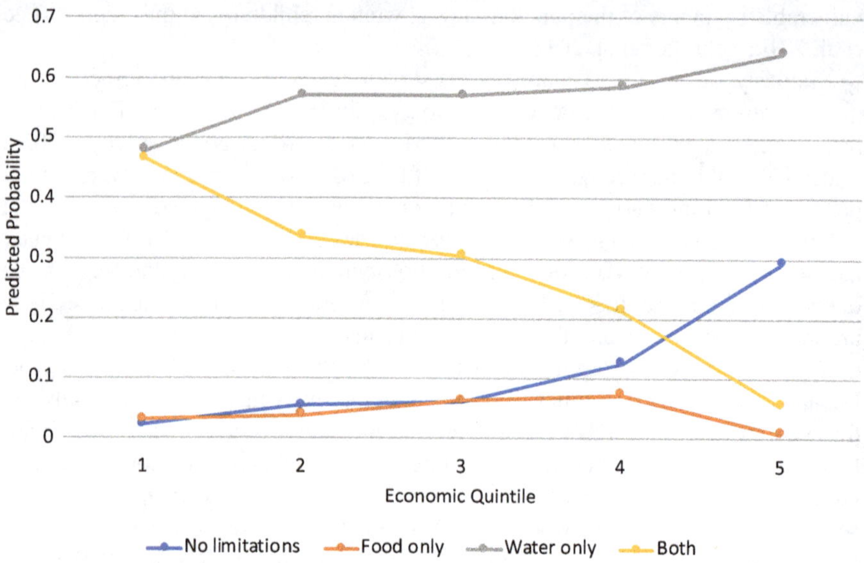

Fig. 7.4 Probability of a household experiencing water and food limitations alone or jointly by economic quintile. (Data from ENSANUT-ECU 2012)

Climate-Related Vulnerability and Health: Lessons from Galapagos

In summary, the geographic isolation of the Galapagos, the limited land available for farming, lack of freshwater sources, and the economic reliance on tourism currently make residents vulnerable to household food insecurity and poor water quality. In turn, these limitations in food and water resources contribute to poor diet quality, infections and, increasingly, the triple burden of infection, overweight and NCDs, and mental health disorders. Further, our research also shows that households that are most at risk for both being water and food insecure, and, also, suffering from disease, are those who are poorer, indigenous, and rural.

These geographic, economic, and social determinants of the current burden of disease in the Galapagos are the same factors that underlie vulnerability to the health impacts of climate change on the islands and more broadly in SIS. Geographically, the low-lying topography and limited freshwater resources of the Galapagos and many SIS already put residents at risk for poor water availability and quality. With increasing temperatures and rising sea waters, these resources may be further threatened by flooding and/or salinization of ground water. Contaminated water sources and standing water will increase disease risk from a number of infectious illnesses including diarrheal diseases and vector-borne diseases like dengue or chikungunya (Akpinar-Elci and Sealy 2014). Increasing salinization may also contribute to malnutrition. Increasing salinity in the soil may reduce productivity and limit the crops that can be grown, contributing to food insecurity and undernutrition. At the same time, consumption of drinking water with higher salinity may also contribute to NCDs by increasing the risk of salt-sensitive hypertension (Khan et al. 2020). Along with these impacts on water sources, higher temperatures and accompanying droughts will threaten food systems both locally and globally, increasing the reliance on processed foods and exacerbating overweight and NCD risk (Frumkin and Haines 2019). Together these limitations and their health sequela will place strain on communities and health care systems. Inequalities in the distribution of limited resources and the increased stress on individuals and households trying to meet their water and food needs may contribute to household dysfunction, such as inter-partner violence (Nunbogu and Elliott 2021), and social schisms at the community level (Wutich and Ragsdale 2008; Brewis et al. 2021) worsening mental health. The greater burden of physical and mental health conditions will place increasing strain on already under-resourced health care systems.

The recognition that these impacts of climate on human health and well-being are multiplicative and interactive has led researchers to label the co-occurrence of climate change, undernutrition, and obesity a "Global Syndemic" (Swinburn et al. 2019). This syndemic framework views the clustering of two or more health conditions as stemming from the interaction between them and detrimental social and physical environments (Mendenhall 2017). In the case of climate change, undernutrition and overnutrition, this syndemic framework is used to highlight their co-occurrence, interaction, and common underlying causes. As the example of the Galapagos has shown, the dual burden of disease, i.e., the syndemic of under- and over-nutrition, is driven by common limitations in food and water environments that, in turn, stem from geographic, economic, and social constraints. Climate change is likely to heighten these negative interactions since it shares many of the same underlying drivers already contributing to vulnerability and may limit resilience and adaptive capacity (Savage et al. 2021).

However, the syndemic nature of climate change and human health also present opportunities for "double-duty" and "triple-duty" actions to address undernutrition, NCDs, and climate change simultaneously. In 2019, the *Lancet Commission* defining the Global Syndemic (Swinburn et al. 2019) offered numerous suggestions for actions at the global, national, and community levels, including offering subsidies to farmers to grow fruits and vegetables to alleviate malnutrition and shift production from greenhouse gas emitting animal husbandry; changes to nutrition education and school lunches to promote environmental sustainability, prevent undernutrition, and enhance human development by keeping children in school; and urban redesign to promote walking or cycling, disincentivizing driving, and improving public transport availability and cost to increase physical activity, reduce overweight and limit greenhouse gas emissions. While these interventions will need to be place-based and appropriate for the local context of islands like the Galapagos and SIS, addressing food sufficiency, providing access to adequate and safe drinking water, and improving health care infrastructure are key human health and climate change adaptation strategies (Hanna and McIver 2014; McMichael et al. 2006; Swinburn et al. 2019). Identifying current sources of vulnerability is important for developing strategies to mitigate these risks now in the face of climate change and is critical for improving physical and mental health in the Galápagos, SIS and LMIC more broadly.

References

Akpinar-Elci M, Sealy H (2014) Climate change and public health in small Island states and Caribbean countries. In: Pinkerton KE, Rom WN (eds) Global climate change and public health. Springer, New York, NY, pp 279–292

Badhwa N, Fejfar D, Pozo R, Nicholas K, Grube A, Stewart J, Thompson A, Ochoa-Herrera V (2022) Water quality and access in Isabela: Results from a household water survey. In: Water, food and human health in the Galapagos, Ecuador. Springer, Cham, pp 57–74

Brewis, A., Workman, C., Wutich, A., Jepson, W., Young, S., Household Water Insecurity Experiences–Research Coordination Network (HWISE-RCN), Adams, E., Ahmed, J.F., Alexander, M., Balogun, M., & Boivin, M. (2020). Household water insecurity is strongly associated with food insecurity: Evidence from 27 sites in low-and middle-income countries. American Journal of Human Biology, 32(1), e23309.

Brewis A, Roba KT, Wutich A, Manning M, Yousuf J (2021) Household water insecurity and psychological distress in Eastern Ethiopia: Unfairness and water sharing as undertheorized factors. SSM Ment Health 1:100008

Brown J, Cairncross S, Ensink JH (2013) Water, sanitation, hygiene and enteric infections in children. Archives of Disease in Childhood 98(8):629–634

Burbano DV, Valdivieso JC, Izurieta JC, Meredith TC, Ferri DQ (2022) "Rethink and reset" tourism in the Galapagos Islands: Stakeholders' views on the sustainability of tourism development. Ann Tour Res Empiric Insights 3(2):100057

Burke A (2021) The crossroads of ecotourism dependency, food security and a global pandemic in Galápagos, Ecuador. Sustainability 13:13094. https://doi.org/10.3390/su132313094

Connor R (2015) The United Nations world water development report 2015: Water for a sustainable world (Vol. 1). UNESCO publishing

Donowitz JR, Petri WA Jr (2015) Pediatric small intestine bacterial overgrowth in low-income countries. Trends in Molecular Medicine 21(1):6–15

Freire W, Ramirez-Luzuriaga M, Belmont P (2015) Tomo I: Encuesta Nacional de Salud y Nutrición de la población ecuatoriana de cero a 59 años, ENSANUT-ECU 2012. Revista Latinoamericana de Políticas y Acción Pública 2(1):117

Freire WB, Waters WF, Román D, Jiménez E, Burgos E, Belmont P (2018) Overweight, obesity, and food consumption in Galapagos, Ecuador: A window on the world. Globalization and Health 14(1):93

Frumkin H, Haines A (2019) Global environmental change and noncommunicable disease risks. Annual Review of Public Health 40:261–282

Granda M, Sandra G, Calvopiña V (2012) Medición de pobreza en Galápagos. Informe Galápagos, Puerto Ayora, Galápagos, Ecuador, pp 2011–2012

Grube A, Stewart J, Ochoa-Herrera V (2020) The challenge of achieving safely managed drinking water supply on San Cristobal Island, Galapagos¨. International Journal of Hygiene and Environmental Health 228:113547

Guyot-Tephany J, Grenier C, Orellana D (2013) Usos, percepciones y manejo del agua en Galapagos. Dirección del Parque Nacional Galápagos, Consejo de Gobierno del Régimen Especial de Galápagos, Fundación Charles Darwin, Galapagos Conservancy, Puerto Ayora, Galapagos, Ecuador

Hanna EG, McIver L (2014) Small Island states - canaries in the coal mine of climate change and health. In: Butler CD (ed) Climate change and global health. CABI, UK, pp 181–192

Herforth AW, Wiesmann D, Martínez-Steele E, Andrade G, Monteiro CA (2020) Introducing a suite of low-burden diet quality indicators that reflect healthy diet patterns at population level. Curr Dev Nutr 4(12), nzaa168:4012003

Hickey GM, Unwin N (2020) Addressing the triple burden of malnutrition in the time of COVID-19 and climate change in Small Island developing states: What role for improved local food production? Food Secur 12(4):831–835

Himmelgreen DA, Romero-Daza N, Amador E, Pace C (2012) Tourism, economic insecurity, and nutritional health in rural Costa Rica: Using syndemics theory to understand the impact of the globalizing economy at the local level. Ann Anthropologic Pract 36(2):346–364

Houck KM, Terán E, Ochoa J, Zapata GN, Gomez AM, Parra R, Dvorquez D, Stewart JR, Bentley ME, Thompson AL (2020) Drinking water improvements and rates of urinary and gastrointestinal infections in Galápagos, Ecuador: Assessing household and community factors. American Journal of Human Biology 32(1):e23358

Jahnke JR, Archer NT, Thompson AL, Ocampo J, Bentley ME (2022) Health-seeking behavior and perceptions of care: A qualitative study investigating the underutilization of public health care on the Galápagos Islands. In: Water, food and human health in the Galapagos, Ecuador. Springer, pp 257–271

Jones AD, Mundo-Rosas V, Cantoral A, Levy TS (2016) Household food insecurity in Mexico is associated with the co-occurrence of overweight and anemia among women of reproductive age, but not female adolescents. Maternal & Child Nutrition

Jones AD, Hoey L, Blesh J, Janda K, Llanque R, Aguilar AM (2018) Peri-urban, but not urban, residence in Bolivia is associated with higher odds of co-occurrence of overweight and anemia among Young children, and of households with an overweight woman and stunted child. Journal of Nutrition 148(4):632–642

Kau AL, Ahern PP, Griffin NW, Goodman AL, Gordon JI (2011) Human nutrition, the gut microbiome and the immune system. Nature 474(7351):327–336

Khan JR, Awan N, Archie RJ, Sultana N, Muurlink O (2020) The association between drinking water salinity and hypertension in coastal Bangladesh. Global Health J 4(4):153–158

Lee J, Houser R, Must A, de Fulladolsa P, Bermudez O (2010) Disentangling nutritional factors and household characteristics related to child stunting and maternal overweight in Guatemala. Economics and Human Biology 8(2):188–196

Mateus C, Quiroga D (2022) Galapagos' water management evaluation under a changing climate and the current COVID-19 pandemic. In: Water, food and human health in the Galapagos, Ecuador. Springer, pp 29–55

McIver L, Kim R, Woodward A, Hales S, Spickett J, Katscherian D, Hashizume M, Honda Y, Kim H, Iddings S, Naicker J (2016) Health impacts of climate change in Pacific Island countries: A regional assessment of vulnerabilities and adaptation priorities. Environmental Health Perspectives 124(11):1707–1714

McMichael AJ, Woodruff RE, Hales S (2006) Climate change and human health: Present and future risks. The Lancet 367(9513):859–869

Mendenhall E (2017) Syndemics: A new path for global health research. The Lancet 389(10072):889–891

Miller JD, Workman CL, Panchang SV, Sneegas G, Adams EA, Young SL, Thompson AL (2021) Water security and nutrition: Current knowledge and research opportunities. Advances in Nutrition 12(6):2525–2539

Nicholas K, Bentley M, Terán E, & Thompson A (2020) Water security in the Galápagos: Socioecological determinants and health implications. EcoHealth 17:111–124

Nunbogu AM, Elliott SJ (2021) Towards an integrated theoretical framework for understanding water insecurity and gender-based violence in low-and middle-income countries (LMICs). Health & Place 71:102651

Pera MF, Katz BN, Bentley ME (2019) Dietary diversity, food security, and body image among women and children on San Cristobal Island, Galapagos. Maternal Child Health J 23:1–9

Pérez-Escamilla R, Villalpando S, Shamah-Levy T, Méndez-Gómez Humarán I (2014) Household food insecurity, diabetes and hypertension among Mexican adults: Results from Ensanut 2012. Salud Pública de México 56:s62–s70

Popkin BM, Adair LS, Ng SW (2012) Global nutrition transition and the pandemic of obesity in developing countries. Nutrition Reviews 70(1):3–21

Prüss-Ustün A, Wolf J, Bartram J, Clasen T, Cumming O, Freeman MC, Gordon B et al (2019) Burden of disease from inadequate water, sanitation and hygiene for selected adverse health outcomes: An updated analysis with a focus on low- and middle-income countries. International Journal of Hygiene and Environmental Health 222(5):765–777

Reyes MF, Trifunovic N, Sharma S, Kennedy M (2015) Water supply assessment on Santa Cruz Island: A technical overview of provision and estimation of water demand. In: Galapagos Report 2013–2014. GNPD, GCREG, CDF and GC. Puerto Ayora, Galapagos, Ecuador, pp 46–53

Rosinger AY, Young SL (2020) The toll of household water insecurity on health and human biology: Current understandings and future directions. Wiley Interdiscip Rev Water 7(6):e1468

Ruiz E, Himmelgreen DA, Romero Daza N, Peña J (2014) Using a biocultural approach to examine food insecurity in the context of economic transformations in rural Costa Rica. Ann Anthropologic Pract 38(2):232–249

Sampedro C, Pizzitutti F, Quiroga D, Walsh SJ, Mena CF (2018) Food supply system dynamics in the Galapagos Islands: Agriculture, livestock and imports. Renewable Agriculture and Food Systems 35:1–15

Savage A, Bambrick H, McIver L, Gallegos D (2021) Climate change and socioeconomic determinants are structural constraints to agency in diet-related non-communicable disease prevention in Vanuatu: A qualitative study. BMC Public Health 21(1):1231

Scheyvens RA, Movono A, Auckram S (2021) Pacific peoples and the pandemic: Exploring multiple well-beings of people in tourism-dependent communities. Journal of Sustainable Tourism 31:1–20

Schmeer KK, Piperata BA (2017) Household food insecurity and child health. Maternal & Child Nutrition 13(3):e12301

Schnitter R, Verret M, Berry P, Tiam C, Fook T, Hales S, Lal A, Edwards S (2018) An assessment of climate change and health vulnerability and adaptation in Dominica. International Journal of Environmental Research and Public Health 16(1):70

Stoler J, Pearson AL, Staddon C, Wutich A, Mack E, Brewis A, Rosinger AY et al (2020) Cash water expenditures are associated with household water insecurity, food insecurity, and perceived stress in study sites across 20 low- and middle-income countries. The Science of the Total Environment 716:135881

Swinburn BA, Kraak VI, Allender S, Atkins VJ, Baker PI, Bogard JR, Brinsden H et al (2019) The global Syndemic of obesity, undernutrition, and climate change: The lancet commission report. The Lancet 393(10173):791–846

Thompson AL, Nicholas KM, Watson E, Terán E, Bentley ME (2020) Water, food, and the dual burden of disease in Galápagos, Ecuador. Am J Human Biol 32(1):e23344

Thompson AL, Teran E, Bentley ME (2022) Syndemic water and food insecurity: Impacts on the dual burden of disease in Galapagos. In: Water, food and human health in the Galapagos, Ecuador. Springer, Cham, pp 91–105

Tzioumis E, Adair LS (2014) Childhood dual burden of under- and overnutrition in low- and middle-income countries: A critical review. Food Nutr Bull 35(2):230–243

Vásquez WF, Raheem N, Quiroga D, Ochoa-Herrera V (2021) Household preferences for improved water services in the Galápagos Islands¨. Water Res Econ 34:100180

Walsh SJ, Mena CF (2016) Interactions of social, terrestrial, and marine sub-systems in the Galapagos Islands, Ecuador. Proceedings of the National Academy of Sciences. 201604990

Walsh SJ, McCleary AL, Heumann BW, Brewington L, Raczkowski EJ, Mena CF (2010) Community expansion and infrastructure development: Implications for human health and environmental quality in the Galápagos Islands of Ecuador. Journal of Latin American Geography 9(3):137–159

Watson E (2019) Social predictors of diet quality in Galapagos, Ecuador. (BA). University of North Carolina at Chapel Hill, Chapel Hill, NC

Watts N, Amann M, Arnell N, Ayeb-Karlsson S, Beagley J, Belesova K, Boykoff M, Byass P, Cai W, Campbell-Lendrum D, Capstick S (2021) The 2020 report of the lancet countdown on health and climate change: Responding to converging crises. The Lancet 397(10269):129–170

Weigel MM, Armijos MM (2015) Food insufficiency in the households of reproductive-age Ecuadorian women: Association with food and nutritional status indicators. Ecology of Food and Nutrition 54(1):20–42

Weigel MM, Armijos RX, Racines M, Cevallos W (2016) Food insecurity is associated with undernutrition but not overnutrition in Ecuadorian women from low-income urban neighborhoods. Journal of Environmental and Public Health 2016:8149459. https://doi.org/10.1155/2016/8149459

Weir T, Pittock J (2017) Human dimensions of environmental change in small Island developing states: Some common themes. Regional Environmental Change 17(4):949–958

World Bank (2022). https://www.worldbank.org/en/country/smallstates/overview

Wutich A, Ragsdale K (2008) Water insecurity and emotional distress: Coping with supply, access, and seasonal variability of water in a Bolivian squatter settlement. Social Science & Medicine 67(12):2116–2125

Wutich A, Budds J, Jepson W, Harris LM, Adams E, Brewis A, Cronk L, DeMyers C, Maes K, Marley T, Miller J (2018) Household water sharing: A review of water gifts, exchanges, and transfers across cultures. Wiley Interdisciplinary Reviews: Water 5(6):e1309

Young SL, Boateng GO, Jamaluddine Z, Miller JD, Frongillo EA, Neilands TB, Collins SM, Wutich A, Jepson WE, Stoler J (2019) The household water InSecurity experiences (HWISE) scale: Development and validation of a household water insecurity measure for low-income and middle-income countries. BMJ Global Health 4(5):e001750

Amanda L. Thompson is Professor of Anthropology and Nutrition at the University of North Carolina at Chapel Hill and a fellow at the Carolina Population Center. She has been working on human health issues in the Galapagos since 2014. Amanda is the interim co-director of the Center for Galapagos Studies and Galapagos Science Center.

Jill R. Stewart, Co-Guest Editor, is the Philip C. Singer Distinguished Professor, Department of Environmental Sciences and Engineering, University of North Carolina at Chapel Hill, USA. She develops noveal techniques to detect and track pathogens in water. She is particularly interested in linkages between ecosystems and human health and well-being.

Margaret Bentley is Distinguished Professor Emeritus in the Department of Nutrition at the Gillings School of Global Publich Health at the University of North Carolina at Chapel Hill.

Jaime Eduardo Ocampo Trujillo is a Professor in the School of Business Administration. In the past 3 years, Jaime has worked on a project of telemedicine with Hospital Oscar Jundl on San Cristobal Island in the Galapagos. Thus far, the program has helped more than 500 patients in securing a second opinion in cooperation with physicians of USFQ`s medical system page.

Enrique Teran is a medical doctor and holds a PhD in Pharmacology. He is a professor at the School of Medicine at the Universidad San Francisco de Quito.

Valeria Ochoa Dr. Ochoa-Herrera is a Professor in the Environmental Engineering Department at Universidad San Francisco de Quito USFQ, Ecuador, and an Adjunct professor in Environmental Sciences and Engineering Department, Gillings School of Global Public Health, University of North Carolina – Chapel Hill, USA. Dr. Ochoa-Herrera research areas involve the impacts of climate change in freshwater quality, wastewater treatment, bioremediation, and sustainability; she has contributed to promoting technological innovation and action against water pollution in Ecuador and Colombia.

Chapter 8
Improvements in the Galapagos Health System: Telemedicine, Research & Medical Assistance

Jaime Eduardo Ocampo Trujillo and María Emilia Menoscal Coello

Introduction

Prior to the opening of the Oskar Jandl Hospital (HOJ) in San Cristobal Island, Galapagos Archipelago of Ecuador, the community only operated a small medical center with 15 beds, built in the late 1960s. Ever since HOJ's inauguration in December 2014, the 7500 inhabitants of the island population use HOJ's public services that includes 23 beds for hospitalization, an obstetric center with operating room, a physiatry area, neonatology area, a unit of intermediate care, among others (Public Health Ministry 2014). According to the Ecuadorian Health Ministry, the inauguration of this hospital sought to benefit the entire population of the archipelago, thereby eliminating the necessity of moving to the continent for access to health services. In addition, HOJ sought to have a positive impact on tourism, since national and international tourists could access its services when arriving to the island.

While the HOJ has significantly added value to public health in the archipelago, more still needs to be accomplished. Since 2015, under a joint alliance with Sistemas Médicos de la USFQ, the School of Public Health of the Universidad San Francisco de Quito (USFQ) joined forces through a cooperative agreement with the hospital and the Ecuadorian Public Health Ministry. This agreement is divided into three areas of emphasis: research, training, and medical assistance. The research branch has several stakeholders involved, including the Ecuadorian Public Health Ministry, University of North Carolina at Chapel Hill (UNC), hospital authorities, USFQ professors and researchers, among other international academic institutions. The focus of the numerous research papers published thus far have centered on exploring the needs of the island regarding the general state of health care of the population. In

J. E. Ocampo Trujillo (✉) · M. E. M. Coello
Universidad San Francisco de Quito, Pichincha, Ecuador
e-mail: jocampo@usfq.edu.ec; memenoscal@gmail.com

S. J. Walsh et al. (eds.), *Island Ecosystems*, Social and Ecological Interactions in the Galapagos Islands, https://doi.org/10.1007/978-3-031-28089-4_8

addition, training that has been accomplished for health professionals and administrative staff of the hospital, which has focused on innovation in various branches of health administration and in several medical specialties. Finally, this project has aimed to fill the medical specialties that the islands lack.

Over the years, the cooperative agreement has encompassed several knowledge areas of USFQ, such as, the School of Public Health, Nutrition and Diet, School of Business and Economics, among others. It also has counted on the support of several international universities, such as the University of North Carolina at Chapel Hill (UNC), Texas A & M University, and Hofstra University. The aim of this project is centered on the development of approaches to provide an active support mechanism for the Ecuadorian Health Ministry to effectively deliver integral access to medical attention for the people of the Galapagos in a variety of medical specialties that were previously unavailable to the population, enhanced through medical brigades with national and international professionals. Regarding the training branch, the HOJ sought to improve the prevention, rehabilitating, recovery, and promotion of health care for the Galapagos population.

All activities have been duly studied with respect to the context of the island. Training programs have been carried out and the population has been assisted in different medical specialties. Progress has been made in conjunction with different actors in the investigations, which has allowed for a deeper understanding of the causes of different diseases in the islands. For this reason, efforts have focused on mitigating the presence of COVID-19 on the islands, always within the main objective: provide the population of Galapagos with a broader range of medical care.

Description of the Environment

Puerto Baquerizo Moreno, the capital of the province, is also one of the most populated sectors of San Cristóbal Island. It is located 960 km from continental Ecuador, with coordinates 0051.30 degrees South Latitude and 08937.60 degrees West Longitude, with an area of 381 km2. Its privileged location engenders high biological diversity, countless endemic species of flora and fauna, and several marine currents that offer diverse habitats for plant and animal species. Among the varieties of its flora, the cactus, mangroves, carob trees, among others, stand out. As for its fauna, sea lions, iguanas, blue-footed boobies, royal frigatebirds are most noteworthy (EcuRed 2021). The climate of Puerto Baquerizo Moreno is characterized by being quite stable throughout the year. There is a temperature constant within the sector, as for every 100 m of elevation, the temperature decreases 1 °C. Regarding precipitation, it varies from 350 to 1800 mm per year, at altitudes from 0 to the highest point above sea level on the island, 700 meters above mean sea level (Directorate of Productive and Sustainable Development XE "Sustainable development" of the Decentralized Autonomous Government of Santa Cruz 2020).

The Galapagos Population and Housing Census indicates that in San Cristóbal, 7199 people were registered in 2015. As for Puerto Baquerizo Moreno, 6533 people were registered. The island shows a rate of annual population growth of 0.8% in the 2010 census. Moreover, 85% of the island's population identifies as mestizo, while 15% say they belong to different ethnic groups, whether indigenous, Afro-Ecuadorian, Montuvio, white, or another (National Institute of Statistics and Censuses 2015). As noted, there is a varied population in the Galapagos Islands, as such, individuals have different health needs. For this reason, it was expected that the project would benefit the total population, not only through medical attention, but also through improvements in health practices of health professionals. Also, the project facilitated a greater depth of understanding of the medical needs of the local people.

Impact on Society

This project seeks to create an ongoing level of support with the HOJ. The hospital has four specialties, however, the needs and demands of the inhabitants of the archipelago extend beyond the hospital's capacity and supply. The USFQ representatives, with the support of other actors mainly from the University of North Carolina at Chapel Hill, seek to strengthen the availability of professionals in different areas of health, both permanently and through medical brigades. Regarding the training branch of the project, the aim is not only to support medical care, but also to contribute to improvements in administrative matters. The training given since 2016 have covered various topics: emergency care, training in disasters, pediatric intensive care, patient safety, customer service, patient experience, neonatology, nursing, medical administration, among others. The impact is intended to be more holistic, since it is not only about innovating in medical care, but also supporting the management and planning of the hospital, which will indirectly contribute to improvements in medical services in the institution.

The impact of the research topic of the project corresponds to achieving an improved understanding of the island's medical needs. In conjunction with hospital authorities and the Ministry of Public Health, the investigations were carried out on issues of obesity, nutrition problems, and other topics related to the threats to the Galapagos' public health. The conclusions of these investigations were very relevant, since they serve to highlight the current diagnoses of the islands, in terms of, their needs that could lead public health authorities to take the necessary measures to deal with patient needs. The impact of the investigative branch has been evidenced in the different actions of the project, which has joined forces to carry out activities in accordance with the conclusions seen in the investigations. Until August 2021, work has been carried out on four research projects: *Overweight, Obesity, and Food Consumption in Galapagos, Ecuador: Window on the World* (Wilma Freire,

Diana Román and others)*; Water, Food, and the Triple Burden of Disease in the Galapagos (Amanda* Thompson, Kristopher Nicholas and others)*; Socio-Ecological Factors Associated with Dengue Risk and Aedes aegypti Presence in the Galapagos Islands (*Ortega, Leon and others); and *Ecuador and Breastfeeding Practices and Complementary Feeding in Ecuador: Implications for Localized Policy Applications and Promotion of Breastfeeding, a Pooled Analysis* (Wilma Freire, Diana Román and others).

Since the opening of USFQ's Galapagos campus on San Cristobal Island, the university has sought to establish an ongoing relationship with local people. Through this project and its positive impact on society, the intention is to strengthen genuine and supportive ties between the academia and the community of the islands. In addition, this project intends to bring complementary medical solutions that local health authorities are unable to provide.

Objective

As mentioned, the objective of this project has focused on the three areas mentioned above. The focus of the project has varied over the years, depending on what the island's population has needed. Regarding the research area, this activity has been carried out based on what has been known about the health problems of the Galapagos population. For medical assistance, both telemedicine and brigades have been developed. Both methods have involved national and international health professionals from USFQ, UNC, SIME-USFQ, Hofstra University, among others. Finally, the training has responded to the level of support that the HOJ has required to improve the quality of patient care. For this reason, the staff has been trained, not only in medical matters but also in administrative matters. It is important to note that, starting with the health emergency, the linkage took a gradual turn. It has prevailed in working from three branches already mentioned, however, the priorities to support the immunization of the island have been taken into consideration. As of 2021, the objective of the project changed towards supporting the health authorities and joining efforts to protect the inhabitants of the island against COVID-19, that is, to achieve total immunization. Based on these objectives, Tables 8.1, 8.2, 8.3, 8.4, and 8.5 show the activities that have been carried out for the project, along with a description of each activity.

Table 8.1 Activities (2016–2017)

Name	Activity Summary
Telemedicine assistance	221 medical appointments made via telemedicine
Research program on Obesity in the archipelago	Research about the factors on Obesity in San Cristobal Island
Results of the program of research about obesity	Presentation about the results on the factors of obesity, with government officials

Table 8.2 Activities (2017–2018)

Name	Activity Summary
Medical brigade	8 health professionals
Telemedicine: Virtual medical Assistance	Transfer of 8 health professionals to the Islands for 3-months
Nursery training	Training in Nursery: Visits of UNC Nurses, July 2017
Results presentation: Research about Obesity in San Cristobal Island	Presentation of results regarding the factors of Obesity on the Islands with government officials
ATLS training of doctors and HOJ nurses by professionals of the Norwell University School of Medicine	Training in Medical urgencies of HOJ doctors and nurses by professors of the USFQ School of Medicine.
Research about breastfeeding and nutrition	Analysis of patterns of breastfeeding customs upon polls made in different sectors of Ecuador: Cumbayá, Morona Santiago, and the Galapagos Islands
Research with UNC	USFQ professors and members of SIME USFQ dictated trainings approaches on health administration for HOJ administration

Table 8.3 Activities (2018–2019)

Name	Activity Summary
Training in management	USFQ professors and SIME USFQ members trained the health administration of HOJ
Telemedicine	36 medical appointments via telemedicine
Nursery training	UNC nurses trained HOJ nurses on relevant nursing topics and innovations
Aedes aegypti research in the Galapagos	USFQ professors conducted research about dengue and derivatives in the islands.
Ecological hospital intervention	A proposal was supported from USFQ graduate students for the Transformation of the HOJ into an ecological hospital
Training in emergency responses by professionals of Hofstra university, Zucker School of Medicine	Physicians, nurses, paramedics and firefighters received emergency response training taught and funded by physicians from Hofstra University's Zucker School of Medicine.
IV research and Conservation symposium in the Galapagos	During the IV Galapagos research and Conservation symposium, a range of national and international professors from USFQ and UNC, and HOJ officials linked to the project and presented the results of the investigations conducted in the islands.

Table 8.4 Activities (2019–2020)

Name	Activity Summary
Presentation of results: Dengue and breastfeeding	Presentation of results to authorities of the Galapagos Islands of both investigations carried out
Training to HOJ administration	Training for administrative personnel of the HOJ in human resource issues (online training)

Table 8.5 Activities (2020–2021)

Name	Activity Summary
Ictus - angels training	The Boehringer platform was provided to train health professionals
Training in health administration and leadership	Training on issues related to improving the quality of services. Not only did HOJ staff attend, but also employees from zone 5 hospitals, who were interested in participating.
Finance training	Training on personal finance topics for HOJ and zone 5 health professionals.
Cardiology training	Cardiologists from USFQ medical systems provided cardiology Services at the Hospital
Training in research methodology	Training on research methodology for health professionals from HOJ and zone 5
Vaccination brigade	Professionals from USFQ, students from the Master's in public health, and students from the School of Medicine worked for 28 days in the COVID-19 vaccination brigade in the Galapagos Islands
Training in corporate communication	Training hospital employees on issues related to corporate communication and customer service
Medical brigade	Transfer of SIME-USFQ personnel in different medical specialties to care for the island's inhabitants in dermatology, dentistry and cardiology specialties
Cardio-Pulmonar resurrecting, Norwell University	Training for HOJ staff in cardiopulmonary resuscitation by Norwell University staff

Methods & Results

To start the project, a collaborative agreement was signed between the HOJ and USFQ, with the support of regional and national health authorities. The cooperative project was coordinated through the assignment of Dr. Jaime Ocampo as the lead delegate responsible for project operations. Sistemas Médicos de la USFQ, on behalf of the university, gave the hospital equipment for telemedicine assistance, for San Cristobal and for Santa Cruz Islands. These pieces of equipment included a Dell laptop, camera, digital stethoscope, and a portable electrocardiograph. To move forward with telemedicine assistance, 17 health professionals from San Cristóbal Island were trained in the use of the equipment. The provision of services to the inhabitants of the island was verified in the project, which was evidenced in the 221 cases attended as a second opinion through telemedicine.

The overweight and obesity work in the Galapagos project was led by Dr. William F. Watters and Dr. Wilma B. Freire from USFQ. Their work was conducted in the Galapagos from October fifth to ninth, 2016. Under this initiative, awareness was raised regarding the problem of overweight and obesity in islands through interviews and structured observations. For the project, there were two coordinators, one on each populated island of the archipelago to conduct focus groups as part of the study. As part of and in support of the research, UNC professors visited the islands and trained on the telemedicine approaches.

The results found in the research on *Overweight, Obesity, and Food Consumption in Galapagos, Ecuador: Window on the World* were very fruitful in determining the factors that cause this disease in the islands. Through the discussions, it was discovered that the inhabitants of San Cristóbal face several obstacles when it comes to buying and consuming fresh vegetables and fruit. The qualitative component of this study indicates that fresh products in Galapagos are expensive, unavailable, and of medium quality (Freire et al. 2018). The investigation resulted in the generation of vital information to improve nutrition in society by local administrators.

The results for the years 2017–2018 vary according to the activities carried out. As for the administrative training that began on August third, 2018, USFQ professors remained on the island for two to three weeks to meet their training objectives in medical administration. Regarding the investigative nature, progress continued with the research project on obesity. However, work also began on studies related to vectors in the transmission of Dengue and Zika, the results of which were presented in October 2018. Studies continued on breastfeeding focused on nutrition, led by professors from the USFQ School of Public Health. Within the context of this research, the results indicated that the rates of early breastfeeding and breastfeeding at an appropriate age are significantly higher in the urban and rural area of Morona Santiago province than in Cumbayá (Quito) or in the Galapagos Islands (Freire et al. 2020). This indicates that it is necessary to develop effective policies and promotion strategies based on factors such as early breastfeeding and breastfeeding practices at middle-ages (Freire et al. 2020).

This year, it was not possible to support the transfer of doctors to the island. However, the telemedicine program continued, in specialties such as dermatology and imaging and cardiology, conducted by USFQ health professionals. Between the second half of 2017 and the first quarter of 2018, 500 telemedicine sessions were carried out. By April, the program was suspended due to changes in authorities within the hospital. To align the parties to the agreement, various periodic meetings were held with national, district, and local authorities to learn in-depth about the progress of the activities. Notifications were received from the Public Health Ministry, in which the benefits of this agreement were disclosed.

This year, five trainings were provided, with 40 direct beneficiaries of the HOJ in different subjects. Two trainings in nursing, administrative issues for health personnel, and health management and emergency responses. The 15 trainers came from different institutions, including USFQ, UNC, Hofstra University, and SIME-USFQ. Professionals from UNC, USFQ and HOJ continued with two investigations: *Water, Food, and the Triple Burden of Disease in the Galapagos; and Aedes aegypti (Dengue, Zika, and Chikungunya) in Galapagos*. In the first investigation, it was found that obesity and the risk of diseases were not easily detected due to an unbalanced diet and the presence of poor water quality (Freire et al. 2018). On the other hand, regarding the research on *Aedes aegypti*, it was found that Dengue infections were reported in more homes in Puerto Baquerizo Moreno (28%) than in Puerto Ayora (20%), although the population in Puerto Ayora is double the size of Puerto Baquerizo Moreno. In addition, the *Aedes aegypti* is concentrated in the water containers of the homes of the Galapagos population (Ryan et al. 2019).

During this period, 36 consultations were carried out within the framework of telemedicine. It was not possible to continue with this branch of the project due to technical problems. For this reason, the transfer of patients in need from different medical branches could not be significantly reduced. However, it is expected to continue with this part of the project in subsequent years.

According to the 2018 HOJ accountability, hospital occupancy this year increased from 23% to 40% (Parra 2018). This was due to many factors, including the training and telemedicine care that could directly contribute to this improvement. As part of the project, it was ensured that the trainees felt comfortable and confident with the knowledge they received.

The most important result was the renewal of the agreement between USFQ and Zone 5 Hospitals. This allowed the project to continue, with the activities already planned and the support of the Galapagos health authorities assured. During the signing of the agreement, the invited authorities were able to document, through the presentation of a short summary, the past and future projects that have been carried out thanks to this agreement. Due to the health emergency from COVID-19, some activities were limited. However, based on that, the opportunity was seen to rethink its objectives based on this new reality.

The cardiology brigade was carried out effectively, and 235 people with different cardiovascular conditions were treated. A follow-up was carried out for each patient, thanks to the support of doctors specialized in cardiology at USFQ. Regarding the training carried out by USFQ School of Business professors, six programs were offered for HOJ staff, in addition to the staff of the hospitals in Zone 5. The training focused mainly on issues related to quality of service, internal communication, and leadership. The main activity of the 2021 was the vaccination brigades against COVID-19 for the population over 18 years of age. USFQ received a special request from the Vice President of Ecuador and the health authorities to lead these brigades, so that this is the first province in the country to achieve mass immunization. Medical staff and graduates from the USFQ School of Medicine moved to the islands for a period of 28 days, to vaccinate 20,122 older adults. From this, a decrease of COVID cases was achieved. With this, tourism was significantly reactivated within the archipelago. Before USFQ School of Medicine led the vaccination brigade in Galapagos, at the beginning of the pandemic, there was no ICU in any of the two hospitals in the province, including HOJ. Although the islands already had centers with intermediate care units, with the help of the university, sections for intensive care units were enabled in the HOJ. USFQ, by leading the vaccination brigades of the archipelago, was an essential actor in immunizing the entire island, as such, the Galapagos became the first archipelago in Latin America immunized from COVID-19 (Universidad San Francisco de Quito 2021). The project achieved this result because of the merged mutual work of the Ecuadorian private and public sector, who had a common goal of immunizing the area (Graph 8.1).

Throughout these six years, with the presence of USFQ, more than 20,000 people have benefited both directly and indirectly through this project in different ways. In each case, there has been an expected result, according to the three approaches of the project.

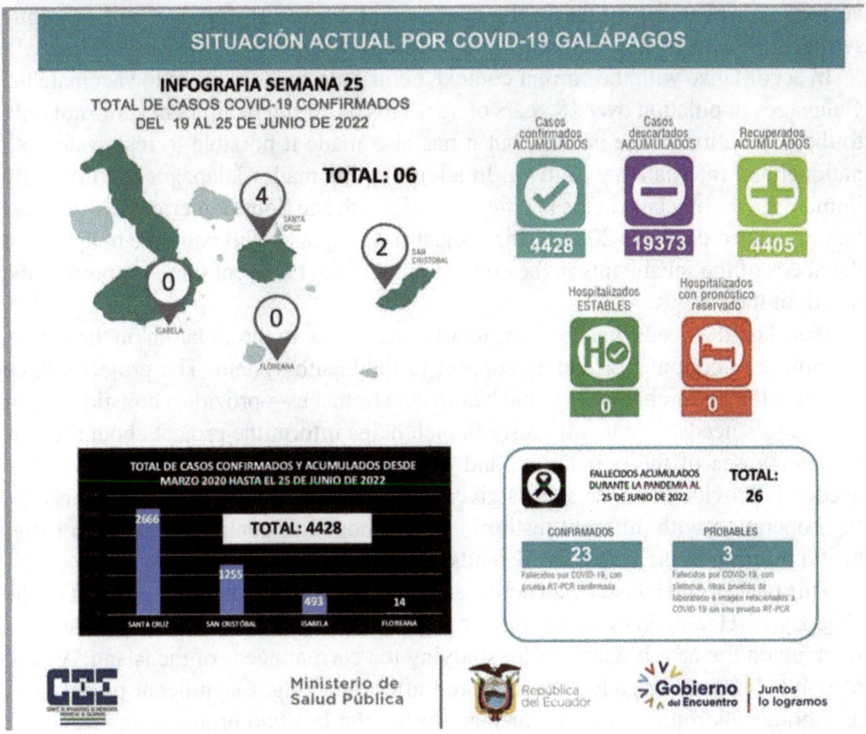

Graph 8.1 Confirmed casos of COVID-19 until June 2022 (Ecuadorian Public Health Ministry 2022)

Discussions and Conclusions

Without a doubt, the USFQ-HOJ cooperative agreement has brought great improvements in the public health of the archipelago. Not only have the inhabitants been able to access a wider range of medical specialties, but they have also been able to become aware of the needs of the population of the islands. First, significant findings were found on the most lethal diseases in the archipelago. For example, regarding research on obesity, the factors that make this disease a threat to the islands are now known in greater depth, so that future actions can be taken.

By providing the archipelago with national and international professionals to train the different stakeholders that are part of the islands'health system, many long-term benefits have been achieved. Not only has there been a commitment to better quality medical care, but the best health administration strategies have also been sought, thus optimizing the available resources, which are still limited. This means that there may be a better flow of attention to the public, increase the availability and improve the quality of public health in the islands. The medical care provided by this project hastened the move of the inhabitants to be treated in different medical specialties that were not available on the islands. This, in turn, indirectly causes the acceleration of tourism, due to the evidence of a general improvement in public services.

In accordance with the current context, contributions were made to vaccinate the Galapagos population over 18 years of age. This has brought direct benefits not only to the population of the islands, but it has also made it possible to reactivate both national and international tourism. In addition, this made Galapagos the first fully immunized archipelago in the Pacific coast of North and South America (Universidad San Francisco de Quito 2021). This project in Galapagos will continue to appeal to the needs of the inhabitants in the context of COVID-19, to continue supporting the health of the islands.

Based on the needs that continue to arise according to the situation on the island, the aim is to continue providing support to the health system. The project's three factors—the research, training, and health care branches—provide a holistic view of the island's needs. The investigative branch helps inform the project about the current diagnoses of the population, and according to that, the medical care that is needed is provided and the staff is given a continuing education in health services. In cooperation with different institutions, it is hoped to continue building a better health offering in the Galapagos Islands, an important heritage for Ecuador.

This project intends to be an active and positive agent to the improvement of the Galapagos' Health System for the coming years. That is why the project team is working on the next brigades, after studying the current needs of the island. A need of ophthalmology brigades has appeared after detecting a significant presence of metabolic syndrome in the archipelago during the last two brigades, in December 2021. A more in-depth study is to be conducted through a brigade focused on ophthalmology. However, the project aims to expand its horizons and start inquiring about health needs in Santa Cruz Island. After attempting all these years to be an active agent for the improvement of Galapagos' public health, for the next upcoming activities such as these, it was learned that for the project is to have an even more direct impact, the USFQ School of Nutrition, Education and the UNC School of Social Work will also intervene to transfer awareness to the inhabitants regarding better practices and reducing the effects of the metabolic syndrome (diabetes, obesity, blood pressure).

Acknowledgements The development of this project would not have been possible, first, without the trust of the Puerto Baquerizo Moreno Zonal Coordination of the Health Minister, who has, over the years, given us the responsibility to be involved actors in the improvement of the public health of the Galapagos Islands.

A special thanks is also offered to the Universidad San Francisco de Quito (USFQ) and many of its administrators, leaders, and professors who have, in some way, been involved in this project. Time, expertise, and resources of university participants have directly impacted the well-being of the Galapagos population. In that sense, the active participation of the ophthalmologists and cardiologists was vital to the discoveries made and documented in the several research papers. Thank you for your support.

Lastly, we would also like to thank our academic international partners who have also collaborated in this project from the University of North Carolina at Chapel Hill, Hofstra University, and Norwell University. Thank you for your selfless support for our country.

References

Directorate of Productive and Sustainable Development of the Decentralized Autonomous Government of Santa Cruz (2020) Puerto Baquerizo Moreno. https://bit. ly/2WFdLLq

EcuRed (2021) Puerto Baquerizo Moreno (Ecuador). EcuRed. https://bit. ly/3BvgHJW

Freire W, Waters W, Roman D, Jiménez E, Burgos E, Belmont P (2018) Overweight, obesity, and food consumption in Galapagos, Ecuador: A window on the world. BMC Global Health 93:9. https://doi.org/10.1186/s12992-018-0409-y

Freire W, Waters W, Román D, Belmont P, Wilkinson-Salamea E, Díaz A, Palacios I, Bucheli E (2020) Breastfeeding practices and complementary feeding in Ecuador: Implications for localized policy applications and promotion of breastfeeding: A pooled analysis. International Breastfeeding Journal 10(1186):1–9. https://doi.org/10.1186/s13006-020-00321-9. International Journal of Environmental Research and Public Health 682, 1–16. 10.3390/ijerph16050682

National Institute of Statistics and Censuses (2015) Main results of the Galapagos Population and Housing Census 2015. [PDF file]. https://bit.ly/3jxlaWn

Parra F (2018) Accountability 2018: Oskar Jandl Hospital. [PDF file]. Ryan S, Lippi A, Nightingale R, Hamerlinck G, Borbor-Cordova M, Cruz M

San Francisco de Quito University USFQ (2021) Making Galapagos the First Immunized Archipelago In Latin America. https://bit.ly/3kJUNM4

Thompson A, Nicholas M, Watson E, Terán E, Bentley M (2019) Water, food and the dual burden of disease of Galápagos, Ecuador. American Journal of Human Biology 10(1002):1–20. https://doi.org/10.1002/ajhb.23344

Jaime Eduardo Ocampo Trujillo founded SIME Sistemas Médicos USFQ more than 25 years ago. He had previously graduated as a Doctor of Medicine in Central University of Ecuador, graduated from his MBA in Loyola University New Orleansand obtained his PhD in Capella University, after more than 5 years of research. In his free time, he enjoysexercising and furthering his knowledge in healthcare management. He has been leadering this project for more than 7 years.

María Emilia Menoscal Coello is currently graduating as a finance student from San Francisco University USFQ. She started publishing her articles in 2020 in Aula Magna University, a magazine from USFQ. Her works covered a variety of topics, such as business management, human resources, economics, sustainable projects, among others. She currently works with Jaime as research assistant.

Chapter 9
Social Issues in the Galapagos Islands: A Participatory and Exploratory Study

Gina Chowa, Cynthia Fraga Rizzo, Amanda Thompson, Margaret Bentley, and Mimi Chapman

Introduction

In its landmark *Our Common Future* (1987) report, the Brundtland Commission defined sustainable development as "development that meets the needs of the present without compromising the ability of future generations to meet their own needs" (World Commission on Environment and Development [WCED] 1987, p. 41). This report's definition has been carried forward into the present day and future via the United Nations' 2030 Agenda for Sustainable Development and this Agenda's constitutive Sustainable Development Goals (SDGs), which are undergirded by five key foci: people, planet, prosperity, peace, and partnership. Of central interest to this paper are the first three foci: people, planet, and prosperity. The emphasis on people (SDGs 1–6) pertains to ending poverty and hunger, in all their forms and dimensions, and ensuring that human beings can fulfill their potential with dignity and equality in a healthy environment. Prosperity entails ensuring that all human beings can enjoy prosperous and fulfilling lives and that economic, social, and

G. Chowa (✉) · C. F. Rizzo
School of Social Work & Global Social Development Innovations Center,
University of North Carolina at Chapel Hill, Chapel Hill, NC, USA
e-mail: chowa@email.unc.edu

A. Thompson
Department of Anthropology & Center for Galapagos Studies,
University of North Carolina at Chapel Hill, Chapel Hill, NC, USA

M. Bentley
Department of Nutrition, University of North Carolina at Chapel Hill, Chapel Hill, NC, USA

M. Chapman
School of Social Work & Global Social Development Innovations Center,
University of North Carolina at Chapel Hill, Chapel Hill, NC, USA

S. J. Walsh et al. (eds.), *Island Ecosystems*, Social and Ecological Interactions in the Galapagos Islands, https://doi.org/10.1007/978-3-031-28089-4_9

technological progress occurs in harmony with nature (SDGs 7–11). The third element, planet, centers on protecting the planet from environmental degradation, including by ensuring sustainable consumption and production, sustainably managing natural resources, and taking urgent action on climate change, so that the planet can support the needs of the present and future generations (SDGs 12–15). Collectively, these undergirding foci constitute a clarion call to balance the needs of human beings with the imperative of protecting the environment.

A substantial scholarly consensus has shown that privileging the desires and economic prosperity of people to the detriment and harm of the planet yields global consequences (e.g., climate change) that will ultimately diminish the quality of life of all people, a message that media outlets are increasingly highlighting to audiences worldwide. At the same time, environmental conservation efforts should seek to preserve the dignity, well-being, and fulfillment of the people already living in those environments. The tension between these two imperatives is playing out in communities and environments worldwide, with varying results.

The Ecuadorian government has done a commendable job to ensure that the physical environment in the Galapagos Islands is protected. The Galapagos National Park (GNP) is a living sanctuary for endemic species, pristine landscapes, virgin forests, and rare flora and fauna (UNESCO Center 2019). The national park maintains 97% of the land area of the Galapagos Islands, which is preserved against human habitation (Lu et al. 2013); the other 3% of the islands is used for human uses. The Galapagos is attractive to the people on mainland Ecuador as a place to work and live and, as a result, the Galapagos has a population growth of around 6.4% per year since the 1990s (Walsh and Mena 2016).

At the same time, Galapagos residents face many economic challenges, including stark, growing inequalities between the very rich who own tourism businesses and those who work in these businesses; limited economic opportunities resulting in unemployment; and, for the many who depend on fishing for their livelihood, having to compete with commercial fishing companies. In 2021, President Guillermo Lasso increased the minimum wage for the private sector for 2022 by 6.5%. The current minimum wage is $425 USD per month (Content Engine 2021). However, many people working in the tourism industry in the Galapagos earn less than the minimum wage because they are hired on an hourly basis without any contractual, legally binding agreements. Exacerbating these economic issues are a slew of social problems common to low-resource communities across the world, including a poor educational system that makes it difficult for young people to qualify for universities, and relational and domestic issues that result in higher rates of violence in general and domestic violence. Galapagos residents also face many health issues including obesity, communicable diseases (partly due to lack of access to clean drinking water), and non-communicable diseases such as diabetes and high blood pressure (partly due to the lack of availability of healthy foods).

Stakeholders in the Galapagos are committed to addressing the islands' social, economic and health issues. Yet, while several recent studies have been conducted to address the health issues faced by residents (Freire et al. 2018; Houck et al. 2020; Pera et al. 2019; Waldrop et al. 2016), comparatively little research has sought to identify or investigate the causes of and potential interventions for the islands' social issues, which do not exist in isolation from the economic and health issues that residents face. This study uses participatory research methods to provide a much-needed foundation for the literature on the social issues affecting the well-being of people on the Galapagos Islands. In other words, the research question being addressed is *What are the social issues that affect the well-being of the people of the Galapagos Islands*?

Intrinsic, Contextual, and Structural (ICS) Analytical Framework

The intrinsic, contextual, and structural (ICS) analytical framework originally proposed by Chowa et al. (2021) is a useful tool for understanding the vulnerability of people to social challenges. Grounded in socio-ecological model (a model that considers the interplay between the micro, mezzo, and macro factors that impact people's lives), ICS helps identify contextually- and culturally-centered meanings of vulnerability by gathering data on the intrinsic factors (e.g., individual physiological and psychological characteristics), contextual factors (e.g., family, school, and peer groups), and structural factors (e.g., a country's political or economic climate) that contribute to people's vulnerability. ICS enables communities to define their social problems rather than requiring researchers to impose preformed ideas of existing problems. This community participation approach logically interacts with a strengths-based approach that frames communities as contributors to their well-being and sources of pre-existing resources and knowledge, rather than a deficit-based approach that depicts communities as sources of the problems that people face.

Materials and Methods

Study Design

This study employs qualitative methods with a cross sectional design. We use purposive sampling due to the nature of the research question of the study i.e., *What are the social issues that affect the well-being of the people of the Galapagos Islands?*

We analyze data using an *Intrinsic, Contextual, and Structural (ICS) Analytical Framework* (Chowa et al. 2021) that examines the intrinsic, contextual, and structural issues affecting residents of the Galapagos. Participants are selected for in-depth interviews and meetings by the staff of the Galapagos Science Center. All meetings are conducted and recorded in Spanish and subsequently translated to English.

Study Participants

Over 30 key stakeholders participated in this study. Stakeholders included elected officials, social workers, members of the inter-institutional committee (a committee that has membership of representatives from all government units), school officials, community-based organizations, Ministry of Health officials, Universidad San Francisco de Quito (USFQ) faculty, and Galapagos Science Center staff. These participants were selected because of the information and knowledge they have regarding the human dimensions of the Galapagos.

Study Settings

The human population in the Galapagos is primarily limited to four islands. The study was conducted on three islands: San Cristobal, Santa Cruz, and Isabela.

Data Collection

Individual interviews were conducted with key stakeholders in October 2019 before the onset of the COVID-19 pandemic. In addition, we conducted focus group discussions (FGDs) with the Galapagos' governing council, some community members, and community-based organizations. The research team also held meetings with school authorities and legal agencies. Some of these meetings were organized as a result of recommendations the team received during the initial meetings.

A discussion guide with the following questions was generated by the research team to gather data on social issues affecting local people: *What are the pressing social issues in the Galapagos? What is being done to address these issues? What are the gaps in addressing these issues, and who else should we talk to understand the social issues in the Galapagos?* Asking these questions allowed for flexibility in the discussion for follow-up and probing questions.

Analysis

Analysis followed Braun and Clarke's (2006) step-by-step guide to reflexive thematic analysis, an approach for analyzing qualitative data to better answer questions about people's experiences, views, perceptions, and representations of a given phenomenon. The process includes reading the notes from all meetings, Key Informant interviews, FGDs for emergent themes, assigning thematic codes, and identifying convergence across interviews around common themes as well as potential

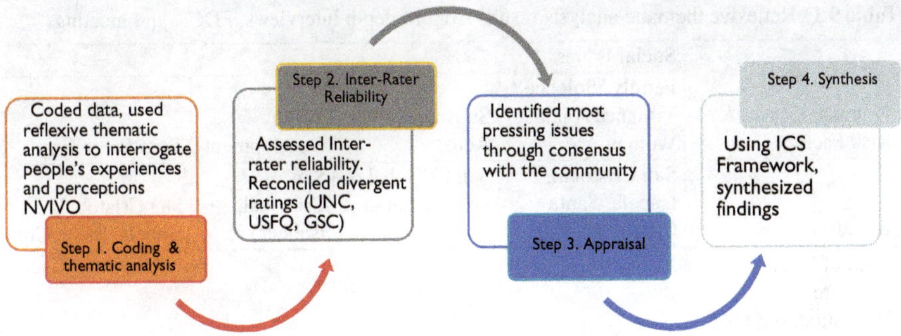

Fig. 9.1 Analysis Phase

differences in viewpoints. We used matrices and tables to display qualitative data and identify key themes. We also used the ICS framework to identify and examine challenging experiences mentioned by community members and the support levels and resources that exist to address these challenges.

The analysis phases included: coding and thematic analysis, assessing inter-rater reliability, appraisal, and synthesis. Figure 9.1 shows the activities performed in each phase.

Findings

Findings from the thematic analysis are represented in Table 9.1.

To provide context, we highlight the most often discussed underlying contributors to these social issues that participants perceived to be driving the escalation of social challenges in the Galapagos. The four social issues that formed themes were: family violence, substance use, youth development and teenage pregnancy. The contributing factors to these issues cut across the four social issues and are a beginning point for understanding the social issues and may provide potential intervention leverage points as well.

Tension Between Human and Animal Needs

Stakeholders and community members reported perceiving tension between the needs of humans and animals on the Galapagos (e.g., restrictions by conservations and the national park on livelihoods such as agriculture and fishing that cause a problem for plant and animal survival). Across the board, participants reported feeling that the needs of human beings on the islands came second to those of the animals. As one participant put it,

Table 9.1 Reflexive thematic analysis results from in-depth interviews, FDGs, and meetings

	Social Issues			
Potential Causes / Risk Factors	Family Violence / Violence Against Women	Substance Abuse	Youth Development	Teen Pregnancy
Island(s)	San Cristobal, Isabela, Santa Cruz	San Cristobal, Isabela, Santa Cruz	San Cristobal, Isabela	San Cristobal, Isabela, Santa Cruz
Lack of legal structure (very new legislation, but not yet enforced)	X			
Lack of resources or access issues	X (examples: legal, victim support/mental health services, shelter, perpetrator services)	X (examples: substance abuse treatment options)		X
Limited attention to prevention efforts	X (example: prevention department in hospital but no shelters)	X		X
Limited knowledge of available resources	X			X (example: birth control is available but not publicized)
Norms, values, beliefs	X (examples: normalization of violence, machismo)	X (examples: normalization of alcohol use by parents)		
Concerns/fear regarding confidentiality	X	X	X	X (regarding use of family planning services)
Small community (everyone knows each other)	X	X	X	
Isolation (isolated from mainland and family)	X	X	X	
Substance use/abuse	X		X	
Lack of father involvement in education of older children	X			
Disintegration of families (note: high divorce rate)	X	X		X

(continued)

Table 9.1 (continued)

Potential Causes / Risk Factors	Social Issues Family Violence / Violence Against Women	Substance Abuse	Youth Development	Teen Pregnancy
Island(s)	San Cristobal, Isabela, Santa Cruz	San Cristobal, Isabela, Santa Cruz	San Cristobal, Isabela	San Cristobal, Isabela, Santa Cruz
Instability of government priorities/previous governmental influence	X			X (example: government stopped family planning services for some time and recently restarted; prior government socialist and Catholic)
Lack of informal social support	X			
Lack of central cultural identity of local people	X	X	X	
Cramped and less than optimal housing	X			
Route of narcotics trafficking		X		
Ships refuel in the islands		X		
Self-medication for depression and anxiety		X		
Lack of recreation activities		X	X	
Limited perception of economic opportunities			X	
Challenges to financing higher education			X	
Limited possible selves			X	
Lack future orientation			X	
Youth with behavioral problems sent to live with family on the island and then these youth negatively influence others			X	

(continued)

Table 9.1 (continued)

	Social Issues			
Potential Causes / Risk Factors	Family Violence / Violence Against Women	Substance Abuse	Youth Development	Teen Pregnancy
Island(s)	San Cristobal, Isabela, Santa Cruz	San Cristobal, Isabela, Santa Cruz	San Cristobal, Isabela	San Cristobal, Isabela, Santa Cruz
Amount of unsupervised time			X	
Engagement in sex work			X	
Preparation for higher education			X	
Lack of open communication with parents about sex				X
Use of morning after pill as a preventative measure against pregnancy				X
Religious units pushing for abstinence				X
Heavy reliance on tourism industry	X	X		
Limited number of professionals/limited professional training	X (i.e., no consistency of care providers because they are sent from brigades)			X
Limited involvement of parents/caregivers		X		X

> We don't see conservation as a threat, but we have problems that obligate us to survive. We have needs that we wish to resolve to conserve and love the islands. We cannot love and conserve the island if we are struggling. Social issues must be addressed and not ignored.

Lack of a Cultural Identity of Local People

Due to the immigration of people from the mainland seeking opportunities of employment on the Galapagos, many participants reported that there was a lack of a positive and coherent cultural identity of local people. Due to this sentiment, participants felt that no overarching identity or culture was uniting the residents of the islands, as most of them identified with their home communities on the mainland. One participant commented:

> Many people on these islands come from multiple places and have different ways of thinking that are tied to their cultural background. On the mainland, people engage in social and cultural events that provide an identity for them. However, because people here do not have common cultures, they do not have those cultural events to bring them together. When there is an issue to be resolved, no one culture can be used.

Exacerbating this lack of cultural identity was longer-term residents' negative perspective of those perceived to be outsiders. In the words of one participant,

> People in the islands came to the islands at different times. Those who were here before, feel that they belong here more than the late comers. Therefore, there is a perception that the early comers belong to the island and are indigenous, and the late comers are the outsiders.

Lack of Confidentiality and Confidence Using Healthcare Services

Residents also reported a general lack of use of health care services, including both mental health services and the hospital. Participants explained that this was in part due to the lack of privacy and confidentiality on the islands due to the small population and physical size of the community. For example, several residents reported that young people seeking reproductive health services often feared that their parents would learn about their visit. One participant said,

> The communities here are very small and everyone knows everyone so no one goes to the psychologist, because they are afraid to ask for help, because other people may either see them go to the psychologist and tell everyone their business.

Another participant said,

> Many young ladies use the morning after pill because they are afraid to use the reproductive health services. Everyone will know that they are using family planning and if the parents find out they will be in trouble. So, they turn to the morning pill because it is the last resort.

Residents also reported avoiding seeking services at the hospital due to a general mistrust of the Cuban doctors hired at the hospital. Many participants sought healthcare from the mainland because of this mistrust of Cuban doctors. As one participant noted,

> There is a new hospital on the island called the Oskar Jandl Hospital with new staff, but local people do not use the hospital. They would rather go to Guayaquil for their health care because they do not trust the doctors at the hospital who are of Cuban origin. It is also because the student physicians used to attend to patients on the island and the local people do not trust the student physicians due to limited practice and experience.

Notably, this viewpoint was communicated by different study participants across the segments of the study, suggesting it is a widely held belief among residents. At the same time, telehealth options were also introduced in recent years, which also reduced the burden of travelling to the mainland for healthcare and offered greater privacy – potentially a contributing factor to residents' avoidance of the local hospital.

Cultural Norms Regarding Violence and Corporal Punishment

Participants reported high levels of intra-family violence on the islands, while they also noted that this issue was not discussed among residents. One participant explained the reasons for this lack of discussion:

> There are high levels of intra family violence on the islands, but culturally, that is not something anyone is supposed to talk about. It is like washing your dirty laundry in public. For example, someone can experience violence and if they report the violence to the authorities, everyone will be talking about that person and what shame they have brought upon their family. On the other hand, authorities are not equipped to handle cases of violence and are influenced by the culture, so they will tell the victims to solve it at home with their partner.

Gender dynamics were also cited as a reason for the silence surrounding intimate partner violence:

> Families typically support the male abusive partner and women are disempowered by mothers-in-law who are perpetuating the violence in the communities. From a young age, a woman is disempowered and disenfranchised to parent children and take care of themselves, so they depend on their husbands later to fend for them.

Other participants noted that corporal punishment was common on the islands because, among residents, it was expected that some violations of expected behaviors e.g., when a father feels that their child has disobeyed them, should be addressed with corporal punishment. As one participant noted,

> In many families, fathers are very authoritarian with families and impose their power with violence. Violence is later replicated in the schools and the cycle continues.

Another participant discussed how *machismo* i.e., a strong sense of male pride that comes from the assertion that males should dominate in everyday life, contributes to the violence on the islands:

> There is a lot of drug and alcohol abuse including cocaine, hallucinogens, marijuana, etc. This contributes to the violence on the islands. Not only within the household, but also violence in general. One contributing factor is machismo where men feel that they must show their strength through this violence, but also because there is high drug use among the fishermen, and illegal activities such as selling gas to the drug lords.

Participants also acknowledged that due to lack of employment opportunities on the islands, men feel emasculated, and they turn to drugs to numb themselves from the shame of unemployment. Compounding this issue is the fact that the Galapagos are on the route of narcotics traffic from Peru, which both ensures supplies of drugs to the islands and provides unemployed men with an alternative means of generating income.

Researchers Seen as Extractive and Not Giving Back to the Communities

Participants discussed prior experiences of researchers interviewing them, collecting data, and then not giving back to the community. Many expressed feeling research fatigue and that they were being ignored, as research findings were not shared in a timely and accessible (non-print) manner and did not lead to on-the-ground changes. As one participant bluntly put it:

> Enough of research already! We need interventions.

Another participant spoke to the community's feeling of being excluded from the research process after data collection:

> There is a lot of research that has taken place on the islands. We are constantly interacting with researchers, but we are not aware of what happened to the research. The researchers do not come back to tell us what happened.

Although this sentiment was shared by many, some participants described an annual event at the Galapagos Science Center when researchers interact with the community to discuss their research. However, they also noted that very few community members can access these findings because they are communicated in ways that are inaccessible to the average community member.

ICS Analytical Framework Findings

Qualitative findings from participant interviews provided a rich context for the ICS analysis. The thematic analysis and the ICS analysis identified similar themes, particularly regarding the contextual and the structural factors that affect the vulnerability of residents of the Galapagos.

Our ICS analysis found that at the individual or intrinsic level, the risk factors of Galapagos residents include gender (i.e., females at more risk because of discrimination, violence, ridicule because of social norms and expectations of a young women) and age (i.e., the younger population does not have a good education, employment opportunities, recreation activities; are at risk of sexual abuse and exploitation; fear accessing reproductive health care services due to lack of confidentiality; and face mental health issues that are compounded by having to cope with all the aforementioned issues affecting the younger population). Individuals' physical and mental health were also notable risk factors. Participants reported high levels of mental health issues on the islands, but no mental health services, and that

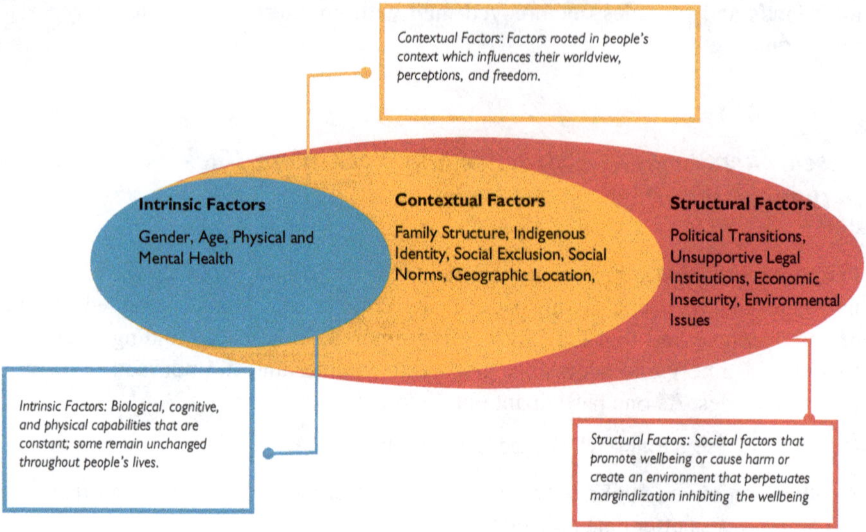

Fig. 9.2 Intrinsic, Contextual, and Structural Analysis of the Risk Factors for Social Issues in the Galapagos

the one resident psychologist rarely consulted due to confidentiality concerns. For physical health, participants from the Ministry of Health and USFQ reported high levels of obesity, high blood pressure, and diabetes among Galapagos residents.

Contextual risk factors included residents' isolation from their families, the disintegration of families, unhealthy interpersonal relationships, and the lack of a cultural identity of local people. At the structural level, ICS analyses revealed a lack of sustainable electoral leadership, which meant that support programs were constantly changing along with elected officials, and a lack of centralized systems to provide data and spearhead correctional services for civil violations on the islands. See Fig. 9.2 for a detailed list of risk factors at all levels.

Current Efforts to Address Social Issues

Current efforts are underway to address these issues identified on the three islands covered by the current study. Table 9.2 presents the data on the current efforts to date to address the identified issues. While several issues remain unaddressed, these also constitute opportunities for developing collective interventions to address these issues.

Table 9.2 Current Efforts to Address the Social Issues Identified

	Violence	Substance Use	Youth Development	Teen Pregnancy
Institutions and Ministries either interested in or currently working to address the issue:	Ministry of Justice (Ministry of Social and Economic Inclusion)	Ministry of Health and Oskar Jandl Hospital – offer services	Galapagos Governing Council – Political Chief Office (Authority for a Day project with high school student government dignitories)	Municipalities throughout Ecuador (Vistazo magazine article, November edition; we are unsure if there are current activities in the islands.)
	Galapagos Governing Council – Political Chief Office			Ministry of Health - services
Organizations and agencies either interested in or currently working to address the issue:	Guicell Gonzalez – Women Warriors Group in Santa Cruz	Municipalities of San Cristobal and Santa Cruz - projects	(?)	(?)
USFQ faculty interested in or currently working to address the issue:	María Amelia Viteri Cristen Dávalos (based in Quito, study began in 2019)	(?)	Diana Pazmiño Andrés Pazmiño Marjorie Riofrío Diego Paez (Community Outreach Project in Galapagos to support IB students prepare honors thesis begun in 2019)	(?)

Discussion

This study was conducted before the onset of the COVID pandemic in 2019. Thus, some study findings may have changed both due to the myriad disruptions that resulted from COVID and due to the efforts, that stakeholders on the islands, the government of Ecuador, and USFQ may be undertaking to address issues that were amplified by COVID. The following discussion of our study findings should be interpreted with this context in mind.

The four key social issues that emerged from the thematic analysis were domestic violence, substance use, youth development, and teen pregnancy. Because we have already discussed the contributing factors for these issues in our Findings section, this section potential starting points to address these issues. These starting points are broad and may seem not to provide detailed guidance for interventions designed to address the identified social issues. However, the purpose of this study was not to provide solutions, but to understand the salient social issues facing local people and their underlying factors. As such, our discussion offers guidance on potential approaches to implement to address the identified social issues.

The overarching theme of this study is the social issues that people of the Galapagos are facing. Clearly, conserving the environment where people live must be done with the people who live in the environment in mind. Ensuring a balance between environmental conservation and meeting the needs of people is key to the success of the conservation movement and requires including communities in the process of conserving the environment in or near which they live. Addressing human needs on the Galapagos Island may benefit from *community-based conservation* models, or "efforts to protect biodiversity in which the local community participates as much as possible" (Vimal et al. 2018. P334). Although community-based conservation has its critics (Noe and Kangalawe 2015; Larson et al. 2016), a recent systematic study has shown the ability to successfully safeguarding endangered species and habitats while alleviating poverty and ensuring the well-being of people (Brichieri-Colombi et al. 2018). Our findings indicate that the lack of livelihoods was driving most of residents' discontentment with the conservation efforts taking place in the Galapagos.

Developing demand-driven, context-specific interventions will be essential to address several noted social issues, such as residents' negative perceptions of the Cuban doctors at their local hospital. Community mobilization and information dissemination is key to developing interventions that are acceptable to the community and that will thus increase their use of important services (e.g., health care services). Similarly, interventions, services, and programs designed to address violence on the islands must be created based on discussions with community members about what they believe would most successfully mitigate violence in their communities. This work must also engage the authorities that oversee law-enforcement institutions to ensure that victims of violence will receive tailored, trauma-informed services. Indeed, in their recent study of violence in the Galapagos, Davalos & Zaragocin (2022) found that 74% of victims of physical violence and 67% victims of sexual violence reported this violence to the authorities, only to find that no legal action was ultimately taken against the perpetrators. Interventions that education law enforcement services would begin to change how victims are treated when they report their experiences to law enforcement.

Violence can also be addressed by drawing on indigenous knowledge and cultural wisdom related to wellness. For example, in other parts of the world, group-based approaches have successfully addressed violence against women by working with men using local knowledge of domestic relational interactions (Gibbs et al. 2022). In some contexts, social norms can normalize violence, but imposing fully foreign perspectives and customs may make it difficult for a community to embrace new norms. Instead, exploring local ways of relating that may highlight healthy relationships could have more success in addressing violence directed against vulnerable community members.

Coordinating efforts with institutions already working on salient social issues is essential to ensure the sustainability of programs and, more importantly, their positive outcomes. To this end, Table 2 lists the government institutions, organizations, and USFQ faculty currently working to address key social issues in the Galapagos. In particular, the inter-institutional committee (with representation across all ministries and units of government) provides an ideal platform for collaboration and exchanging information in pursuit of effective cross-sectoral interventions. Cross-sectoral approaches are ideal for developing interventions to address social issues such as those found in the Galapagos, as they entail purposeful programming at both community and structural levels (Melinkas et al. 2019, Stark et al. 2018). For instance, given that interventions designed to promote sexual and reproductive health, mental health, and physical and psychological safety must account for the intrinsic, contextual, and structural level factors detailed in Fig. 9.2, a cross-sectoral approach that addresses the educational, social, economic, and cultural aspects of these issues simultaneously will optimize these interventions' effectiveness. At the same time, interventionists must ensure that these interventions align with community values by fully engaging the community and local expertise in their development and implementation.

Conclusion

This formative study sought to identify the salient problematic social issues that exist in the Galapagos and the key drivers of these social issues. Because this was an exploratory study with no preconceived agenda to address these social issues, it lays out its findings to inform future work designed to address the issues it identifies. To this end, our discussion describes possible next steps toward developing effective interventions that address these issues via participatory approaches that strike the right balance between addressing the needs of community members and furthering conservation efforts on the Galapagos.

References

Braun V, Clarke V (2006) Using thematic analysis in psychology. Qual Res Psychol 3(2):77–101. https://doi.org/10.1191/1478088706qp063oa

Brichieri-Colombi TA, McPherson JM, Sheppard DJ, Mason JJ, Moehrenschlager A (2018) Standardizing the evaluation of community-based conservation success. Ecol App 28(8):1963–1981. https://doi.org/10.1002/eap.1788

Chowa, G., Masa, R., Manzanares, M., Bilotta, N., & Barrington, C. (2021). A systematic literature review of positive youth development impacts on marginalized and vulnerable youth.. USAID. https://www.youthpower.org/resources/systematic-literature-review-positive-youth-development-impacts-marginalized-and-vulnerable-youth

Content Engine LLC (2021) President Ecuador raises minimum wage by 6.25% by 2022, Despite criticism from businessmen [Translated from Spanish]. CE Noticias Financieras. https://www.proquest.com/wire-feeds/president-ecuador-raises-minimum-wage-6-25-2022/docview/2610341991/se-2

Dávalos C, Zaragocín S (2022) Island feminism meets feminist geopolitics: The spatial dynamics of gender-based violence in the Galapagos Islands. Area 54:313–321. https://doi.org/10.1111/area.12783

Freire WB, Waters WF, Román D, Jiménez E, Burgos E, Belmont P (2018) Overweight, obesity, and food consumption in Galapagos, Ecuador: A window on the world. Global Health 14(1):93–93. https://doi.org/10.1186/s12992-018-0409-y

Gibbs A, Mkhwanazi S, Sikweyiya Y (2022) Stepping-stones and creating futures: A group-based approach to addressing violence against women through working with men. J Clin Psychol 78:26–37. https://doi.org/10.1002/jclp.23293

Houck KM, Terán E, Ochoa J, Zapata GN, Gomez AM, Parra R, Dvorquez D, Stewart JR, Bentley ME, Thompson AL (2020) Drinking water improvements and rates of urinary and gastrointestinal infections in Galápagos, Ecuador: Assessing household and community factors. Am J Human Biol 32(1):e23358. https://doi.org/10.1002/ajhb.23358

Larson LR, Conway AL, Krafte KE, Hernandez SM, Carroll JP (2016) Community-based conservation as a potential source of conflict around a protected area in Sierra Leone. Environ Conserv 43:242–252

Lu F, Valdivia G, Wolford W (2013) Social dimensions of 'nature at risk' in the Galapagos Islands, Ecuador. Conserv Soc 11(1):83–95. https://doi.org/10.4103/0972-4923.110945

Melinkas AJ, Saul G, Singh SK, Mkandawire J, Gueye M, Diarra A, Amin S (2019) More than brides alliance: midline evaluation report. Population Council, New York

Noe C, Kangalawe RYM (2015) Wildlife protection, community participation in conservation, and (dis) empowerment in southern Tanzania. Conserv Soc 13:244

Pera MF, Katz BNH, Bentley ME (2019) Dietary diversity, food security, and body image among women and children on San Cristobal Island, Galapagos. Maternal Child Health J 23:830–838. https://doi.org/10.1007/s10995-018-02701-4

Stark L, Asghar K, Seff I, et al (2018) Preventing violence against refugee adolescent girls: findings from a cluster randomised controlled trial in Ethiopia. BMJ Global Health 3:e000825

UNESCO Centre (2019) Galápagos Islands. https://whc-unesco-org.libproxy.lib.unc.edu/en/list/1

Vimal R, Khalil-Lortie M, Gatiso T (2018) What does community participation in nature protection mean? The case of tropical national parks in Africa. Environ Conserv 45(4):333–341. https://doi.org/10.1017/S0376892917000583

Waldrop JB, Page RA, Bentley ME (2016) Perceptions of body size in mothers and their young children in the Galapagos Islands. Maternal Child Health J 20:2012–2018. https://doi.org/10.1007/s10995-016-2022-0

Walsh SJ, Mena CF (2016) Interactions of social, terrestrial, and marine sub-systems in the Galapagos Islands, Ecuador. Proc Nat Acad Sci USA 113(51):14536–14543. https://www.jstor.org/stable/26472918

World Commission on Environment and Development [WCED] (ed.) (1987) Our common future: The world commission on environment and development. Oxford University Press

Gina Chowa is the Johnson-Howard-Adair Distinguished Professor and the Founding Director for the Global Social Development Innovations in the School of Social Work at the University of North Carolina at Chapel Hill. Chowa conducts research on global social development, particularly, the intersection of economic security, workforce development, social protection and financial inclusion and its impact on the well-being of vulnerable and marginalized populations in the Global South. She develops and evaluates interventions that aim to promote gainful employment, meaningful financial inclusion, long-term economic security, and improved health and education outcomes. Her research focuses on employing rigorous methodology to examine causal relationships between holistic economic security interventions and well-being of children, youth, and their families.

Cynthia Fraga Rizo has practice experience providing services to survivors of intimate partner violence and their children. Dr. Rizo has worked on a number of projects in the area of interpersonal violence, including intimate partner violence, human trafficking, and sexual assault.

Amanda Thompson is Professor of Anthropology and Nutrition at the University of North Carolina at Chapel Hill and a fellow at the Carolina Population Center. She has been working on human health issues in the Galapagos since 2014. Amanda is the interim co-director of the Center for Galapagos Studies and Galapagos Science Center.

Margaret Bentley research focuses on women and infant's nutrition, infant and young child feeding, behavioral research on sexually transmitted diseases, HIV, and community-based interventions for nutrition and health. She has particular expertise in qualitative research methods and the application of these for program development and evaluation.

Mimi Chapman joined the faculty in 2001. Her research interests span child maltreatment and child and adolescent well-being, particularly among new immigrant families. She has developed arts-based interventions aimed at decreasing implicit and explicit bias among "high intensity professionals" such as public-school teachers and health care providers, work that has garnered media attention in outlets such as National Public Radio and the New York Times. Her global work in China used photovoice to understand mother's experiences of in-country migration. She worked with Chinese colleagues to examine the re-emergence of social work and currently collaborates with colleagues to understand the perspectives of youth in the Galapagos Islands to inform interventions for risk behavior.

Chapter 10
Towards Increased Island Food System Resilience: Lessons Learned from the COVID-19 Pandemic

Khristopher M. Nicholas, Margaret E. Bentley, Claire Barrington, and Amanda L. Thompson

Introduction

Island ecosystems are often described as facing two types of barriers: "classical" challenges related to sustaining rapidly growing populations with limited natural resources and "emerging" challenges related to import dependence and exposure to global markets (Schwarz et al. 2011). Complicating this distinction, however, are challenges imposed by climate change and the recent coronavirus pandemic (Farrell et al. 2020; Syddall et al. 2022). Climate change is expected to drastically reduce the fishing and agriculture yield for many small island developing states (SIDS) with implications for food security and economic development (Bell et al. 2021). Because many island nations are dependent on food imports, supply chain stagnation during COVID-19 pandemic contributed to income loss and food insecurity (Farrell et al. 2020).

Given the growing threat of external stressors such as climate change and global pandemics, increased attention has been given to understanding adaptive capacity and resilience. A community's social adaptive capacity reflects its ability to limit exposure to stressors and foster resilience through adaptation (Walker et al. 2004).

K. M. Nicholas (✉) · M. E. Bentley
Department of Nutrition, University of North Carolina at Chapel Hill, Chapel Hill, NC, USA
e-mail: knicholas@hsph.harvard.edu

C. Barrington
Department of Health Behavior, University of North Carolina at Chapel Hill, Chapel Hill, NC, USA

A. L. Thompson
Department of Nutrition, University of North Carolina at Chapel Hill, Chapel Hill, NC, USA

Department of Anthropology, University of North Carolina at Chapel Hill, Chapel Hill, NC, USA

S. J. Walsh et al. (eds.), *Island Ecosystems*, Social and Ecological Interactions in the Galapagos Islands, https://doi.org/10.1007/978-3-031-28089-4_10

A recent study in Fiji showed that resilience to climate stressors was bolstered by knowledge sharing, adequate government support, and the capacity of communities to rally resources for collective use (Medina Hidalgo et al. 2021).

This chapter explores resilience among residents of the Galapagos islands, a setting that is subject to the geographic, economic, and social pressures shared by other SIDS. Specifically, we explore resilience in the context of food acquisition during the height of the COVID-19 pandemic in 2020. The Galapagos Islands of Ecuador represent a unique opportunity to understand the interaction between systemic and community factors that drive dietary practices. Of the approximately 30,000 residents, 75% of adults experience overweight or obesity, compared to the national average of 65% (Freire et al. 2018, INEC (Instituto Nacional de Estadística y Censos del Ecuador 2015). Local food production is limited by restrictions on the use of pesticides, fertilizers, gas-operated farming equipment to lessen the impact on local ecosystems. Located 1000 km away from the continent, only 3% of terrestrial land is available for human use which contributes to a food economy that is dependent almost entirely on imports every 15 days by boat (Page et al. 2013; Sampedro et al. 2018). Although some markets sell food imported by airplane, these goods can be costlier than goods imported on the boat. Local food markets are the primary source of food for residents on the island. However, food prices are not externally regulated and price volatility across markets for the same goods is common.

Recent research has highlighted the presence of the double burden of nutrition among (and within) Galapagos households, where prevalence of morbidities of undernutrition (e.g., iron-deficient anemia) co-occur alongside overweight and obesity (Thompson et al. 2019). Residents in the Galapagos Islands make food acquisition decisions under a variety of constraints. These constraints include financial limitations, time constraints, nutrition knowledge constraints, and the unique case of COVID-19 and food availability, which has had stark impacts on food availability and coping strategies globally (O'Meara et al. 2022).

COVID-19 in the Galapagos

This study took place during January and February 2022, not long after the lifting of a strict quarantine period in the Galapagos due to the coronavirus pandemic (Galapagos Government Council 2021). In San Cristobal Island, there was a three-month period during 2020 where residents were only allowed outside their homes from 6 AM to 12 PM to obtain food and other essentials. Tourism, the biggest source of income in the Galapagos, came to a standstill during the pandemic, affecting the livelihoods of many residents. Food supply chains from mainland Ecuador to the Galapagos were disrupted and this unsteady flow of food imports coupled with economic hardship left few families unaffected. Social life was also disrupted by the pandemic, with 33% of students reporting absence from school due to lack of internet access and a reported increase in the number of domestic violence cases (Galapagos Government Council 2021).

Objective

This study's primary objective was to understand vulnerability to external stressors related to food acquisition practices among community members in San Cristobal Island, Galapagos, Ecuador. A secondary objective was to identify community-sourced solutions towards increased resilience. Although prior research has reported on the health landscape in the Galapagos, market inventory and pricing, and food purchasing strategies (Nicholas et al. 2022), little is known about perceived food system vulnerabilities or potential community-sourced solutions. Specifically, we sought to (a) identify perceived points of vulnerability between individuals and their food environment, (b) understand the effects of the COVID-19 pandemic on food acquisition and consumption in the Galapagos, and (c) identify community-sourced solutions towards increased resilience to environmental stressors.

Methods

Research Team and Reflexivity

Focus group moderators consisted of two locally born residents of San Cristobal Island (hereafter M1 and M2). US-based researchers attended focus groups via Zoom, recorded each session, and took written notes which were discussed with the moderators after sessions. We were aware of implicit power dynamics between ourselves as researchers in the United States, especially given the potentially sensitive discussions. Accordingly, US-based researchers did not actively moderate focus group discussions and deferred to the leadership of the moderators. Weekly meetings between the moderators and US-based research team were used to discuss key insights, strategies for probing, and to discuss specific nuances characteristic of colloquial Galapagos Spanish that arose during interview transcription and translation.

Participants and Procedures

Participants were recruited using a convenience sampling approach. Participant inclusion criteria were status as native Galapagueños and having a minimum age of 16 years. We conducted four sex-specific focus groups (two for men, two for women). Each focus group had 5–6 participants. Sex-specific focus groups were used to reduce the tendency of male participants to center their own perspectives and to foster safe atmosphere during conversations about healthy eating, diet habits, body size, and stress during the coronavirus pandemic. Each focus group took place in a conference room at the Galapagos Science Center on San Cristobal Island. Focus groups were audio-recorded to facilitate speaker identification. Upon arrival, written consent was obtained for each participant. All participants and facilitators wore masks in compliance with coronavirus safety protocols in the Galapagos Science Center. Each session took an average of 1 h and 30 min. At the end of the session, each participant received 15 USD.

We created a semi-structured focus group guide to address the following topics: (1) common food shopping habits, (2) priority setting when food shopping (e.g., deciding between cost and health), (3) definitions of healthy eating, (4) perceived barriers to healthy eating during the COVID-19 pandemic, and (5) suggestions for the local government to enable consumption of healthy diets.

Data Analysis

We used thematic analysis to identify key codes and themes in the data (Braun et al. 2019). We created a baseline codebook to map participant responses within the food environment and supply chain framework. We also used analytic memos and discussions among the research team to identify convergent codes and nascent themes. Next, we identified participant-centered actions, feelings, and beliefs, using verbatim codes where possible. This process of iterative coding allowed us to navigate between *a priori* assumptions and knowledge of Galapagos food environment and *de novo* synthesis of participant responses on the lived experiences of food acquisition, preparation, and consumption. Qualitative data analysis was conducted in Atlas.ti (version 9.1.3).

Results

Participant Summary

There were 11 male participants and 12 female participants across four focus groups (Table 10.1). The mean participant age was 31.2 years old, and 10 participants were under 30 years of age. Occupations included students (n = 6), individuals employed in the public sector (n = 6), individuals privately employed (n = 7), and full-time parents (n = 4).

"If I had money, there was no food. And if there was food, there was no money": COVID-19 and healthy eating.

Table 10.1 Summary characteristics of focus group participants including sex, age range and occupation

Participant ID	Sex	Age Range	Occupation
P1	F	30–35	Stay at home parent
P2	F	35–40	Public employee
P3	F	16–20	Student
P4	F	25–30	Public employee
P5	F	25–30	Unemployed
P6	F	16–20	Student
P7	M	30–35	Privately employed
P8	M	16–20	Student
P9	M	16–20	Student
P10	M	30–35	Privately employed
P11	M	25–30	Public employee
P12	M	30–35	Public employee
P13	F	25–30	Privately employed
P14	F	35–40	Stay at home parent
P15	F	25–30	Stay at home parent
P16	F	35–40	Stay at home parent/private
P17	F	40–45	Public employee
P18	F	45–50	Privately employed
P19	M	40–45	Public employee
P20	M	25–30	Privately employed
P21	M	20–25	Student
P22	M	45–50	Privately employed
P23	M	20–25	Student

Three codes were applied that categorize the variety of participant responses to food acquisition during the coronavirus pandemic: (1) *Coping via supplemental income, home gardens, and food sharing*, (2) *Stress (eating) and sedentarism*, and (3) *Reducing to the essentials.*

Coping via Supplementary Income, Home Gardens, and Food Sharing

During the pandemic, participants expressed that many people became unemployed and had to find alternative ways to generate income. Selling homemade food and homegrown produce from home gardens was a commonly reported strategy. Participants discussed these goods in a positive light or a negative light:

Positive light

In my house, more than anything else, people came, uh, selling their products, the sale was super informal, but they brought from their farms and we, just the same, tried to negotiate. So, yes, there were many purchases that saved us, literally, because we got our food in that way. (P21, M, 20–25, Student)

Negative light

M1: At that time during the pandemic, was it easy to buy sweets and snack foods?

> P19: I gave it a four [out of five], "somewhat agree"... there were people who sold tamales, who sold gummies, *hallacas*, cakes, who came by the house or contacted by cell phone and by WhatsApp groups, "Look, someone is selling this, come buy here" and yea.
>
> P20: Right? I also put four ["somewhat agree"].
>
> M2: Why?
>
> P20: Exactly the same reason, in my area there were like six people selling cakes, sweets, and pastries.
>
> (P19, M, 40–56, public employee; P20, M, 25–30, physical trainer)

For some participants, the availability of food, especially produce from home gardens, was viewed as a necessity given the unreliable supply of imported foods during the pandemic. The dominant view among participants across focus groups was slightly tinged by a negative opinion on all the snacks and cakes that were sold. There was general frustration at being stuck inside and only eating unhealthy food for months during the pandemic and some viewed the ubiquity of sellers on *WhatsApp* or coming by the house as having contributed to their diet patterns during quarantine.

Food sharing and social support was a common thread across focus group discussions. During the pandemic, the community of Puerto Baquerizo Moreno organized baskets of food that were shared with members of the tourism association (individuals whose income were most affected by the lockdown) and families in need. Below is an account of the benefit of food sharing from P14. In her focus group, P14 was vocal about the limitations that her budget imposed on feeding her family, even before the pandemic.

> [On the availability of foods to cook healthy meals in her neighborhood] Me too, I put two [out of five], "somewhat in disagreement", because in my personal case, we practically supplied ourselves with the food that the population organized. So many of those foods were not 100% nutritious. Yes, they were healthy, but if you're used to, I don't know, eating a vegetable soup, suddenly that wasn't available...
>
> We lived off what the population organized and helped us. Because just on that date my father-in-law died and 17 people stayed in a house. So those were the most traumatic two and a half months... But good thing there was food. No one starved to death. (P14, F, 35–40, stay at home mom)

In addition to supplementary income and relying on food sharing networks, several participants also named various dimensions of social support as crucial to their coping strategies during the coronavirus pandemic. P14 described the impact of her church community, which provided information on when food arrived on the island, as this was a period of stress for everyone to access food during the small window where lockdown was lifted to obtain food and essential items. Another participant (P16, F, 35–40, stay at home mom/privately employed) shared having family connections to the highland farm as another avenue of social support. She shared that her mother gave her family a hen which provided eggs and "fruits of the farm" including yucca, sweet potato, pepper, and onion.

Stress (Eating) and Sedentarism

Stories of stress and anxiety during the pandemic recurred across the focus groups. Some stress related to misinformation at a time when no one knew exactly how the coronavirus was contracted. For example, one participant (P17, F, 45–50, public employee) shared that during the pandemic, she wanted to buy fresh fruits and vegetables, but was told that they might contract COVID from handling infected produce and expressed, above all, confusion at that time. There was also stress related to food acquisition:

> You had to be on the lookout for the ship to arrive and go to the-to the store… And some didn't respect the 6-foot distance, so you had to – apart from the fact that there weren't enough vegetables – deal with people like that. (P15, F, 25–30, stay at home mom)

Most commonly, stress and anxiety during the pandemic was associated with unhealthy eating habits:

> No, I think that-I think that an important factor there in the-the pandemic was stress and that stress made you have, um, a lot of anxiety. So, anxiety makes you eat a lot of things that are not healthy. That's why I think that in some cases weight was gained and it can't be lost.
>
> [*P10 raises his hand as if to say, "yep, that's me"*]
>
> (P12, M, 30–35, public employee)

This exchange follows the focus group discussion around definitions of healthy eating when M1 asked the group if they thought that the types of diets they described as healthy were possible during the pandemic. This acknowledgement of weight gain, due to unhealthy eating and sedentarism during the pandemic was a theme across focus groups. In one focus group among women, a participant laughingly said, "We came out chubby" and later joked that they were going to die, "not from COVID, but from stress eating." A levity to these discussions underlies what was also a traumatic experience during the pandemic.

Reducing to the Essentials

> M1: What about snack and junk foods during the pandemic, was it easy to find? What answer did you write down?

> P16: I really didn't know what to answer. I put, I put four [somewhat agree]. But the truth is that at home in that time of the pandemic I spent as little as possible, because I did buy only what was necessary, which were fruits, vegetables, and bread.
>
> …Because as I told you, we rely 100% on tourism. We closed the agency and closed-and closed the boat. Vacations stopped-stopped, so our resources were extremely limited. And I have three children and there are five of us in the house, so what I did was try to reduce expenses. No, they understood.
>
> What I bought them were the biscuits from – ***Las Universal***, which they could have with a coffee, with cheese as well as for a snack, and some saltine crackers. Soda we left out because it's $3. I mean, I really didn't know what to answer. (P16, F, 35–40, stay at home mom/tourism)

Reducing to the essentials was a strategy adopted during the period of COVID-related food scarcity. **P16** described having to limit food purchases in response to the pandemic, because their primary source of income was from tourism. **P16** also reported having relied significantly on the food sharing organized by the community. **P14** shared her own views on food acquisition during the pandemic as a balance between being limited by variable monetary resources and variable food access, stating, "If I had money, there was no food. And if there was food, there was no money."

Participant-Sourced Ideas for Municipal Assistance

The last component of the thematic analysis is the theme on *municipal assistance*. One of the last questions asked was what ideas participants would suggest to the municipality to improve healthy eating. The array of responses fell within five categories coded as *eat local – Galapagos pride, nutrition education, exercise infrastructure and incentives, more government regulation,* and *helping farmers helps everyone*.

Eat Local – Galápagos Pride

For some participants, "eating local" is the whole point; it is the silver bullet that both supports the local economy and fills the gap of food scarcity due to reliance on imports. For others, "eating local" is a barrier to having good diets. Many participants fell to one side or the other – they strongly viewed supporting local farmers as a community imperative or they believed local goods were of mediocre quality and excessively expensive. The below quote exemplifies both perspectives in a single participant, P14:

> P14: There are us, there are people who live with $10 a day, like that, $10 in your pocket for the day. From there you have to buy bread, from there you have to buy lunch. Then, they should raise a little awareness of the prices they are putting on the products. Because it is

> true, what comes from abroad is expensive in itself and what is consumed here is also expensive…
>
> I like to go to the-to the farmer's market that they do in *Los Algarrobos* because the product is very fresh. Buying a cauliflower there compared to buying one in the store, the one from there lasts me up to two weeks in the refrigerator, but the one from the store in three days it's done, it rots. But I have to have at least $40 in my pocket to be able to stock up on the essentials, the essentials.
>
> P16: Yes it's true. I went on the weekend and with exactly $40 I bought the essentials.
>
> P14: The essentials [only], it's not that you also buy a chicken, a piece of meat, and I don't know what else; yucca – No, you get only the essentials for the week. Then, and apart from that, you end up lacking salt, you lack oil… (P14, F, 35–40, stay at home mom; P16, F, 35–40, stay at home mom/tourism affiliated)

P14 discussed the difficulties of making decisions given finite resources. She likes locally grown foods and enjoys going to the local weekend farmer's market yet acknowledges that shopping at the farmer's market depletes her food budget. In one focus group, opposition arose between P13 (F, 25–30, self-employed) and P16 (F, 35–40, stay at home mom/tourism affiliated) on the importance of locally grown food in the Galapagos:

> Because, excuse me, because if prices are already rising here in Galapagos, what's the difference between buying a product from here, and one from abroad? I mean, I prefer to buy foods from abroad. (P13, F, 25–30, self-employed)

This question came as an intercession in the conversation at the time between M1 and P16, who are both supportive of eating local. In her elicited response, P16, begins defending local producers, saying, "If [people] pay that value for goods from abroad, why not pay that to us?" As she describes, local farmers have to do all the work as farmers abroad except without the aid of pesticides or fertilizers, which are strongly regulated to preserve Galapagos ecosystem stability:

> …The coffee process [entails] taking out those husks, drying, from there grinding and the selection of the coffee, that selection from bean to bean is a super tough process. After it is already selected in that round, it is selected again, it is roasted and it is packed.
>
> That's the whole process… and [they] are right now selling [it] at $6 and it's not-and it's not a value that reflects all the work that has been done. So, that's why [they] were seeing if [they] would go up a little [in price]… And it's a coffee that people are coming to order precisely because it's such good quality and highly reviewed. (P16, F, 35–40, stay at home mom/tourism affiliated)

As with price, not all participants agree on the quality or trustworthiness of locally grown foods. P16 lauds locally grown coffee and reports that it is grown without any additional chemicals. However, in a separate focus group, two participants raised the issue over both of these points. In one instance, P22 (M, 50–55, accountant) discusses his experience trying to support locally grown coffee farmers. He shares that he spent $15 but the coffee beans were bitter and burnt and, ultimately, said "I was sorry I paid it." P22 frequently referenced the difficulty of providing enough healthy food for his family, underscoring his frustration at his disappointing experience with local goods. Another participant in the same focus group is P19 (M, 40–45, public employee) who shares his own doubts on the production process of local goods: "We say that in Galápagos they are organic products, but in reality,

I would not put my hands in the fire for them, because I do not know what is being put in our products up in the highlands."

There is a divergence among participants on what ultimately prevents eating locally grown foods. For some participants like P22, P19, and P14, external factors such as cost and quality is paramount. For others, like P16, the barriers are community members themselves:

> Sorry, I believe that there is no real barrier [to healthy eating]. Because you will find healthy products here. The barrier is ourselves as people, that's the barrier. Because we have products, we have fishing… I think that rather, the only barrier is us. (P16, F, 35–40, stay at home mom/tourism affiliated)

Ultimately, participant responses converged on the resolution that the solution cannot simply be to eat local. Whether education, consumer assistance, or farmer subsidies, external support is needed:

> That's why I say, it would be for the institutions to make a plan together. Not only, 'Let them consume what is local', no. 'Look, we are going to help the vendors so that this product is not so expensive. We are going-to invest.' (P14, F, 35–40, stay at home mom)

Nutrition education evokes the desire to learn more about healthy eating. Participants acknowledge the desire to learn more about how to feed themselves given unique dietary and body needs. An important dimension of *nutrition education* was the desire for education programs in schools to combat the conflation of "dieting" with "healthy eating".

> Workshops for children when they go to school, healthy meals, what kind of diet to eat. Because-because it's not just dieting. No, "I stop eating and go on a diet", I mean, no. I mean, the-the diet is also eating as-as I mentioned a moment ago, things in their proper proportion. So, I would tell the municipality that, talk to the children, create projects in the schools, so that they can strengthen from-from home and from education, in the schools, a healthy diet. (P11, M, 25–30, public employee)

P11 was a participant that was more willing to spend more money for good food. Throughout his comments in the focus group, he highlighted the importance of doing what it takes (such as walking to multiple stores) to find fresh produce and was a strong supporter of eating local foods. Above, he acknowledges the conflation of "dieting" with "healthy eating" as a poor view on healthy eating. Yet this acknowledgement of *nutrition education* is not the entire picture. Throughout her focus group, P14 discussed her experience having difficulty paying for enough healthy food for her family. On the subject of nutrition education, P14 says,

> What's the point of giving me the nutrition talk if I don't have the money to be able to buy what the nutritionist is telling me? Everything goes hand in hand. (P14, F, 35–40, stay at home mom)

The implication is that knowledge is important, but financial support to leverage this knowledge is imperative.

Exercise infrastructure and incentives refers to investment from the municipality in better public exercise equipment and finding a way to incentivize participants to be active. Despite the small size of Puerto Baquerizo Moreno, many participants prefer not to walk even short distances. This code applied to only two participant responses, underscoring the general lack of emphasis on physical activity in the Galápagos.

More government regulation referred to the desire for market oversight. The two types of oversight were for food safety and cost. Regarding food safety, several participants discussed shopping at stores and finding food that was already spoiled. Most commonly, this applied to fresh produce, but it also applied to processed foods. One participant (P21, M, 20–25, student) recounts his experience when buying yogurt at the market, only to find out that it was already spoiled. He links this poor food safety oversight with large tourism companies that bring food (for wealthy tourists) to the island and offload the leftovers on local markets. The other main category of oversight relates to the variable cost of the same good across markets, especially goods from abroad. A suggestion to the municipality is to ensure price matching across markets.

Helping farmers helps everyone applies to participants who wanted to advocate for government support as a means of supporting local farmers, which would lower the cost of local goods. The two most common ideas related to government-assisted transportation and required local sourcing of food by tourism companies:

Government-assisted transportation

P21: Yes, I also think that the municipality could intervene in the transportation, because it is true, it is too expensive for the farmer to go down [from the highlands] and have to – I mean, like neglecting production there to have to come down here to sell. There must be a mechanism, a way, in which a car, like a car going through all the farms, picking up all the products, and bring them down here so that it is not each farmer paying individually for transport, but there is one-one collective –

> P20: That would be excellent.
> (P21, M, 20–25, student; P20, M, 25–30, physical trainer)
> *Mandated local food sourcing among tourism companies.*
> So, I think that the authorities should, eh-eh, by legal means, ensure that [local] products are consumed mostly by tour operators, right? So that they can improve the income of the farmer and at a given time make sure the costs are accessible to the local consumer.
> For example, for all tour operators at least 60% or 50% of organic products must be purchased here and not brought from the mainland.
> (P22, M, 50–55, accountant)

Transportation is acknowledged as a challenge for local farmers, not only for consumers. P21's idea would be to provide transportation to all farmers, negating their need to raise costs (good for the consumer) and providing them with reliable, affordable market access (good for local farmers). Similarly, P22 suggests requiring tour operators to supply their food needs using locally grown products. In this way, farmers have another avenue for reliable market share, and consumers are less subject to price hikes.

Conclusions

This study identified perceived food supply challenges within Galapagos food environments and explored the case of healthy eating during the height of the COVID-19 pandemic. Focus group participants identified food acquisition challenges that impede access to having enough healthy foods, especially fresh fruits and vegetables. On the production side, restrictions to local farming limit the domestic supply of food thereby inducing reliance on food imports. Related to food processing, concern over food additives and chemicals are a large concern. Despite reliance on food imports, participants expressed concern about the chemicals necessary to preserve food during transport. Transportation was one of the largest concerns among participants. Because imported foods travel 1000 km, there are frustrations at its inflated cost and unpredictable quality. Lastly, challenges for food acquisition relate to markets being frequently out of stock and lack of price matching across markets leading to perceived price gauging.

During the coronavirus pandemic, these challenges were greatly exacerbated. The global economic downturn led to an unstable food environment wherein food imports were less frequent, and food stock was unpredictable. The vulnerabilities to food supply challenges were further increased when the tourism-dependent Galapagos economy came to a standstill with quarantine restrictions. Participant narratives demonstrated several examples of social resilience. At an individual level, some residents supplemented lost income by selling homegrown food. At a community level, baskets of essential foods were organized and distributed to residents most in need. Several participants reported that these food baskets enabled them to survive the pandemic.

The confluence of a global pandemic in an import-dependent remote island ecosystem highlighted the Galapagos islands' vulnerabilities to inadequate food supply. Participants suggested several ideas to foster increased community resilience. Increased nutrition education for both adults and children would reduce the perceived guesswork participants report in trying to eat enough healthy food. Though somewhat debated among participants, there is a consensus that increased consumption of local foods would decrease dependence on food imports. However, to achieve this, municipal investment is necessary to support local farmers and local consumers. Providing transportation infrastructure, instituting pricing regulations, and stricter policies on tourism companies' roles in the food environment were examples of increased governmental oversight.

Island ecosystems such as the Galapagos are vulnerable to climate change, economic fluctuations, and tourism disruptions. The coronavirus pandemic demonstrated the risks posed to community wellbeing when these vulnerabilities are left unaddressed. Community members themselves, however, demonstrated the capacity for social adaptation towards increased resilience. To improve resilience to external stressors in the Galapagos, future policies would benefit from public engagement to implement the community-sourced solutions outlined here.

References

Bell JD, Senina I, Adams T, Aumont O, Calmettes B, Clark S, Dessert M, Gehlen M, Gorgues T, Hampton J, Hanich Q, Harden-Davies H, Hare SR, Holmes G, Lehodey P, Lengaigne M, Mansfield W, Menkes C, Nicol S, Williams P (2021) Pathways to sustaining tuna-dependent Pacific Island economies during climate change. Nat Sustain 4(10):900–910. https://doi.org/10.1038/s41893-021-00745-z

Braun V, Clarke V, Hayfield N, Terry G (2019) Thematic analysis. In: Handbook of research methods in health social sciences

Farrell P, Thow AM, Wate JT, Nonga N, Vatucawaqa P, Brewer T, Sharp MK, Farmery A, Trevena H, Reeve E, Eriksson H, Gonzalez I, Mulcahy G, Eurich JG, Andrew NL (2020) COVID-19 and Pacific food system resilience: opportunities to build a robust response. Food Secur 12(4):783–791. https://doi.org/10.1007/s12571-020-01087-y

Freire WB, Waters WF, Román D, Jiménez E, Burgos E, Belmont P (2018) Overweight, obesity, and food consumption in Galapagos, Ecuador: a window on the world. Glob Health 14(1):93. https://doi.org/10.1186/s12992-018-0409-y

Galapagos Government Council (2021) Galapagos 2030 Plan

INEC (Instituto Nacional de Estadística y Censos del Ecuador) (2015) Censo de Población y Vivienda-Galápagos

Medina Hidalgo D, Nunn PD, Beazley H (2021) Uncovering multilayered vulnerability and resilience in rural villages in the Pacific: a case study of Ono Island, Fiji. Ecol Soc 26(1):art26. https://doi.org/10.5751/ES-12197-260126

Nicholas K, Delamater P, Thompson A (2022) Spatial and social dimensions of Galápagos food environments. In: Thompson AL, Ochoa-Herrera V, Teran E (eds) Water, food and human health in the Galapagos, Ecuador. Springer, pp 121–141. https://doi.org/10.1007/978-3-030-92411-9_8

O'Meara L, Turner C, Coitinho DC, Oenema S (2022) Consumer experiences of food environments during the Covid-19 pandemic: global insights from a rapid online survey of individuals from 119 countries. Glob Food Sec 32:100594. https://doi.org/10.1016/j.gfs.2021.100594

Page R, Bentley M, Waldrop J (2013) People live here: maternal and child health on Isla Isabela, Galapagos. In: Walsh CF, Stephen J, Mena (eds) Science and conservation in the Galapagos Islands: frameworks & perspectives. Springer, pp 141–153. https://doi.org/10.1007/978-1-4614-5794-7_4

Sampedro C, Pizzitutti F, Quiroga D, Walsh SJ, Mena CF (2018) Food supply system dynamics in the Galapagos Islands: agriculture, livestock and imports. Renew Agric Food Syst 35(3):234–248. https://doi.org/10.1017/S1742170518000534

Schwarz A-M, Béné C, Bennett G, Boso D, Hilly Z, Paul C, Posala R, Sibiti S, Andrew N (2011) Vulnerability and resilience of remote rural communities to shocks and global changes: empirical analysis from Solomon Islands. Glob Environ Chang 21(3):1128–1140. https://doi.org/10.1016/j.gloenvcha.2011.04.011

Syddall VM, Fisher K, Thrush S (2022) Collaboration a solution for small Island developing states to address food security and economic development in the face of climate change. Ocean Coast Manag 221:106132. https://doi.org/10.1016/j.ocecoaman.2022.106132

Thompson AL, Nicholas KM, Watson E, Terán E, Bentley ME (2019) Water, food, and the dual burden of disease in Galápagos, Ecuador. Am J Hum Biol. https://doi.org/10.1002/ajhb.23344

Walker B, Holling CS, Carpenter S, Kinzig A (2004) Resilience, adaptability and transformability in social–ecological systems. Ecol Soc 9(2). https://doi.org/10.5751/ES-00650-090205

Khristopher M. Nicholas is focused on research that explores the relationship between behavior and environment as contributors to diet. Originally from Trinidad and Tobago, he is particularly interested in food environments within climate-vulnerable island communities.

Margaret E. Bentley is a Distinguished Professor Emeritus of Nutrition at the The University of North Carolina at Chapel Hill. She is a medical and nutritional anthropologist whose expertise includes maternal and child nutrition and health, with a focus on the first 1000 days of life. She has been working in the Galapagos since 2009.

Claire Barrington is a Professor in the Department of Health Behavior at the University of North Carolina Gillings School of Global Public Health. She is the Latin American Programs Director for the Institute for Global Health and Infectious Disease and a fellow at the Carolina Population Center. Dr. Barrington conducts mixed-methods research on social and structural influences on health, in particular HIV and type-2 diabetes, with a geographic focus in Latin America.

Amanda L. Thompson is Professor of Anthropology and Nutrition at the University of North Carolina at Chapel Hill and a fellow at the Carolina Population Center. She has been working on human health issues in the Galapagos since 2014. Amanda is the interim co-director of the Center for Galapagos Studies and Galapagos Science Center.

Chapter 11
Understanding the Impacts of a Natural Disaster: Evidence from the 2004 Indian Ocean Tsunami

Elizabeth Frankenberg, Cecep Sumantri, and Duncan Thomas

Introduction

Across the globe, human and animal populations and the ecosystems in which they reside are experiencing pressures from climate change, both from slow-onset gradual changes and from rapid-onset events that are often large-scale and more intense. These pressures affect the health and resources of those exposed to them, but at present our knowledge of which outcomes are affected, the magnitude of the effects, and their longevity is limited.

For scientists who study human and natural systems and the links between them, one key challenge to understanding the impacts of a changing climate is the lack of "pre-onset" information against which to compare the evolution of various phenomena over time and thereby quantify the effects of climate change on outcomes of interest. Samples constructed without regard to the populations in place before a change, that rely only on data from those still in place after a change, or that focus on residents displaced only to organized highly visible camps and shelters, risk mischaracterizing an event's impact on the entire population. From a methodological perspective, it is important to understand how estimates of the impacts of contextual change may be biased by samples that miss particular population sub-groups. On one hand, it may include people who are most able to adapt to the change and move away before or immediately after the event. On the other hand, those most

E. Frankenberg (✉)
University of North Carolina at Chapel Hill, Chapel Hill, NC, USA
e-mail: e.frankenberg@unc.edu

C. Sumantri
SurveyMETER, Yogyakarta, Indonesia

D. Thomas
Duke University, Durham, NC, USA

S. J. Walsh et al. (eds.), *Island Ecosystems*, Social and Ecological Interactions in the Galapagos Islands, https://doi.org/10.1007/978-3-031-28089-4_11

deleteriously impacted by the event, including those who die, may be missed. The magnitude and direction of this bias is not obvious.

This paper investigates the nature and importance of these biases for understanding the impacts on well-being in both the short- and longer-term of a large-scale natural disaster, the 2004 Indian Ocean earthquake and tsunami. We use data from a large-scale household survey collected on the island of Sumatra, Indonesia. Importantly, the survey is longitudinal, spanning the period before and after the tsunami. The disaster, which killed an estimated quarter of a million people worldwide, is one of the most devastating natural disasters in recorded history. Nowhere was hit harder than the coastline of the Indonesian province of Aceh. The tsunami completely destroyed some communities but left other comparable, nearby coastal communities relatively untouched. Although the tsunami was not a consequence of climate change, the waves flooded coastal areas and pushed water up river basins, generating a surge of saltwater over land in the way that storm surges often accompany tropical cyclones.

The Study of the Tsunami Aftermath and Recovery (STAR) is ideally-suited for this research. In 2004, ten months before the tsunami, Statistics Indonesia conducted a large socioeconomic survey (SUSENAS) throughout Indonesia, as a part of an annual survey that is population-representative at the *kabupaten* (district) level. After the tsunami inundated the western coastline of the province of Aceh and, to a lesser extent, North Sumatra, we worked with Statistics Indonesia to field a longitudinal follow-up. The goal was to recontact every surviving SUSENAS respondent who was living in any of the 11 districts that had a potentially vulnerable coastline in the provinces of Aceh and North Sumatra.

Two features of STAR are unusual but critically important for this research investigating biases that arise from using non-representative samples to understand the effects of disasters. First, our pre-tsunami baseline is representative of the entire at-risk population and thus an ideal vehicle for a longitudinal study that tracks the lives of survivors in the aftermath of the tsunami. Second, we continue to follow survivors to this day. In each survey round, we track each target respondent to their location of residence, including those who were displaced or chose to move after the tsunami. Many have moved multiple times to places in Aceh, North Sumatra and other provinces across Indonesia. We attempt to interview every respondent in every follow-up. We are, therefore, able to provide scientific evidence on the value of this design for drawing conclusions about the short and longer-term impacts of exposure to a large-scale natural disaster.

Context

At 8 a.m. on Sunday, December 26, 2004, one of the most powerful earthquakes in recorded history occurred 150 miles from the coast of the island of Sumatra, Indonesia. The earthquake displaced a trillion tons of water, which formed a series of tsunami waves that hit the coast of Sumatra about 15 minutes later, eventually reaching across the Indian Ocean. The tsunami was completely unexpected. Geological records indicate that the last tsunami to hit mainland Sumatra was over 600 years ago (Monecke et al. 2008).

Aceh, the northern most province on Sumatra, was hardest hit. Along 800 kilometers of coastline communities experienced varying degrees of inundation, resulting in destruction of the built and natural environment and the deaths of more than 170,000 people.

Impacts varied considerably even between coastal communities that were otherwise similar and close to one another. The water's height and inland reach were a function of slope, water depth, and coastal topography (Ramakrishnan et al. 2005). Along parts of Aceh's west coast, trees up to 13 meters tall lost their bark (Borrero 2005). At the beachfront in Banda Aceh, the province's capital and largest city, the water was as deep as 9 meters; though rarely exceeded the height of a two-story building (Borrero 2005). Low-lying communities within a few kilometers of the coast were largely destroyed and many of their residents perished. River basins allowed the waves to move inland as much as 9 kilometers in some areas, whereas in other locations they encroached only 3–4 kilometers (Kohl et al. 2005; Umitsu et al. 2007). Areas sheltered by altitude, distance from the coast, or other topographical features sustained damage to structures and deposition of sediment and debris, but larger proportions of the population survived. For some communities the tsunami had few if any direct effects, although the earthquake was felt throughout Aceh and damaged property and infrastructure in some areas that the water never reached. The tsunami affected the transportation network along the coast. Some communities were cut off from the main roads connecting major population centers. In some cases, residents of communities that were not directly impacted by the tsunami saw increased demand for their goods and services, particularly food and housing — a benefit for those who sell food or housing but not for net food purchasers or renters.

The tsunami was followed by an unprecedented outpouring of financial support from governments, aid agencies, international and domestic NGOs, and private citizens. Pledges to Indonesia totaled more than US$7 billion (Nicol 2013). Of the total amount committed to Indonesia, US$1.5 billion were in excess of the estimated cost of the reconstruction, which allowed the Indonesian government to set the goal of "building back better."

The tsunami resulted in the destruction of livelihoods and tremendous economic stresses for many along with disruption to their social networks (Frankenberg et al. 2008; Gillespie et al. 2014). Many of the people living in the hardest hit areas moved away to temporary housing in barracks or camps, for example, or to private homes (Gray et al. 2014). Some of those people returned to their pre-tsunami communities, particularly after a massive housing reconstruction was launched (Laurito et al. 2022). Studies have also established some individuals and families displayed resilience and navigated the trauma of the disaster with modest impacts on well-being, others were able to recover much of the economic losses and others have fallen permanently behind (Lawton et al. 2022).

Data

Working with Statistics Indonesia to select 11 districts in the province of Aceh and two in North Sumatra whose coastlines were potentially vulnerable to inundation by a tsunami. Within each selected district we included all SUSENAS enumeration areas, regardless of distance from the coast. All members of all households enumerated in these districts in the 2004 SUSENAS were selected to form the STAR baseline study population.

The February/March 2004 SUSENAS, which was conducted 10 months before the tsunami, provides the population-representative baseline for STAR and covers communities in all coastal districts in Aceh and North Sumatra that would have been at-risk of being directly affected by the tsunami. We conducted the first follow-up between May 2005 and July 2006, at a time when the full impact of the tsunami and where it had hit was not well-known. The fieldwork was extremely challenging. We remained permanently in the field and completed four annual follow-ups and then completed additional follow-ups at roughly 10 and 15 years after the disaster.

In the communities that were hardest hit, 80% of the respondents died in the tsunami (Frankenberg et al. 2011). We triangulated across multiple sources of information to establish survival status. We are confident it is accurate for 99% of the pre-tsunami baseline respondents. Information comes from interviews with household and family members (whose reports we consider most reliable), community leaders, and neighbors. Information from the latter two sources is critical for households in which no members could be located. In each follow-up, every household member is interviewed. Parents or caregivers provide information about children age 11 years or younger, proxy respondents provide information for adults unable to answer for themselves. The first two follow-up surveys collected detailed information on experiences at the time of the tsunami from each respondent. All surveys include questions on physical health, psycho-social well-being, and behavioral responses to the event, including displacement and migration, as well as information about individual and household demographics and socioeconomic status.

In this paper we analyze 16,342 baseline respondents who survived the tsunami and were 15 or older at that time. Half are male. The average respondent was age 35-years at the time of the tsunami and had completed 8 years of schooling (just shy of completing junior secondary school in Indonesia). Our first measure of well-being is completed schooling of each respondents.

STAR is designed to collect data at the household and individual levels. A key respondent in each household provides information about every household member along with household-level measures of social and economic well-being including expenditures. We draw on these data to trace the evolution of a key marker of economic status, household per capita expenditure (PCE), in the aftermath of the tsunami. We have information on PCE for between 91% and 94% of the baseline respondents in each of the follow-ups. This is a high rate of follow-up in any large-scale longitudinal survey. It is unprecedented in a follow-up after a large-scale natural disaster that caused enormous damage to the built and natural environment and resulted in large-scale displacement and migration.

Every member of a STAR household age 15-years or older is eligible for an interview that collects more detailed information about their health, economic status and perceptions of well-being. We draw on these interviews to investigate the evolution of an indicator of overall measure of well-being in the aftermath of the tsunami. The indicator is measured for 80–85% of the respondents for whom PCE is measured in each follow-up.

Methods and Measures

Our primary question revolves around the ways in which members of our baseline sample redistribute themselves in the aftermath of a major disaster. We begin by analyzing how pre-tsunami characteristics are associated with location 1 year after the tsunami, then turn to how location 1 year after the tsunami is related to economic resources in that year and at 2 and 5 years after the tsunami. The question is important for the design of studies that aim to understand well-being associated with the environmental and contextual changes that will accompany global warming, where the goal is characterizing outcomes for a population as a whole, rather than solely those found in a particular location or type of housing in the aftermath of change.

To characterize location after the tsunami, we create a variable that assumes four values. Those who are in the same location at the first follow up, roughly 1 year after the tsunami, as they were in the baseline survey are distinguished from those living in a different location but within the same neighborhood, those living in a different neighborhood but within the same village or township, and those living in a different village or township. For the analytical sample 53% remained in the same home at the first follow up interview, 13% had changed residences but were in the same neighborhood, 8% had changed neighborhoods but remained in the same village, and 14% had moved to a new village (11% were not interviewed).

For individuals who cannot or do not want to remain in their original location, another factor in relocation is the type of housing available to them. In the aftermath of a disaster, securing, safe, stable shelter is a core component of recovery, but it is challenging when significant property destruction has occurred. For those living somewhere other than their pre-tsunami location, we classify individuals based on whether they were in emergency temporary housing (tents, camps, or barracks, 4.4%) or private housing (31%).

Where people live, the type of housing in which they live, and how they fare after a major event depends in part on the degree of damage to which they were exposed. We designed STAR to include communities along a continuum ranging from destruction of almost all buildings and vegetation to no direct damage from the tsunami waves (though some communities sustained damage from the earthquake).

For our analyses we classify communities into three groups with respect to level of damage in the community (heavy, moderate, and not directly damaged). The damage measure is based on remote sensing measures of damage, direct observations from our team supervisors, and reports from community leaders.[1] This measure is closely correlated with levels of tsunami mortality and other outcomes for individuals. About 20% of respondents analyzed here resided in communities heavily damaged by the tsunami, 58% were in moderately damaged communities, and 22% were in communities with light to no damage.

We use three measures to characterize human capital and well-being before and after the tsunami. The first measure is years of educational attainment, which for most study subjects was established before the tsunami. Education is widely regarded as a good indicator of long-run economic well-being and has the advantage, in this context, of being fixed over time.

The second measure is monthly household PCE which varies over time. PCE is calculated from questions about spending on 7 food and 12 non-food categories of goods. Examples of food categories are rice; meat and fish; fruits and vegetables; non-food examples are clothing; personal goods; and energy. For those who do not pay rent, housing expenditures are imputed based on the rental value of the home. In general, PCE is thought to be a good measure of resource availability and thus economic well-being (Deaton 2016), particularly in settings of substantial temporal variation in income because of seasonality or the nature of work. This is important in the context of a disaster that destroyed livelihoods and income-earning capacity. During the 1998 financial crisis when incomes collapsed in Indonesia, households adjusted their spending patterns and drew on their savings, support from family and friends as well as public programs (Frankenberg et al. 2003). Since household expenditures increase with household size, we standardize by household size which is a crude way to take this into account. PCE is measured in real terms taking into account local area price variation. The distribution of PCE is skewed to the right and so we use the natural logarithm of PCE, lnPCE, in the analyses. In the year after the tsunami, the average household spent Rp 1.4 million per month which is approximately US$150 (or US$40 per person).

We complement lnPCE with a global indicator of well-being that is measured at the individual level using responses to a Cantril ladder-type question. Specifically, are asked to imagine a six-step ladder where on the bottom (the first step), stand the poorest people, and on the highest step (the sixth step), stand the richest people. They are then asked to locate where they feel they are at the time of the survey. The question has been used extensively as a source of information on perceptions of well-being around the world. At the first post-disaster interview, 51% of

[1] Our satellite-based damage measures come from three publicly-available damage products produced after the tsunami and a measure we constructed using data from NASA's MODIS sensor. Images from December 17, 2004 and December 29, 2004 were geographically linked using the MODIS reprojection tool.

respondents reported themselves on the first or second step, and just under 8% reported themselves on the fourth, fifth, or sixth step.

This third measure of well-being has three advantages relative to lnPCE. First, PCE is measured at the household level and every household member is assigned the same level of PCE. This does not take into account potential within-household variation in access to resources which may be especially important after a disaster if some members tighten their belts more for the sake of others in the household. This pattern was documented in the context of the 1998 financial crisis in Indonesia where older women apparently reduced their own consumption in favor of grand-children in the household (Frankenberg, Smith, and Thomas 2003). Second, and related, when individuals split off from a household and form a new household or live alone, changes in PCE may not accurately reflect changes in well-being. This is also important in the context of a the tsunami as large numbers of people were displaced and many of the affected household split up. Third, it is complicated to deal with price heterogeneity in surveys and this concern is side-stepped by the ladder question.

The ladder question has two disadvantages. First, the definition of poorest and richest is likely to vary with socio-economic status which affects interpretation of the indicator. It is advantageous, therefore, to simultaneously investigate patterns in lnPCE and the ladder question to interpret the results with these caveats in mind. The second disadvantage is that since the question reflects each individual's perception of well-being, it imposes a higher burden on the survey as each respondent has to be individually interviewed. Completion rates of individual assessments are lower than household-level assessments.

Results

To set the stage for descriptions of the evolution of well-being in the aftermath of the tsunami and how it varies with community-level damage and post-event location, Fig. 11.1 displays the distribution of respondents across damage level and residential arrangement at the time of the first post-tsunami interview. Three points are important. First, across the three damage levels there is substantial variation with respect to the proportion of respondents who remain in the same location as before the tsunami, but even in undamaged locations 20% of respondents have relocated by the first follow up (in heavily damaged areas fully 76% of respondents are elsewhere). Second, substantial proportions of respondents relocate to private homes. Even among those from moderately and heavily damaged areas, post-disaster residence at 1 year is dominated by private residences rather than temporary shelter. Third, a lot of movers leave their communities to settle in new places. This is true regardless of level of damage to the community but is particularly dominated by movers from areas without damage.

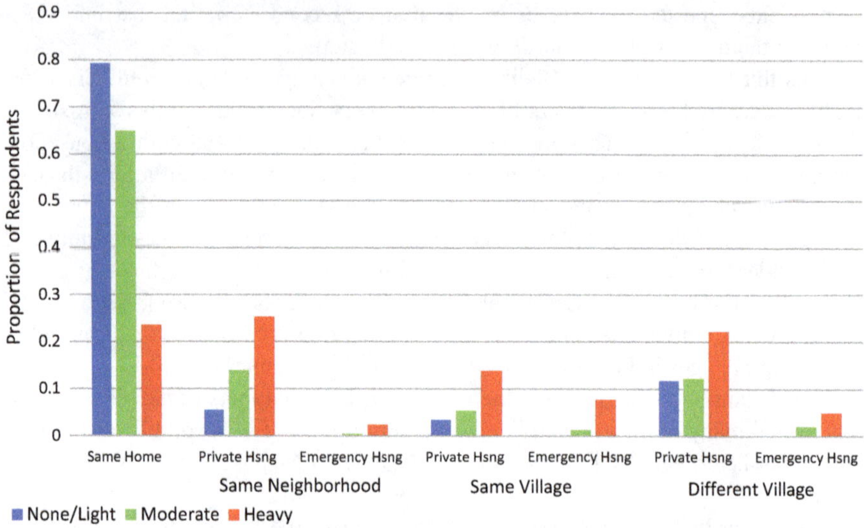

Fig. 11.1 Respondents' locations one year Post-Tsunami by community damage level

Pre-Tsunami Characteristics and Post-Tsunami Changes in Location and Living Arrangements

Table 11.1 presents summary statistics for educational attainment (in years) and household PCE, both measured before the tsunami. Means are presented for individual respondents, stratified by where they were living at the time of the first post-tsunami follow up (on average 1 year after the tsunami), relative to their pre-tsunami location.

Education levels are lowest, at just under 8 years, for those who, 1 year after the tsunami, are still located at the site of their pre-tsunami residence. Individuals in different locations all have on average about a half a year more education. The sizes of the gaps, which range from 0.539 to 0.642, are presented in column 2. These differences are all statistically significant, indicating that movers in the aftermath of the disaster differ from stayers on educational attainment, a key component of human capital.

Means of lnPCE are reported in column 3 of Table 11.1. The gaps in column 4 can be interpreted as approximately percentage differences. Relative to individuals who remain in their pre-tsunami location, the PCE of those who are in a new location, but within the same neighborhood, is about 14% higher before the tsunami—a gap that is statistically significant. No substantively important or statistically significant differences exist between stayers and those who are found outside of their original neighborhood.

Table 11.1 Pre-tsunami characteristics and subsequent residential mobility

	Years of education		lnPCE	
	Means	Gaps	Means	Gaps
	[1]	[2]	[3]	[4]
Pre-tsunami location	7.97		2.82	
	[0.12]		[0.02]	
Same neighborhood	8.59	0.62	2.95	0.14
	[0.23]	[0.22]	[0.04]	[0.04]
Same desa	8.51	0.54	2.84	0.03
	[0.19]	[0.20]	[0.03]	[0.03]
Different desa	8.61	0.64	2.83	0.01
	[0.20]	[0.20]	[0.03]	[0.03]
No. respondents	16,342		16,342	
R^2	0.825	0.006	0.964	0.007
p: F(all locations)	0.000	0.001	0.000	0.013

Years of education and ln(per capita expenditure) measured pre-tsunami
Location measured in year after tsunami relative to location pre-tsunami
Notes: Standard errors in parentheses below means and gaps take into clustering at level of baseline enumeration area. p: F(…) is p-value of F test for joint significance of all locations in year after tsunami

The results of this table provide initial evidence that individuals who move from their pre-event place of residence in the aftermath of a major change differ from those who do not move. These results illustrate the importance of tracking movers in the aftermath of an event, rather than simply interviewing those who remain in their original location.

In Table 11.2 we again consider years of education and economic resources measured before the tsunami, but we stratify by level of tsunami damage in the community. For each level of damage, evidence emerges that respondents' educational levels and household spending levels differ by individuals' post-tsunami locations relative to their pre-tsunami locations.

In areas without direct tsunami damage, for example, those who relocated within the same village but outside of their original neighborhood have about a year and a half more education than their counterparts who remain in the same location. These individuals also have higher levels of per capita spending. In areas of moderate damage there differences between those who remain in their pre-tsunami location and those who move, but for both measures the advantages appear for those who shift residences within the same neighborhood.

The results differ for those from areas that were heavily damaged by the tsunami. In these areas levels of education and per capita spending are significantly lower for those who relocate within their village than for those who stay in the same place.

These results confirm that not only do movers differ from stayers on important dimensions of human capital and economic resources, but the direction of the differences varies by extent of damage from the disaster.

Table 11.2 Tsunami exposure, pre-tsunami education and lnPCE, and residence in year after tsunami

	Years of education			lnPCE		
Extent of damage:	No direct [1]	Moderate [2]	Heavy [3]	No direct [1]	Moderate [2]	Heavy [3]
A. Location of residence in year after tsunami						
Pre-tsunami location	8.03	7.71	9.36	2.77	2.79	3.08
	[0.19]	[0.15]	[0.36]	[0.04]	[0.03]	[0.07]
Relative to pre-tsunami location						
Same neighborhood	−0.55	0.65	−0.11	0.05	0.13	−0.05
	[0.52]	[0.30]	[0.41]	[0.07]	[0.06]	[0.08]
Same desa	1.53	0.54	−0.80	0.18	−0.01	−0.20
	[0.41]	[0.30]	[0.40]	[0.08]	[0.05]	[0.07]
Different desa	0.48	0.40	0.14	−0.08	−0.07	0.00
	[0.42]	[0.25]	[0.41]	[0.04]	[0.04]	[0.07]
B. Location and type of residence in year after tsunami						
Relative to pre-tsunami location						
Same neighborhood - emergency		−0.35	−1.34		−0.18	−0.47
		[1.02]	[0.59]		[0.21]	[0.15]
Same neighborhood - private		0.67	0.01		0.14	−0.01
		[0.29]	[0.41]		[0.05]	[0.08]
Same desa – Emergency		−0.80	−1.20		−0.10	−0.22
		[0.69]	[0.53]		[0.11]	[0.08]
Same desa - private		0.85	−0.58		0.02	−0.20
		[0.28]	[0.39]		[0.05]	[0.08]
Diffferent desa - emergency		−0.51	−1.73		−0.22	−0.17
		[0.42]	[0.72]		[0.09]	[0.09]
Different desa – Private		0.56	0.56		−0.05	0.04
		[0.25]	[0.40]		[0.03]	[0.08]
No. respondents	3660	9508	3174	3660	9508	3174
p: F(location)	0.002	0.066	0.071	0.017	0.010	0.003
p: F(location+type res)		0.001	0.002		0.00705	0.000415

Notes: Standard errors in parentheses below coefficient estimates take into clustering at level of baseline enumeration area. p: F(…) is p-value of F test for joint significance of covariates in (…)

The bottom panel examines patterns by both location of post-tsunami residence and type of housing that individuals were living in at the first follow up interview. These results are presented only for individuals who originated in areas that sustained moderate or heavy tsunami damage because the small number of individuals from areas without direct damage precludes analysis.

In this panel we see that individuals interviewed in emergency housing are systematically less educated and from households with fewer resources before the tsunami. This is true for those from moderately damaged and for those from heavily damaged areas. In other words, one year after the tsunami, the individuals who are in emergency shelters were poorer and are more poorly educated than individuals who were not displaced. In areas of heavy damage the educational disparity is over one year for those in emergency housing relative to those who remained in their origin location. With respect to education, for those from moderately damaged areas, movers who are living in private homes are significantly better educated than those who remain in the same location. In areas of both moderate and heavy damage, the pattern of relative disadvantage (pre-tsunami) for those in emergency housing is present with respect to per capita spending levels but the differences are less consistently statistically significant.

These results further enrich the narrative. It is not just where one moves that varies by important pre-tsunami characteristics. In addition the type of housing that movers find themselves living within varies as well. To capture a full picture of how individuals fare after a disaster, tracking movers across the myriad types of locations and living arrangements is important, because pre-event characteristics are associated with what happens after the event.

Post-Tsunami Locations and the Evolution of Economic Resources

We turn now to the ladder question to investigate how one's individual perception of economic circumstances evolves after a disaster, and how this varies by living situation in the first year after the event (Table 11.3).

In areas without tsunami damage relocating out of one's original neighborhood in the year after the tsunami is associated with significantly higher per capita spending than remaining in place. This same relationship holds for individuals from moderately and heavily damaged areas, where relocation to a private home (but not to emergency housing) is associated with higher per capita spending. Two years after the tsunami, those who have relocated beyond their original neighborhood (but not those who relocated within it) have significantly higher spending than those who remained in place. By 5 years after the tsunami this relationship has weakened, regardless of damage zone.

Although residence in emergency housing 1 year after the tsunami was significantly more likely for those from households with fewer resources before the tsunami, living in emergency housing is not associated with lower spending levels relative to those of non-movers, in any of the post-tsunami waves.

As mentioned above, our per capita spending measure reflects household-level economic resources, the allocation of which may vary across households. The question on perception of socioeconomic status, as represented by steps on a ladder, is a measure of economic resources that is specific to individuals rather than to households. In Table 11.4 we present results from assessing the relationship of this ladder measure with location of residence 1 year after the tsunami.

Table 11.3 Tsunami exposure, lnPCE and location and type of housing 1, 2 and 5 years after tsunami

	1 year after			2 years after			5 years after		
Extent of damage:	No direct [1]	Moderate [2]	Heavy [3]	No direct [4]	Moderate [5]	Heavy [6]	No direct [7]	Moderate [8]	Heavy [9]
Same neighborhood - emergency		−0.20	0.15		−0.08	−0.03		−0.22	−0.05
		[0.29]	[0.12]		[0.23]	[0.13]		[0.28]	[0.16]
Same neighborhood - private	0.06	0.18	0.24	−0.16	0.09	0.18	−0.05	0.04	0.06
	[0.08]	[0.07]	[0.09]	[0.08]	[0.07]	[0.10]	[0.04]	[0.06]	[0.08]
Same desa - emergency		−0.07	0.10		−0.09	0.06		0.04	0.06
		[0.11]	[0.11]		[0.07]	[0.12]		[0.10]	[0.09]
Same desa - private	0.24	0.11	0.17	0.24	0.11	0.20	0.16	0.03	0.08
	[0.10]	[0.07]	[0.09]	[0.09]	[0.05]	[0.09]	[0.09]	[0.07]	[0.08]
Diffferent desa - emergency		0.13	0.23		0.07	0.16		0.05	0.03
		[0.09]	[0.14]		[0.08]	[0.12]		[0.06]	[0.09]
Different desa - private	0.28	0.42	0.59	0.20	0.24	0.47	0.08	0.12	0.19
	[0.13]	[0.05]	[0.09]	[0.11]	[0.05]	[0.10]	[0.07]	[0.04]	[0.08]
Reference (pre-tsunami	3.28	3.22	3.50	3.47	3.48	3.80	3.93	3.94	4.14
Location)	[0.04]	[0.03]	[0.08]	[0.04]	[0.03]	[0.09]	[0.03]	[0.02]	[0.08]
No. respondents	3487	8995	2879	3464	8652	2824	3537	8964	2933
p: F (All location/ residence)	0.024	0.000	0.000	0.003	0.000	0.000	0.025	0.036	0.078

Notes: Standard errors in parentheses below coefficient estimates take into clustering at level of baseline enumeration area. p: F(…) is p-value of F test for joint significance of all locations and residence types relative to pre-tsunami location

In the year after the tsunami, those from heavily damaged areas who are living in emergency housing position themselves on a substantially and significantly lower step than do individuals who remain in the same residence. Regardless of location relative to their pre-tsunami home, those in emergency housing report being half a step lower on the ladder. Also the relationship is strong in year 1, it is not detectable in year 2 or in year 5.

Table 11.4 Tsunami exposure, step on well-being ladder and location and type of housing 1, 2 and 5 years after tsunami

	1 year after			2 years after			5 years after		
Extent of damage:	No direct [1]	Moderate [2]	Heavy [3]	No direct [4]	Moderate [5]	Heavy [6]	No direct [7]	Moderate [8]	Heavy [9]
Same neighborhood – Emergency		0.01	−0.45		−0.08	−0.11		−0.23	−0.10
		[0.20]	[0.17]		[0.11]	[0.11]		[0.17]	[0.10]
Same neighborhood - private	0.07	0.15	0.14	−0.24	0.06	0.17	−0.08	0.14	0.13
	[0.08]	[0.06]	[0.08]	[0.12]	[0.05]	[0.06]	[0.09]	[0.05]	[0.07]
Same desa - emergency		−0.09	−0.39		−0.14	0.05		0.05	−0.01
		[0.12]	[0.11]		[0.08]	[0.09]		[0.08]	[0.10]
Same desa – Private	0.09	−0.01	−0.22	−0.09	0.01	0.07	0.16	0.03	−0.03
	[0.09]	[0.06]	[0.10]	[0.12]	[0.05]	[0.07]	[0.08]	[0.05]	[0.08]
Diffferent desa - emergency		−0.13	−0.50		−0.01	0.01		0.32	−0.02
		[0.08]	[0.11]		[0.05]	[0.09]		[0.07]	[0.09]
Different desa - private	−0.09	0.07	−0.05	0.10	0.08	0.10	−0.04	0.12	0.08
	[0.08]	[0.04]	[0.10]	[0.07]	[0.03]	[0.08]	[0.09]	[0.04]	[0.07]
Reference (pre-tsunami Location)	2.54	2.43	2.62	2.67	2.66	2.74	2.94	2.86	2.99
	[0.04]	[0.03]	[0.08]	[0.04]	[0.02]	[0.06]	[0.03]	[0.02]	[0.06]
No. respondents	3001	7521	2461	2733	6898	2212	3033	7665	2492
p: F (all location/ residence)	0.386	0.014	0.000	0.078	0.051	0.024	0.204	0.000	0.136

Notes: Standard errors in parentheses below coefficient estimates take into clustering at level of baseline enumeration area. p: F(…) is p-value of F test for joint significance of all locations and residence types relative to pre-tsunami location

For those from moderately damaged areas residence in emergency housing does not have the same dramatic relationship with perception of ladder step. In these areas, there are no statistically significant differences between those in emergency housing and those who did not move in either in the first year or second year after the disaster. By year five, those who were in emergency housing in a different village report an improvement in status of about third of a step, whereas those in emergency housing within the same community at year 1 report about a quarter of a step drop in status—a strong difference in outcomes.

For those who moved but to private homes, differences in step relative to those in the same location tend to be positive, but small and not statistically significant. One exception to this pattern is that those from heavily damaged areas who moved to private homes in the same village report a significantly lower ladder position than those who remained in the same location.

Among those from undamaged areas some movers report a higher step position than stayers, whereas some report a lower step position. No clear pattern emerges for those from areas that were undamaged in the tsunami.

Although the focus of this paper is on the methodological importance of harnessing baseline data and following up respondent who move, we note in passing that although first post-tsunami locations vary widely, these locations do not dictate degree of recovery in the subsequent period. We have documented considerable resilience even after such a devastating disaster but, for some population subgroups the effects are long-lasting, particularly for economic and health-related indicators of well-being (Frankenberg et al. 2016; Frankenberg et al. 2017; Frankenberg et al. 2019; Lawton et al. 2021; Lawton et al. 2021; Thomas et al. 2018).

Discussion and Conclusions

As extreme events increase in frequency and severity, there is a need for data on how people fare during and recover from these threats, based on representative samples of the population at risk. Because constructing such a sample can be a complicated endeavor, much of the work on the impacts of disasters relies on convenience samples of one form or another, such as people remaining in their pre-event residence rather than moving, or people who have relocated to official shelters or camps for the displaced.

Our results provide scientific evidence of the importance of samples that represent the population before the event, rather than studying only individuals present after an event. We have shown that pre-event characteristics condition where and in what circumstances people live after an event, that post-event well-being is associated with post-event living conditions in the short-term, and that over time, the link weakens between short-term living arrangements and post-event well-being. Were one to interpret the results for a subset of those we interview as the outcomes for the entire population, one would misrepresent population outcomes both at points in time as well as with respect to the evolution of outcomes over time.

The STAR project uses a pre-tsunami survey as a baseline, which was possible both because Statistics Indonesia regularly conducts large high quality cross-sectional surveys and because of their collaboration in the aftermath of the devastation of the tsunami. Other studies have repurposed survey or census data in order to construct a baseline (see for example the RISK project in New Orleans), relied on records from databases of mailing addresses or phone numbers, or used pre-event satellite imagery to build a frame of structures from which a sample can be drawn.

As environmental pressures mount, designs for data collection that produce unbiased estimates of the impact and evolution of population well-being are imperative. These designs may draw on tools social scientists have used for decades, supplemented with novel methods harnessing new technologies or new administrative data streams. Investments in establishing observatories where data collection can occur regularly and at relatively high frequency may have important pay-offs with respect to developing a better understanding of the impacts of climate change.

Acknowledgements Support from the National Institute for Child Health and Human Development (R01HD052762, R01HD051970, R03HD071131, P2C HD050924), the National Institute on Aging (R01AG031266), the National Science Foundation (CMS-0527763), the Hewlett Foundation, the World Bank, and the MacArthur Foundation (05-85158-000) is gratefully acknowledged.

References

Borrero J (2005) Field data and satellite imagery of tsunami effects in Banda Aceh. Science 308:1596

Deaton A (2016) Measuring and understanding behavior, welfare, and poverty. Am Econ Rev 106(6):1221–1243

Frankenberg E, Smith JP, Thomas D (2003) Evidence from the Indonesia family life survey. J Hum Resour 38(2):280–321

Frankenberg E, Friedman J, Gillespie T, Ingwersen N, Pynoos R, Rifai IU, Sikoki B, Sumantri C, Suriastini W, Thomas D (2008) Mental health in Sumatra after the tsunami. Am J Public Health 98(9):1671–1677

Frankenberg E, Gillespie T, Preston S, Sikoki B, Thomas D (2011) Mortality, the family, and the Indian Ocean tsunami. Econ J 121:F162–F182

Frankenberg E, Sumantri C, Thomas D (2016) The evolution of well-being in the aftermath of a disaster: evidence from Aceh and North Sumatra. In: James H, Paton D (eds) The consequences of disasters: demographic, planning and policy implications. C. C. Thomas

Frankenberg E, Friedman J, Ingwersen N, Thomas D (2017) Linear child growth after a natural disaster: a longitudinal study of the effects of the 2014 Indian Ocean tsunami. Lancet 389:S21. https://doi.org/10.1016/S0140-6736(17)31133-9

Frankenberg E, Lawton R, Seeman T, Sumantri C, Thomas D (2019) The causal impact of stress on inflammation over the long-term. Manuscript

Gillespie T, Frankenberg E, Chum KF, Thomas D (2014) Nighttime lights time series of tsunami damage, recovery, and economic metrics in Sumatra, Indonesia. Rem Sens Lett 5:286–294

Gray C, Frankenberg E, Sumantri C, Thomas D (2014) Studying displacement after a disaster using large scale survey methods: Sumatra after the 2004 tsunami. Ann Assoc Am Geogr 104(3):594–612

Kohl P, O'Rouke A, Schmidman D, Dopkin W, Birnbaum M (2005) The Sumatra-Andaman earthquake and tsunami of 2004: the hazards, events, and damage. Prehosp Disaster Med 20(6):355–363

Laurito M, Frankenberg E, Thomas D (2022) Effects of housing aid on psychosocial health after a disaster. Int J Environ Res Public Health 19(12):7302

Lawton R, Frankenberg E, Sumantri C, Seeman T, Crimmins E, Thomas D (2021) Cortisol levels 15 years after a natural disaster. Manuscript

Lawton, R., Frankenberg, E., Friedman, J. & Thomas, D. (2022). Long-term economic Well-being after the 2004 Indian Ocean tsunami. Manuscript

Monecke K, Finger W, Klarer D, Kongko W, McAdoo B, Moore A, Sudrajat S (2008) A 1,000-year sediment record of tsunami recurrence in northern Sumatra. Nature 453:1232–1234

Nicol W (2013) The tsunami chronicles: adventures in disaster management, 1st edn. CreateSpace Independent Publishing Platform

Ramakrishnan D, Ghosh S, Raja V, Chandran R, Jeyram A (2005) Trails of the killer tsunami: a preliminary assessment using satellite remote sensing technique. Curr Sci 88(5):709–711

Thomas D, Frankenberg E, Seeman T, Sumantri C (2018) Effect of stress on cardiometabolic health 12 years after the Indian Ocean tsunami: a quasi-experimental longitudinal study. Lancet Planet Health 2:S8

Umitsu M, Tanavud C, Patanakanog B (2007) Effects of landforms on tsunami flow in the plains of Banda Aceh, Indonesia, and Nam Khem, Thailand. Mar Geol 242:141–153

Elizabeth Frankenberg is the Cary C. Boshamer Distinguished Professor of Sociology and Public Policy and Former Director of the Carolina Population Center at the University of North Carolina at Chapel Hill. Frankenberg's research focuses on three thematic areas: the ways in which the health and social service shape the environmental well-being of individuals, the ways that interactions among family members influence well-being, and how individuals respond to changes induced by unexpected events.

Cecep Sumantri is the Director of Capacity Building at SurveyMETER, a research organization based in Yogyakarta, Indonesia. Sumantri's research focuses on human capital outcomes in Indonesia. He has directed many large-scale longitudinal survey projects.

Duncan Thomas is Norb F. Schaefer Distinguished Professor of International Studies and a Professor of Economics at Duke University. Thomas's research investigates the interrelationships between health, human capital and socio-economic status with a focus on the roles that individual, family and community factors play in improving levels of health and well-being across the globe.

Part V
Island Ecosystems: Terrestrial Sub-systems

Chapter 12
Unravelling the Interactions Between Endemic and Invasive Plant Species in the Galapagos Islands

María de Lourdes Torres, Diego Urquía, Leonie Moyle, Matt Gibson, Todd Vision, and Bryan Reatini

Introduction

Invasive species are currently considered one of the major threats for global biodiversity, especially in isolated ecosystems such as oceanic archipelagos (Primack 2014).

Currently, the number of introduced plant species in the Galapagos Islands exceeds that of native and endemic species combined. While the last two do not total more than 800 (Galapagos Conservancy 2022), 879 introduced plants were reported as of 2010 in the archipelago (Brewington 2011). Of these, 37 are considered highly aggressive invasive species, including guava (*Psidium guajava*), cascarilla (*Cinchona pubescens*), tupirrosa (*Lantana camara*), blackberry (*Rubus*

M. de L. Torres (✉)
College of Biological & Environmental Studies and USFQ-UNC Galapagos Science Center, Universidad San Francisco de Quito, Quito, Ecuador
e-mail: ltorres@usfq.edu.ec

D. Urquía
College of Biological & Environmental Studies and USFQ-UNC Galapagos Science Center, Universidad San Francisco de Quito, Quito, Ecuador

L. Moyle · M. Gibson
Department of Biology, Indiana University, Bloomington, IN, USA

T. Vision
Galapagos Science Center, USFQ-UNC, San Cristóbal, Galápagos, Ecuador

Department of Biology, University of North Carolina, Chapel Hill, NC, USA

B. Reatini
Department of Biology, University of North Carolina, Chapel Hill, NC, USA

S. J. Walsh et al. (eds.), *Island Ecosystems*, Social and Ecological Interactions in the Galapagos Islands, https://doi.org/10.1007/978-3-031-28089-4_12

niveus), elephant grass (*Pennisetum purpureum*), and Cuban cedar (*Cedrela odorata*) (Tye et al. 2007). These species began to disperse and spread uncontrollably, especially on the populated islands —Santa Cruz, Isabela, San Cristóbal and Floreana (Whittaker 2007; Tye et al. 2007). Here, invasive species have been displacing several endemic plant species including native ferns, plants of the genus *Scalesia*, *Miconia robinsoniana* and *Psidium galapageium* (Galapagos National Park Directory 2007). Thus, the presence of invasive plant species in the Galapagos has been linked to a reduction in the richness and abundance of native/endemic species, and has prevented the regeneration of the original vegetation in disturbed areas (Weber 2003).

The impacts of invasive plants can be direct, as is the case when they compete with and displace native species in the face of limited resources and space; they can also bear pathogens and parasites to which native species usually have little or no resistance (Whittaker 2007). Invasive species can also have an indirect impact on local biota by altering the physical characteristics of the habitat and biological communities, and by cascading alterations in the trophic interactions of the invaded ecosystem (Sakai et al. 2001; Jäger et al. 2007). Of particular concern are those invasive species that are also ecosystem engineers, since the introduction of one or a couple of specimens is sufficient to have a large-scale alteration in the invaded ecosystem (Rilov et al. 2012). Thus, invasive species, especially invasive plants, may alter ecological processes and disrupt ecosystem services (Tye 2001).

Invasive species can also affect native biota at the genetic and evolutionary level, especially when the invasive species alters the physical environment of the habitat it arrives in (Crooks 2002; Jäger et al. 2007). The worst of these cases occurs when there is hybridization between closely related introduced and native/endemic species (Sakai et al. 2001). Here, the latter are in imminent danger of rapid extinction via genetic erosion, outbreeding depression, and/or genetic swamping (López-Caamal et al. 2014; Ellstrand and Rieseberg 2016; Chafin et al. 2019). In addition, through introgression, the native species can pass genes to the introduced species, increasing the adaptability of the latter to the invaded ecosystem (Huxel 1999; Ellstrand and Schierenbeck 2006). Hybridization between native/endemic and introduced species has already been observed in several cases, being more common in insular plants (Ellstrand and Schierenbeck 2006; López-Caamal et al. 2014). For this phenomenon to occur, it is required that introduced species coexist with a closely related native or endemic species in sympatry; moreover, these should have similar flower morphologies, and to share common pollinators (Chamorro et al. 2012; Abdallah et al. 2021). In Galapagos, there are some introduced-endemic species pairs meeting these requirements that may potentially lead to a hybridization scenario. These include *P. guajava* (guava-introduced) and *P. galapageium* (guayabillo-endemic), *L. camara* (tupirrosa-introduced) and *L. peduncularis* (endemic), *Gossypium barbadense* (cotton-introduced) and *G. darwinii* (cotton-endemic), and even a more complex case involving four tomato species —*Solanum lycopersicum* and *S. pimpinellifolium* (both introduced), and *S. cheesmaniae* and *S. galapagense* (endemic).

In this chapter we discuss two of these cases in more detail —the complex tomato, and the guava-guayabillo cases— to explore how invasive and endemic plant species might be interacting in the Galapagos Islands, from the genetic to the ecological levels.

For both examples we present highlights of the research we have conducted using modern tools of population genetics and genomics, combined with observational and experimental fieldwork, and the review of historical sources. Although many cases remain to be studied yet, the two presented here give us an initial insight on the magnitude of the threats invasive plants pose over endemic species, including potential hybridization.

The Complex Tomato Case

Wild tomatoes are one case in which we have combined field work with population genomics to examine the complex outcomes of interactions between Galapagos endemic and invasive species. The wild tomato group (Solanum section Lycopersicum) includes two Galapagos endemic species—*S. cheesmaniae* and *S. galapagense*—that differ on the basis of habitat, leaf, and fruit characters (Darwin et al. 2003). Both species are unique among the wild tomatoes in having orange or yellow fruits when ripe (Darwin et al. 2003). Genetic data indicates that these two species are descended from a single ancestral lineage on the Galapagos (e.g., they share unique variants not present in any other wild tomato species; Pease et al. 2016; Pailles et al. 2017; Gibson et al. 2021). From genomic data, we estimate the timing of dispersal to the Galapagos to be ~770–1850 generations ago (Gibson et al. 2021). Therefore, these species are likely younger than the archipelago itself.

Two close relatives of these endemic species have been recently introduced to the Galapagos: *Solanum lycopersicum* L. (domesticated tomato) and *Solanum pimpinellifolium* L. (a wild species). Both species appear to have been introduced after human colonization (Darwin et al. 2003; Darwin 2009). Domesticated tomato is still primarily known from cultivated (farm and garden) sites on the islands, unlike *S. pimpinellifolium* whose recent demography is consistent with invasion (Darwin et al. 2003; Nuez et al. 2004; Gibson et al. 2020). *S. pimpinellifolium* is likely the closest wild relative to the two Galapagos endemic species (Pease et al. 2016); its native range is coastal to mid-elevation Andean sites in mainland Ecuador and Peru (Gibson and Moyle 2020). Confirmed *S. pimpinellifolium* records appear in Galapagos collections dating back to at least 1985 (Darwin et al. 2003), however field observations from the last 30 years have documented a rapid increase in the abundance of this species on human-inhabited islands (Darwin et al. 2003; Nuez et al. 2004; Gibson et al. 2020). For example, over the period 2018–2021 we identified 18 previously undescribed *S. pimpinellifolium* sites (Gibson et al. 2020, 2021, unpubl. data), mostly on San Cristobal and Santa Cruz. Causes for this rapid demographic expansion remain speculative, but likely include spread by humans and by non-human animal dispersers. While some of these invasive populations are close to human habitation, approximately half are more distant (>1 km) and located in arid/transitional sites that partially overlap with the range of endemic *S. cheesmaniae* (Darwin 2009; Nuez et al. 2004; Gibson et al. 2020). In contrast to introduced *S. pimpinellifolium*, populations of the two endemic tomato species appear to be experiencing a demographic decline. For instance, of 24 sites previously documented to have endemic populations on San Cristobel, Santa Cruz, or Isabela, we found that only four sites still had endemic individuals in the period 2018–2021 (Gibson et al. 2020, unpubl. data). Endemic populations on uninhabited islands might be more persistent over time (unpubl. data), although a more systematic assessment of these remote sites is needed.

With >5700 SNPs generated from reduced representation (ddRADSeq) genotyping of >300 mainland and Galapagos collections, we have used population genomics to identify the number and timing of introductions of *S. pimpinellifolium* onto the Galapagos, and their possible consequences for interactions with endemic tomato species (Gibson et al. 2021). We find evidence for at least three separate introductions of *S. pimpinellifolium* onto the Galapagos: one major event from central Ecuador, that is responsible for nearly all sampled invasive populations, as well as two minor events from Peru and Ecuador, each resulting in a single historical or extant location (Gibson et al. 2021) (Fig. 12.1). Our point estimate for the major introduction is 200 generations ago, consistent with human-mediated dispersal; the other two introductions are likely even more recent, based on their close genetic similarity to mainland *S. pimpinellifolium* collections (Gibson et al. 2021).

Our analysis also confirms that hybridization and introgression is occurring among introduced and endemic species, especially between *S. pimpinellifolium* and endemic *S. cheesmaniae*. These two species can now be found within meters of each other at several locations within the arid/transitional zone on Santa Cruz Island. We and others have documented phenotypically mixed individuals at these sites (Darwin 2009; Gibson et al. 2020, unpubl. data), and our genomic data confirms that these intermediate individuals are early generation (F1 and F2) hybrids (Gibson et al. 2021). Our analyses also detects evidence for older gene flow between *S. cheesmaniae* and *S. pimpinellifolium*, including genomic signatures of hybridization that are shared across invasive populations collected from both Santa Cruz and San Cristobal; this gene flow presumably pre-dates the dispersal and differentiation between invasive populations on these two islands (Gibson et al. 2021) (Fig. 12.2). In most cases, the recipient *S. pimpinellifolium* populations show no obvious phenotypic evidence of this hybrid ancestry, but there is one important exception. Since 2018, we have identified several *S. pimpinellifolium* populations on Santa Cruz that have a novel fruit color polymorphism, where up to 40% of individuals have orange instead of their usual red fruits (Gibson et al. 2020). Ancestry assignment across the genomes of individuals from two of these polymorphic populations indicates that the orange fruit phenotype is uniquely associated with *S. cheesmaniae* ancestry at two genes known to be involved in carotenoid biosynthesis (Gibson et al. 2021) (Fig. 12.2). In one population, orange fruits are associated with the *S. cheesmaniae* allele for *CYC-B*—a lycopene beta cyclase previously shown to underlie orange fruit color in the Galapagos endemic species (Rick 1956). In the second polymorphic population,

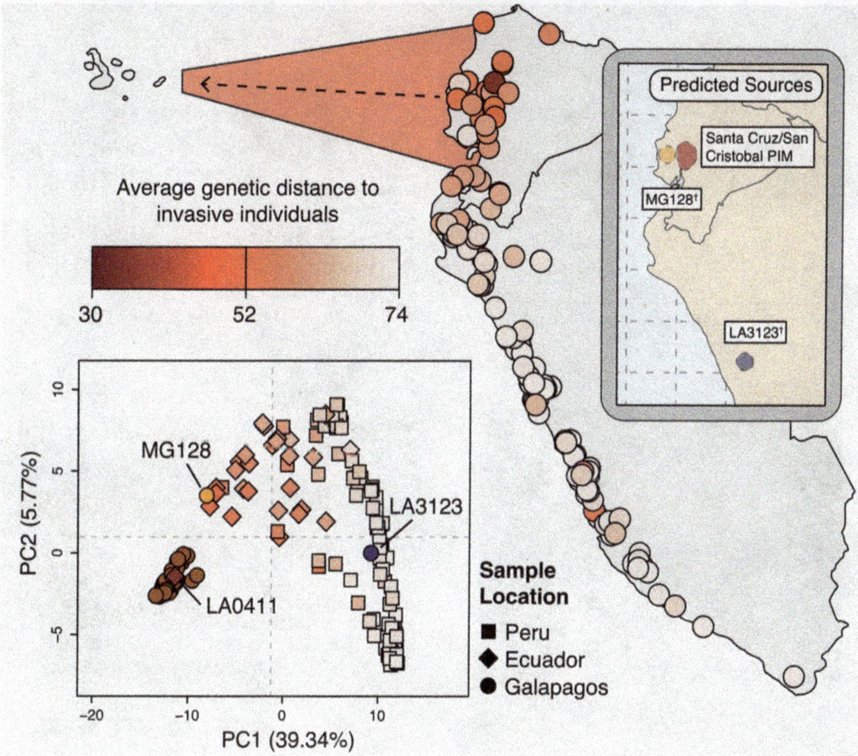

Fig. 12.1 Genomic data indicate that Galapagos *S. pimpinellifolium* (PIM) is the result of a recent invasion from Ecuador. The source of >95% of Galapagos PIM is central Ecuador, with two minor introductions from elsewhere. The map shows the average genetic distance between Galapagos PIM collections and each of the 132 mainland accessions. Inset left Plot: A multi-locus principal components analysis (PCA) of these *S. pimpinellifolium* samples. Squares, diamonds, and circles indicate Peruvian, Ecuadorian, and Galapagos collections, respectively. Inset right plot: Predicted continental origins for Galapagos PIM collections using ML inference. Colors match those shown in the multi-locus PCA. (Results from a single run are shown). (Figure reproduced from Gibson et al. 2021)

orange fruit color is associated with endemic ancestry at a different carotenoid locus—*PSY1* (Phytoene synthase 1). The amount (3.6–4.4%) and distribution of *S. cheesmaniae* ancestry in both cases is consistent with gene flow in the last 4–12 generations (Gibson et al. 2021). Together, our data indicate that at least two separate loci underlie the novel origin of orange fruit color in invasive *S. pimpinellifolium* — both of which appear to be derived via introgression from *S. cheesmaniae* in the recent past.

While most of our evidence for admixture indicates introgression from endemic into invasive populations, we also detect gene flow in the reciprocal direction —from *S. pimpinellifolium* into *S. cheesemaniae*— near their contact sites on Santa Cruz. In addition, our genomic data confirm hybridization between the two Galapagos endemic species at a contact site described on Isabela since 2000 (Darwin 2009;

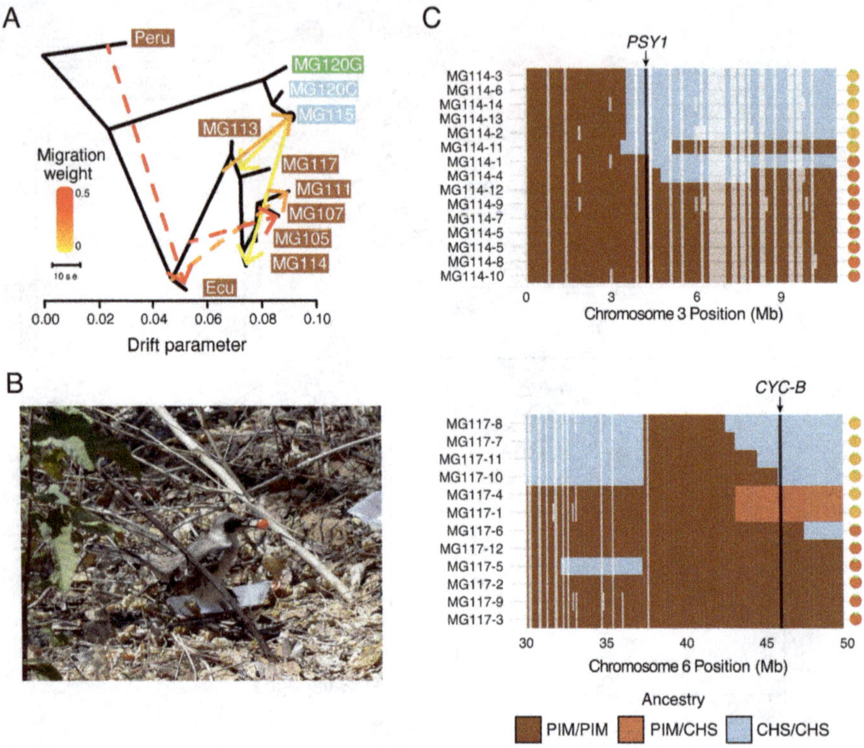

Fig. 12.2 Patterns of population genetic structure and admixture between *S. pimpinellifolium* and endemic Galapagos tomatoes: A. Treemix analysis summary (m = 6; ln[L] = 395.08) of inferred admixture events in the history of collected populations of *S. pimpinellifolium* (brown), *S. cheesmaniae* (blue), and *S. galapagense* (green). Solid lines indicate inferred interspecific admixture events and dashed lines indicate intraspecific events. B. Galapagos mockingbird at an admixed site on Santa Cruz, consuming a red fruit of *S. pimpinellifolium*. C/D. Patterns of local ancestry across focal chromosome regions in two polymorphic populations of *S. pimpinellifolium* (MG114 and MG117). Rows indicate individual genotypes, colored by inferred ancestry at each chromosomal position. Fruit color for each individual is shown at far right. C. CHS ancestry at carotenoid biosynthesis gene PSY1 on chromosome 3 correlates with observed fruit color variation (orange versus red) in MG114. D. CHS ancestry at carotenoid biosynthesis gene CYC-B on chromosome 6 correlates with fruit color variation (orange versus red) in MG117. (Panels A, C, and D reproduced from Gibson et al. 2021)

Gibson et al. 2020); morphologically intermediate individuals were identified as either backcross or F2 generation hybrids (Gibson et al. 2021). Other less resolved cases of hybridization involve possible gene exchange between one or both endemic species and domesticated tomato *S. lycopersicum*, also on Isabela (Gibson et al. 2021).

Together our data clearly show that recent demographic expansion has brought invasive populations into contact with endemics, resulting in ongoing hybridization and gene flow, and the evolutionary origin—via introgression of endemic alleles—of at least one novel trait in some invasive populations. The frequency and outcome of this contact is likely to be amplified by reproductive and ecological similarities between native and introduced species. All four species are very closely related (<0.5MYA; Pease et al. 2016) and largely crossable in the greenhouse (Rick 1963, and unpubl. data) indicating that intrinsic (post-pollination) reproductive barriers between them are incomplete. All four species have similar yellow buzz-pollinated flowers, and the principal native pollinator—the Galapagos carpenter bee *Xylocopa darwini*—moves directly between them in the field (Darwin 2009; pers. obs.). These observations suggest there are little to no floral isolating barriers. We have also observed both endemic and introduced (*S. pimpinellifolium*) fruits being eaten by native frugivorous birds, including Galapagos mockingbirds (Fig. 12.2). Our current hypothesis is that these native fruit dispersers are contributing to the geographical expansion of invasive populations, including dispersal to sites that bring them into sympatry with endemics. Overall, these close evolutionary relationships, low intrinsic barriers, and shared mutualists likely all contribute biologically to increasing the frequency of both demographic and reproductive contact between invasive and endemic species, and the likelihood and persistence of hybridization in these contexts.

The future consequences of this contact for invasive populations remains unclear, including whether hybridization (and the introgression of orange fruit color) might further facilitate the invasion of *S. pimpinellifolium.* The future consequences of invasive-endemic contact is even less resolved for endemic tomato species. If endemics and invasives compete for similar habitats or for mutualists, increasing contact between them threatens the demographic stability of endemic populations. Moreover, although we detect introgression of invasive (*S. pimpinellifolium*) alleles into endemic (*S. cheesmaniae*) genomes (Gibson et al. 2021), we do not yet understand the consequences of this gene flow for genomic integrity of these recipient populations or for functional trait changes that might result.

The Guava-Guayabillo Case

Guava invasion and Human History in the Galapagos

Guava (*Psidium guajava*) is one of the most aggressive invasive plants in the Galapagos (Fig. 12.3a). It is widely distributed in all the four populated islands of the archipelago (Tye et al. 2007). Guava gathers several characteristics that make it an ideal invasive species. Firstly, it is generalist and adaptable, capable of growing in any type of soil and light conditions (Somarriba 1986; Binggeli et al. 1998); second, it can reproduce both sexually and asexually (Loh and Rao 1989; Reichard and Hamilton 1997), letting it spread fast while conserving genetic diversity (Urquía et al. 2019); third, guava has small seeds that are easily dispersed by introduced and native animals (Blake et al. 2012), which can also germinate in a short time (Nava et al. 2014); finally, guava seedlings grow fast, and they can grow at the same rate regardless of the season and climatic conditions (Baker 1974; Rejmanek and Richardson 1996; Nava et al. 2014).

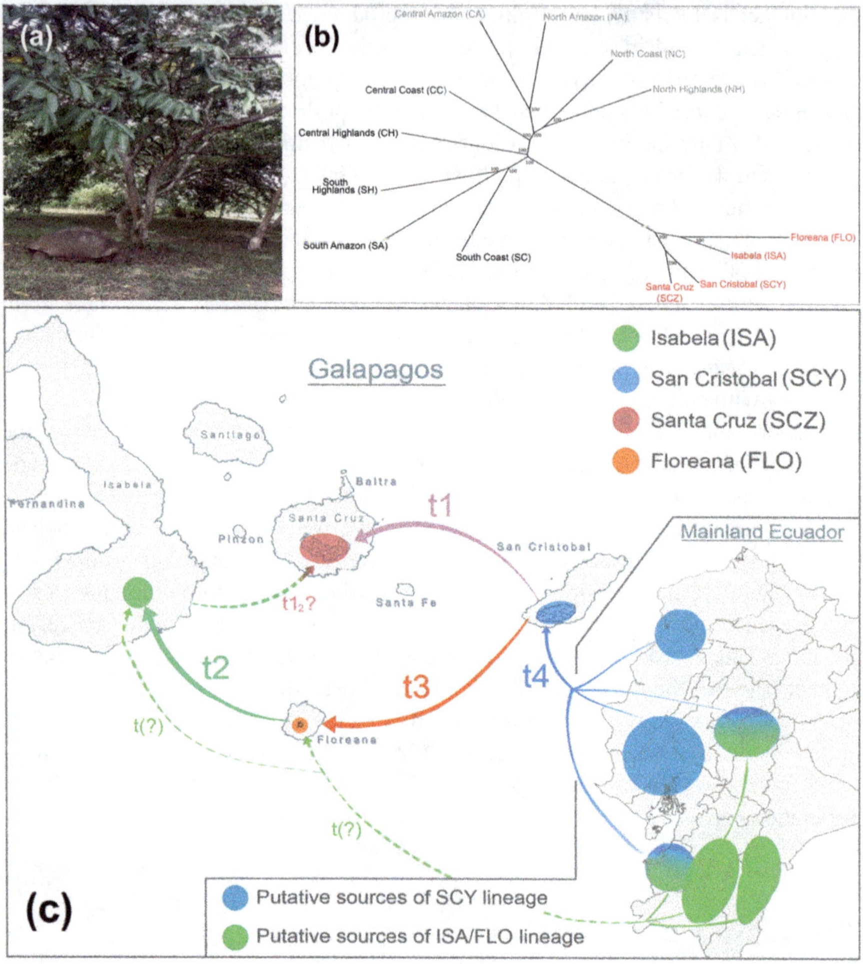

Fig. 12.3 (**a**) A *Psidium guajava* tree in Santa Cruz Island, next to an endemic Galapagos giant tortoise. (**b**) Neighbor-joining (NJ) tree illustrating Cavalli-Sforza & Edwards genetic distances (DCH) among all the mainland Ecuador regions and Galapagos populations of guava; bootstrap values are shown for each node. Galapagos populations are labeled in red, and mainland regions are labeled in a gray scale, with darker levels showing a greater proximity to the Galapagos populations. (**c**) Model proposed for the introduction and colonization history of guava in the Galapagos Islands, as inferred from the ABC analysis: From the mainland (particularly from the CH and coastal regions), an initial introduction would have occurred in San Cristobal Island (t4), from where guava passed then to Floreana Island (t3), and from this one, later to Isabela (t2); finally, from San Cristobal, guava would have passed to Santa Cruz, the last island colonized by this invasive species (t1). An admixture of the Isabela and San Cristobal guavas in Santa Cruz (t1 & t12) was also suggested from genetic data. A second, independent introduction from the mainland (likely from the southern regions) to Floreana and Isabela islands, may contribute to the origin of the Isabela/Floreana lineage as well (t(?)). Note: Mainland Ecuador and Galapagos maps are not at scale in this figure

For finding out the origin of the Galapagos guava, and for reconstructing the colonization pathway of this invasive plant after its arrival in the archipelago, a total of 11 microsatellite (SSR) markers were analyzed on 96 guava samples from mainland Ecuador, and 280 from Galapagos. The Galapagos samples came from all four islands this invasive species is distributed in: San Cristobal, Santa Cruz, Isabela, and Floreana (Urquía et al. 2019; Urquía et al. 2021). Then, genetic distances among Galapagos and mainland guavas were visualized (Fig. 12.3b), and different colonization models were proposed and compared via Approximate Bayesian Computation (ABC) analyses in order to get the most likely colonization scenario (Fig. 12.3c).

The Central-Highlands (CH) region of mainland Ecuador was the most likely origin of the Galapagos guava populations according to genetic distances (Fig. 12.3b). In addition, the guavas of Isabela and Floreana showed possible genetic input from southern mainland Ecuador, while the San Cristóbal population was linked to the coastal regions of the mainland (Urquía et al. 2021). Interestingly, these proposed origins of the Galapagos guava coincided with the origin of the human settlers in the archipelago. The CH region includes the provinces of Tungurahua and the northern portion of Azuay (Fig. 12.1b); in this regard, the Galapagos Islands received numerous immigrants from Tungurahua during their first human settlements, including indigenous people of the Salasaca culture, who have established communities in the archipelago (Grenier 2007; Wogan 2009; Granda and Chóez 2013); northern Azuay includes the areas surrounding the city of Cuenca, which is the birthplace of Manuel J. Cobos; Cobos is one of the most famous settlers of the Galapagos Islands, who established a massive sugarcane plantation on San Cristobal Island —Hacienda El Progreso— and a dye processing company in Floreana (Astudillo 2018; Lundh 2004). The contribution from the Ecuadorian coast to the Galapagos guava, notably from the Guayaquil area, is also reasonable. Guayaquil is home to Ecuador's most important port and serves as the main hub between the mainland and the Galapagos Islands. In the 19th and 20th centuries, almost everything and everyone intending to reach the Galapagos, including emigrants, livestock, and agricultural products, had to stop at the port of Guayaquil. Likewise, several Galapagos settlers and landowners were originally from Guayaquil and the surrounding area; in fact, Cobos brought a significant number of workers and crops from this region to San Cristobal (Latorre 1997; Lundh 2004). The strong link between the Galapagos guavas and those of the mainland coast could also be reinforced by human migration from the province of Manabí (Granda and Chóez 2013; Wogan 2009). Finally, a contribution from southern Ecuador to the populations of Isabela and Floreana is also consistent with the colonization history of Galapagos, since the archipelago received several immigrants from this region (which includes the provinces of Loja and Azuay), historically affected by severe droughts that forced its inhabitants to migrate (Granda and Chóez 2013).

Through an ABC analysis based on the genetic data, a likely colonization model was proposed (Urquía et al. 2021). This model shows that San Cristobal was the first island to be colonized by guava, from where it would have spread to Floreana then, and finally to Santa Cruz; Isabela would have been invaded from Floreana. An independent, parallel introduction —from southern Ecuador— could also have contributed to the invasion of Floreana and Isabela (Fig. 12.3c). The pathway shown is consistent with the history of human colonization of the Galapagos Islands. San Cristobal was not the first island to be colonized by humans in the archipelago, yet the first large-scale, permanent settlements in the Galapagos were founded here. Notably, the earliest physical evidence of guava found in San Cristobal (charred seeds underground) dates from the same time Cobos began raising sugarcane at Hacienda El Progreso (Astudillo 2018). Furthermore, according to Lundh (2004), the earliest report of guava in Galapagos dates from around 1889–1890, with three individuals planted on the Cobos property in San Cristobal. Regarding Floreana, it was the first island in the archipelago to be colonized by humans, in the early nineteenth century (Lundh 2004, 2006). Therefore, as shown in the suggested colonization model, an early invasion of guava in this island is also consistent with the history of the Galapagos. Moreover, during most of the nineteenth century, San Cristobal and Floreana were the only Galapagos islands that hosted permanent human settlements, and sustained activities such as trade and exchange of products and workers between them; therefore, it is very likely that guava arrived in Floreana during this period (Lundh 2004; Urquía et al. 2019). From here, guava was introduced to Isabela, the third island to be formally colonized by humans. A colonist named Antonio Gil may have played a role in this event, as he brought cattle (and perhaps guava) from Floreana to Isabela, where he erected a large farm in 1897 (Latorre 1997; Lundh 2004). Santa Cruz was the last island to be invaded by guava according to the proposed model, coinciding with the fact this island was the last to be colonized by humans, about 100 years ago. Cobos' workers originally living on San Cristobal made their way to Santa Cruz in the late 19th and early 20th centuries, and Cobos himself established several small farms there. There was also constant trade between these two islands shortly after the settlement of Santa Cruz, between the 1920s and 1930s (Lundh 2004); therefore, guava could have arrived from San Cristobal to Santa Cruz at this time. Genetic data also suggested that Santa Cruz guavas had a genetic input from the lineages that settled in Isabela and Floreana, but in a lower extent than the contribution from San Cristobal.

The history of the guava introduction into the Galapagos Islands was either driven by a single event, or by a series of introduction events in rapid succession —from different sources in the mainland. Despite the lower genetic diversity of guava island populations compared to their mainland counterparts (Urquía et al. 2019; Urquía et al. 2021), the contribution of multiple mainland sources to the gene pool of the invasive population would have led it to display high adaptability and success in invading the archipelago (Hughes et al. 2008; Xu et al. 2015). Thus, the importance of genetic research complemented by independent data sources, such as historical records (Astudillo 2018), is reinforced in order to gather key information to explain the characteristics of an invasion process. These characteristics include the adaptability of the invasive species and its success in spreading, as well as the aggressiveness of the invasion and the factors and events driving it. In particular, immigration, colonization, and past human activities may have been important factors that help explain the arrival of guava in the Galapagos Islands and its spread to at least four islands in the archipelago. Such information, in turn, may be a crucial first milestone in the development of a roadmap to address the guava problem in the Galapagos Islands.

Consequences of Secondary contact in Psidium

The Galapagos endemic tree guayabillo (*Psidium galapageium*) is in the same genus as guava. Though it is found on many islands in the archipelago, including the four inhabited ones (McMullen 2018), and is an important component of the native forest, it is classified as Near Threatened in the Red Book of endemic plants of Ecuador (Kawasaki et al. 2017). Threats to guayabillo are thought to include specific ones, such as exploitation of its highly-valued wood, and more generic ones such as habitat loss and herbivory from introduced livestock (Kawasaki et al. 2017; Urquía et al. 2020). In addition, there has long been concern about hybridization between guava and guayabillo (Tye 2001), reinforced by the fact that hybridization is relatively common among *Psidium* species (Urquía et al. 2020; Reatini 2021). Guayabillo can be found at lower elevations than the zone at which guava is most abundant, but the two do co-occur at intermediate elevations on the four inhabited islands (Valdebenito 2018). As discussed above, hybridization between these two species in this zone of secondary contact could potentially introduce traits that would allow guava to spread more aggressively in the archipelago, or lead to the genetic erosion of guayabillo populations, or both.

Prezygotic mating barriers between the two species are weak. Both species flower together for several months of the year and the most common pollinator observed for both species is the ubiquitous Galápagos carpenter bee (*Xylocopa darwinii*, Valdebenito 2018; Reatini 2021). Furthermore, pollen grains from guava can successfully germinate and develop pollen tubes that reach the ovary of guayabillo flowers (Reatini 2021). Suspected hybrid plants were identified by our colleague Hugo Valdebenito from the Universidad San Francisco de Quito, and by Galápagos National Park personnel, at a site on San Cristóbal. This site includes phenotypically stereotypical individuals of both species but also nine individuals of uncertain affinity. Morphological analysis found these putative hybrids to have intermediate leaf traits while tree architectural traits more closely resembled guayabillo (Reatini 2021).

However, molecular genetic analysis using genome-wide markers (Reatini 2021, and Reatini et al. in prep) assigned these individuals to guava, with no detectable contribution from guayabillo. In fact, no individuals with a high probability of hybrid ancestry could be found in a wider microsatellite sample. This sample included 269 guava and 135 guayabillo individuals from San Cristobal, Santa Cruz, and Isabela islands, analyzed using nine of the microsatellite loci described in Urquía et al. (2019). Nor were examples of hybrid ancestry found from a genome-wide sequencing sample of over 100,000 markers. These markers were genotyped in the nine individuals of uncertain affinity, plus 19 guava and 20 guayabillo individuals from sympatry from the same inhabited islands, a guayabillo from uninhabited Santiago, and a guava from mainland Ecuador.

The genetic results are consistent with evidence that, although prezygotic barriers are low, postzygotic barriers to mating are high. Reatini (2021) performed reciprocal crosses between guava and guayabillo and found that interspecific crosses had reduced fruit set in both species relative to intraspecific crosses and self-fertilization. Out of approximately 30 interspecific crosses in both directions, only one fruit reached maturity when guayabillo was the maternal parent and two reached maturity when guava was the maternal parent. The hybrid fruit produced when guayabillo was the maternal parent yielded no viable seeds, and all seeds were shriveled and apparently inviable. The seed set observed for the hybrid fruit produced when guava was the maternal parent was approximately halved. This high barrier to production of hybrids is consistent with the two species being more distantly related than in the case of invasive and endemic tomatoes (Proença et al. 2022). Additionally, both the microsatellite and sequencing data suggest that guava and guayabillo may differ in ploidy (Urquía et al. 2020; Reatini 2021), which would lead to chromosomal incompatibility.

Although viable hybrids appear to be rare, the lack of low premating barriers raises the possibility that guava poses a direct threat to the reproduction of guayabillo in zones of secondary contact, either by clogging guayabillo's flowers with foreign pollen or via gametic wastage due to low hybrid viability. If so, this would be an instance of the more general phenomenon of reproductive interference, in which there is a reduction of fitness for at least one of the species involved due to a negative sexual interaction (Kyogoku 2020). One prediction of the reproductive interference hypothesis is that there would be lower reproduction of guayabillo in areas of secondary contact. To test this, Reatini (2021) investigated fruit set and seed set at two sympatric sites and one allopatric site for each species on San Cristóbal. As predicted by reproductive interference, both fruit and seed set were substantially and significantly reduced in sympatry relative to allopatry for guayabillo. As a second test, Reatini (2021) investigated seedling recruitment, measuring the height of all guayabillo individuals as a proxy for age in three quadrants of 10 m^2 at each site. Again, consistent with the prediction of reproductive interference, there were only two guayabillo individuals <2 m in height across both sympatric sites, whereas there were over 30 individuals <2 m at both sites in which guava was absent. While both these results are consistent with reproductive interference, we cannot exclude other factors being at play. For instance, the sites where guava is now growing in proximity to guayabillo may be affected by ecological disturbance in ways that independently reduce guayabillo's reproduction and favor guava's establishment. Similarly, resource competition between the two species could contribute to lower reproductive output of guayabillo.

Whatever the cause, there is evidence that populations of guayabillo have been steadily disappearing from areas now occupied by guava. Guayabillo is now absent from the highlands of San Cristóbal, where the guava invasion began and where guava dominates today (Laso et al. 2020), although it is present in the highlands of all other islands on which it occurs (Reatini 2021). To see if there was further evidence for historical range contraction of guayabillo, Reatini (2021) constructed species distribution models (Phillips et al. 2006) for both guayabillo and guava and compared these to contemporary occurrence records for both species. While there was a large degree of overlap between the predicted ranges of both species on San Cristóbal, all extant guayabillo populations were found to be located near the outer edge of predicted sympatry, which was also the lower limit of guava dominated forest as determined by remote sensing (Laso et al. 2020). Applying this same analysis to the other inhabited islands, the distance of currently known guayabillo populations from the center of their predicted geographical distribution was inversely related to the age of the guava invasion as inferred by Urquía et al. (2019). Santa Cruz, where guava was introduced only in the 1950s, showed the least decline in areas of predicted sympatry, while Floreana and Isabela were intermediate between Santa Cruz and San Cristóbal, where guava was introduced the earliest.

This finding suggests that guayabillo populations in areas of sympatry, may, without human intervention, be on a path toward extirpation. If reproductive interference is the main factor contributing to the decline in guayabillo's range, then we would expect those populations in closest proximity to high densities of guava are those at most risk, but that archipelago-wide extinction would be unlikely unless guava expanded its range to lower elevations and to the uninhabited islands. However, there may still be threats to these allopatric populations of guayabillo from other factors (wood harvesting, disturbance, climate change, introduced herbivores and pathogens). In any event, the case of guayabillo and guava reminds us that secondary contact between invasive and native species may have negative consequences even in the absence of gene flow.

Common Themes and Future Directions

Despite some differences, these two case studies underscore several shared features of Galapagos plant invasions, and their potential impact on endemic communities. Guava is among the most aggressive Galapagos plant invasions, with introduction records dating to the nineteenth century, and a widespread current distribution on populated islands within the archipelago. In comparison, the tomato/*Solanum* invasion is less demographically dramatic and possibly more recent, with a more restricted current occurrence. Nonetheless, both the guava and tomato invasions are dynamic and ongoing, with evidence for multiple historical introductions and continued demographic spread, especially in association with humans. Both of these invasions also appear to have been facilitated by biological features of the invading species: Each invasion is derived from populations of a widespread, genetically diverse, mainland species that has broad ecological tolerance and physiological mechanisms (such as animal dispersal and rapid germination) that facilitate rapid dispersal and establishment.

These case studies also clearly demonstrate several predicted consequences of invasions for endemic communities, including close endemic relatives. Guava's invasion is associated with community-wide demographic effects, such as displacement of endemic plant communities (Tye et al. 2007), including its close endemic relative guayabillo. *Solanum*'s invasion has had clear demographic and reproductive consequences specifically for close endemic relatives, including genetic and evolutionary effects resulting from hybridization. Both invasions also appear to have influenced or modified ecological processes within Galapagos communities, including interactions with mutualists such as seed dispersers.

Our present understanding of these processes has relied critically on the use of diverse, independent, and complementary sources of data. For instance, the combination of historical records and contemporary molecular genetic analyses has been a powerful tool for tracking genealogical relationships and reconstructing the history of invasion of Guava onto the islands. Further, in situ crossing experiments, demographic sampling, and comparative niche modeling suggest this invasion has had both demographic and reproductive consequences for endemic guayabillo. Similarly, in *S. pimpinellifolium*, only by pairing genomic analyses with field sampling and observations were we able to uncover the timing and history of species invasion, confirm evidence for invasive-endemic species contact and hybridization, and reveal introgression as responsible for phenotypic evolution in some invasive populations. In both cases, cutting edge genotyping and population genetic technologies have provided unique insights into these processes, but the full power of these data has relied on its integration with field surveys, observations, and collections, within a well-described historical context.

These diverse and complementary data will be equally critical for monitoring, assessing, and predicting the future consequences of invasive-endemic interactions, in both invasive and endemic species. From the perspective of the invasive species, we still have much to understand about the complex factors shaping their continued demographic expansion, and their contact with endemics. For instance, we do not yet know whether introgression of endemic alleles into invasive populations has or will increase their adaptability within Galapagos ecosystems, as has been predicted for endemic-invasive introgression (Huxel 1999; Ellstrand and Schierenbeck 2006). Understanding the fitness consequences of fruit color evolution in *S. pimpinellifolium* populations —via introgression of endemic alleles—, and how much it could affect future invasiveness, will require additional ecologically-informed observations and experiments *in situ*. For example, if this reflects a color preference in fruit dispersers (such as Galapagos mockingbirds), trait introgression might amplify the future spread of this invasive species.

Our current understanding of the longer-term demographic and genetic prospects of endemic populations is even more limited. Both guava and *Solanum* invasions have the potential to disrupt ecological interactions in endemic communities, either through direct physical displacement or via disrupted ecological services, such as pollinator service or animal-mediated seed dispersal. Our data currently suggest, for example, that reproductive interference from invasive guava could be a substantial threat to endemic guayabillo. These factors could have future cascading effects on the demographic and reproductive stability and long-term population sustainability of endemics (Tye 2001; Sakai et al. 2001; Jäger et al. 2007). Assessing these possibilities will require more data from endemic populations, including those that are less vulnerable to contact with invasives, such as on uninhabited islands in the archipelago. The same is true of the potential genetic threats to endemic species because of introgression from invasives, including increased vulnerability to extinction via outbreeding depression or via genetic erosion and/or swamping (Huxel 1999, López-Caamal et al. 2014, Ellstrand and Rieseberg 2016). Our current genomic data indicates that the amount of historical gene flow from invasive into endemic populations is fairly limited. However, these patterns reflect past generations of contact and hybridization that may change with the recent sharp increase in frequency and density of invasive *S. pimpinellifolium*.

Regardless, even with incomplete information, our analyses already indicate that a complex set of demographic and genetic factors shapes the origin and progression of these two Galapagos plant invasions, and their subsequent interactions with endemic species. In both cases, understanding the future trajectory of these endemic-invasive interactions will also likely require examining how they feed back to affect each species' individual interactions with components of their abiotic and biotic environment, including mutualists like pollinators and dispersers.

Invasive species pose a major threat to biodiversity worldwide, especially in island ecosystems such as the Galapagos Islands. Scientific data, such as those generated in the present research, allow us to understand the mechanisms by which invasive species outcompete, displace, and may ultimately drive native and endemic species to extinction, as well as the genetic, ecological, and even sociological factors that drive and trigger such mechanisms. By having this type of information at hand and raising awareness of the threats posed by invasive species, we can come up with ideas for controlling them and preventing the displacement of endemic species. Thus, with this type of studies we will be able to contribute to the conservation of island ecosystems' biodiversity in this challenging context of imminent environmental change.

References

Abdalla M, Hervías-Parejo S, Traveset A (2021) Low pollinator sharing between coexisting native and non-native plant pairs: The effect of corolla length and flower abundance. Front Ecol Evol. https://doi.org/10.3389/fevo.2021.709876

Astudillo FJ (2018) Environmental and historical archaeology of the Galápagos islands: Archaeobotany of Hacienda El Progreso, 1870–1920. Veget Hist Archaeobot 27(5):737–751. https://doi.org/10.1007/s00334-018-0668-9

Baker HG (1974) The evolution of weeds. Ann Rev Ecol Syst 5(1):1–24. https://doi.org/10.1146/annurev.es.05.110174.000245

Binggeli P, Hall JB, Healey J (1998) An overview of invasive woody plants in the tropics. School of Agricultural and Forest Sciences

Blake S, Wikelski M, Cabrera F, Guezou A, Silva M, Sadeghayobi E, Yackulic CB, Jaramillo P (2012) Seed dispersal by Galápagos tortoises. J Biogeogr 39(11):1961–1972. https://doi.org/10.1111/j.1365-2699.2011.02672.x

Brewington L (2011) The politics of invasion: Defining and defending the natural, native and legal in the Galápagos Islands of Ecuador. In partial fulfillment of the requirements for the degree of Doctor of Philosophy in the Department of Geography. University of North Carolina at Chapel Hill

Chafin TK, Douglas MR, Martin BT, Douglas ME (2019) Hybridization drives genetic erosion in sympatric desert fishes of western North America. Heredity 123(6):759–773. https://doi.org/10.1038/s41437-019-0259-2

Chamorro S, Heleno R, Olesen J, McMullen C, Traveset A (2012) Pollination patterns and plant breeding systems in thc Galápagos: a review. Ann Bot 110(7):1489–1501. https://doi.org/10.1093/aob/mcs132

Crooks JA (2002) Characterizing ecosystem-level consequences of biological invasions: The role of ecosystem engineers. Oikos 97(2):153–166. https://doi.org/10.1034/j.1600-0706.2002.970201.x

Darwin S (2009) The systematics and genetics of tomatoes on the Galápagos Islands (Solanum, Solanaceae). A thesis submitted for the degree of Doctor of Philosophy at University College London, Department of Genetics, Evolution and Environment University College London

Darwin SC, Knapp S, Peralta IE (2003) Taxonomy of tomatoes in the Galápagos Islands: Native and introduced species of Solanum section Lycopersicon (Solanaceae). Syst Biodiverse 1(1):29–53. https://doi.org/10.1017/S1477200003001026

Ellstrand NC, Rieseberg LH (2016) When gene flow really matters: Gene flow in applied evolutionary biology. Evolution App 9(7):833–836. https://doi.org/10.1111/eva.12402

Ellstrand NC, Schierenbeck KA (2006) Hybridization as a stimulus for the evolution of invasiveness in plants? Euphytica 148(1):35–46. https://doi.org/10.1007/s10681-006-5939-3

Galápagos Conservancy (2022) Biodiversity. In: Galápagos Conservancy. https://www.galapagos.org/about_galapagos/biodiversity/. Accessed 21 Aug 2022

Gibson MJS, Moyle LC (2020) Regional differences in the abiotic environment contribute to genomic divergence within a wild tomato species. Mol Ecol 29(12):2204–2217. https://doi.org/10.1111/mec.15477

Gibson MJS, de Lourdes TM, Moyle LC (2020) Local extirpation is pervasive among historical populations of Galápagos endemic tomatoes. Evolution Ecol 34(2):289–307. https://doi.org/10.1007/s10682-020-10035-3

Gibson MJ, de Torres ML, Brandvain Y, Moyle LC (2021) Introgression shapes fruit color convergence in invasive Galápagos tomato. eLife 10:e64165. https://doi.org/10.7554/eLife.64165

Granda M, Chóez G (2013) Sistemas Humanos: Población y Migración en Galápagos

Grenier C (2007) Conservación contra natura. Las Islas Galápagos. Editorial Abya Yala

Hughes AR, Inouye BD, Johnson MTJ, Underwood N, Vellend M (2008) Ecological consequences of genetic diversity. Ecol Lett 11(6):609–623. https://doi.org/10.1111/j.1461-0248.2008.01179.x

Huxel GR (1999) Rapid displacement of native species by invasive species: Effects of hybridization. Biol Conserv 89(2):143–152. https://doi.org/10.1016/S0006-3207(98)00153-0

Jäger H, Tye A, Kowarik I (2007) Tree invasion in naturally treeless environments: Impacts of quinine (Cinchona pubescens) trees on native vegetation in Galápagos. Biol Conserv 140(3):297–307. https://doi.org/10.1016/j.biocon.2007.08.014

Kawasaki L, Holst B, Bazante G (2017) *Psidium galapageium*. In: León-Yánez S, Valencia R, Pitmam N, Endara L, Ulloa C, Navarrete H (eds) Libro Rojo de Plantas Endémicas del Ecuador. Publicaciones del Herbario QCA, Pontificia Universidad Católica del Ecuador. https://bioweb.bio/floraweb/librorojo/

Kyogoku D (2020) When does reproductive interference occur? Predictions and data. Popul Ecol 62(2):196–206. https://doi.org/10.1002/1438-390X.12041

Laso FJ, Benítez FL, Rivas-Torres G, Sampedro C, Arce-Nazario J (2020) Land cover classification of complex agroecosystems in the non-protected highlands of the Galapagos Islands. Remote Sens 12(1):65. https://doi.org/10.3390/rs12010065

Latorre O (1997) Galapagos: Los primeros habitantes de algunas islas. Noticias de Galápagos 56–57:62–66

Loh CS, Rao AN (1989) Clonal propagation of guava (*Psidium guajava* L.) from seedlings and grafted plants and adventitious shoot formation in vitro. Scientia Horticulturae 39(1):31–39. https://doi.org/10.1016/0304-4238(89)90035-6

López-Caamal A, Cano-Santana Z, Jiménez-Ramírez J, Ramírez-Rodríguez R, Tovar-Sánchez E (2014) Is the insular endemic *Psidium socorrense* (Myrtaceae) at risk of extinction through hybridization? Plant Syst Evol 300(9):1959–1972. https://doi.org/10.1007/s00606-014-1025-9

Lundh JP (2004) Galápagos: A brief history. Human and cartographic history of the Galápagos Islands

Lundh JP (2006) The farm area and cultivated plants on Santa Cruz, 1932–1965, with remarks on other parts of Galapagos. Galapagos Res 64:12–25

McMullen CK (2018) Flowering plants of the Galápagos. Cornell University Press

Nava AD, Hernández VAG, Jaimes MN, Castro EH, Álvarez DV, Villaseñor GD, Alberto FP, Lagunas BC (2014) Growth kinetics of vegetative and reproductive organs of Guava (*Psidium guajava* L.) in Iguala Guerrero, Mexico. Agricul Sci 5(14):1468–1475. https://doi.org/10.4236/as.2014.514157

Nuez F, Prohens J, Blanca JM (2004) Relationships, origin, and diversity of Galapagos tomatoes: Implications for the conservation of natural populations. Am J Bot 91(1):86–99. https://doi.org/10.3732/ajb.91.1.86

Pailles Y, Ho S, Pires IS, Tester M, Negrão S, Schmöckel SM (2017) Genetic diversity and population structure of two tomato species from the Galapagos Islands. Front Plant Sci 8:138. https://doi.org/10.3389/fpls.2017.00138

Pease JB, Haak DC, Hahn MW, Moyle LC (2016) Phylogenomics reveals three sources of adaptive variation during a rapid radiation. PLoS Biol 14(2):e1002379. https://doi.org/10.1371/journal.pbio.1002379

Phillips SJ, Anderson RP, Schapire RE (2006) Maximum entropy modeling of species geographic distributions. Ecol Model 190(3):231–259. https://doi.org/10.1016/j.ecolmodel.2005.03.026

Primack RB (2014) Essentials of conservation biology.

Proença CEB, Tuler AC, Lucas EJ, da Costa Vasconcelos TN, de Faria JEQ, Staggemeier VG, de Carvalho PS, Forni-Martins ER, Inglis PW, da Mata LR, da Costa IR (2022) Diversity, phylogeny and evolution of the rapidly evolving genus *Psidium* L. (Myrtaceae, Myrteae). Ann Bot 129(4):367–388. https://doi.org/10.1093/aob/mcac005

Reichard SH, Hamilton CW (1997) Predicting invasions of Woody Plants introduced into North America. Conserv Biol 11(1):193–203

Rejmanek M, Richardson DM (1996) What attributes make some plant species more invasive? Ecology 77(6):1655–1661. https://doi.org/10.2307/2265768

Rick CM (1956) Genetic and systematic studies on accessions of Lycospersicon from the Galapagos Islands. Am J Bot 43(9):687–696. https://doi.org/10.2307/2438834

Rick CM (1963) Biosystematic studies on galapagos tomatoes by Rick, Charles M.: Collectible; Very Good Pamphlet (1961) | Robinson Street Books, IOBA. Occasional Papers of the California aacdemy of Sciences (44):59–77

Rilov G, Mant R, Lyons D, Bulleri F, Benedetti-Cecchi L, Kotta J, Queirós AM, Chatzinikolaou E, Crowe T, Guy-Haim T (2012) How strong is the effect of invasive ecosystem engineers on the distribution patterns of local species, the local and regional biodiversity and ecosystem functions? Environ Evid 1(1):10. https://doi.org/10.1186/2047-2382-1-10

Sakai AK, Allendorf FW, Holt JS, Lodge DM, Molofsky J, With KA, Baughman S, Cabin RJ, Cohen JE, Ellstrand NC, McCauley DE, O'Neil P, Parker IM, Thompson JN, Weller SG (2001) The population biology of invasive species. Ann Rev Ecol Syst 32(1):305–332. https://doi.org/10.1146/annurev.ecolsys.32.081501.114037

Somarriba E (1986) Effects of livestock on seed germination of guava (*Psidium guajava* L.). Agroforestry Syst 4(3):233–238. https://doi.org/10.1007/BF02028357

Tye A (2001) Invasive plant problems and requirements for weed risk assessment in the Galápagos Islands. In: Groves R, Panetta F, Virtue J (eds) Weed risk assessment. CSIRO Publishing, Melbourne, pp 153–175

Tye A, Atkinson R, Carrion V (2007) Increase in the number of introduced plant species in Galapagos. DPNG, GCREG, FCD, GC, Puerto Ayora, Galapagos

Urquía D, Gutierrez B, Pozo G, Pozo MJ, Espín A, de Lourdes Torres M (2019) *Psidium guajava* in the Galapagos Islands: Population genetics and history of an invasive species. PLoS One 14(3):e0203737. https://doi.org/10.1371/journal.pone.0203737

Urquía D, Pozo G, Gutierrez B, Rowntree JK, de Lourdes Torres M (2020) Understanding the genetic diversity of the guayabillo (*Psidium galapageium*), an endemic plant of the Galapagos Islands. Global Ecol Conserv 24:e01350. https://doi.org/10.1016/j.gecco.2020.e01350

Urquía D, Gutierrez B, Pozo G, Pozo MJ, de Torres ML (2021) Origin and dispersion pathways of guava in the Galapagos Islands inferred through genetics and historical records. Ecol Evol 11(21):15111–15131. https://doi.org/10.1002/ece3.8193

Valdebenito H (2018) A study contrasting two congene plant species *Psidium guajava* (introduced guava) and *P. galapageium* (Galapagos guava) in the Galapagos Islands. In: de Lourdes Torres M, Mena CF (eds) Understanding invasive species in the Galapagos Islands. Springer, pp 47–68

Weber E (2003) Invasive plant species of the world: A reference guide to environmental weeds, 2nd edn. CABI Publication

Whittaker RJ, Whittaker S of GRJ, Fernandez-Palacios JM, Fernández-Palacios D de EF de BJM (2007) Island biogeography: Ecology, evolution, and conservation. OUP Oxford

Wogan P (2009) Magical writing in Salasaca: Literacy and power in Highland Ecuador. Routledge

Xu C-Y, Tang S, Fatemi M, Gross CL, Julien MH, Curtis C, van Klinken RD (2015) Population structure and genetic diversity of invasive *Phyla canescens*: implications for the evolutionary potential. Ecosphere 6(9):art162. https://doi.org/10.1890/ES14-00374.1

María de Lourdes Torres is the Head of the USFQ Department of Biotechnology and Director of the Plant Biotechnology Lab. Her main research interests include genetics and conservation of local species of mainland Ecuador and the Galapagos Islands.

Diego Urquía Graduate student at the Biology Department of UNC-Chapel Hill. His research interests lie in using molecular tools to answer questions about the ecology, demography, and evolution of wildlife and plant populations, especially in the Galapagos Islands.

Leonie Moyle Professor of Biology at Indiana University Bloomington, studies the evolution and genetics of speciation and adaptation, including in the plant group Solanum.

Matt Gibson now a Statistical Geneticist at Gencove, did his doctoral research at Indiana University Bloomington, on the ecological genomics of natural variation within and between wild tomato species.

Todd Vision Associate Professor of Biology at the University of North Carolina at Chapel Hill, is an evolutionary geneticist interested in the forces that shape the genomes, functional diversity and biogeographical distribution of plants.

Bryan Reatini postdoctoral associate at the University of Arizona, is a conservationist, ecologist, and evolutionary biologist researching the evolutionary causes and ecological consequences of biological invasions.

Chapter 13
Galapagos Land Snails and Environmental Sustainability

Stella de la Torre and Isabel Villarruel-Oviedo

Introduction

Land use change is severely affecting soil ecosystems globally. Decreases in the diversity of the soil biota caused by anthropogenic disturbances are known to alter decomposition and nutrient cycling of organic matter contributing to eutrophication of water bodies, reduced aboveground biodiversity, and global warming (Coleman et al. 2004; Bender et al. 2016). These facts evidence the importance of soil management and conservation to maintain above-ground biodiversity and ecosystem services and, therefore, to achieve environmental sustainability. A necessary first step in soil management and conservation is to understand the structure of its diverse biological community that includes bacteria and fungi of thousands of taxa, and a huge variety of macro-invertebrates (e.g., nematodes, earthworms, mollusks, arthropods) (Bardgett and van der Putten 2014). Among this last group, macro-detritivores that feed on leaf litter, like land snails, play an important role in decomposition processes, directly influencing soil properties (Astor et al. 2015; Handa et al. 2014).

Galapagos land snails are an endemic and highly diverse group of soil macro-detritivores in the Galapagos islands. They have been negatively affected by anthropogenic disturbances including the loss of native vegetation, the introduction of exotic species like black rats, fire ants and introduced snails, and climate change (Coppois and Wells 1987; Trueman et al. 2011; Villarruel-Oviedo and de la Torre 2014). Although the role of land snails in litter decomposition, retention of soil calcium and trophic webs has been studied in several terrestrial ecosystems, from deserts to temperate and tropical forests (Astor et al. 2015; Graveland and van der

S. de la Torre (✉) · I. Villarruel-Oviedo
College of Biological & Environmental Studies and USFQ-UNC Galapagos Science Center, Universidad San Francisco de Quito, Quito, Ecuador
e-mail: sdelatorre@usfq.edu.ec

S. J. Walsh et al. (eds.), *Island Ecosystems*, Social and Ecological Interactions in the Galapagos Islands, https://doi.org/10.1007/978-3-031-28089-4_13

Wal 1996; Jones and Shachak 1994; Oli and Gupta 2000; Schilthuizen 2011), almost nothing is known about the role of Galapagos land snails on the functioning of the terrestrial ecosystems in the archipelago.

Soils in the inhabited islands of the Galapagos archipelago have been affected by agricultural activities (Gerzabek et al. 2019). Differences in soil properties such as organic carbon stocks and total nitrogen between arable and natural forest areas have been attributed to the use of agrochemicals and changes in pH (Gerzabek et al. 2019). However, to our knowledge, there is no information about how changes in the soil biota, specifically, in land snails, could be related to the observed differences in soil properties.

In this paper we describe the diversity of land snails in three sites with different land use patterns in the island of San Cristobal, correlating it with the concentrations of soil organic carbon and nitrogen to assess the possible influence of snail diversity on soil properties. In addition, we present the results of a pilot field experiment comparing the consumption rates of two native and one introduced snail species in the two climatic seasons in a preliminary assessment of their role in plant litter decomposition.

Study Areas

We selected three study sites with different land use patterns: an Organic Agriculture site, a Restoration site, and a site inside the Galapagos National Park (Fig. 13.1).

The Organic Agriculture site was in Hacienda El Cafetal, at 260 m elevation, near El Progreso town. Being in the south-facing slopes of the island, this site is more humid and has wetter soils than the other two (Snell and Rea 1999). Vegetation in this site was dominated by shrubs of coffee *Coffea* cf. *arabica*, but other introduced tree species were also present (e.g., *Cedrela odorata, Cinchona pubescens*). Ferns (cf. *Polypodium* sp.) and vines (*Momordica charantia*) occurred in the undergrowth.

The Restoration site was in Hacienda La Tranquila, at 370 m elevation. It was formerly an area of pasture infested with guava (*Psidium guajava*) and raspberry (*Rubus niveus*); few individuals of these two species were still present in the area. The reforested native species included *Lecocarpus darwinii* and *Scalesia pedunculata*.

The site inside the Galapagos National Park (GNP) was close to La Soledad settlement, at 370 m elevation. Introduced plant species were abundant in this area, including agaves (*Furcraea hexapetala*), guava and raspberry. Among the native species, small patches of manzanillo trees (*Hippomane mancinella)* were conspicuous along the study transects.

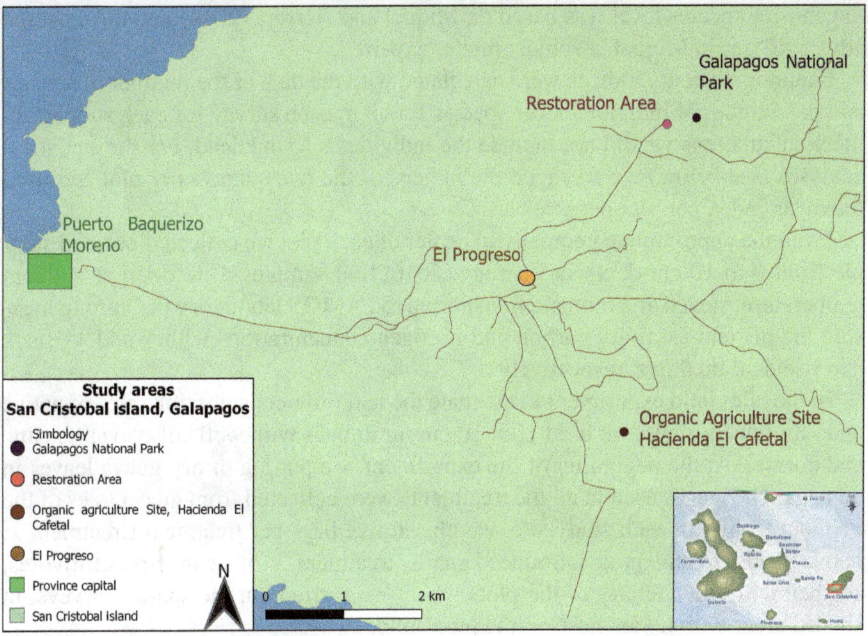

Fig. 13.1 Study areas in San Cristobal, Galapagos

Methods

Field work was carried out from August through September 2015, and from March through April 2016. These two periods were selected as representative of the dry and rainy seasons of the islands, respectively (Trueman and d'Ozouville 2010). However, 2016 was a year of La Niña so the rainy season was unusually dry, with monthly precipitations below the average (INAMHI 2016; Bellavista (darwinfoundation.org)).

In the Organic Agriculture and Restoration sites, we built five randomly located 50 m-transects. In each transect we placed two 1 m^2-plots, separated from each other by at least 10 m for a total of 10 plots per study site. In the GNP site, we built the five transects in areas where the abundance of native plant species was higher than that of introduced plants. The location of the two 1 m^2-plots per transect was not random neither, in each transect we placed the plots in microhabitats with fewer or no introduced plants.

We assessed the diversity of land snails through surveys in one subplot of 25 cm^2 in each of the 1m^2-plots in the study areas. In each subplot we conducted two surveys in different days, from the soil surface to 5 cm depth, in each climatic season. We photographed and identified snails using guides (Correoso 2008; Villarruel-Oviedo and de la Torre 2014). Shells of individuals found dead that could not be identified in the field, were taken to the Terrestrial Ecology Laboratory of the Galapagos Science Center for a more careful examination. Taxonomic identification to genus or species level was based on Miquel and Herrera (2014) and the personal advice of Sergio Miquel, a subject matter expert.

Shannon diversity indices were calculated with the data of the number of species and the number of individuals per species found in each survey for each subplot. In these calculations we did not include the individuals found dead. For the statistical analyses (see below) we averaged the indices of the two surveys per plot and used the mean index per plot per season.

From the approximate geometrical center of each plot we collected one soil sample from 0 to 10 cm depth in February 2016. Soil samples were dried at ambient temperature, sieved at 2 mm and transported to FAMOS laboratory in Quito to measure the pH and assay for carbon and nitrogen concentrations with Walkley-Black and Kjeldahl methods, respectively.

In the pilot field experiment to estimate the leaf litter consumption rates of native and introduced snails, we used 10 × 10 cm mesh bags with wefts of two sizes (fine and coarse). At the beginning of the experiment, we put 5 g of dry guava leaves in all bags. The leaves used in all the treatments were collected from guava trees of the PNG area only. In each study site, we placed five bags per treatment (treatment 1: native snails, treatment 2: introduced snails, treatment 3: other macro-detritivores, see below) in the vicinity of the plots where we carried out the snails' surveys, in shaded areas with no ant nests in the proximity.

For treatments 1 and 2 we used bags with the fine mesh, introducing10 individuals of a native or introduced snail species. In each study area we only used snails collected alive in that area. After the experiment ended, individuals were released in the same sites where they were collected. For treatment 3, we used bags with the coarse mesh since the mesh size allowed the entry and exit of several groups of macro-detritivores. Millipeds, ants and isopods were frequently recorded inside these bags during the experiment monitoring; snails were rare, we recorded only two individuals in two different bags in the dry season experiment of the El Cafetal study area. Thus, we used these coarse mesh bags to estimate the ingestion rate of other macro-detritivores in each area.

The native species were different between sites since we selected the native species that was more abundant in each site based on the diversity surveys we carried out. The native species used in the Organic Agriculture site was *Bulimulus* cf. *curtus,* the native species used in the Restoration site was *Succinea* sp. We found very few individuals of native species in the GNP site so in this area we carried out the experiment with two, instead of three treatments. We selected *Subulina octona* as the introduced species in the three study sites since it was abundant in all of them.

In each site, at the beginning of the experiment, we weighed the bags and left them for 21 days and 27 days in the dry and rainy season, respectively. At the end of each period, we weighed the bags again. We calculated the daily leaf litter consumption rate by subtracting the final weight from the initial weight per bag and dividing the result by the number of days of the experiment in each season.

We used repeated-measures ANOVA to compare the Shannon diversity indices of the dry and wet season surveys among sites. We used one-way ANOVAs to compare the soil pH, transformed (ln and log) percentages of soil organic carbon and nitrogen, and C:N ratios. We carried out Pearson correlations between each soil variable and the average Shannon diversity indices per site. Finally, we compared the daily leaf litter consumption rates between seasons and treatments with a multifactorial ANOVA after checking for normality.

Results

Snail Diversity

We recorded a total of 11 snail species, including the introduced *Subulina octona* and *Zonitoides arboreous*, which were present in all three study sites (Table 13.1). Among the native species, three occurred in the three study sites (e.g., *Helicina nesiotica*) whereas others were only found in one of the areas, usually in the Organic Agriculture site (e.g., *Euconolus galapaganus*).

In the dry season surveys, 41% of the individuals of all species were found dead (n = 796). The percentage of dead individuals in the rainy season increased to 50% (n = 500).

The Shannon diversity index, calculated with live individuals only, was significantly higher in the Organic Agriculture site than in the other two study sites ($F_{2, 57} = 21.06$; $p < 0.001$). In the rainy season we found a decrease in snail diversity in two of the sites (GNP and Organic Agriculture), and an increase in the Restoration site (Table 13.2).

Table 13.1 Snail species recorded (X) in the three study sites

Species	Restoration site	GNP site	Organic Agriculture site
Helicina nesiotica	X	X	X
Tornatellides chathamensis	X	X	
Euconulus galapaganus			X
Succinea sp.	X	X	X
Retinella sp.	X		X
Habroconus (*Pseudoguppya*) aff. *pacificus*	X	X	X
Bulimulus cf. *curtus*			X
Bulimulus cf. *galapaganus*			X
Bulimulus cf. *canaliferus*	X		
Subulina octona	X	X	X
Zonitoides arboreus	X	X	X

Table 13.2 Shannon diversity indices (H) in the three study sites (mean plot diversity ± st. dev.), dry and rainy seasons

Sites	H dry season (Sept. 2015)	H rainy season (March 2016)
Restoration	0.132 ± 0.227	0.202 ± 0.230
GNP	0.264 ± 0.342	0.092 ± 0.201
Organic Agriculture	0.974 ± 0.292	0.462 ± 0.291

Table 13.3 Soil pH, organic carbon, nitrogen and C/N ratio in the three study areas (mean ± st. dev)

Site	pH	C (%)	N (%)	C/N
Restoration	5.75 ± 0.12	4.25 ± 0.60	0.98 ± 0.11	4.40 ± 0.70
GNP	5.91 ± 0.09	5.27 ± 0.52	0.99 ± 0.14	5.37 ± 0.75
Organic Agriculture	5.68 ± 0.28	4.62 ± 0.42	0.92 ± 0.10	5.08 ± 0.63

Soil Characterization

The soil samples of the study sites differ significantly in pH ($F_{2,27} = 4.213$; $p = 0{,}025$), organic carbon concentration ($F_{2,27} = 9.045$; $p = 0{,}001$) and C/N ratio ($F_{2,27} = 5.378$; $p = 0{,}018$) (Table 13.3). Soils in the Organic Agriculture site were more acidic, whereas soils in the GNP site had the higher organic carbon concentration and C/N ratio. Differences in nitrogen concentration were not significant.

We found high negative correlations between snail diversity and soil nitrogen content in the two climatic seasons (dry season $r = -0.93$, rainy season $r = -0.99$). Correlations between snail diversity and soil pH were also moderately high and negative (dry season $r = -0.63$, rainy season $r = -0.89$). Due to the small sample size ($n = 3$ in all correlations), these correlations were not statistically significant.

Consumption Rates Experiment

In the consumption rates experiment, we found significant differences in the daily consumption rates among treatments ($F_{2,76} = 22.43$; $p < 0.0001$) and between seasons ($F_{1,76} = 19.78$; $p < 0.0001$). The interaction between treatment and season was also significant ($F_{2,76} = 19.78$; $p < 0.0001$). In the dry season, the daily consumption rates of the introduced snail *S. octona* were significantly higher than the rates of the native species (*Bulimulus* cf. *curtus* and *Succinea* sp.) and of other macro-detritivores. In the rainy season, the daily consumption rates were significantly lower than in the dry season, but the daily consumption rates of the native species were significantly higher than those of the introduced species and of the other macro-detritivores (Fig. 13.2).

Discussion

We found differences in snail diversity among the study sites. These differences could be partly explained by differences in humidity and soil moisture, since the Organic Agriculture site had the highest snail diversity and is the most humid habitat. Previous studies have also reported more diverse snail communities in habitats with high air humidity and soil moisture (e.g. Čejka and Hamerlík 2010). The importance of humidity for land snails in San Cristobal is also suggested by the

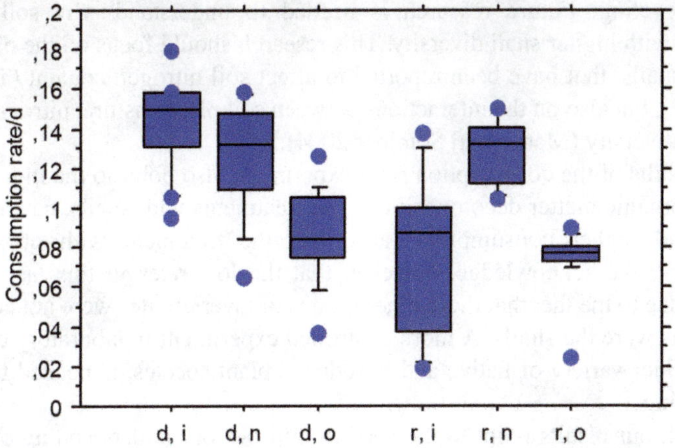

Fig. 13.2 Daily consumption rates in the dry (d) and rainy (r) seasons of introduced (i), native (n) and other macro-detritivores (o)

reduction in snail diversity and the increased number of dead snails we found in the rainy season of 2016, when monthly precipitations were below the average (INAMHI 2016).

In addition to the influence of humidity, our results suggest that land use change is also affecting snail diversity. Although the GNP and Restoration sites are at the same altitude in the north-facing slopes of the island, have similar humidity and are relatively close to each other (aprox. 1 km), the abundance of native snails in the GNP site was low compared to the Restoration site. This difference could be due to a reduced availability of native plants in the GNP site –dominated by introduced plants–, and points to the importance of native plants for native snails, as reported by Coppois and Wells (1987).

Decreases in the populations of native snails may also be caused by the potential competition of the two introduced snail species, *S. octona* and *Z. arboreus*. Our surveys of snail diversity suggest that these two introduced species are widespread and abundant in the highlands of San Cristobal. In addition, in the consumption rates experiment, the introduced *S. octona* had higher consumption rates than native snails in periods with average precipitations. This higher consumption rate may give this introduced species a competitive advantage (Spiller 1984). Interestingly, this advantage may be lost in unusually dry periods, like the rainy season of 2016. The higher consumption rates of native snails in this period, suggest that native species are better adapted to extreme dry conditions. Future studies are needed to complement these findings and assess the effects of competition of introduced snails, including the giant African snail *Lissachatina fulica*, that is present in Santa Cruz (Darwin Foundation n.d.) and may eventually colonize San Cristobal, although it was not recorded in our surveys.

The high negative correlation between snail diversity and soil nitrogen content suggests that snails have an important role in nutrient dynamics, in particular, in nitrogen cycling. Future research is needed to understand why soil nitrogen decreases with higher snail diversity. This research should focus on the diet preferences of snails, that have been reported to affect soil nitrogen content (Thompson et al. 1993), but also on the interactions between soil pH, moisture, nitrogen content and snail diversity (Martin and Sommer 2004).

The results of the consumption rates experiment also point to the importance of snails in organic matter decomposition. The treatments with snails, native or introduced, had higher consumption rates than the treatment with other macro-detritivores. We acknowledge, however, that the low rates in this last treatment could be due to the fact that these other groups of invertebrates were not confined to the bags as were the snails. A more controlled experiment in laboratory conditions using a wider variety of native and introduced plant species, is needed to confirm these findings.

Overall, our results point to the need to improve our understanding of the ecological role of native snails and to identify those environmental conditions that may enhance individual survival and population growth. Although preliminary, our results suggest that preserving native snails in Galapagos is key to maintain soil ecological integrity and to assure environmental sustainability. Galapagos land snails are iconic not only for their diversity and endemism but also for their role in soil dynamics and the conservation of the terrestrial biodiversity of the islands. Efforts should be made to increase awareness among scientists, visitors and local people on the importance of studying and conserving these fascinating animals. To enhance visitors' awareness, a chapter on the evolution, ecological role and conservation of Galapagos land snails could be included in the contents of the training courses for tourist guides so that they could share this information with their tourists. Increasing awareness of local people is even more important since they could actively engage and benefit from snail conservation actions that may eventually improve soil quality and other ecosystem services. We are currently working with the Galapagos Science Center to develop an outreach program for the local community to educate them in the conservation of native snails and their habitats. Educational activities, including citizen science, may allow us not only to share with them updated information about the ecological importance of land snails but also to engage them in research to better understand their ecological requirements (e.g., diet, tolerance limits) and their main conservation threats.

Acknowledgements This research was funded by a GAIAS Grant from Universidad San Francisco de Quito. The Galápagos National Park allowed us to conduct this research with the permits PC-63-15 and PC-20-16. We deeply acknowledge all the team of the Galapagos Science Center (USFQ/UNC) and the personal of the GNP, in particular, Maryuri Yépez and Jeffreys Málaga for their support to our research. Our species thanks to Sergio Miquel, Angel Correoso, Heinke Jäger, Carlos Valle and Esteban Suárez for their valuable scientific advice. Geovanny Sarigu from Hacienda La Tranquila, and Nicolás Balón and Edgar Román from Hacienda El Cafetal kindly allowed us to carry out the field research in those areas.

References

Astor T, Lenoir L, Berg M (2015) Measuring feeding traits of a range of litter-consuming terrestrial snails: leaf litter consumption, faeces production and scaling with body size. Oecologia 178. https://doi.org/10.1007/s00442-015-3257-y

Bardgett RD, van der Putten WH (2014) Below ground biodiversity and ecosystem functioning. Nature 515:505–511

Bender SF, Wagg C, van der Heijden MGA (2016) An underground revolution: biodiversity and soil ecological engineering for agricultural sustainability. Trends Ecol Evol 31(6):440–452. https://doi.org/10.1016/j.tree.2016.02.016

Čejka T, Hamerlík L (2010) Land snails as indicators of soil humidity in Danubian woodland (SW Slovakia). Pol J Ecol 57:741–747

Coleman DC, Crossley DA Jr, Hendrix PF (2004) Fundamentals of soil ecology, 2nd edn. Elsevier Academic Press, San Diego, CA

Coppois G, Wells S (1987) Threatened Galápagos snails. Oryx 21(4):236–241

Correoso M (2008) Los Moluscos terrestres y fluviales del Ecuador Continental. La Biodiversidad desconocida. SIMBIOE, Quito, Ecuador

Darwin Foundation (n.d.) Galapagos Species Checklist *Lissachatina fulica*(darwinfoundation.org). Accessed 24 Aug 2022

Gerzabek MH, Bajraktarevic A, Keiblinger K, Mentler A, Rechberger M, Tintner J, Wriessnig K, Gartner M, Valenzuela X, Troya A, Couenberg PM, Jäger H, Carrión JE, Zehetner F (2019) Agriculture changes soil properties on the Galápagos Islands – two case studies. Soil Res 57:201–214. https://doi.org/10.1071/SR18331

Graveland J, van der Wal R (1996) Decline in snail abundance due to soil acidification causes eggshell defects in forest passeries. Oecologia 105(3):351–360

Handa IT, Aerts R, Berendse F, Berg MP, Bruder A, Butenschoen O, Chauvet E, Gessner MO, Jabiol J, Makkonen M, McKie BG, Malmqvist B, Peeters ETHM, Scheu S, Schmid B, van Ruijven J, Vos VCA, Hättenschwiler S (2014) Consequences of biodiversity loss for litter decomposition across biomes. Nature 509(7499):218–221

INAMHI (2016) Boletín Climatológico Semestral – Boletín de vigilancia climática del Ecuador 1. Instituto Nacional de Meteorología e Hidrología, Quito, Ecuador

Jones CG, Shachak M (1994) Desert snail's daily grind. Nat Hist 103(8):56–61

Martin K, Sommer M (2004) Relationships between land snail assemblage patterns and soil properties in temperate-humid forest ecosystems. J Biogeogr 31:531–545. https://doi.org/10.1046/j.1365-2699.2003.01005.x

Miquel S, Herrera H (2014) Catalogue of terrestrial gastropods from Galápagos (except Bulimulidae and Succineidae) with description of a new species of *Ambrosiella* Odhner (Achatinellidae) (Mollusca: Gastropoda). Archiv für Molluskenkunde Int J Malacol 143. https://doi.org/10.1127/arch.moll/1869-0963/143/107-133

Oli BP, Gupta PK (2000) Land mollusc fauna of the Kumaon Himalayan Forest and the role of snail in plant litter decomposition. Trop Ecol 41(2):1–5

Schilthuizen M (2011) Community ecology of tropical forest snails: 30 years after Solem. Contrib Zool 80(1):1–15

Snell H, Rea S (1999) The 1997–98 El Niño in Galápagos: Can 34 years of data estimate 120 years of pattern? Noticias de Galápagos: a biannual news publication about science and conservation in Galápagos, Galápagos National Park Service and the Charles Darwin Research Station: 11

Spiller DA (1984) Seasonal reversal of competitive advantage between two spider species. Oecologia 64(3):322–331. http://www.jstor.org/stable/4217468

Thompson L, Thomas CD, Radley JM, Williamson S, Lawton JH (1993) The effect of earthworms and snails in a simple plant community. Oecologia 95(2):171–178. https://doi.org/10.1007/BF00323487

Trueman M, d'Ozouville N (2010) Characterizing the Galapagos terrestrial climate in the face of global climate change. Galapagos Res 67:26–37

Trueman M, Hanna L, d'Ozouville N (2011) Terrestrial ecosystems in Galápagos: potential responses to climate change. In: Larrea I, Di Carlo G (eds) Climate change vulnerability assessment of the Galápagos Islands. WWF and Conservation International, pp 29–46
Villarruel Oviedo I, de la Torre S (2014) Estudio preliminar de caracoles terrestres en la isla San Cristóbal, Galápagos. Avances en Ciencias e Ingenierías 6(2):B20–B25

Stella de la Torre is a professor and researcher at the USFQ School of Biology. She obtained her PhD in Zoology at the University of Wisconsin, Madison. Stella's research interests are in population ecology, ecosystem ecology, and behavioral ecology. In Galapagos, her research focuses on the impacts of land use change on community diversity and ecosystem functions.

Isabel Villarruel-Oviedo is a researcher and conservationist. She obtained a BS in Biology and a MS in Applied Ecology at USFQ. Her undergraduate and graduate research projects focused on the ecology of Galapagos land snails.

Chapter 14
Galapagos Petrels Conservation Helps Transition Towards a Sustainable Future

Leo Zurita-Arthos, Carolina Proaño, Jonathan Guillén, Sebastián Cruz, and David Wiedenfeld

Introduction

Seabirds of the Procellariiformes order are among the most threatened group of birds in the world. Their declines are related to both sea-based and land-based threats (Borrelle et al. 2015). The Galapagos Petrel (*Pterodroma phaeopygia*) is a seabird that breeds only in the Galapagos Archipelago. Threats to the petrel at sea, as for other species that share this niche, include water pollution and commercial fisheries. Some individuals are killed as bycatch in fisheries, while others are losing their prey as a result of overfishing (Croxall et al. 2012). On land, petrels face threats including predation of eggs, chicks, and adults, destruction of nests from livestock trampling, and habitat lost to agricultural conversion and introduced plants (Raine et al. 2017; Duffy 1984), and these are not recent threats, but long-term problems in all the islands where they nest.

The introduced species affecting the Galapagos Petrel on land during the breeding season are dogs (*Canis familiaris*), cats (*Felis catus*), pigs (*Sus scrofa*), and rats (*Rattus* spp.). All prey on petrel eggs, chicks, and occasionally, on adult individuals (Personal observations, 2017–2020). Domestic livestock, such as cattle (*Bos*

L. Zurita-Arthos (✉)
Universidad San Francisco de Quito USFQ. Cumbaya, Quito, Ecuador

Galapagos Science Center GSC, Puerto Baquerizo Moreno, Galápagos, Ecuador
e-mail: lzurita@usfq.edu.ec

C. Proaño · S. Cruz
Tierramar Research Group, Puerto Ayora, Galápagos, Quito, Ecuador

J. Guillén
Naturalist Guide of Galapagos, Galápagos, Ecuador

D. Wiedenfeld
American Bird Conservancy ABC, Washington, DC, USA

S. J. Walsh et al. (eds.), *Island Ecosystems*, Social and Ecological Interactions in the Galapagos Islands, https://doi.org/10.1007/978-3-031-28089-4_14

taurus), horses (*Equus ferus*), donkeys (*E. asinus*), and even the eradicated goats (*Capra hircus*) will collapse petrel nest burrows by tramping on them, causing loss of the contents and nest site (Valarezo and Wiedenfeld 2005). This is particularly harmful since the petrels show a pattern of philopatry and return each season to the same breeding site, and even to the same nest (Cruz Delgado 2005; Warham 1996).

Introduced plants also threaten the petrel nests. Introduced blackberries (*Rubus* spp.) grow rapidly and densely, thus they can overgrow nest burrow entrances during the nonbreeding season, making the nest inaccessible to the returning petrels (Cruz Delgado 2005). Furthermore, introduced red cinchona (*Cinchona pubescens*), a low-stature tree, grows thickly and creates root networks that the petrels cannot penetrate when excavating nest burrows, rendering large areas unusable to the petrels for nesting.

These issues are not new and have been addressed at different levels since the 1980s, when the population decline reached rock bottom (Cruz and Cruz 1987). At some point, all the nesting sites, in the five islands where they breed (namely Santa Cruz, San Cristóbal, Isabela, Floreana, and Santiago) were under different levels of threat (Harris 1970; Castro and Phillips 1996). Since then, the conservation programs have achieved important milestones, although they have not completely secured the population numbers for the long-term (Coulter et al. 1985; Valarezo and Wiedenfeld 2005; Gummer et al. 2015). In fact, the International Union for the Conservation of Nature has kept the Galapagos Petrel on the Critically Endangered List due to their limited population number and geographic range (Granizo et al. 2002; BirdLife International 2018). Aiming to bring some of these issues to light and to discuss the steps towards the conservation of the species, a historic workshop was held in 2019 with all the interested actors at the local and regional levels. The main result of this gathering of experts was the formation of the Conservation Action Plan for the Galapagos Petrel. Furthermore, a Galapagos Petrel Research Group was created to secure the conservation of the species in the long-term, working with local stakeholders where the nesting areas are located and to bring science and new techniques to better understand the role and importance of the species to the ecosystems of the archipelago. The present project aims to implement some of the strategies to enhance the protection and enhancement of petrel nesting areas developed in the Action Plan. Our objectives are ambitious, but at the same time feasible:

- Develop methods to monitor petrel nests on private lands.
- Evaluate the understanding of farmers and landowners, as well as their attitudes toward petrel conservation on Santa Cruz and San Cristóbal islands.
- Provide information about the breeding season and guidance for protecting petrel burrows.
- Monitor a set of petrel nest burrows on private lands to determine predation.
- Trial artificial burrows to increase nest site availability and replace nests lost to livestock trampling

Nest Mapping

We began the project by mapping the locations of nests on private lands on Santa Cruz and San Cristóbal islands (Figs. 14.1 and 14.2). This process was based on the initial interviews with farmers and landowners about whether they knew or thought there were nests on their properties, with follow-up to confirm and locate actual nest

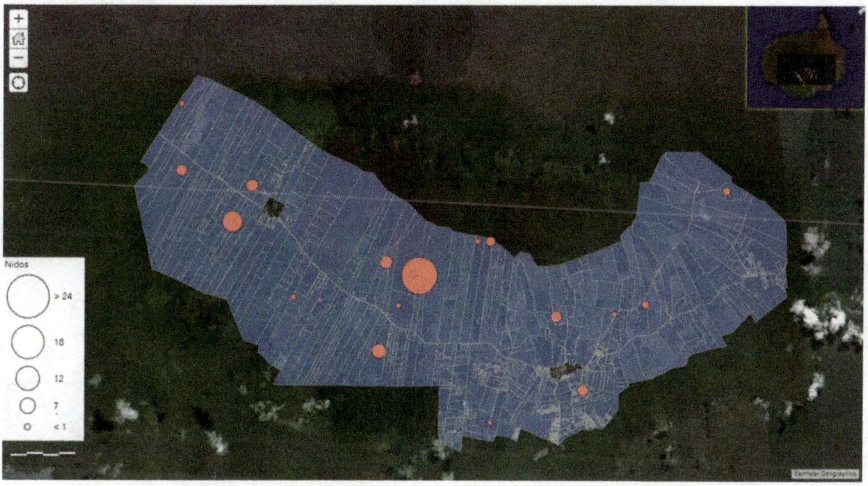

Fig. 14.1 Highlands of Santa Cruz Island. The agricultural zone is highlighted in blue. Farms with reported petrel nests show the light-red circles

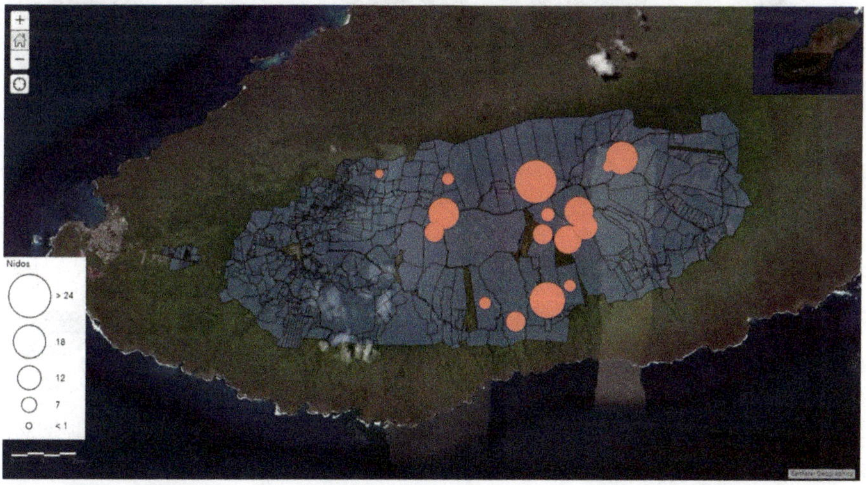

Fig. 14.2 Highlands of San Cristóbal Island. The agricultural zone is highlighted in blue. Farms with reported petrel nests show the light-red circles

burrows. A total of 39 farmers reported the presence of nests on their land or properties, although actual breeding activity was not verified. Nevertheless, the ease of the process and the understanding of the local actors about the species showed that the petrel is an integral part of the *galapagueño* culture.

Interviews and Outreach

Petrel nests were found on the properties of 22 landowners on Santa Cruz Island and 17 on San Cristóbal Island. During interviews, all the landowners (100%) expressed interest in helping with petrel conservation on their lands. These landowners were contacted up to three times during the year for separate interviews/interactions, each lasting about 10–15 min (planned as an initial conversation prior to nesting season, during nesting season providing outreach materials, and exit/post-breeding season). All landowners that expressed interest in maintaining and protecting the petrel nests on their property were provided with handouts about ways to protect petrels by season (Fig. 14.3).

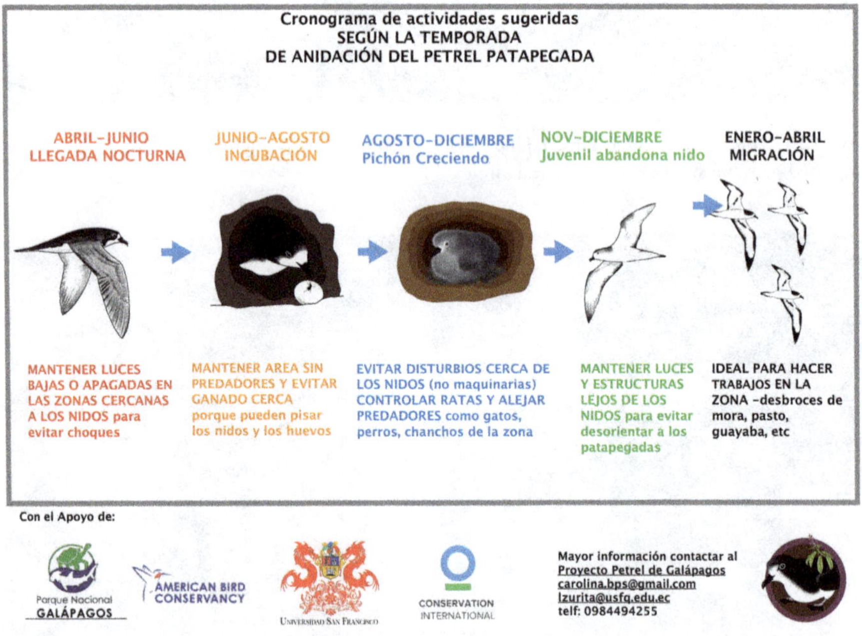

Fig. 14.3 Handout provided to landowners, showing conservation actions they could take to protect petrels and petrel nests, by season

Nest Monitoring

Nest monitoring using "burrowscopes" (video cameras using a flexible fiber-optic cable to peer into natural burrows) showed occupied nests (Fig. 14.4). This monitoring contributes to the collection of data and information for the permanent monitoring program that the Galapagos National Park leads every year for the past decade. In fact, only by joining forces with the park, other institutions, and local farmers, it would be possible to produce vital information about the breeding season and guidance for protecting petrel burrows. This is particularly important when considering the challenges that climate change is already bringing to the islands and the quest for sustainable solutions to these challenges.

Six artificial nests were built using basic gardening materials (flowerpots and drain channels) – three on San Cristobal and three on Santa Cruz Island (Fig. 14.5). One artificial burrow on Santa Cruz Island was constructed at the exact same place where a natural burrow had been damaged and collapsed, in hopes that the breeding pair would occupy it. Two more artificial burrows were installed in the vicinity of this original nest. In San Cristóbal, the artificial nests were located at a small breeding site, where more activity was observed, since the land offers a safe place for long-term conservation. It is understood that the process of getting the petrels to use the artificial nests for breeding might take a few years, as it was already tested in the sister species, the Hawaiian Petrel, where just recently they had young individuals coming back to the artificial nests where they were relocated (ABC 2021).

Fig. 14.4 Galapagos Petrel in burrow on San Cristóbal Island

Fig. 14.5 Artificial burrow construction on San Cristóbal Island

Fig. 14.6 Images from the camera traps showing (**a**) Galapagos Petrel investigating and entering an artificial burrow on Santa Cruz Island. (**b**) Galapagos Petrel using an artificial burrow on Santa Cruz Island. (**c**) Rat, probably Black Rat (*Rattus rattus*), investigating an artificial nest burrow. (**d**) Cat investigating an active nest burrow on San Cristóbal Island

Camera traps installed in the nests' vicinity showed petrels visiting and entering the artificial burrows. We also identified predators investigating the same burrows (Fig. 14.6). Predators captured in the pictures include rats, feral cats, and dogs. This is promising and at the same time worrisome, since the petrels are facing the dangers of these introduced predators.

Conclusions: Achievements and Next Steps

Only the artificial burrow that replaced the collapsed burrow was occupied successfully. The breeding pair carried out some modifications, deepening the floor of the nest. An egg was detected in August of 2021. Following this, the nest was constantly monitored with camera traps and careful inspection. We placed rat poison baits (Brodifacoum) and a Goodnature A24 trap near the nest to control rodents and predators in the vicinity. The petrel chick successfully fledged in December 2021 (Fig. 14.7).

We believe that artificial burrows can provide a stable and protective nest for Galapagos Petrels and thus aid in their conservation in the long-term. They can also make monitoring and data gathering easier and more effective for researchers and rangers from the Galapagos National Park. In fact, by providing safe access to the chick via the removable top, these nests can be potentially used to relocate the fledging chicks to a safe location free of predators. All farm owners consulted for this project are willing to collaborate by accepting visits and monitoring of nests on their property. Furthermore, there is interest in controlling introduced species around nests and understanding the value of these actions.

Further efforts need to be made to secure more nesting sites, and trial of artificial burrows needs to be scaled up to properly increase nest site availability, particularly, necessary in some areas where there is more impact from the introduced species, both animal and plant species. Finally, we believe that it would be useful to perform

Fig. 14.7 Galapagos Petrel chick hatched in artificial nest burrow on Santa Cruz Island

soil analysis in the nesting areas to prove the importance of the long-term presence of the petrels and their contribution as vectors of nutrients from sea to land, especially trace elements.

Acknowledgments We thank the David and Lucille Packard Foundation for financing this project, the Galapagos National Park Service for their support and engagement in the conservation of Galapagos, and especially the farmers and landowners for allowing us to carry out the research activities with them in their land.

References

American Bird Conservancy, ABC (2021) After Years at Sea, First Hawaiian Petrel Pair Nests at Nihokū. https://abcbirds.org/news/hawaiian-petrel-nest-nihoku-2021/. Accessed on 26 February 2023

BirdLife International (2018) *Pterodroma phaeopygia*. The IUCN Red List of Threatened Species 2018. Accessed on 26 February 2023

Borrelle S, Jones H, Salguero-Gomez R, Bollard-Breen B, Towns D (2015) Are we rearranging the deckchairs on the Titanic? Evaluating multiple marine threats to Procellariiformes. In: Conference: world seabird conference. Cape Town, South Africa

Castro I, Phillips A (1996) A guide to the birds of the Galapagos Islands. Christopher Helm, A & B, London

Coulter M, Cruz F, Cruz J (1985) A programme to save the Dark-rumped Petrel, Pterodroma phaeopygia, on Floreana Island, Galapagos, Ecuador. International Council For Bird Preservation Technical Publication no 3, pp 177–180

Croxall JP, Butchart SH, Lascelles BEN, Stattersfield AJ, Sullivan BEN, Symes A, Taylor PHIL (2012) Seabird conservation status, threats and priority actions: a global assessment. Bird Conserv Int 22(1):1–34

Cruz J, Cruz F (1987) Conservation of the dark-rumped Petrel Pterodroma phaeopygia in the Galapagos Islands, Ecuador. Biol Conserv 42:303–311

Cruz Delgado F (2005) Biología Reproductiva de *Pterodroma phaeopygia* (Procellariidae) Isla San Cristóbal – Galapagos. Tesis de Biología. Universidad Central del Ecuador

Duffy D (1984) The endangered petrel of the Galapagos volcanoes. Noticias de Galápagos 39, pp 24–27

Granizo T, Pacheco C, Ribadeneira MB, Guerrero M, Suárez L (eds) (2002) Libro rojo de las aves del Ecuador. SIMBIOE/ Conservación Internacional/ EcoCiencia/ Ministerio de l Ambiente/ UICN. Serie Libros Rojos del Ecuador, tomo 2. Quito, Ecuador

Gummer H, Taylor G, Wilson KJ, Rayner MJ (2015) Recovery of the endangered Chatham petrel (Pterodroma axillaris): a review of conservation management techniques from 1990 to 2010. Glob Ecol Conserv 3:310–323

Harris MP (1970) The Biology of an Endangered Species, the Dark-Rumped Petrel (*Pterodroma phaeopygia*), in the Galapagos Islands. The Condor 72(1):76–84

Raine AF, Holmes ND, Travers M, Cooper BA, Day RH (2017) Declining population trends of Hawaiian Petrel and Newell's shearwater on the Island of Kaua 'i, Hawaii, USA. Condor 119(3):405–415

Valarezo JC, Wiedenfeld DA (2005) Estado actual y distribución del Petrel de Galápagos (*Pterodroma phaeopygia*) en la isla Santa Cruz. Reporte técnico. Fundación Charles Darwin. Galápagos

Warham J (1996) The behaviour, population biology and phisyology of the Petrels. Academic, London

Leo Zurita-Arthos is a geographer, expert on GIS and Geospatial Sciences, with a special interest in conservation, ecology, and sustainability. His team's work in Galapagos has helped to better understand the geographic distribution of the nesting areas, as well as the conflicts and opportunities they represent towards a more sustainable future.

Carolina Proaño is an Ecuadorian ecologist with vast experience in petrels and other seabirds in the Galapagos. She is a PhD student at University of Konstanz in Germany, precisely researching the ecology of the Galapagos petrel, as part of the Max Planck Institute for Ornithology.

Jonathan Guillén is a naturalist guide, born and raised in the Galapagos. He is an expert in the field and a real adventurer in finding known and new petrel nests. Starting as a volunteer, he has more than a decade of experience working with birds and other endemic species on the islands.

Sebastián Cruz is a permanent resident of the Galapagos who has worked in conservation of the islands' biodiversity since a young age. He holds a PhD. In Biology from the University of Konstanz in Germany where he studied the movement ecology of seabirds in the Galapagos.

David Wiedenfeld is a Senior Conservation Scientist at ABC. His research focuses on bird population ecology and conservation biology. He was Head of the Department of Vertebrate Ecology at the Charles Darwin Research Station in the Galapagos Islands. His current work is primarily with bird populations in the Americas.

Chapter 15
Impact of Weathering and Mineralogy on the Chemistry of Soils from San Cristobal Island, Galapagos

Xiao-Ming Liu, Heather D. Hanna, and Julia G. Barzyk

Introduction

Chemical weathering of silicate rocks plays a vital role in global processes such as oceanic nutrient fluxes and atmospheric CO_2 regulation (Misra and Froelich 2012; Penniston-Dorland et al. 2017). Chemical weathering of basalt, in particular, is an important contributor to atmospheric CO_2 regulation. Despite representing only ~8% of exposed silicate rock (Gaillardet et al. 1999), basalt chemical weathering is responsible for at least 30% of global CO_2 consumption (Dessert et al. 2003). Thus, characterizing the complexities of basalt weathering can provide an important foundation for global weathering studies (Liu and Rudnick 2011). Economically, basalt weathering is important to study as extreme weathering forms laterites, which are important iron ores. Finally, progressive weathering of basalt results in elemental loss, influencing the soil fertility (Kronberg and Nesbitt 1981).

This research aims to understand how the different climate zones on the San Cristobal Island, Galapagos, impact weathering of the underlying basalt and its resulting soil mineralogy and chemistry. The Galapagos Islands were established as a UNESCO World Heritage site in 1978, and have since been identified as one of the world's 100 most irreplaceable sites (Le Saout et al. 2013). However, increasing

---

Supplementary Information The online version contains supplementary material available at https://doi.org/10.1007/978-3-031-28089-4_15.

X.-M. Liu (✉) · H. D. Hanna
Department of Earth, Marine and Environmental Sciences,
University of North Carolina at Chapel Hill, Chapel Hill, NC, USA
e-mail: xiaomliu@unc.edu

J. G. Barzyk
Army Research Office, Research Triangle Park, NC, USA

S. J. Walsh et al. (eds.), *Island Ecosystems*, Social and Ecological Interactions in the Galapagos Islands, https://doi.org/10.1007/978-3-031-28089-4_15

immigration and tourism on the islands are posing ever greater threats to the delicate ecosystems of the archipelago (Gonzalez et al. 2008). To aid conservation efforts, high-priority questions whose answers would significantly benefit conservation and sustainability efforts have been identified. How soils vary across the Galapagos Islands has been identified as one of those questions in need of answering (Izurieta et al. 2018).

To date, research into Galapagos soils has focused heavily on Santa Cruz Island, while few studies have been conducted on the San Cristobal soils (Stoops 2014). San Cristobal is older than Santa Cruz and soils have been characterized as Alfisols instead of the Mollisols and Inceptisols reported for Santa Cruz (Adelinet et al. 2008; Stoops 2014). Thus, research on Santa Cruz soils is not necessarily applicable to San Cristobal soils. Additionally, San Cristobal has the longest history of agriculture among the Galapagos Islands (Stoops 2014) with agriculture occupying 17.7% of the island's total area (Lasso and Espinosa 2018). The most detailed published studies to date have been by Adelinet et al. (2008), who examined soil mineralogy and hydrodynamic properties, and a more vegetation-focused soil phytolith study by Astudillo (2018). Additionally, Franz (1980) provided an overview of soil types in some areas of San Cristobal, while Lasso and Espinosa (2018) summarized some findings in a Spanish-language publication by Wicknell (1997). Since sustainable agriculture is an important topic in the Galapagos, an improved understanding of the impact of weathering on soil chemistry could also help inform these practices in accordance with the Galapagos National Park's management plan.

Geologic Setting and Samples

San Cristobal Island, located on the easternmost side of the Galapagos archipelago (Fig. 15.1a) where the Nazca plate has moved off of the hotspot (Geist et al. 2008), provides an ideal location to study the impact of climate on basalt weathering. Morphologically, San Cristobal is composed of two separate volcanoes, but geochemical similarities in the lavas suggest they come from the same volcanic plumbing system (Geist et al. 2008). An extinct shield volcano forms the southwestern portion of San Cristobal and provides the island's topographic high (Fig. 15.1b). In contrast, the low elevations to the northeast are dominated by a newer series of fissure eruptions (Geist et al. 1986). Compositionally, the island is almost entirely basalt, which ranges in age from 2.33 ± 0.13 Ma to an estimated <1 ka (Geist et al. 1986). Soils are best developed on the windward (southeast) side of the shield volcano (Adelinet et al. 2008) where most lavas have been determined to be of Brunhes age by magnetic polarity measurements (Group 3 in Fig. 15.1b) with two K-Ar ages measured at 0.66 Ma and 0.89 Ma (Geist et al. 1986).

The trade winds and ocean currents that control the Galapagos climate are strongly influenced by the Inter Tropical Convergence Zone (ITCZ) and the El Niño Southern Oscillation (ENSO; Trueman and D'Ozouville 2010), resulting in a colder and drier climate than is generally found at the equator (Adelinet et al. 2008). The ITCZ

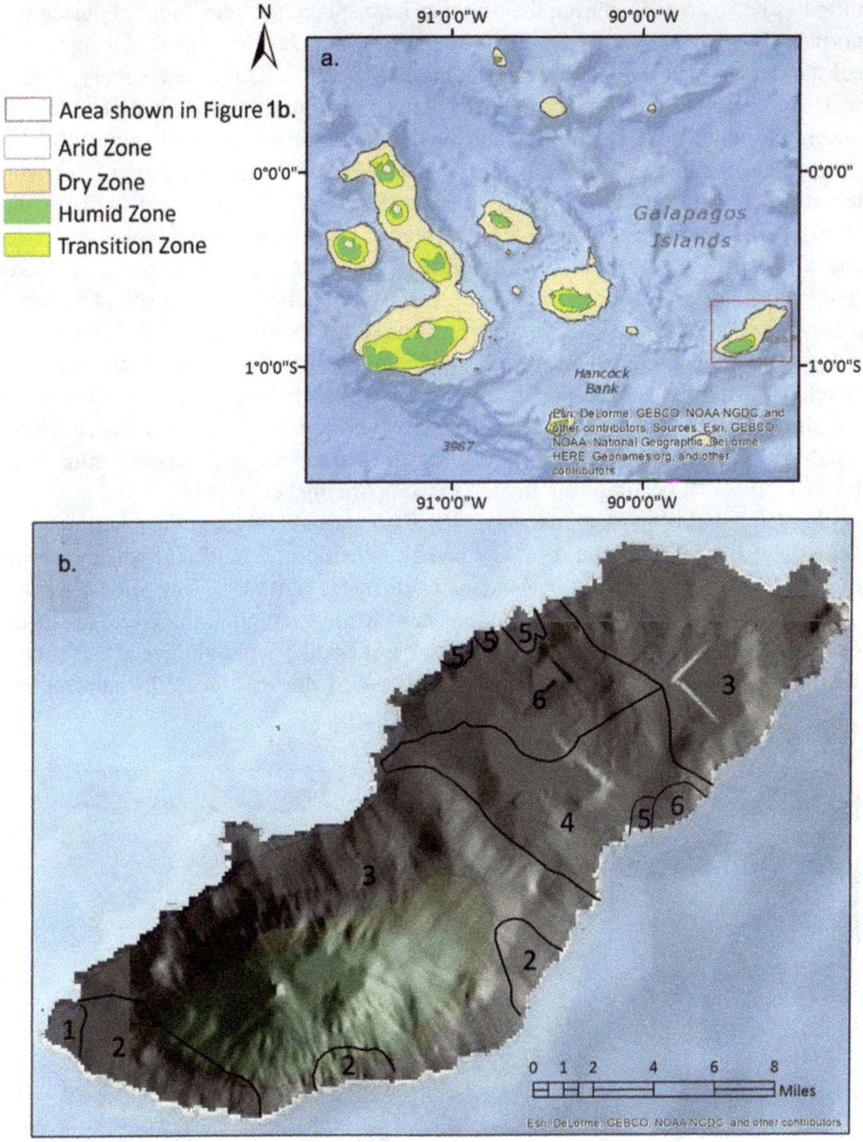

Fig. 15.1 (**a**) Map of climate zones on the Galapagos islands after Huttel (1986) and (**b**) San Cristobal island with LiDAR base to show topography. Location of San Cristobal island shown in red box. Climate zones from part a are overlain on LiDAR base with 70% transparency. Black lines and corresponding numbers denote age distribution of lava flows as determined using flow morphology and paleomagnetism by Geist et al. (1986). Ages range from Group 1 lavas, inferred to be the oldest with K-Ar age of 2.32 Ma, to Group 6 lavas, inferred to be the youngest based on plant development and surface weathering. See Fig. 15.1 of Geist et al. (1986) for more information. Lavas to be sampled in this study are mostly Group 3 lavas with site in the Group 2 lavas. (Esri, DeLorme GEBCO, NOAA NGDC, and other contributor. Sources: Esri, GEBCO, NOAA, National Geographic, DeLorme, HERE, Geonames.org, and other contributors. Esri, DeLorme, GEBCO, NOAA NGDC, and other contributors)

migrates between 10°N during the northern hemisphere summer and 3°N during the northern hemisphere winter resulting in a hot season (January–May), during which substantial rain can fall on lowlands, and a cold season (June–December), during which a temperature inversion produces abundant stratus clouds that increase moisture in the highlands (Trueman and D'Ozouville 2010; Lasso and Espinosa 2018). The presence of these stratus clouds has caused elevation-related climate zonation up the steep topography of the shield volcano (Fig. 15.2; Trueman and D'Ozouville 2010). An arid zone receiving <400 mm/year of precipitation is present at the coast and a dry zone receiving < 800 mm/year of precipitation occurs in the lowlands (Huttel 1986). A transition zone is located mid-way up the side of the shield volcano, receiving 800–1100 mm/year of precipitation (Huttel 1986). Finally, the humid zone is located at higher elevations, receiving annual precipitation up to 1500–2000 mm (Huttel 1986; Adelinet et al. 2008). While these climate zones occur on all sides of the shield volcano, the north-facing leeward side is drier than the southern-facing windward side (Huttel 1986) in part because the former only receives rain from heavy storms that occur during the hot season (Adelinet et al. 2008).

San Cristobal is an ideal place to study basalt weathering and the resulting soils because of the relatively uniform age and composition of the shield volcano, combined with the different climate zones at different altitudes. This allows for the source material to be held relatively constant while examining the effect of differences in rainfall and, by extension, the degree of weathering. Samples of fresh rock, weathered rock, saprolite, and soil were collected from four locations representing

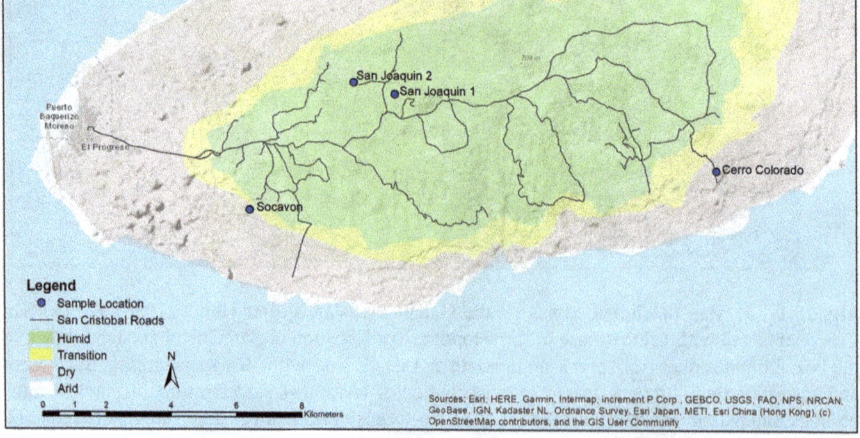

Fig. 15.2 Map of study area on San Cristobal Island showing climate zones and sample locations. (Sources: Esri, HERE, Garmin, Intermap, increment P Corp., GEBCO, USGS, FAO, NPS, NRCAN, GeoBase, IGN, Kadaster NL, Ordnance Survey, Esri Japan, METI, ESri China (Hong Kong). (c) OpenStreetMap Contributors,and the GIS User Community)

different climate zones in the southern portion of San Cristóbal (Fig. 15.2). Details of samples are reported in Hanna (2020) and a brief description is provided here.

The Cerro Colorado site (Fig. 15.2) is located in an abandoned quarry in the dry zone at ~120 m elevation, sampled a 160 cm-thick profile which consisted of a thin (<20 cm) soil horizon underlain by cohesive rock. The soil horizon was sampled at 0 cm and 10 cm while the cohesive rock was sampled in 20 cm intervals from 20 cm to 160 cm depth due to visual homogeneity of the profile and difficulty of collecting samples from the well-indurated rock. Photographs of the Cerro Colorado site from before and after sampling are shown in Fig. 15.3a. Fresh rock was not present at the Cerro Colorado site, so potential "parent" samples were not collected in this location.

A pit was dug for sampling at the dry to transition zone Socavòn site (~180 m elevation; Fig. 15.2), which contained ~40 cm of soil underlain by saprolite. Sample locations were scored into the pit wall every 10 cm from the top down, including a saprolite sample at the bottom of the pit (46 cm). Samples were then collected from the bottom of the pit upwards to prevent contamination of lower samples by falling debris during upper profile sampling. Since digging the pit contaminated the surface with soil from deeper in the hole, a 0 cm sample was collected a few feet away in an uncontaminated area. Finally, a corresponding rock sample was collected from an outcrop at that site. Photographs of the Socavòn site from before and after sampling are shown in Fig. 15.3b.

The San Joaquin 2 site (Fig. 15.2), located in a humid zone road cut at ~470 m elevation, was the longest profile collected at 170 cm. The San Joaquin 2 site was composed of soil for the top ~55 cm, which was underlain by saprolite. Soil and saprolite samples were collected every 10 cm, and a corresponding rock sample was collected from an outcrop adjacent to the soil profile in the road cut. Photographs of the San Joaquin 2 site from before and after sampling are shown in Fig. 15.3c.

A 120-cm profile was sampled from the humid zone San Joaquin 1 site at ~540 m elevation (Fig. 15.2), which was composed of ~90 cm-thick soil underlain by saprolite. Soil and saprolite samples were collected every 10 cm starting at the surface. Photographs of the San Joaquin 1 site from before and after sampling are shown in Fig. 15.3d. A corresponding rock sample was collected from the nearest accessible outcrop, located ~100 m away.

Methods

Soil and saprolite samples were baked at 200 °C for 4 min according to the United States Department of Agriculture's Animal and Plant Health Inspection Service permit requirements at the Galapagos Science Center, San Cristobal, Galapagos. All subsequent sample preparation and elemental concentration analyses were conducted at the University of North Carolina at Chapel Hill. Soil samples were ground using an agate mortar and pestle or a shatterbox. Weathered areas of rock samples were first removed using a rock saw, they rock samples were cleaned and powdered using a shatterbox.

Fig. 15.3 Before and after sampling photographs of (**a**). Cerro Colorado and (**b**) Socavòn profiles. Debris fell into the hole and obscured view of 46 cm sample in Socavòn after the 46 cm sample was collected. Before and after sampling photographs of (**c**). San Joaquin 2 and (**d**) San Joaquin 1 profiles

Fig. 15.3 (continued)

Powdered soil, saprolite, and rock samples were dissolved for elemental analysis using a protocol modified from Li et al. (2019). Approximately one hundred milligrams of each sample were weighed and transferred to 50 ml Teflon beakers for sample digestion. Three milliliters of H_2O_2 were immediately added to the sample to oxidize organics. The beakers were placed on a hotplate at 60 °C overnight and were then dried down at 50 °C the next day. Once H_2O_2 was evaporated, 3 ml of Aqua Regia and 0.5 ml of concentrated HF were added to the beakers, which were then placed on a 180 °C hotplate in a fume hood until total sample dissolution. Then the samples were dried down at 130 °C and fluxed on a hot plate with 5 ml of HNO_3 at 180 °C for 24 h. After complete dissolution, samples were dried down at 130 °C and then re-dissolved in 2 ml of concentrated HCl for future elemental analysis using an Agilent™ 7900 Quadrupole Inductively Coupled Plasma Mass Spectrometer (Q-ICP-MS). The accuracy and reproducibility of the Q-ICP-MS analyses were evaluated using the BHVO-2 and SBC-1 standards (Table S1).

Aliquots of select powdered soil, saprolite, and rock samples were also analyzed using X-Ray Diffraction (XRD) at the Chapel Hill Analytical and Nanofabrication Lab (CHANL) at the University of North Carolina at Chapel Hill. Diffraction patterns were collected on randomly oriented powder samples using the Rigaku SmartLab theta-theta diffractometer using CuKa(alpha) radiation (40 kV, 44 mA) with Bragg-Brentano focusing and a K beta filter. Scans were conducted from 5° to 80° with a scan rate of 1° 2 (theta)/min, and quantitative mineralogical percentages were obtained using the Whole Powder Pattern Fitting function of the Rietveld Analysis, built into the Rigaku PDXL software. Percentages given represent percentage of crystalline phases since the program cannot quantify amorphous phases.

Neodymium (Nd) isotope data were collected on select soil, saprolite, and rock powders from the Cerro Colorado, Puerto Chino, Socavòn and San Joaquin 2 sites, and were analyzed at the State Key Laboratory of Isotope Geochemistry, Guangzhou Institute of Geochemistry, Chinese Academy of Sciences. Isotope separation was conducted using the two-column protocol of Ma et al. (2013). After two-step column chemistry, the resulting solutions were analyzed using a Nu Plasma 1700 multi-collector–inductively coupled plasma–mass spectrometer (MC-ICP-MS). Neodymium mass bias was corrected using sample-standard bracketing, and triplicate analysis of the JNDi Nd standard during the run yielded an average $^{143}Nd/^{144}Nd$ value of 0.512101. This agrees with the average GEOREM $^{143}Nd/^{144}Nd$ value of 0.512115, ranging from 0.51109 to 0.51295 based on 583 values (Jochum et al. 2005).

Strontium (Sr) isotope analysis was performed at the State Key Laboratory of Isotope Geochemistry of the Guangzhou Institute of Geochemistry, Chinese Academy of Sciences. The Sr isotope separation followed the procedure of Zhu et al. (2018). Strontium mass bias was corrected using $^{86}Sr/^{88}Sr = 0.1194$, and repeat analysis of the SRM987 and BHVO-2 standards yielded average values of $^{87}Sr/^{86}Sr = 0.710252$ and 0.703455, respectively. Our NIST SRM987 value agrees with the certified $^{87}Sr/^{86}Sr$ values of 0.71034 ± 0.00026. Our BHVO-2 standard $^{87}Sr/^{86}Sr$ average is consistent with the GeoREM database, which ranges from 0.703404 to 0.7037 (128 values).

Results

X-Ray Diffraction

XRD analysis indicates that crystalline phases in dry zone Cerro Colorado samples are mainly composed of the primary igneous minerals anorthite (~63% to ~81%), forsterite (~7% to ~22%), and augite (~7% to ~19%; Table S2). The only alteration mineral detected is ferrihydrite ranging from ~1% to ~9%. However, ferrihydrite content may be underrepresented since poor crystallinity can make it difficult to detect with an XRD. A small amount of cristobalite (<1%) was detected in the 160 cm sample, however, this may be an artifact caused by the peak interference (Nelson et al. 2017). Moreover, the presence or absence of smectites could not be confidently determined with the Rigaku SmartLab since the sample holder interferes with analyses at angles below ~8°, probably missing an important smectite peak.

XRD analysis of dry to transition zone Socavòn samples indicates that all primary igneous minerals have been altered to clay minerals and Fe oxides and oxyhydroxides (Table S2). The clay minerals are composed of kaolinite and halloysite, representing between 45% and 58% of the crystalline phases, while Goethite (~16% to ~45%), and hematite (~1% to ~5%) make up the iron-bearing crystalline phases (Table S2). Additionally, between ~18% and ~23% of cristobalite is present in the top 30 cm of the profile.

Two sites San Joaquin 2 and San Joaquin 1 are in the area mapped as the humid zone. All San Joaquin 2 samples contain a mixture of primary and secondary minerals (Table S2). Anorthite (~1% and ~40%), Augite (~15% to ~38%), and forsterite (~0% to ~11%; Table S2) are the primary igneous minerals identified. The clay minerals detected by XRD are kaolinite and halloysite (~3% to ~36%), while the Fe-bearing crystalline phases are goethite (~3% to ~46%) and hematite (~2% and ~11%; Table S2). The San Joaquin 1 samples do not contain any primary igneous minerals (Table S2). Clay minerals in the San Joaquin 1 profile are composed of kaolinite and halloysite (~7% to ~56%), and gibbsite (~10% and ~60%), while goethite (~<1% to ~62%), and hematite (~<1% to ~24%) comprise the oxide and oxyhydroxide minerals (Table S2).

Chemical Indices of Weathering

The Chemical Index of Alteration (CIA) was calculated for San Cristobal samples using major element data in Table S3. CIA is a weathering index that primarily reflects feldspar dissolution and the resulting loss of mobile CaO, Na_2O, and K_2O relative to Al_2O_3, the latter of which is presumed to be immobile due to its incorporation into pedogenetic clay minerals (Babechuk et al. 2014). CIA is calculated as the molar ratios of $[Al_2O_3/(Al_2O_3+CaO^*+Na_2O+K_2O)]x100$, with CaO*

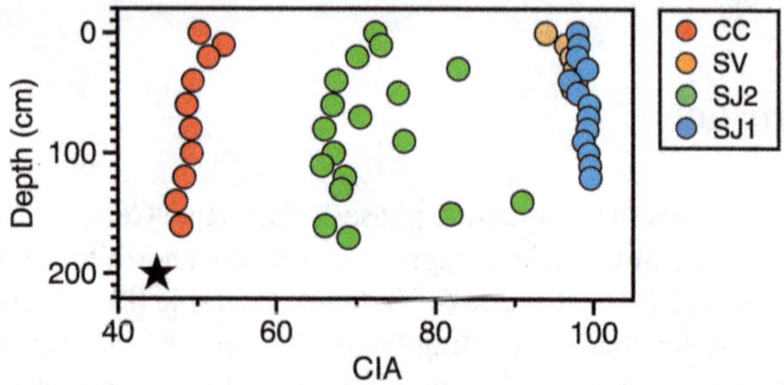

Fig. 15.4 Chemical Index of Alteration (CIA) versus depth for San Cristobal samples. The black star indicates the average parent, while filled circles of different colors represent corresponding weathered regolith/soil samples: Cerro Colorado (CC), Socavòn (SV), San Joaquin 2 (SJ2), and San Joaquin 1 (SJ1)

representing the CaO in silicate phases after the contribution from carbonates and apatite has been removed (Nesbitt and Young 1982). The fresh basalt CIA range is 30–45 (Nesbitt and Young 1982), while idealized montmorillonites and illite have CIA values between 75 and 85, and idealized kaolinite with CIA value close to 100 (Nesbitt and Young 1982). Due to the low CaO concentration in the samples and the lack of evidence for carbonates or apatite in the XRD data, we do not make the CaO* correction for our CIA calculation.

All fresh rock samples in this study plot at the upper end of the fresh basalt CIA range with an average value of 45, while CIA values for the soil sites show elevated values of various degrees (Fig. 15.4). The Cerro Colorado samples in the arid zone have slightly evaluated CIA values (47–53) with minor amount of ferrihydrite. The transition zone Socavòn soil and saprolite samples have CIA values ranging from 94 at the surface to 98 near the bottom of the profile. For the humid zone sites, San Joaquin 2 soil and saprolite samples display a jagged pattern encompassing a wide range of CIA values (66–91), with most samples exhibiting CIA values from 66 to ~75. Samples collected at 30 cm (CIA = 83), 140 cm (CIA = 91), and 150 cm (CIA = 82) displaying noticeably higher values than the other San Joaquin 2 samples. The San Joaquin 1 CIA values are the highest (97–99.6) among these soils.

Sr and Nd Isotopes

All isotope data collected in this study are reported in Table S4. All samples from soil profiles show similar Sr isotope values similar to the average parent (Fig. 15.5a) with the one exception – Socavòn samples. The Socavòn saprolite sample collected from the bottom of the profile (46 cm) also falls within the Sr isotope range exhibited by San Cristobal parent basalts, however, the top 40 cm of the profile have significantly more radiogenic ratios than any of the studied samples.

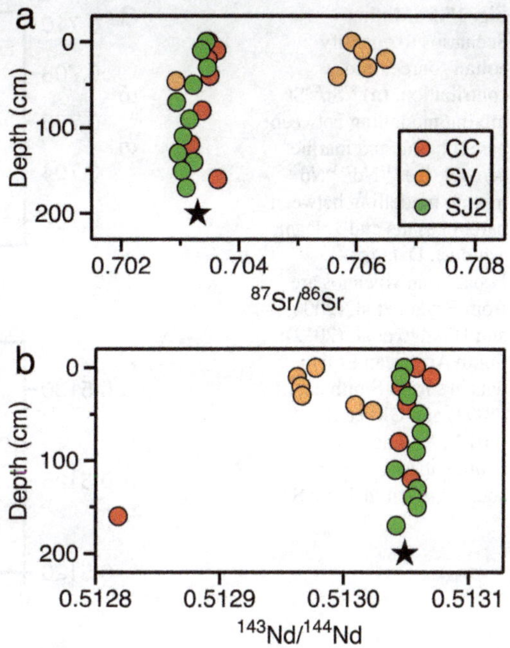

Fig. 15.5 (**a**) $^{87}Sr/^{86}Sr$ and (**b**) $^{143}Nd/^{144}Nd$ versus depth for San Cristobal samples. The black star indicates the average parent basalt and color symbols show soils from different profiles. Error bars for Sr ($2\sigma = \pm 0.000012$) and Nd ($2\sigma = \pm 0.00001$) isotopes are smaller than the symbols

Most soil samples are within the Nd isotope range of the San Cristobal parents, except that the sample from 160 cm of the Cerro Colorado profile has the least radiogenic value ($^{143}Nd/^{144}Nd = 0.512828$) (Fig. 15.5b). Moreover, the Socavòn soil samples from 0 cm to 30 cm exhibit significantly less radiogenic values than the lower Socavòn samples, the San Joaquin 2 samples, or the Cerro Colorado samples from the top 120 cm (Fig. 15.5b).

Discussion

Eolian Contributions to San Cristóbal Samples

Marine Aerosol Influence

Sr isotope ratios of Socavòn samples from the top 40 cm of the profile plot to significantly heavier values than the Cerro Colorado and San Joaquin 2 samples (Fig. 15.5a). Thus, atmospheric deposition of marine aerosols formed by evaporation of hydrated and/or dissolved ocean-derived salts can be important sources of some cations in soil profiles (Derry and Chadwick 2007). These marine aerosols can be approximated using seawater (Vitousek et al. 1999), which has a heavier Sr isotopic signature and lower Sr concentration than Socavòn rocks (Paytan et al. 1993).

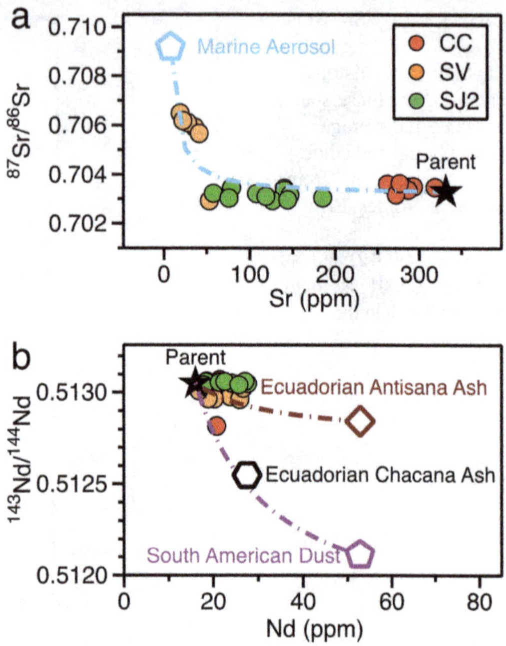

Fig. 15.6 Mixing Scenarios to quantify eolian sources and contribution. (**a**) ^{87}Sr/^{86}Sr mixing modelling between parent basalt and marine aerosol; (**b**) ^{143}Nd/^{144}Nd mixing modelling between parent basalts and volcanic ash/dust. Data for Ecuadorian volcanos are from Bryant et al. (2006) and Hidalgo et al. (2012); South American Eolian data are from Smith et al. (2003) and Gili et al. (2017). Mixing endmember values and sources given in Table S5

Thus, Sr concentration versus ^{143}Nd/^{144}Nd ratio mixing models are shown for the top 30 cm of the Socavòn samples in Fig. 15.6a and endmember compositions are given in Table S5.

Volcanic Ash and Eolian Dust Influence

Samples from the top 30 cm of the Socavòn profile, and the 160 cm Cerro Colorado sample display significantly lower ^{143}Nd/^{144}Nd ratios compared to the other San Cristóbal samples with isotopic data from this study and White et al. (1993) (Fig. 15.5b). Thus, mixing of an eolian component with the Socavòn parent rock is required to explain the Nd isotopic compositions of the Socavòn samples from the top 30 cm and the Cerro Colorado sample from 160 cm. The marine aerosols that dominated Sr isotopic signatures are not suitable for Nd isotope mixing endmembers due to the extremely low Nd concentration in seawater.

The volcanoes from the other Galapagos islands, including the currently active volcanoes of Isabela Island, are not likely contributors due to the southeasterly trade winds, which are the dominant winds in the Galápagos. Thus, the predominant southeasterly wind direction along with San Cristobal's location as the eastern-most island, can be used to exclude other Galapagos volcanoes as likely contributors of ash to San Cristobal while supporting mainland Ecuadorian volcanoes as potential sources.

South American eolian dust is another potential endmember for the lower $^{143}Nd/^{144}Nd$ samples. While it may be possible that dust from Australia, the Sahara, or Asia is reaching San Cristóbal Island during interglacial periods (Xie and Marcantonio 2012), Nd isotope values of these areas largely overlap with those of South American loess. Thus, we use South American loess from Gili et al. (2017) for mixing models.

The Nd concentration versus $^{143}Nd/^{144}Nd$ ratio mixing model is shown for Socavòn samples from the top 30 cm in Fig. 15.6b, with endmember compositions given in Table S5. The mixing scenarios use an ash composition from the Antisana volcano in Ecuador (Bryant et al. 2006; Hidalgo et al. 2012), suggesting ~20% of ash mixing with the Socavòn basalt endmember. The values for the top 30 cm are within the range of Liu et al. (2013) who noted 20–60% dust addition to bauxites developed from the Columbia River basalts. A dust-derived component in the Socavòn profile is also supported by the presence of cristobalite, representing ~18% to ~23% of crystalline phases in samples from the top 30 cm. XRD analysis indicates cristobalite is not present in the parent rock samples or in the soil samples with Nd isotopic values similar to those of the rocks. Additionally, attributing up to ~23% quartz from eolian sources is consistent with the findings of Kurtz et al. (2001), who attributed up to 30% quartz in their Hawaiian samples to Asian dust. The majority of dust compositions from Gili et al. (2017) do not, as a pure endmember, explain the lower $^{143}Nd/^{144}Nd$ Socavòn samples, however, they cannot be ruled out as a mixed contribution with dominantly volcanic ash.

The 160 cm Cerro Colorado sample also displays notably lower $^{143}Nd/^{144}Nd$ values than the rest of the San Cristóbal samples. While the Cerro Colorado site is composed mostly of coherent rock, the 160 cm sample was collected from an area that looks like a contact between two lava flows. Mixing models suggests $^{143}Nd/^{144}Nd$ ratios in these samples can be explained by mixing of basalt with ~35% ash of similar composition to the pre-caldera eruption phase of Ecuadorian Chacana volcano, or ~10% input from South American Eolian deposits (Fig. 15.6b). Unlike the upper Socavòn samples, the 160 cm Cerro Colorado sample does not have significant quartz. However, due to the low melting point of quartz relative to the temperature of basaltic lavas, it is possible emplacement of the overlying lava flow melted and incorporated any quartz that was previously present on the surface of the lower flow.

We note that the amounts of eolian addition calculated above serve as a first-order estimate than an exact percentage, especially given the complexity of the system and the challenges involved in pinpointing values for endmembers. For example, loss of Sr and Nd from basalt during weathering would shift concentrations to lower values, probably underestimating the eolian inputs. Additionally, although mixing models have been calculated using pure ash or dust endmembers, it is likely that both contribute in some proportion to the lower $^{143}Nd/^{144}Nd$ Cerro Colorado and top Socavòn samples. Nonetheless, this research suggests that the dust and volcanic ash from the South American mainland can adequately explain Nd isotopic ratios in San Cristóbal soil samples without input from Australia and the northern hemisphere.

Weathering Intensity and Mineralogy Control in San Cristóbal Soils

We note that clay mineral content (kaolinite + Halloysite) positively correlates with soil depth in San Cristobal soils (Fig. 15.7), demonstrating that the weathering intensity of our soil profiles generally increases toward the surface for each individual profile. Overall, CIA values in soil sites show elevated values of various degrees compared to the parent basalt, reflecting their weathering intensity difference with potential source variation (Fig. 15.4).

Cerro Colorado samples exhibit low CIA values and minimal presence of alteration minerals compared to samples from other sites, suggesting only incipient weathering has occurred. This is consistent with the site's location in the dry zone, where there is limited precipitation to alter primary minerals. Surprisingly, Socavòn samples show very high CIA values, and all primary igneous minerals have been altered to kaolinite and Fe oxides and oxyhydroxides. These suggest high-intensity weathering has occurred, which was unexpected for the dry to transition zone Socavòn samples. However, the Socavòn site is located near the area mapped as Group 2 Matuyama lavas (~1 Ma to ~0.7 Ma; Geist et al. 1986) and may be older than the Brunhes age of most Group 3 lavas. An older age for Socavòn lavas relative to the other sites may explain the high intensity of weathering despite the Socavòn site's dry to transition zone location. Alternatively, Socavòn soils show significant input from eolian sources as discussed in the previous section (>20% ash). Therefore, high CIA and clay mineral content may result from easily weathered volcanic ash instead of basaltic lava.

Both San Joaquin 1 and 2 are located on the leeward side of the island in what has been mapped as the humid zone, however, they exhibit significant differences in the extent of weathering. San Joaquin 2 samples have retained some primary igneous minerals, and most samples exhibit moderate CIA values suggesting a moderate degree of weathering. Samples from the San Joaquin 1 site have high CIA values, and are composed of kaolinite and halloysite, goethite, hematite, and gibbsite. The highly weathered San Joaquin 1 site is located in the highlands (>500 m), and thus is likely influenced by rainfall and fog from the extensive cold-season stratus clouds

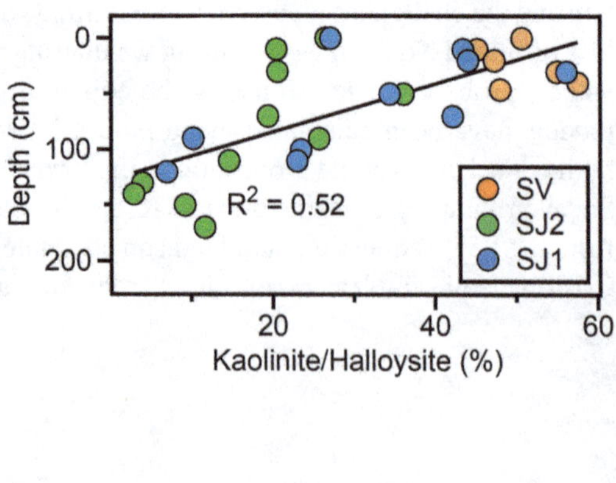

Fig. 15.7 Major secondary mineral content (Kaolinite + Halloysite, %) versus depth for San Cristobal samples. Cerro Colorado (CC) does not have detectable secondary minerals

that serve as an important source of precipitation in the highlands (Percy et al. 2016). This could result in high humidity, therefore higher weathering intensity at the site. In contrast, the relatively lower elevation of the San Joaquin 2 site suggests it would be less impacted by the stratus clouds. Thus, the San Joaquin 2 site would be drier, and therefore less weathered, than the San Joaquin 1 site.

Conclusions

The San Cristobal Island soils provide insight into how mineralogical controls on weathering may progress as climate changes, as well as how weathering in drier areas may progress with time. When age is held constant to examine the impact of climate, congruent weathering dominates in minimally weathered basalt in a dry climate, with ferrihydrite formation and incongruent weathering of plagioclase playing a lesser role. As humidity and weathering intensity increase, primary igneous minerals become secondary minerals. Crystalline secondary phases become important, but do not become dominant until conditions of high humidity and weathering intensity are reached. At this point, gibbsite becomes dominant along with kaolin minerals and Fe oxides and oxyhydroxides.

Radiogenic isotopes indicate parent basalt weathering is not the only contributor to soils from the Socavòn site. $^{87}Sr/^{88}Sr$ ratios suggest the parent rock-derived Sr from the top 40 cm of the Socavòn site has been depleted by weathering and significantly influenced by marine aerosol. The lowest $^{143}Nd/^{144}Nd$ value from a deep Cerro Colorado sample also suggests an eolian influence. Mixing models suggest Nd isotopic compositions of the Cerro Colorado sample can be explained by ~35% contribution from the ash of similar composition to the Ecuadorian Chacana volcano, and/or ~10% input from South American dust deposits. This suggests South American mainland sources alone can explain Nd isotope signatures in San Cristobal soils, without needing northern hemisphere and Australian dust deposition in the Eastern Equatorial Pacific.

Acknowledgements This research was supported by the US Army Research Office under grant W911NF-17-2-0028. The views and conclusions contained in this document are those of the authors and should not be interpreted as representing the official policies, either expressed or implied, of the sponsors, including the Army Research Laboratory or the U.S. Government. The U.S. Government is authorized to reproduce and distribute reprints for Government purposes notwithstanding any copyright notation herein. XRD analysis was performed at the Chapel Hill Analytical and Nanofabrication Laboratory, CHANL, a member of the North Carolina Research Triangle Nanotechnology Network, RTNN, which is supported by the National Science Foundation, Grant ECCS-2025064, as part of the National Nanotechnology Coordinated Infrastructure, NNCI. A small grant from CHANL was received to support XRD analysis. Liu and Hanna would like to acknowledge funding support from the University of North Carolina at Chapel Hill. Finally, the authors thank the Parque Nacional Galápagos and the Galápagos Research Center for their roles in facilitating field work on San Cristóbal Island.

References

Adelinet M, Fortin J, D'Ozouville N, Violette S (2008) The relationship between hydrodynamic properties and weathering of soils derived from volcanic rocks, Galápagos Islands (Ecuador). Environ Geol 56:45–58. https://doi.org/10.1007/s00254-007-1138-3

Astudillo FJ (2018) Soil phytoliths as indicators of initial human impact on San Cristóbal Island, Galápagos. Palaeogeogr Palaeoclimatol Palaeoecol 490:522–532. https://doi.org/10.1016/j.palaeo.2017.11.038

Babechuk MG, Widdowson M, Kamber BS (2014) Quantifying chemical weathering intensity and trace element release from two contrasting basalt profiles, Deccan Traps, India. Chem Geol 363:56–75. https://doi.org/10.1016/j.chemgeo.2013.10.027

Bryant JA, Yogodzinski GM, Hall ML, Lewicki JL, Bailey DG (2006) Geochemical constraints on the origin of volcanic rocks from the Andean Northern volcanic zone, Ecuador. J Petrol 47:1147–1175. https://doi.org/10.1093/petrology/egl006

Derry LA, Chadwick OA (2007) Contributions from earth's atmosphere to soil. Elements 3:333–338. https://doi.org/10.2113/gselements.3.5.333

Dessert C, Dupré B, Gaillardet J, Francois LM, Allegre CJ (2003) Basalt weathering laws and the impact of basalt weathering on the global carbon cycle. Chem Geol 202:257–273. https://doi.org/10.1016/j.chemgeo.2002.10.001

Franz H (1980) Old soils and land surfaces on the Galápagos Islands. GeoJournal 4(2):182–184

Gaillardet J, Dupré B, Louvat P, Allègre CJ (1999) Global silicate weathering and CO2 consumption rates deduced from the chemistry of large rivers. Chem Geol 159:3–30. https://doi.org/10.1016/S0009-2541(99)00031-5

Geist DJ, McBirney AR, Duncan RA (1986) Geology and petrogenesis of lavas from San Cristóbal Island, Galápagos Archipelago. Geol Soc Am Bull 97:555–566. https://doi.org/10.1130/0016-7606(1986)97<555:GAPOLF>2.0.CO;2

Geist D, Diefenbach BA, Fornari DJ, Kurz MD, Harpp K, Blusztajn J (2008) Construction of the Galápagos platform by large submarine volcanic terraces. Geochem Geophys Geosyst 9:1–27. https://doi.org/10.1029/2007GC001795

Gili S, Gaiero DM, Goldstein SL, Chemale F, Jweda J, Kaplan MR et al (2017) Glacial/interglacial changes of Southern Hemisphere wind circulation from the geochemistry of South American dust. Earth Planet Sci Lett 469:98–109. https://doi.org/10.1016/j.epsl.2017.04.007

González JA, Montes C, Rodriguez J, Tapia W (2008) Rethinking the Galápagos Islands as a complex social-ecological system: implications for conservation and management. Ecol Soc 13. https://doi.org/10.5751/ES-02557-130213

Hanna, H. (2020). Geochemical investigation of fluid-rock interactions during chemical weathering and subduction. https://doi.org/10.17615/5w3w-k022

Hidalgo S, Gerbe MC, Martin H, Samaniego P, Bourdon E (2012) Role of crustal and slab components in the Northern Volcanic Zone of the Andes (Ecuador) constrained by Sr-Nd-O isotopes. Lithos 132–133:180–192. https://doi.org/10.1016/j.lithos.2011.11.019

Huttel C (1986) Zonificación bioclimatológica y formaciones vegetales en las Islas Galápagos. Cult Del Banco Cent Ecuador 24:1–13

Izurieta A, Delgado B, Moity N, Calvopiña M, Cedeño I, Banda-Cruz G et al (2018) A collaboratively derived environmental research agenda for galápagos. Pac Conserv Biol 24:168–177. https://doi.org/10.1071/PC17053

Jochum KP, Nohl U, Herwig K, Lammel E, Stoll B, Hofmann AW (2005) GeoReM: a new geochemical database for reference materials and isotopic standards. Geostand Geoanal Res 29:333–338. https://doi.org/10.1111/j.1751-908X.2005.tb00904.x

Kronberg BI, Nesbitt HW (1981) Quantification of weathering, soil geochemistry and soil fertility. J Soil Sci 32:453–459. https://doi.org/10.1111/j.1365-2389.1981.tb01721.x

Kurtz AC, Derry LA, Chadwick OA (2001) Accretion of Asian dust to Hawaiian soils: isotopic, elemental, and mineral mass balances. Geochim Cosmochim Acta 65:1971–1983. https://doi.org/10.1016/S0016-7037(01)00575-0

Lasso L, Espinosa J (2018) Soils from the Galápagos Islands. In: Espinosa J, Moreno J, Bernal G (eds) The soils of Ecuador. Springer, pp 139–150

Le Saout S, Hoffmann M, Shi Y, Hughes A, Bernard C, Brooks TM et al (2013) Protected areas and effective biodiversity conservation. Science (80–) 342:803–805. https://doi.org/10.1126/science.1239268

Li W, Liu X-M, Godfrey L (2019) Development of a matrix independent lithium chromatography and isotopic analysis in geostandards by MC-ICP-MS. Geostand Geoanal Res 43:261–276

Liu X-M, Rudnick RL (2011) Constraints on continental crustal mass loss via chemical weathering using lithium and its isotopes. Proc Natl Acad Sci 108:20873–20880. https://doi.org/10.1073/pnas.1115671108

Liu X-M, Rudnick RL, McDonough WF, Cummings ML (2013) Influence of chemical weathering on the composition of the continental crust: Insights from Li and Nd isotopes in bauxite profiles developed on Columbia River Basalts. Geochim Cosmochim Acta 115:73–91. https://doi.org/10.1016/j.gca.2013.03.043

Ma J, Wei G, Liu Y, Ren Z, Xu Y, Yang Y (2013) Precise measurement of stable neodymium isotopes of geological materials by using MC-ICP-MS. J Anal Spectrom 28:1926–1931. https://doi.org/10.1039/c3ja50229e

Misra S, Froelich PN (2012) Lithium isotope history of cenozoic seawater: changes in silicate weathering and reverse weathering. Science (80) 335:818–823. https://doi.org/10.1126/science.1214697

Nelson ST, McBride JH, Sowards KF, Bickmore BR, Tingey DG, Rey KA et al (2017) Incorrect info, actually sowards article early weathering of Ocean Islands: a synthesis of geochemistry and geophysics, Kohala Peninsula, Hawaii, USA, vol. 14, pp 1324–1342. https://doi.org/10.1130/abs/2017am-300124

Nesbitt HW, Young GM (1982) Early proterozoic climates and plate motions inferred from major element chemistry of lutites. Nature 299:715–717

Paytan A, Kastner M, Martin EE, MacDougall JD, Herbert T (1993) Marine barite as a monitor of seawater strontium isotope composition. Nature 366:445–449. https://doi.org/10.1038/366445a0

Penniston-Dorland SC, Liu X-M, Rudnick RL (2017) Lithium isotope geochemistry. Rev Mineral Geochem 82:165–217. https://doi.org/10.2138/rmg.2017.82.6

Percy MS, Schmitt SR, Riveros-Iregui DA, Mirus BB (2016) The Galápagos archipelago: a natural laboratory to examine sharp hydroclimatic, geologic and anthropogenic gradients. Wiley Interdiscip Rev Water 3:587–600. https://doi.org/10.1002/wat2.1145

Smith J, Vance D, Kemp RA, Archer C, Toms P, King M et al (2003) Isotopic constraints on the source of Argentinian loess – with implications for atmospheric circulation and the provenance of Antarctic dust during recent glacial maxima. Earth Planet Sci Lett 212:181–196. https://doi.org/10.1016/S0012-821X(03)00260-7

Stoops G (2014) Soils and paleosoils of the Galápagos Islands: what we know and what we don't know, a meta-analysis. Pac Sci 68:1–17. https://doi.org/10.2984/68.1.1

Trueman M, D'Ozouville N (2010) Characterizing the Galápagos terrestrial climate in the face of global climate change. Galápagos Res 67:26–37. Available at: http://datazone.darwinfoundation.org/darwin_weather/pdf/trueman_dozouville2010_vol67pp26-37.pdf%5Cnpapers2://publication/uuid/EC3E489C-7997-4635-B74D-3A68BA180A77

Vitousek PM, Kennedy MJ, Derry LA, Chadwick OA (1999) Weathering versus atmospheric sources of strontium in ecosystems on young volcanic soils. Oecologia 121:255–259. https://doi.org/10.1007/s004420050927

White WM, McBirney AR, Duncan RA (1993) Petrology and geochemistry of the Galápagos Islands: portrait of a pathological mantle plume. J Geophys Res 98:19533–19563

Wicknell A (1997) Los paisajes de las Islas Galápagos. In: Winckell A, Zebrowski C, Sourdat M, (CEDIG – IPGH – IRD, Quito) (eds) Los paisajes naturales del Ecuador: Las regiones y paisajes del Ecuador

Xie RC, Marcantonio F (2012) Deglacial dust provenance changes in the Eastern Equatorial Pacific and implications for ITCZ movement. Earth Planet Sci Lett 317–318:386–395. https://doi.org/10.1016/j.epsl.2011.11.014

Zhu H, Liu F, Li X, Wang G, Zhang Z, Sun W (2018) Calcium isotopic compositions of normal mid-ocean ridge Basalts from the Southern Juan de Fuca Ridge. J Geophys Res Solid Earth 123:1303–1313. https://doi.org/10.1002/2017JB014699

Xiao-Ming Liu is an Associate Professor at the Department of Earth, Marine and Environmental Sciences, the University of North Carolina at Chapel Hill. Dr. Liu's research group at UNC-Chapel Hill focuses on understanding fundamental questions using geochemistry, including chemical weathering and critical zone processes; evolution of Earth's surface environments; and environmental geochemistry. Liu has won the 2021 Early Career Award by the Geological Society of America (GSA), Mineralogy, Geochemistry, Volcanology & Petrology Division. Liu is also a 2019 recipient of the NSF CAREER Award.

Heather D. Hanna owns Hanna Forensics, LLC, and specializes in forensic geology and geochemistry consulting for law enforcement agencies. She also consults for geological surveys, providing geochemical expertise for evaluating potential domestic sources of critical minerals. She received her Ph.D. from the University of North Carolina at Chapel Hill in 2020.

Julia G. Barzyk is the Founder of Wise Investigator LLC. Prior to going into business, she served for 7 years as program manager for the Earth Materials and Processes program at the United States Army Research Office. She began her career working in geochemistry and cosmochemistry, receiving her Ph.D. from the University of Chicago in 2007.

Chapter 16
Mapping Narratives of Agricultural Land-Use Practices in the Galapagos

Francisco J. Laso and Javier A. Arce-Nazario

Introduction

Mapping plays a role in producing and disseminating knowledge about drivers of landscape change in agricultural landscapes, informing policy and farmers' practices. In the Galapagos, agricultural landscape narratives engage with interconnected themes including farmer livelihoods, preservation of native ecosystems as a resource for global science and local ecotourism, and health and food security. Scientists can use quantitative methods to reveal changes in land cover and compare land use practices and their effects across the landscape. The latter can be incredibly challenging when mapping a large area because individual farmers often have particular ways of managing their land depending upon unique environmental and socioeconomic conditions.

Using quantitative data entails making choices about how data are collected and interpreted. The choices we make in mapping land cover and land use affect how useful the results are for different audiences. It is essential to acknowledge the choices we make in the mapping process and other limitations common to quantitative land cover change methodologies. This includes practical obstacles and biases in data collection and categorization, technical constraints associated with remote sensors, and the reliability of algorithmic approaches to image classification when extrapolating beyond potentially limited training examples.

F. J. Laso (✉)
Environmental Studies, Western Washington University, Bellingham, WA, USA
e-mail: lasof@wwu.edu

J. A. Arce-Nazario
Department of Geography, University of North Carolina at Chapel Hill, Chapel Hill, NC, USA

S. J. Walsh et al. (eds.), *Island Ecosystems*, Social and Ecological Interactions in the Galapagos Islands, https://doi.org/10.1007/978-3-031-28089-4_16

Participatory mapping and critical cartography offer frameworks that accommodate these limitations (Colloredo-Mansfeld et al. 2020; Crampton 2009; Hauck et al. 2013). These approaches also aim to enhance the usefulness and sustainability of mapping projects and avoid reinforcing asymmetric power dynamics that often disadvantage the people living and working in the mapped areas. This chapter describes an iterative process of mapping that attempts to expose and address several of the limitations of our quantitative methods to derive a narrative of the drivers of land cover change in Galapagos agroecosystems that would be locally relevant to farmers, policymakers, and other stakeholders. We inform our process through participatory and critical approaches and a clear vision of the map's intended audiences and uses. We repeatedly incorporated feedback from farmers and experts to refine the sources and representation of quantitative data in mapping land cover and in mapping land management practices. The participation of farmers and stakeholders and the critical reflexivity in examining and re-examining the priorities and content of the maps were instrumental in making the resulting map narratives useful to various actors in the Galapagos.

The process of iteratively refining land cover and land use data reveals tensions between "ecologically sound" farming and the conservation of "native" species composition. This narrative was complicated by the socio-economically nuanced relationships between farmers and tourism in the Galapagos. We present result into context by documenting the mapping methods that brought these tensions to the foreground. By describing the roles taken by each stakeholder in the mapping process, we provide insights into the perspectives they contribute to the resulting maps. The dynamics between ourselves and our scientific collaborators, the farmers, governmental and non-governmental institutions, and other stakeholders in the mapping process shapes the narratives that emerged. In the framework of critical cartography, this documentation is necessary and integral to mapping and landscape representation (Kim 2015). We conclude by reflecting on the mapping processes and the printed map images that emerged, the social and political role of the resulting maps in the Galapagos, and the practical and impractical elements of critical approaches when mapping for the multifaceted audience of stakeholders in the Galapagos.

Background: Trajectory of Invasive Plants in the Land Cover Composition of the Galapagos Highlands

Shao et al. (2020) use remotely sensed images and represent a common starting point for our exploration of land cover in Galapagos agroecosystems. The striking increase in the regions "Invasive Species"at the borders between "Forest" and "Agriculture" (Shao et al. 2020) provides essential insight into how the land cover has changed through time in Galapagos (Fig. 16.1).

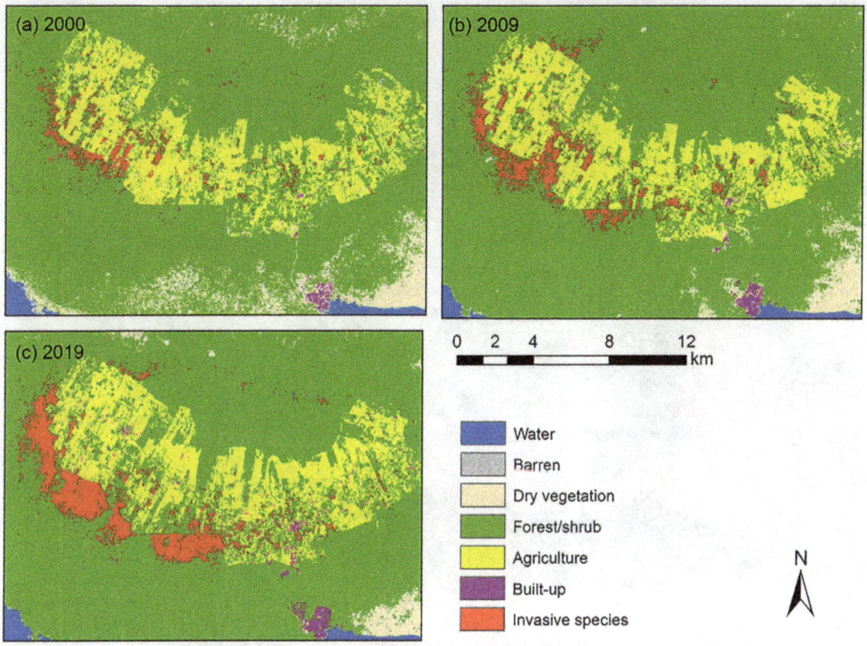

Fig. 16.1 Land cover maps of agricultural areas of Santa Cruz Island for 2000, 2009, and 2019. (Shao et al. 2020)

To understand the dynamics of this trajectory, however, we must confront the many ways that this clear presentation necessarily simplifies reality. Broad terms like "Agriculture" or "Invasive Species" represent the entire land mass of the islands in Fig. 16.1. The omissions that led to the definition of these broad categories depend in part on data and instrumentation limitations. More importantly, the mapmakers' perspectives and the social context of knowledge production shapes these characterizations. To interpret this landscape history, we must consider the narratives these maps depict.

First Iteration: Expanding the Spatial and Categorical Resolution in Land Cover

Our first step towards a more detailed understanding of landscape dynamics was to update the existing land cover maps of Galapagos agroecosystems with the highest spatial resolution land cover classification to date (Laso et al. 2020), using an expanded set of land cover categories. The result (Fig. 16.2) required a tremendous amount of data, collected from remote sensing satellite images and supplemented by high-resolution drone imagery. It also required a significant effort to calibrate

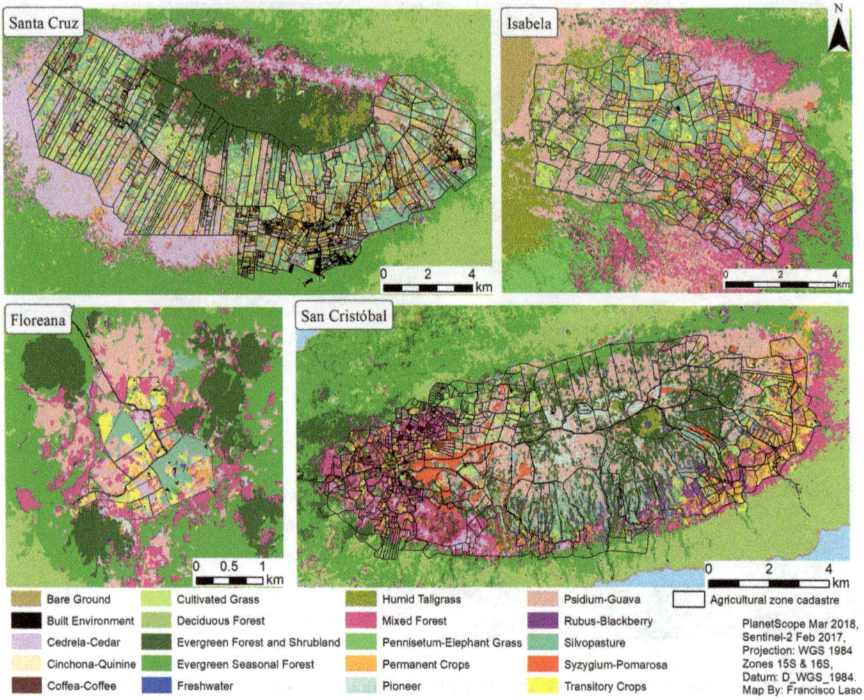

Fig. 16.2 Land cover maps of the humid highlands of the Galapagos islands. (Laso et al. 2020)

and apply an algorithm to identify land cover classes from our high-resolution image data.

This mapping allows higher accuracy since it explodes categories like "Agriculture"and "Invasive Species" into more specific land cover types, even including individual species (Fig. 16.3). Because of this, these maps create a different impression than Shao et al. (2020), depicting a diverse and heterogeneous landscape where agricultural areas are incrusted with patches of invasive plants and native ecosystems. It also illustrates a compatible narrative of invasive species overtaking productive and protected areas, with 30% of the agricultural zones covered by invasive plants (Fig. 16.3) (Laso et al. 2020).

One compelling narrative in Fig. 16.3 is of the dominance of *Psidium guajava* as the most widespread of the invasive plants, covering about 20% of the agricultural areas (Figs. 16.2 and 16.3). This narrative was constructed from algorithmic image classification that simplified the sensor data into categories. However, like any simplification, the results of this classification are an approximation and subject to uncertainty. We used drone images to train the algorithm to classify satellite images, verify predicted land covers, and validate the resulting map and our confidence in the classification. Because of time and resource limitations, the drone images that were used to validate and verify land covers did not cover all agricultural areas equally (Fig. 16.4).

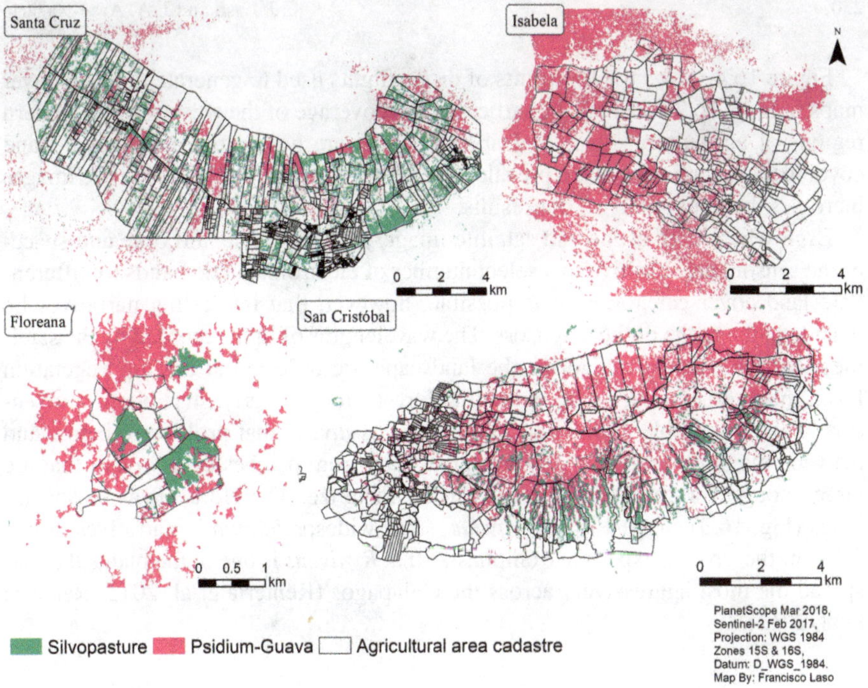

Fig. 16.3 Distribution of invasive species in the humid highlands of the Galapagos islands. (Laso et al. 2020)

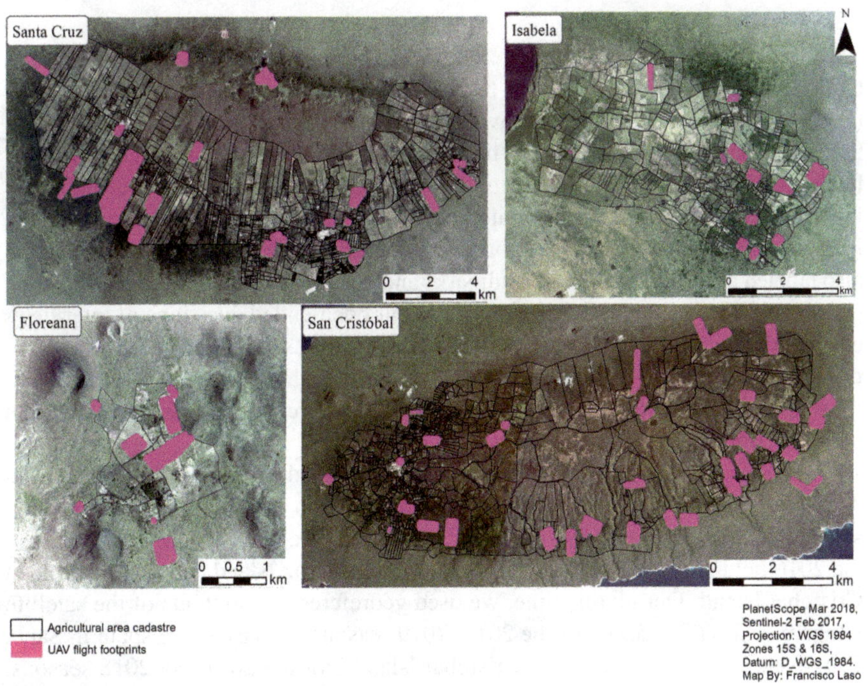

Fig. 16.4 Distribution of drone images collected for training, validation, and verification of land cover maps

Figure 16.4 shows the footprints of drone flights used to generate the land cover maps in Figs. 16.2 and 16.3. In particular, the coverage of the western and southern regions of agricultural areas of Isabela Island were not evenly distributed. Land cover in these regions shows a prevalence of *P. guajava*, but the lack of drone images increases the uncertainty of our results.

Even where both drone and satellite imagery are available, airborne and spaceborne sensors are sensitive to a select number of electromagnetic bands to differentiate land cover categories. It is possible, however, that interesting narratives lie outside of the range of these sensors. The wavelengths of light recorded by the satellites that we used to categorize the landscape are reflected at the top vegetation layer, making anything found below the forest canopy nearly invisible to our sensors. This is a problem with species like *Rubus niveus* that proliferate below and between the cover of trees. Very few areas have pure patches of *R. niveus* that are large enough for spaceborne sensors to characterize. Therefore, invasive species maps (Fig. 16.3) suggest that *R. niveus* is not widespread, while narratives drawn from on-the-ground experience emphasize that *R. niveus* is one of the plants that has spread the most aggressively across the Galapagos (Renteria et al. 2012; Rentería et al. 2012).

Second Iteration: Synthesizing Qualitative Data to Refine and Enhance the Land Cover Narrative

We applied an iterative process for creating and revising our maps to refine the spatially explicit narratives generated by our sensor and algorithm-based approaches (Fig. 16.5). This iterative process involved several rounds of interviewing key stakeholders throughout the agricultural highlands. To be compatible with the traditional landscape mapping process, we spatially referenced the interview data, using mosaics from spaceborne images generated before the field trips (Fig. 5.1), to guide the interviews. The mosaics included roads and boundary lines with protected areas. We also labeled a several prominent landmarks and regions to help interviewees become geographically explicit in their responses to our open-ended questions. If interviewees mentioned an observed change in land cover, we asked them to draw it on the printed map. Some interviewees interacted directly with the paper map. Others narrated approximate locations verbally, in which case, we marked the maps to match their descriptions.

Over 84 semi-structured interviews were conducted during the June, July, and August of 2016–2019. Interviewees were involved in different parts of the food system: food production, distribution, consumption, regulation, and research. For the 2016 field season, 2 weeks were spent in Santa Cruz Island and 3 weeks in San Cristobal Island. During this time, we used georeferenced printouts of the satellite image mosaics (Fig. 5.2). For the 2017–2019 seasons, 2 weeks were spent in Santa Cruz Island and 3 weeks in San Cristobal Island. For the 2017 and 2018 seasons,

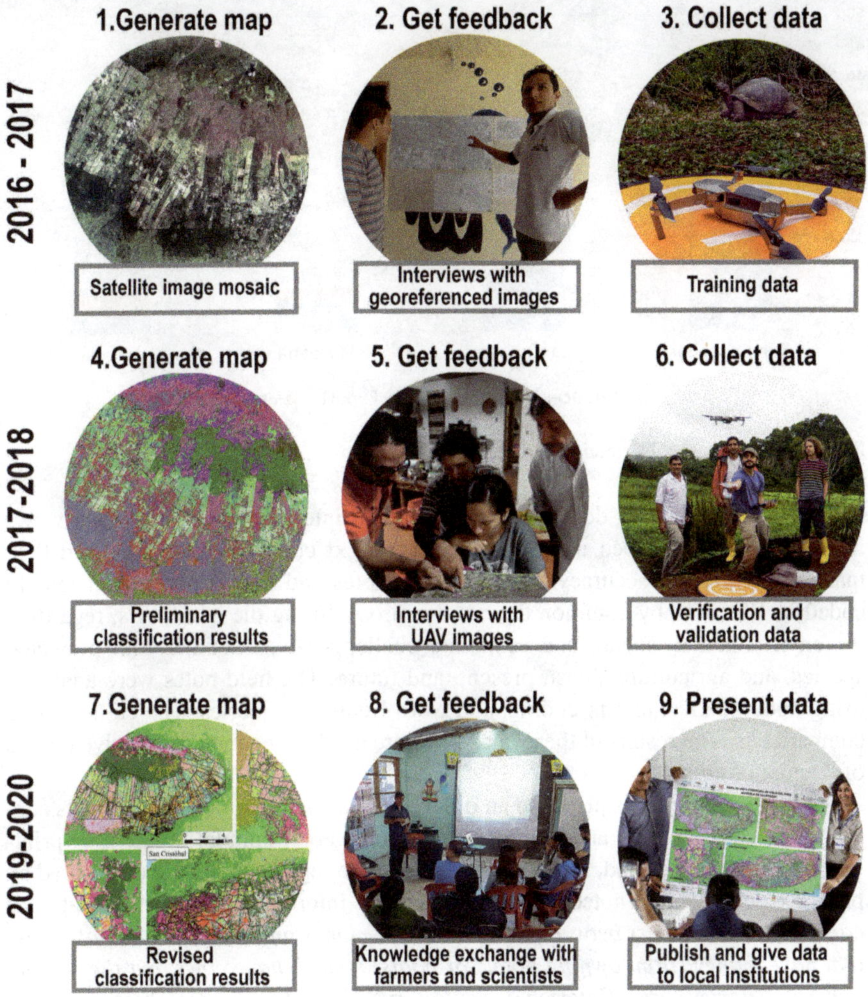

Fig. 16.5 Step-by-step overview of the iterative process of generating maps, obtaining feedback from farmers and specialists, and collecting additional data to improve maps

1 week was spent in Floreana Island. Additionally, a field assistant spent 1 week in Isabela Island in 2017, and another week was spent there in 2018. During this time, we conducted 39 interviews in 2016, 44 in 2017, and the remaining follow-up interviews were conducted during on-site verification for land cover maps in 2018 and 2019. In total, we interviewed 77 individuals during all field seasons.

Figure 16.6 describes the numbers of participants from each island for both field seasons that included authorities (government officials or public servants), experts (researchers, NGOs), producers (landowners, farmers, and cattle ranchers), and vendors (merchants & restaurant owners).

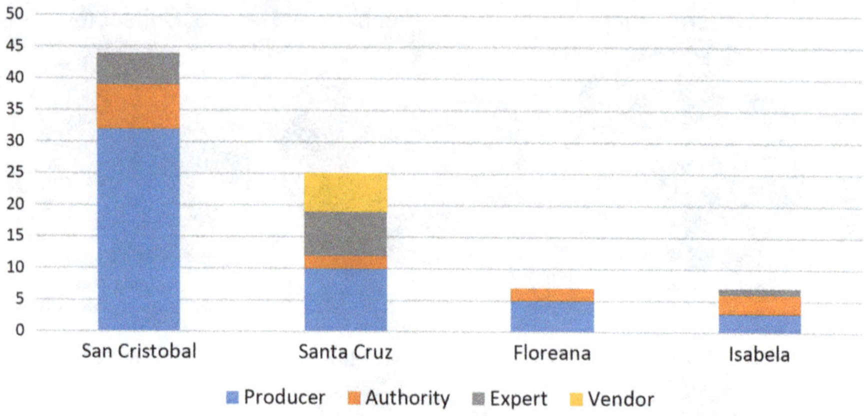

Fig. 16.6 Interviewee distribution per role and per island

We recorded or took detailed notes during all interviews. Recorded interviews were initially transcribed using *Trint* audio-to-text converting software and then manually edited for accuracy. The interview notes and transcripts were manually coded in *NVivo 12* by common themes that arose during the interviews, regarding the environmental, social, and economic challenges, interactions with protected species, and agriculture's past, present, and future. The field notes were analyzed using node queries, matrix coding, and word frequency queries, as well as coding comparisons. The results of these analyses were used to generate descriptive figures of the main findings.

These first interviews provided an overview of a complex food system in its ecological, socioeconomic, and political dimensions. For example, on the map printouts of Santa Cruz Island, interviewees marked the regions that were dominated by pastures for cattle and noted '*power differences*.' Interviewees explained that "*one can categorize farmers between those that have money and those that do not. Those that have money (from owning hotels or boats) tend to have cows, but they do not live from agriculture*" (Interview, Santa Cruz, July 14, 2018). Their access to resources enhanced cattle farmers' ability to cope with hardships and adapt to disturbances. Similarly, most interviewees mentioned 'agua' (water, in Spanish), as seen in the word cloud (Fig. 16.7), as a primary challenge to agriculture. Water scarcity and periodic droughts severely threaten farmer livelihoods in the Galapagos, along with the lack of affordable labor (or 'mano de obra', in Spanish).

Interviews revealed that some of the most significant elements that shape agricultural landscapes in Galapagos are invisible to satellite images of the highlands. For example, the connections to the tourism industry involve power dynamics and connections to locations outside of the agricultural zone. Because they are temporally restricted, droughts are not likely to be discernable from the limited time-sequence of images available for the region. This first stage in the feedback process helped refocus the data collection to represent a varied sample of land uses. We collected training data from all inhabited islands using GPS and a drone in 2017 (Fig. 5.3) and generated preliminary results (Fig. 5.4).

Fig. 16.7 Word cloud visualization to responses to the question "What are the main challenges to Agriculture in the Galapagos?"

Third Iteration: Silvopastures

As described in Colloredo-Mansfeld et al. (2020), a year after collecting the first set of drone images, we returned to a select group of landowners with printed versions of the aerial images of their farms (Fig. 5.5). Depending on the comfort level of farmers, we asked that they cover the image with tracing paper and mark the maps with important elements from their farms (Fig. 16.8) (Colloredo-Mansfeld et al. 2020).

The interviews with UAV images (Fig. 16.8) allowed us to check the accuracy of our preliminary classification results. But most importantly, they helped us revise our classification scheme to include terminology relevant to the farmers themselves. Thanks to interviews with farmers conducted at this stage, we realized the importance of including agriculture-centric categories like "silvopasture" in the land cover classification scheme (Fig. 16.9). Silvopastures are an agroforestal design where pastures for cattle grazing and trees are grown together. Many silvopastures have citrus trees as part of their design, but it can be any tree species, and many farmers combine their pastures with *P. guajava*. It was challenging to train an algorithm to distinguish between silvopastures (maintained by farmers) and *P. guajava*

Fig. 16.8 Examples of drone images that were annotated by farmers to describe their lands in their own terms

growing wild (Fig. 16.8), and including this category increased the risk of underreporting the amount of *P. guajava* in agricultural areas. We decided that these drawbacks were acceptable tradeoffs to allow us to acknowledge this important agricultural practice, as illustrated in Fig. 16.9.

Using feedback from farmers and specialists, we collected more aerial images for verification and validation (Fig. 5.6) and revised the land cover classification (Fig. 5.7). In 2019, Laso shared these results at an interdisciplinary knowledge exchange that he helped co-organize with a farmer cooperative in Santa Cruz Island (Figs. 5.8 and 16.10). Farmers are normally excluded from academic events because of the cost to travel to urban locations for discussions, so the event took place at a small rural school in the highlands of Santa Cruz Island that was regularly used for meetings of an exceptionally active farmer cooperative. The event brought together farmers and scientists in conversation. Three farmers presented their work and their vision for sustainable agriculture in the islands. Then three scientists presented their work that was directly related to agriculture – providing an opportunity to share the ongoing mapping with farmers from Santa Cruz and obtain their feedback. Attendees and presenters discussed each other's work over coffee and tamales.

The event, which was meant to foster communication and collaboration between farmers and conservation scientists, was jeopardized by a boycott by the Ministry of Agriculture, which was unwilling to participate without having a more active role in

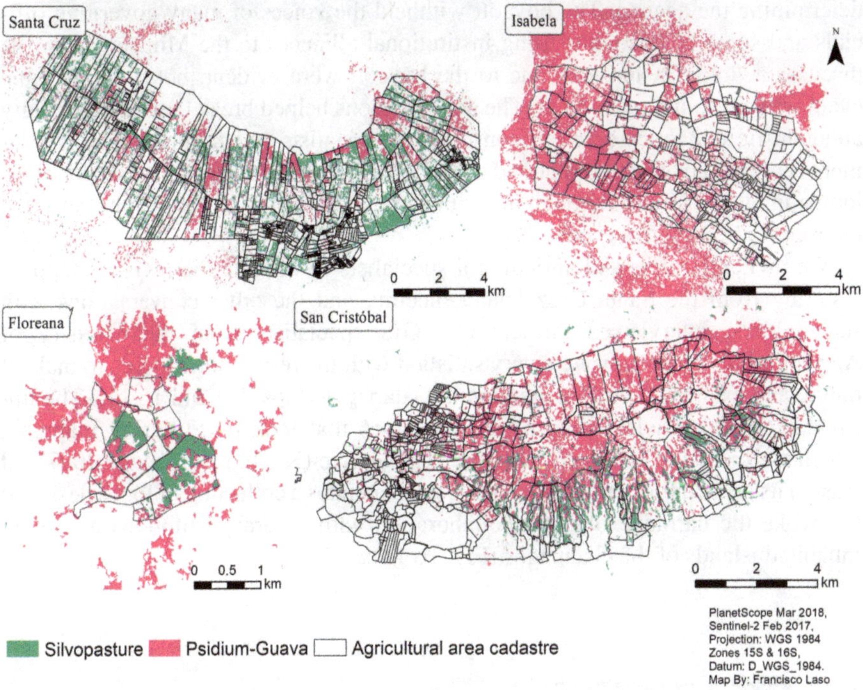

Fig. 16.9 Distribution of *Psidium guajava* and silvopastures in agricultural areas of the Galapagos Islands

Fig. 16.10 Interdisciplinary knowledge exchange between farmers and conservation scientist of Santa Cruz Island. The event took place on July 18, 2019. (Photo by Carter Hunt)

determining the agenda. The boycott withheld the voices of many governing officials and some NGOs with strong institutional alliances to the Ministry from the discussion. Initially, tensions due to the boycott were evident, but ultimately the event was a resounding success. The conversations helped break down traditionally antagonistic dynamics between conservation scientists and farmers. The government representatives that did attend ended up offering their support for similar colloquia in other islands, although unfortunately, the COVID-19 pandemic prevented us from replicating this event.

We invited farmers and agricultural specialists to validate the revised mapping resulting from the Santa Cruz Island meeting and the other conversations with stakeholders. Interviewed farmers and GIS specialists from the Ministry of Agriculture in Galapagos were very satisfied with the result, allowing us to make it public and available to all institutions in Galapagos (Figs. 5.9 and 16.11). The finished product excludes many of the processes that took place during mapping, which are crucial to effectively interpreting the map (Kitchin and Dodge 2007). At least in its printed form, the frame of this map includes a collection of logos designed to invoke the narrative of close collaboration with several institutions across all inhabited islands of the Galapagos (Fig. 16.11).

Fig. 16.11 Formal presentation of the land cover maps of the agricultural areas of the Galapagos during a meeting to celebrate conservation agreements by farmers. From left to right, Francisco Laso (Author), Jimmy Bolaños (Director MAG) and Marilyn Cruz (Director ABG). (Photo: ABG)

Fourth Iteration: Characterizing Agricultural Practices in Galapagos

Given the clear narrative of land use via the categories developed in the mapping, we sought a classification scheme that would describe relationships between humans and their environments in the same spaces. A mixed-methods approach was used (Fig. 16.12) to investigate the intersecting effects of environmental, socioeconomic, and political drivers on farmers' livelihoods and land cover. As before, stakeholder feedback (Fig. 16.12b,c, d) informed the development of land management practice categories and the classification of agricultural production units into a farm typology (Bergman 2008). This land use classification, described by Laso (2021) and

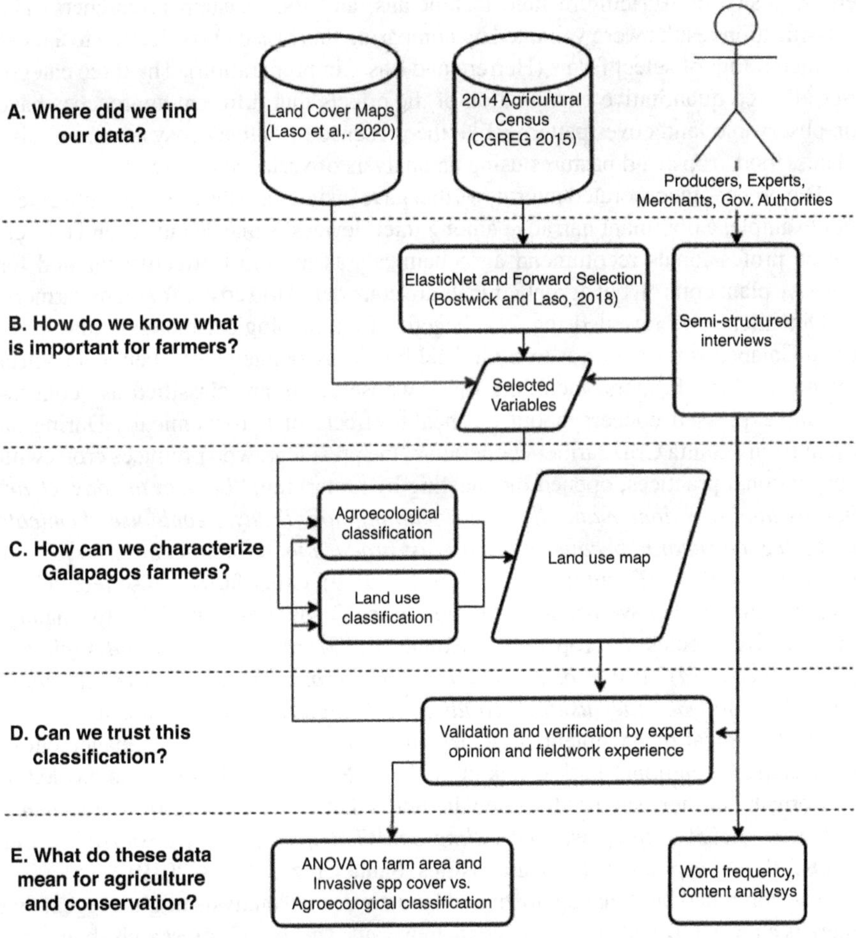

Fig. 16.12 Graphic outline of mixed-methods analysis. Cylinders indicate existing datasets, stick figures represent actors, rectangles are processes, and trapezoids are inputs/outputs

Herrera & Laso (in preparation), sorted farms into one of three general categories: Conventional, Transition, and Agroecological farms. Conventional farms depend on environmentally harmful and industrial practices, typically prioritizing economic profit over environmental and social integrity, and individualistic actions regarding access to information. At the other end of the spectrum, agroecological farms are diversified agricultural systems, nurtured by the exchange of traditional knowledge systems and farmer innovations (Altieri et al. 2011). Interactions between human and ecological values play an important role in their economic decision-making, making them less likely to use environmentally harmful agrochemicals.

Farms were classified based on their responses to the 2014 Agricultural Census for questions about land management practices, source of germplasm, farm inputs, farm culture and identity, and belonging to a farming cooperative. In-depth field experience for farm classification to verify this information came from food producers, Ministry of Agriculture field technicians, and food system researchers. The classification results were validated by comparing this broad classification to a more detailed rating of select farms (Herrera and Laso, in preparation). The three categories allowed quantitative examination of the effects that different approaches have on observable land cover patterns, like the presence of native ecosystems, invasive plants, food crops, and pastures using an analysis of variance (ANOVA).

The process incorporated interviews that gave insight into the dynamics observed. For example, a dominant narrative among interviewees is that scientists and conservation professionals recommend agrochemicals as the most effective method for invasive plant control, but agrochemicals are considered too costly for many farmers.

Our interviews revealed that one benefit of agroecological methods that interested Galapagos farmers was to avoid health risks associated with chemical control methods. Many farmers, including those whose farms are classified as "conventional," expressed concern about the health effects of agrochemicals. During an event from a Santa Cruz farmer cooperative, the president, who produces crops with conventional practices, opened the meeting by remarking, "*Back in the day, clearing the land was done manually with a planting hoe. Today, people use chemicals that affect our health, because we don't just produce food for sale, but also for our own consumption.*" (Event transcript, July 17, 2019). Other farmers also recognized that agrochemicals have become common in the Galapagos, especially for managing invasive species. A crop farmer explained, "*Mora (R. niveus) and supirrosa (Lantana camara) have to be killed using ester herbicides...But we are allergic to herbicides, so when they use it, I go far away*" (Interview, San Cristobal, July 19, 2016). Chemical management strategies are also tools for species management in the Galapagos National Park (Filek et al. 2018; Soria et al. 1999). This has led to concerns by farmers such as the one who noted that "*scientists are spraying chemicals and releasing fungi without telling us. What if that gets transferred to our crops? To native plants?*" (Research Notes, Santa Cruz, July 18, 2019).

The combined land management and land cover map analysis might suggest that there is a need to use chemicals to control invasive species. Our research shows that regardless of land use, agroecological farms have a significantly higher percentages of their surface area covered by invasive plants (Laso 2021). This result suggests a tension between the desire to expand agroecological farming practices in Galapagos and protect the islands' native ecosystems.

Narratives from the interview process offer alternative interpretations. Farmers with agroecological practices may be less likely to control the spread of some invasive plants because they fulfill desirable functions within their farms. For example, a farmer whose farm was classified as agroecological considered that *P. guajava* is not a problem for several reasons. The farmer stated that *P. guajava* had a positive effect on water conservation since the lichen that copiously grows on a guava tree captures water from fog, and grass surrounding guava trees often remains green even during droughts (Laso 2021). Several farmers also noted that guava fruit feeds farm animals, especially during drought events. The strategy of incorporating guava in pastures was also noted by Ministry of Agriculture officials, who observed that farms with a silvopastural design were less affected by the severe 2016 drought (Research notes, Santa Cruz, July 09, 2018). The narratives of these farmers and policymakers help link management choices and land cover in a way that clearly suggests policy considerations: despite their "invasive" status, species like *P. guajava* and *Pennisetum purpureum* fulfill invaluable functions that should be considered for adapting agroecosystems to the more extreme weather events predicted in climatic projections (Mena et al. 2020).

Another reason for the correlation between agroecological farms and invasive plants may be that this incursion does not threaten farmer livelihoods. This latter interpretation accentuates the socioeconomic disparity between farmers with more land than they can manage and likely have additional income sources and farmers with smaller plots of land that represent their entire livelihoods. A producer with both cattle and crops who practices transition farming explained "*I do not practice agriculture to get rich. I do it as a hobby and for my own health*" (Interview, Santa Cruz, July 14, 2018). When asked about invasive plants the farmer responded, "*We have sauco (Citharexylum gentry-verbenaceae), mora (R. niveus), and guayaba (P. guajava). But we have guava under control because it is a small plot of land*" (Interview, Santa Cruz, July 14, 2018). This farm employs a part-time worker who is instrumental in keeping these plants from taking over large sections of the farm. These testimonies and qualitative analyses show most farmers do not make a living out of agriculture, which concurs with the observation of a Ministry of Agriculture technician. His observation is that farmers in the Galapagos can be divided into "*those that have money and those who don't*" (Interview, Santa Cruz, July 14, 2018).

Finally, farmer interviews suggested viable alternatives to agrochemicals. One crop farmers categorized as having transition farming practices advocated for manual control: "*Here we control invasive species manually. Everyone can tell me that it is impossible, but it is more possible to work manually than to throw chemicals all around.... The rocky soils of the low elevation regions make it harder to do this than the soils from the highlands.*" (Interview, San Cristobal, July 17, 2017). This point of view was echoed by another young coffee farmer: "*There is an invasive vine that grows over coffee and can even topple the plants. It is very hard to uproot, and this increases the labor necessary for producing coffee without chemicals. However, over time, it seems to me that this method requires less upkeep, because our*

Fig. 16.13 Alternative strategies to manage *Psidium guajava*

neighbors have to keep applying herbicides periodically to keep their lands clear." (Research notes, Santa Cruz, July 18, 2019). Manually uprooting plants may be more effective than applying chemicals in the long term, but the amount and intensity of this task might only be feasible for younger farmers or those with access to affordable labor. Agroecological farmers who can afford to pay a worker or are young enough to do the work are willing to spend substantial amounts of time and resources in controlling invasive species.

As was the case for the land cover map, the information generated in this mapping is not fully represented or effectively applied by the cartographic artifact alone. To represent the spectrum of responses to invasive species discussed by participants, Laso (2021) generated a brochure (Fig. 16.13) depicting different strategies for invasive species control.

Reflection

The iterative process in which the stakeholders, who we envision as users of our map, are invited to read and critique it creates a valuable feedback loop. This cycle can be 'noise-cancelling,' suppressing irrelevant detail, but it can also be 'feature-enhancing' by bringing hidden features to the foreground. Enhanced features include the important silviculture practices and invasive species management approaches that became visible classification categories and image elements through this feedback process. However, other important features that arose in the transcribed interviews, in the map layers generated by farmers drawing over aerial images, and in the discussions with agencies and stakeholders, are less legible in the narrative offered by the map image. Given these data products emerged in alternative spaces such as farmers' homes and farming cooperatives not fully aligned with existing power dynamics, the intermediate products of the mapping process could contain significant omissions from the final map image. The traditional approaches we used to create our map images do not go far enough to include such polyvocal data. Exploring better ways to communicate what has emerged from this process is a subject of future work.

Nevertheless, traditional mapping is the medium of choice for government actors. We designed the visual products of our mapping to foster communication between groups in the Galapagos, especially farmers and farmer organizations, government ministries, and conservation scientists. We conversed with farmers using UAV images, which were effective because they photo-realistically depicted familiar terrain for the farmers from perspectives familiar to us as GIS professionals. Conventional maps and brochures were similarly effective for the more varied groups of stakeholders interested in developing policies for Galapagos agriculture.

Kim (2015) calls on the next generation of critical cartographers to be attentive to the "social position of the new map and how it engages institutions." During the first iteration of this mapping, the disconnect between researchers and farmers was apparent. In 2016, several farmers individually agreed to attend our focus group about Galapagos food systems scheduled for a Friday morning in the offices of an environmental education organization. Farmers, field technicians from the Ministry of Agriculture, and students were all invited, but only the latter two groups attended: no farmers showed up to the focus group. Farmers, it emerged, were too polite to decline, but unable to prioritize academic endeavors in faraway urban areas over harvesting and preparing for the Saturday market. Effectively engaging farmers required us to visit them on their terms. When we visited farmers for UAV assisted mapping, they became engaged in the broader mapping project, and thus the meetings to discuss and remake the developing land cover maps became popular. By the third iteration of our maps, the project was well-known and farmer turnout to the interdisciplinary knowledge exchange in 2019 was outstanding. The productive conversations that took place during this meeting with the cooperative, in turn, led to greater engagement from government agencies, who expressed interest in organizing similar events on all islands. We think the mapping thus represents a step towards more coordination among institutions, with clear recognition of the authority and relevance of the farmers' perspectives.

The final map image benefits from the process behind the mapping process, not only did the engagement with stakeholders shape its content, but this collective authorship created a broader audience for the map. The land cover image has found its place as an important resource. Since the dataset was published, it has been adopted by local government institutions to evaluate agricultural lands and inform their projects. The map has also been requested by ornithologists, entomologists, agronomists, conservation biologists, city planners, and tourism agencies. Meanwhile, the agroecological classification of farms has been used by academic institutions to allocate funding for regional climate-change adaptation plans. Similarly, the land management brochure of *P. guajava* serves as a visual reminder for researchers to go beyond their notion that chemical and mechanical control of invasive plants is always the most efficient. Thus, the images created through these mappings have a role in multiple new narratives about the islands.

References

Altieri M, Funes-Monzote F, Petersen P (2011) Agroecologically efficient agricultural systems for smallholder farmers: contributions to food sovereignty. Agron Sustain Dev 32(1):1–13. https://doi.org/10.1007/s13593-011-0065-6

Bergman MM (2008) Quality of inferences in mixed methods research: calling for an integrative framework. In: Advances in mixed methods research, pp 1–8. https://doi.org/10.4135/9780857024329

Colloredo-Mansfeld M, Laso FJ, Arce-Nazario J (2020) Drone-based participatory mapping : examining local agricultural knowledge in the Galapagos. Drones 4(62):13. https://doi.org/10.3390/drones4040062

Crampton JW (2009) Cartography: performative, participatory, political. Prog Hum Geogr 33(6):840–848. https://doi.org/10.1177/0309132508105000

Filek N, Cimadom A, Schulze CH, Jäger H, Tebbich S (2018) The impact of invasive plant management on the foraging ecology of the Warbler Finch (Certhidea olivacea) and the small tree finch (Camarhynchus parvulus) on Galápagos. J Ornithol 159(1):129–140. https://doi.org/10.1007/s10336-017-1481-4

Hauck J, Gorg C, Varjopuro R, Ratamaki O, Maes J, Wittmer H, Jax K (2013) "Maps have an air of authority": potential benefits and challenges of ecosystem service maps at different levels of decision making. Ecosyst Serv 4:25–32. https://doi.org/10.1016/j.ecoser.2012.11.003

Kim AM (2015) Critical cartography 2.0: from "participatory mapping" to authored visualizations of power and people. Landsc Urban Plan 142:215–225. https://doi.org/10.1016/j.landurbplan.2015.07.012

Kitchin R, Dodge M (2007) Rethinking maps. Prog Hum Geogr 31(3):331–344. https://doi.org/10.1177/0309132507077082

Laso FJ (2021) Agriculture, wildlife, and conservation in the Galapagos Islands. University of North Carolina at Chapel Hill

Laso FJ, Benítez FL, Rivas-torres G, Sampedro C, Arce-Nazario J (2020) Land cover classification of complex agroecosystems in the non-protected highlands of the Galapagos Islands. Remote Sens 12(65):1–39. https://doi.org/10.3390/rs12010065

Mena CF, Paltán HA, Benitez FL, Sampedro C, Valverde M (2020) Threats of climate change in small Oceanic Islands: the case of climate and agriculture in the Galapagos Islands, Ecuador.

In: Walsh SJ, Riveros-Iregui D, Arce-Nazario J, Page PH (eds) Land cover and land use change on islands. Springer, Cham, pp 119–135. https://doi.org/10.1007/978-3-030-43973-6_5
Rentería JL, Gardener MR, Panetta FD, Atkinson R, Crawley MJ (2012) Possible impacts of the invasive plant Rubus niveus on the native vegetation of the Scalesia forest in the Galapagos islands. PLoS One 7(10):e48106. https://doi.org/10.1371/journal.pone.0048106
Renteria JL, Gardener MR, Panetta FD, Crawley MJ (2012) Management of the invasive hill raspberry (Rubus niveus) on Santiago Island, Galapagos: eradication or indefinite control? Invasive Plant Sci Manag 5(1):37–46. https://doi.org/10.1614/IPSM-D-11-00043.1
Shao Y, Wan H, Rosenman A, Laso FJ, Kennedy LM (2020) Evaluating land cover change on the Island of Santa Cruz, Galapagos Archipelago of Ecuador through cloud-gap filling and multi-sensor analysis. In: Walsh SJ, Riveros-Iregui D, Arce-Nazario J, Page PH (eds) Land cover and land use change on islands. Springer, Cham, pp 167–182. https://doi.org/10.1007/978-3-030-43973-6_7
Soria M, Gardener M,Tye A (1999) Eradication of potentially invasive plants with limited distributions in the Galapagos Islands. In: Turning the tide: the eradication of invasive species, pp 287–292

Francisco J. Laso holds a Ph.D. in Geography from the University of North Carolina at Chapel Hill and an M.Sc. in Ecology and Evolution from Rijksuniversiteit Groningen (Netherlands) and Université Montpellier II (France). Francisco has worked on conservation and education projects in the Galapagos Islands since 2003. His research uses quantitative and qualitative methods to explore human-environment interactions in the Galapagos and the Ecuadorian Amazon.

Javier A. Arce-Nazario is an Associate Professor of Geography at UNC. His work in the Amazon, the Caribbean, and the Galapagos looks for new landscape histories and new ways to explain them. Javier uses ecology, social science, the humanities, and participatory mapping methods to increase the impact that non-academic stakeholders have on landscape change research, and the positive impact that landscape change research has on their communities.

Chapter 17
Land Use and Land Cover Change in the Galapagos: Economic and Natural Drivers

Madeline Giefer

Introduction

Oceanic islands around the world are intense microcosms of global environmental pressures, including land use and land cover (LULC). Small size and isolation make oceanic islands especially vulnerable to environmental change (Alomía Herrera et al. 2022), while complicating socioeconomic challenges brought on by limited land resources and institutional capacities (Benítez et al. 2019). These environmental and social pressures make islands compelling "natural laboratories" for ecological, economic, and sociological research, and few have garnered as much scholarly attention as Ecuador's Galapagos Islands. With acute land limitations, brisk economic growth, and severe exposure to climate change, they are an ideal place to study the ecologies and economics of land use and land cover in a rapidly changing world. Despite strict and extensive environmental protections, land use and land cover are changing more quickly than ever before (Percy et al. 2016), alongside a rapidly growing population and tourism industry. More than a 1000-km from the South American mainland, the archipelago went unseen by humans until 1535, after which inhospitable terrain and limited freshwater staved off permanent settlement until 1832. While any notion of the Galapagos as an "untouched" wilderness is misconceived, the archipelago is indeed closer to its pre-human state than almost any other place on Earth (Orellana and Smith 2016; Izurieta et al. 2018), providing valuable opportunities to study the early effects of human use on a range of ecosystems (Khatun 2018; González et al. 2008).

Land use and land cover changes in the Galapagos reflect both competition and synergy among environmental and social goals. The "Galapagos paradox" (Walsh and Mena 2016) describes this uneasy balance; the tourism-dependent economy

M. Giefer (✉)
Department of Geosciences, Austin Peay State University, Clarksville, TN, USA
e-mail: gieferm@apsu.edu

S. J. Walsh et al. (eds.), *Island Ecosystems*, Social and Ecological Interactions in the Galapagos Islands, https://doi.org/10.1007/978-3-031-28089-4_17

rests heavily on the islands' ecological integrity, both real and perceived, while tourists' presence degrades the ecosystems that attract them. Tourism is the Galapagos' largest driver of land use and land cover change (Benítez et al. 2019; Walsh et al. 2010), as lodging, dining, entertainment, and transportation accommodations expand to meet growing demand. The industry meanwhile siphons labor and population from the agricultural highlands, accelerating the spread of invasive plants that thrive on abandoned farmland (Barrera et al. 2021). This intersects with the ongoing tension between conservation interests and local economic growth and well-being. The Galapagos' current regulatory framework, while extensive, is failing both in sustaining ecosystems and maintaining the faith of local stakeholders. This chapter outlines a web of ecological and human processes to present a comprehensive picture of ongoing land use and land cover challenges and underpin holistic planning that embraces social and economic realities.

Land Cover History and Trajectory

Each Galapagos Island has unique ecological characteristics (Watson et al. 2010), but the islands are unified by "altitudinally compressed" land cover systems wherein disparate vegetation and climate zones exist in close proximity to each other (Laso et al. 2019). The largest islands have six altitudinal zones: the bare zone (lava rock and beaches near the coastline), littoral zone (shrubs, mangroves, and other salt-tolerant species), arid zone (low scrubs, cacti), transition zone, humid zone (scalesia shrubs and trees), and very humid zone (miconia, sedges, and ferns) (Orellana and Smith 2016). These varied land covers support the archipelago's flagship fauna, such as sea lions in the bare zone, land iguanas in the arid zone, and giant tortoises in the humid highlands. Since the first successful human settlement in 1832, land use has entailed the clearing of native vegetation and the spread of plants introduced for food and ornamentation, altering these habitats and shifting balances among plants, humans, and other animals. National and international designations now make the Galapagos one of the most regulated regions in the world in terms of land cover and land use; Galapagos National Park, established in 1959, occupies 79% of land surface (Orellana and Smith 2016). The Galapagos Special Law, passed at the national level in 1998, restricts virtually all immigration, at least in theory, and increased the authority of the national park administration (Hoyman and McCall 2013). However, immigration continues (Villacis and Carrillo 2013), and current policies fail to stem the loss of biodiversity and natural land covers.

Agricultural Highlands

The humid highlands are the archipelago's most biologically productive regions and the most conducive to agriculture (Laso et al. 2019), putting them at the forefront of early land cover changes. First came small-scale farmers, targeting lands with the deepest soils (Alomía Herrera et al. 2022), clearing forest and vegetation and introducing familiar Ecuadorian crops (Astudillo 2018), some of which would become naturalized or invasive (Laso et al. 2019). The 1860s saw the rise of the El Progreso plantation on San Cristobal, along with the first large-scale deforestation powered by hired workers from the mainland, vastly expanding existing cleared spaces (Astudillo 2018). The largest wave of migration came in the 1970s, and although by this point farmers comprised less than one-half of arrivals (Walsh et al. 2010), agricultural land area continued to expand with illegal migration from mainland Ecuador (Pizzitutti et al. 2020). Today, the amount of agricultural land is around 19,000 hectares (Barrera et al. 2021), having contracted from its peak in some regions during the late twentieth century as farmers abandon plots and take other jobs (Alomía Herrera et al. 2022).

While agriculture's tenure as the primary livelihood in the Galapagos was short-lived, its impacts on land cover are extensive and permanent. Invasive plants, the strongest driver of land cover change in the Galapagos (Percy et al. 2016), proliferate on active and abandoned cropland. Areas of human settlement contain more introduced than native plant species, and 42% of humid highland area on the four inhabited islands is altered by agriculture and invasives (Trueman et al. 2010). More than 800 non-native plant species have been documented on the archipelago (Gardener et al. 2013; Trueman et al. 2010; Walsh et al. 2010), most of them deliberately for food or ornamental purposes (Barrera et al. 2021), and most in the latter half of the twentieth century (Gardener et al. 2013). About one-third of these species have naturalized, one-sixth have become invasive, and 3% have transformed within their new environment (Trueman et al. 2010). With less than two hundred years of settlement, the archipelago is at an "early stage" of invasion where most introduced species are confined to farms and gardens (Gardener et al. 2010), but "early stage" should not be interpreted as "mild." Invasive plants are already replacing natural ecosystems far more quickly than current management regimes can contain, and "extinction deficit" may be building up for the coming decades and centuries (Trueman et al. 2010). Key invasive species in the agricultural highlands include guava, blackberry, quinine, supirosa, and pomarrosa (Laso et al. 2019), among which guava is especially prevalent and aggressive. This highly tolerant shrub forms dense thickets that crowd out native vegetation, and it thrives on both abandoned and active agricultural land (Walsh et al. 2010). Invasive plants, especially guava, make cultivation more difficult on remaining cropland, closing a feedback loop in which invasive plants and cropland abandonment exacerbate each other. This may push highland ecosystems to new equilibria where key ecological relationships are damaged beyond repair (Wilkinson et al. 2005).

Despite agriculture's central role in introducing invasive plants, it is also critical in controlling their spread. While cropland retirement contributes to ecologically beneficial reforestation in many regional contexts (Li and Li 2017), in the Galapagos, it accelerates biodiversity loss as invasive species proliferate across abandoned cropland and into naturally vegetated zones (McCleary 2012). Various policy mechanisms may keep farmers on their lands controlling invasive species, including subsidizing local produce to compete with cheaper imports from the mainland (Khatun

2018; Miller et al. 2010), and directly subsidizing farmers' efforts to remove invasive plants (Miller et al. 2010). Other recommendations include re-engaging abandoned agricultural land for crop and livestock production, creating a framework for farmers to rent land to each other (Puente-Rodríguez et al. 2019), improving productivity through technology (Barrera et al. 2021), and strengthening enforcements against removing or harvesting native vegetation (Quiroga et al. 2011). While eradication, when possible, is usually ecologically and economically ideal, it usually proves untenable in the Galapagos (Gardener et al. 2010). An intensive 5-year effort to eradicate raspberry on uninhabited Santiago Island did lead to declines in plant and seed bank densities in managed areas, but new populations continued to crop up on other parts of the island, while collateral damage by herbicides on native plants outweighed conservation benefits (Renteria et al. 2012). Further, eradication is an appropriate goal only when reintroduction is unlikely (Meyer 2014), and frequent foot traffic across the four inhabited Galapagos Islands, and some uninhabited, keeps cross-island reintroduction possible.

Agriculture-related land cover change and invasive species threaten the survival of wildlife in addition to native plants (Khatun 2018), perhaps most visibly the Galapagos giant tortoise. Agricultural land cuts off migration routes (Benitez-Capistros et al. 2019) and removes forage, leading to lower tortoise densities and diets dominated by invasive plants (Laso et al. 2019). The high presence of invasives in tortoise diets further accelerates their proliferation and destruction of native food sources through seed dispersal (Walsh et al. 2010). Meanwhile tortoises sharing land with livestock may harbor and disperse antibiotic resistance, putting themselves, humans, domestic animals, and other wildlife at risk (Nieto-Claudin et al. 2021). While agriculture itself threatens tortoises, the decline in agriculture may be a greater threat; tortoise densities are even lower on abandoned cropland than on active cropland. Some farmers, however, have covered former cropland with semi-natural environments to attract giant tortoises and tourists. While the practice is relatively new and may not fit a purist's definition of conservation, research suggests these "tortoise farms" are successful in attracting the animals (Pike et al. 2022). However, these are profit-driven enterprises that require concerted upkeep, and with a finite market for this tourist experience, "tortoise farms" will probably amount to a very small fraction of the highlands' land use portfolio.

As new plant species shift competitive balances in the ecosystem, climate change introduces new pressures that complicates efforts to preserve natural land cover. The narrow ecological niches that make oceanic islands susceptible to invasion make them especially sensitive to warming temperatures and rising sea levels (Escobar-Camacho et al. 2021; Pizzitutti et al. 2020). The Galapagos National Park Service and Charles Darwin Foundation identify climate change as the main cause of biodiversity loss after invasive species (Dueñas et al. 2021), and the two forces interact at many spatial and temporals scales (Escobar-Camacho et al. 2021). Rising sea surface temperature will increase rainfall in both the humid highlands and arid lowlands, altering plant growth patterns, increasing erosion, and widening the competitive advantages of some already-robust invasive species (Dueñas et al. 2021; Escobar-Camacho et al. 2021). This is especially true for guava, which may expand into new areas as the humid zone grows and tortoises disperse seeds into newly hospitable areas (Ellis-Soto et al. 2017). As more frequent and intense droughts and floods (Izurieta et al. 2018) push more farmers out of agriculture, the archipelago may also lose its first line of defense against invasive species expansion, making the conservation of native vegetation cover even more expensive and untenable. While coastal lowlands are currently less affected by invasive species than agricultural highlands (Watson et al. 2010), increased rainfall will shrink the arid zones and leave them open to some of the same invasive plants that dominate much of the highlands (Ellis-Soto et al. 2017).

Urbanized Lowlands

While demand for agricultural land has plateaued and even contracted in many areas, demand for urban space is growing quickly. Urban space in the Galapagos is generally concentrated along the coast in arid zones (Guézou et al. 2010), driven directly and indirectly by tourism. To accommodate more than 250,000 tourists per year (Escobar-Camacho et al. 2021), built areas are becoming denser and more expansive (McCleary 2012), while jobs in the tourism industry draw farmers from the Galapagos highlands and (illegally) from mainland Ecuador, who further increase demand for permanent built infrastructure. The Galapagos have a population growth rate three times higher than that of the mainland (Escobar-Camacho et al. 2021), and impervious surface on the three main inhabited islands (Santa Cruz, San Cristobal, and Isabela) increasing from 2.2% to 5.7% between 1990 and 2015, putting pressure on the health of humans, wildlife, and endemic vegetation (Benítez et al. 2019). This expansion creates its own demand for freshwater, energy (Percy et al. 2016), material imports, and waste management systems, and few concerted efforts have been undertaken to mitigate these loads (Alava et al. 2022). Traditionally, the land use imprints of water and waste management in the Galapagos have been minimal and handled at the household level, with cisterns and rooftop tanks (Grube et al. 2020), septic tanks (Ragazzi et al. 2016), and most families burying their own trash (Ragazzi et al. 2014). However, as the archipelago's population surpasses 30,000 with more than seven times that many tourists over the course of a year (Mena et al. 2020), municipal treatment plants and landfills have slowly taken root. The first landfill was constructed near Puerto Ayora in 2009 (Ragazzi et al. 2014), and now all four inhabited islands have landfills (Jaramillo et al. 2020). No working wastewater treatment plants existed in the Galapagos as late as 2010 (Walsh et al. 2010). San Cristobal and Isabela islands received wastewater treatment plants in 2012 and 2015 respectively, although Santa Cruz, the archipelago's most populous island, still does not have one (Mateus and Quiroga 2022) due to technological difficulties (Ragazzi et al. 2016). Ongoing intensification and extensification of urban land cover will make centralized water management systems increasingly critical, as demand for safe water rises while loss of forest and wetland reduce the islands' natural stormwater filtration (Mateus and Quiroga 2022). With land prices rising and available space depleting, the provincial and municipal governments will need to act quickly to secure appropriate water management infrastructure, both built and natural. While national park boundaries and rough topography constrain urban growth more strictly than on most oceanic islands, the impacts of urbanization spread far outside town limits. The tourism industry has motivated park management to open previously off-limits land areas to visitors, replacing habitat with built infrastructure, disturbing wildlife, and introducing new species (Orellana and Smith 2016). Meanwhile rising land prices have driven some residents to build homes in unincorporated rural areas, further decreasing natural vegetation cover (Pizzitutti et al. 2020). As on many other oceanic islands, there may also be informal and peri-urban development that puts coastal ecosystems at risk (Sierra and Feng 2018).

An especially compelling consequence of urban transformation in the archipelago famous for inspiring the theory of natural selection is its effect on wildlife's evolutionary trajectories. Evolutionary processes have been less affected by human land use in the Galapagos than on longer-settled islands (González et al. 2008), but there is a growing body of work demonstrating mixed effects on Darwin's finch. The availability of human food in urban areas helps finches survive and reproduce during dry years with limited natural food sources, and urban finches produce more offspring (Harvey et al. 2021) and have higher population densities than rural finches. However, processed human foods may degrade health and overall fitness, and the urbanized niche brings Darwin's finch into more direct competition with other species (De León et al. 2019). Meanwhile finches use human-made debris to build nests, and some die from entanglement (Harvey et al. 2021; Theodosopoulos and Gotanda 2018). Urban expansion also impacts the Galapagos sea lion, which competes closely with humans for space on beaches and streets. Sea lions living on more crowded beaches are less reactive to, and avoidant of, human presence (Pavez et al. 2015), and human presence affects behavior, nursing patterns, and mother-pup recognition (Denkinger et al. 2015). With urban infrastructure and humans increasingly encroaching on sea lion rookeries, this may lead to new selection processes with uncertain long-term impacts on the endangered species' health, reproduction, and survival.

Galapagos coastal towns are among the world's most climate-vulnerable communities, with sea level rise, flooding, and exacerbated ENSO events degrading physical safety and economic security. These risks are only growing as population and infrastructure expand to accommodate more residents and tourists and the built environment replaces natural flood-regulating landscapes (Quiroga et al. 2011). Emerging physical realities will meet with uneven economic geography and force the local, provincial, and national governments to make difficult choices to preserve the Galapagos' social and economic future. With tourism accounting for 80% of the economy (Escobar-Camacho et al. 2021), preserving businesses, infrastructure, and comfort in coastal towns will undoubtedly be a high priority. However, this may create conflicts of equity when public spending disproportionately benefits wealthier coastal populations and tourists at the expense of rural citizens. Such visible disparities in both economic status and public funds may exacerbate the exodus of farmers, whose work is critical for food security and controlling invasive plants. These imbalances may be resolved with international funding for climate resilience throughout the Galapagos as a matter of global environmental heritage, so the burden does not fall entirely on a tiny, remote province or a small, middle-income country. International nongovernment organizations already spend millions of dollars each year on broad conservation efforts in the Galapagos, largely on education and projects that directly conserve habitats and endemic species. The author is unaware of any major international funding efforts to assist Galapagos farmers in continuing agricultural production and invasive species management amid growing economic and environmental pressures. As farmers are the primary custodians of disturbed highland ecosystems, and pillars of the archipelago's long-term economic diversity and food security, environmental groups may be justified in directing some funds toward farmers' success and security. With market forces strongly favoring coastal urban economies, it is up to the public and nonprofit sectors to protect and support the other keepers of this natural laboratory.

Closing Remarks

The Galapagos' millennia-old land cover patterns are permanently altered from a few centuries of human use. When the first boots introduced alien plant species and later the first residents-built farms and towns, the terrestrial ecosystems were sent on a path toward new equilibria that, even now, may not be reached for centuries. While the archipelago is undoubtedly "more pristine" than longer-settled oceanic islands, it is far from untouched, and it is too late to plan for a Galapagos without a bustling population of residents and visitors. Land use and land cover are changing more rapidly than ever before, and it is critical that policies and infrastructure are carefully designed to contain human impacts past and present, namely the replacement of native environments with concrete surfaces and aggressively invasive plants. The economics of the Galapagos push toward rapid coastal development and abandonment of farming, both of which threaten the integrity of their respective ecosystems. It is thus up to policymakers and funders to react to prevailing market forces to preserve the pristineness that remains on the "Enchanted Islands."

References

Alava JJ, McMullen K, Jones J, Barragán-Paladines MJ, Hobbs C, Tirapé A, Schofield J (2022) Multiple anthropogenic stressors in the Galápagos Islands' complex social-ecological system: interactions of marine pollution, fishing pressure and climate change with management recommendations. Integr Environl Assess Manag

Alomía Herrera I, Paque R, Maertens M, Vanacker V (2022) History of land cover change on Santa Cruz Island, Galapagos. Land 11(7):1017

Astudillo FJ (2018) Environmental and historical archaeology of the Galápagos islands: archaeobotany of Hacienda El Progreso, 1870–1920. Veg Hist Archaeobot 27(5):737–751

Barrera V, Monteros-Altamirano Á, Valverde M, Escudero L, Allauca J, Zapata A (2021) Characterization and classification of agricultural production systems in the Galapagos Islands (Ecuador). Agric Sci 12(5):481–502

Benitez-Capistros F, Couenberg P, Nieto A, Cabrera F, Blake S (2019) Identifying shared strategies and solutions to the human–giant tortoise interactions in Santa Cruz, Galapagos: a nominal group technique application. Sustainability 11(10):2937

De León LF, Sharpe DM, Gotanda KM, Raeymaekers JA, Chaves JA, Hendry AP, Podos J (2019) Urbanization erodes niche segregation in Darwin's finches. Evol Appl 12(7):1329–1343

Denkinger J, Gordillo L, Montero-Serra I, Murillo JC, Guevara N, Hirschfeld M et al (2015) Urban life of Galapagos Sea lions (Zalophus wollebaeki) on San Cristobal Island, Ecuador: colony trends and threats. J Sea Res 105:10–14

Dueñas A, Jiménez-Uzcátegui G, Bosker T (2021) The effects of climate change on wildlife biodiversity of the Galapagos Islands. Clim Change Ecol 2:100026

Ellis-Soto D, Blake S, Soultan A, Guézou A, Cabrera F, Lötters S (2017) Plant species dispersed by Galapagos tortoises surf the wave of habitat suitability under anthropogenic climate change. PLoS One 12(7):e0181333

Escobar-Camacho D, Rosero P, Castrejón M, Mena CF, Cuesta F (2021) Oceanic islands and climate: using a multi-criteria model of drivers of change to select key conservation areas in Galapagos. Reg Environ Chang 21(2):47

Gardener MR, Atkinson R, Rentería JL (2010) Eradications and people: lessons from the plant eradication program in Galapagos. Restor Ecol 18(1):20–29

Gardener MR, Trueman M, Buddenhagen C, Heleno R, Jäger H, Atkinson R, Tye A (2013) A pragmatic approach to the management of plant invasions in Galapagos. In: Plant invasions in protected areas. Springer, Dordrecht, pp 349–374

González JA, Montes C, Rodríguez J, Tapia W (2008) Rethinking the Galapagos Islands as a complex social-ecological system: implications for conservation and management. Ecol Soc 13(2)

Grube AM, Stewart JR, Ochoa-Herrera V (2020) The challenge of achieving safely managed drinking water supply on San Cristobal island, Galápagos. Int J Hyg Environ Health 228:113547

Guézou A, Trueman M, Buddenhagen CE, Chamorro S, Guerrero AM, Pozo P, Atkinson R (2010) An extensive alien plant inventory from the inhabited areas of Galapagos. PLoS One 5(4):e10276

Harvey JA, Chernicky K, Simons SR, Verrett TB, Chaves JA, Knutie SA (2021) Urban living influences the nesting success of Darwin's finches in the Galápagos Islands. Ecol Evol 11(10):5038–5048

Hoyman MM, McCall JR (2013) Is there trouble in paradise? The perspectives of Galapagos community leaders on managing economic development and environmental conservation through ecotourism policies and the special law of 1998. J Ecotour 12(1):33–48

Izurieta A, Delgado B, Moity N, Calvopina M, Cedeño I, Banda-Cruz G, Sutherland WJ (2018) Corrigendum to: a collaboratively derived environmental research agenda for Galápagos. Pac Conserv Biol 24(2):207–207

Jaramillo P, Tapia W, Negoita L, Plunkett E, Guerrero M, Mayorga P, Gibbs J (2020) The Galapagos Verde 2050 Project, vol 1. Charles Darwin Foundation, Puerto Ayora, Galápagos-Ecuador, pp 1–130

Khatun K (2018) Land use management in the Galapagos: a preliminary study on reducing the impacts of invasive plant species through sustainable agriculture and payment for ecosystem services. Land Degrad Dev *29*(9):3069–3076

Laso FJ, Benítez FL, Rivas-Torres G, Sampedro C, Arce-Nazario, J (2019) Land cover classification of complex agroecosystems in the non-protected highlands of the Galapagos Islands. Remote Sensing 12(1):65

Li S, Li X (2017) Global understanding of farmland abandonment: a review and prospects. J Geogr Sci 27(9):1123–1150

Mateus C, Quiroga D (2022) Galapagos' water management evaluation under a changing climate and the current COVID-19 Pandemic. In: Water, food and human health in the Galapagos, Ecuador. Springer, Cham, pp 29–55

McCleary AL (2012) Characterizing contemporary land use/cover change on Isabela Island, Galapagos. In: Science and conservation in the Galapagos Islands: frameworks & perspectives. New York, Springer, pp 155–172

Mena CF, Paltán HA, Benitez FL, Sampedro C, Valverde M (2020) Threats of climate change in Small Oceanic Islands: The case of climate and agriculture in the Galapagos Islands, Ecuador. In: Land cover and land use change on islands. Springer, Cham, pp 119–135

Meyer JY (2014) Critical issues and new challenges for research and management of invasive plants in the Pacific Islands. Pac Conserv Biol 20(2):146–164

Miller BW, Breckheimer I, McCleary AL, Guzmán-Ramirez L, Caplow SC, Jones-Smith JC, Walsh SJ (2010) Using stylized agent-based models for population–environment research: a case study from the Galápagos Islands. Popul Environ 31:401–426

Nieto-Claudin A, Deem SL, Rodríguez C, Cano S, Moity N, Cabrera F, Esperón F (2021) Antimicrobial resistance in Galapagos tortoises as an indicator of the growing human footprint. Environ Pollut 284:117453

Orellana D, Smith F (2016) Assessing geographic isolation of the Galapagos Islands. In: International archives of the photogrammetry, remote sensing & spatial information sciences, p 41

Pavez G, Munoz LILY, Barilari F, Sepúlveda M (2015) Variation in behavioral responses of the South American sea lion to tourism disturbance: implications for tourism management. Mar Mamm Sci 31(2):427–439

Percy MS, Schmitt SR, Riveros-Iregui DA, Mirus BB (2016) The Galápagos archipelago: a natural laboratory to examine sharp hydroclimatic, geologic and anthropogenic gradients. Wiley Interdiscip Rev Water 3(4):587–600

Pike KN, Blake S, Gordon IJ, Cabrera F, Nieto-Claudin A, Deem SL, Schwarzkopf L (2022) Sharing land with giants: habitat preferences of Galapagos tortoises on farms. Glob Ecol Conserv 37:e02171

Pizzitutti F, Brewington L, Walsh SJ (2020) Human and natural environments, island of Santa Cruz, Galapagos: a model-based approach to link land cover/land use changes to direct and indirect socio-economic drivers of change. In: Land cover and land use change on islands. Springer, Cham, pp 183–203

Puente-Rodríguez D, Bos AB, Koerkamp PWG (2019) Rethinking livestock production systems on the Galápagos Islands: organizing knowledge-practice interfaces through reflexive interactive design. Environ Sci Policy 101:166–174

Quiroga D, Mena C, Karrer L, Suzuki H, Guevara A, Murillo JC (2011) Dealing with climate change in the Galapagos: adaptability of the tourism and fishing sectors. In: Climate change vulnerability assessment of the Galápagos islands. Conservation International, Quito, p 81

Ragazzi M, Catellani R, Rada EC, Torretta V, Salazar-Valenzuela X (2014) Management of municipal solid waste in one of the Galapagos Islands. Sustainability 6(12):9080–9095

Ragazzi M, Catellani R, Rada EC, Torretta V, Salazar-Valenzuela X (2016) Management of urban wastewater on one of the Galapagos Islands. Sustainability 8(3):208

Renteria JL, Gardener MR, Panetta FD, Crawley MJ (2012) Management of the invasive hill raspberry (Rubus niveus) on Santiago Island, Galapagos: eradication or indefinite control? Invasive Plant Sci Manag 5(1):37–46

Sierra VLA, Feng X (2018) Landscape connectivity approach in oceanic islands by urban ecological island network systems with the case study of Santa Cruz Island, Galapagos (Ecuador). Curr Urban Stud 6(4):573–610

Theodosopoulos AN, Gotanda KM (2018) Death of a Darwin's Finch: a consequence of human-made debris? Wilson J Ornithol 130(4):1023–1028

Trueman M, Atkinson R, Guézou A, Wurm P (2010) Residence time and human-mediated propagule pressure at work in the alien flora of Galapagos. Biol Invasions 12(12):3949–3960

Villacis B, Carrillo D (2013) The socioeconomic paradox of Galapagos. In: Science and conservation in the Galapagos Islands. Springer, New York, NY, pp 69–85

Walsh SJ, Mena CF (2016) Interactions of social, terrestrial, and marine sub-systems in the Galapagos Islands, Ecuador. Proc Natl Acad Sci 113(51):14536–14543

Walsh SJ, McCleary AL, Heumann BW, Brewington L, Raczkowski EJ, Mena CF (2010) Community expansion and infrastructure development: implications for human health and environmental quality in the Galápagos Islands of Ecuador. J Lat Am Geogr:137–159

Watson J, Trueman M, Tufet M, Henderson S, Atkinson R (2010) Mapping terrestrial anthropogenic degradation on the inhabited islands of the Galapagos Archipelago. Oryx 44(1):79–82

Wilkinson SR, Naeth MA, Schmiegelow FKA (2005) Tropical forest restoration within Galapagos National Park: application of a state-transition model. Ecol Soc 10(1)

Madeline Giefer is an assistant professor of Geography at Austin Peay State University. Her research examines land use change and the indirect effects of conservation policies on rural communities.

Part VI
Island Ecosystems: Marine Sub-systems

Chapter 18
Common Oversights in the Design and Monitoring of Ecosystem-Based Management Plans and the Siting of Marine Protected Areas

Sergio A. Navarrete, Christopher M. Aiken, M. Isidora Ávila-Thieme, Daniel Valencia, Alexandre Génin, and Stefan Gelcich

Introduction

Nearly 200 million tons of fish, invertebrate and macroalgae are extracted every year from wild or cultured populations at sea, which is considered essential for human food security and health (FAO 2020), and it is expected to increase in importance for human life in the next decades (Naylor et al. 2021). There is no doubt that the scale of these extractive activities is large enough to be causing negative impacts on marine life. This pressure, coupled with the stress caused by a changing climate, is pushing many ecosystems and their services to the brink of collapse. This is certainly the case for the southeastern Pacific. Since the 1960s, the industrial and coastal fisheries along the nutrient-rich waters of the Humboldt Upwelling Ecosystem (HUE) and the smaller Galápagos upwelling have provided a sizable fraction of the world landings of fish biomass, but many of the fished resources

S. A. Navarrete (✉)
Estación Costera de Investigaciones Marinas, Las Cruces, Departamento de Ecología, Santiago, Chile

Millenium Nucleus for Ecology and Conservation of Temperate Mesophotic Reefs Ecosystems (NUTME), Pontificia Universidad Católica de Chile and Center COPAS-COASTAL, Universidad de Concepción, Concepción, Chile

Instituto Milenio de Socio-Ecología Costera (SECOS), Santiago, Chile

Center for Applied Ecology and Sustainability (CAPES), Pontificia Universidad Católica de Chile, Santiago, Chile
e-mail: snavarrete@bio.puc.cl

C. M. Aiken
CQ University, Coastal Marine Ecosystems Research Centre, Gladstone, Rockhampton, QLD, Australia

S. J. Walsh et al. (eds.), *Island Ecosystems*, Social and Ecological Interactions in the Galapagos Islands, https://doi.org/10.1007/978-3-031-28089-4_18

show signs of over-exploitation or outright collapse (Salas et al. 2011; IFOP 2014; Vinueza et al. 2014). As a result, there is an urgent need for scientists to provide guidelines to preserve ecosystem services, foster sustainability and restore exploited populations.

Unfortunately, the scientific information to achieve this grand goal will never be sufficiently long, extensive or complete to provide a definitive 'how-to' user manual for managers and conservation agencies. Facing the impossibility of scientifically assessing coastal fisheries using traditional fishery protocols and recognizing the inadequacies of the 'resource-focused' approaches, which disregard the wider effects of fisheries, especially the human dimensions of sustainability, scientists and practitioners have in the last decades turned to more integrated and holistic approaches. In this context, the implementation of Marine Protected Areas (MPAs) and the development of Ecosystem Based Management (EBM) programs have been largely endorsed by the scientific community, and are being prioritized around the world (Charles 2001; Douvere and Ehler 2009), and in the Galápagos Islands in particular (Vinueza et al. 2014; Castrejón et al. 2014; Walsh and Mena 2016), where artisanal fisheries exert strong pressure on coastal resources (Fig. 18.1).

It is clear that there will be no sustainability of coastal ecosystems and their services without an effective and explicit inclusion of humans and their social structures as the main players within managed ecosystems. Yet, resource governance structures, whether polycentric or monocentric, even when well-matched to social systems (Cáceres et al. 2022), do not guarantee fishery sustainability, and may still lead to fishery degradation if they don't consider the scales and spatial dependencies

M. I. Ávila-Thieme
Instituto Milenio de Socio-Ecología Costera (SECOS), Santiago, Chile

Facultad de Ciencias Biológicas, Departamento de Ecología, Pontificia Universidad Católica de Chile, Santiago, Chile

Advanced Conservation Strategies, Midway, UT, USA

D. Valencia
Estación Costera de Investigaciones Marinas, Las Cruces, Departamento de Ecología, Santiago, Chile

A. Génin
Estación Costera de Investigaciones Marinas, Las Cruces, Departamento de Ecología, Santiago, Chile

Facultad de Ciencias Biológicas, Departamento de Ecología, Pontificia Universidad Católica de Chile, Santiago, Chile

Copernicus Institute of Sustainable Development, Utrecht University, Utrecht, The Netherlands

S. Gelcich
Instituto Milenio de Socio-Ecología Costera (SECOS), Santiago, Chile

Center for Applied Ecology and Sustainability (CAPES), Pontificia Universidad Católica de Chile, Santiago, Chile

Facultad de Ciencias Biológicas, Departamento de Ecología, Pontificia Universidad Católica de Chile, Santiago, Chile

Fig. 18.1 The fish market on the main street of Santa Cruz Island open to tourists and locals

of metapopulation dynamics. Indeed, a rather stationary approach has dominated thinking when it comes to spatial management and the siting or spatial design of no-take MPAs. Below, we illustrate why the consideration of metapopulation principles is critical if we are to sustainably manage exploited ecosystems using spatial strategies. We highlight the advances in ecological and oceanographic sciences that make possible to realistically model connectivity and provide an example of why they should be implemented in the Galapagos Marine Reserve (GMR).

It is also clear that, if the fabric of nature that maintains fish biomass is slowly degrading, failure to consider the wider effects of fisheries on the ecological web, including non-exploited species, could mean that apparent long-term sustainability of even 'well-managed' fisheries is only a short-term illusion (Travis et al. 2014). Coastal spatial management and integrated coastal management, which are considered as fundamental to Ecosystem Based Management programs (Douvere and Ehler 2009), are conceived as a response to overcome the shortcomings of mono-specific and resource-focused-management. Consequently, well-designed monitoring programs, including baseline information and time series, are essential and integral to any EBM programs (Edgar et al. 2004; Day 2008). Many have criticized varied aspects of EBM approaches (see Murawski et al. 2007, and references therein). Our goal here is not provide another criticism to EBM or MPAs but to note that the state of knowledge and capabilities in ecological and oceanographic sciences today, make it possible to overcome some of the shortcomings.

Including Connectivity in Zoning, Siting of MPAs and Assessment of Performance of EBM Programs

Spatial management approaches and spatially integrated coastal management – the foundations of the EBM approach – focus on managing the multiple human uses of designated areas or 'zones' that are assigned to varied, overlapping or non-overlapping uses. Since the designation of no-take MPAs is another spatial management strategy to improve conservation and increase resilience to fisheries, it has become part of most if not all EBM programs. The approach attempts to best match and find best tradeoffs solutions to stakeholders' spatial priorities and interests. For varied reasons, from the point of view of the species subjected to fisheries the spatial zoning approach has been static, rarely considering the inherent underlying dynamics of spatially structured populations imposed by dispersal in an advective environment.

The absence of connectivity principles in spatial management plans may owe in part to lack of suitable information, but also to unawarness of frameworks that allow metapopulation dynamics to be reconciled with habitat suitability. Advances over the last decade suggest that this oversight may now be overcome. Firstly, non-equilibrium metapopulation theory, which focuses on timescales that approach management and conservation goals, has been well-developed in the last decade and can provide useful guidelines even under limited metapopulation information (Aiken et al. 2007; Aiken and Navarrete 2014; Aiken and Navarrete 2020; Aiken and Navarrete 2011; Williams and Hastings 2013; Mari et al. 2017; Farrell and Ioannou 1996). This body of knowledge comes to complement the more traditional, stability approaches to metapopulations which have been reviewed in the context of spatial management and marine protected areas (Hastings 2014; Botsford et al. 2001; Gaines et al. 2010). Aiken et al. (2023) provide an example of how principles of population persistence and growth in reactive systems, *e.g.* those exhibiting transient dynamics at timescales matching scales of management, can be applied to guide restoration of seagrass *Zoostera mulleri* in Port Curtis, Australia, even with incomplete information about dispersal.

Secondly, our capacity to realistically simulate larval dispersal has improved considerably. Numerical 3D models of the coastal circulation can be run at high enough resolution to capture important meso-scale and sub-mesoscale oceanographic processes, including the complex coastal boundary layer (Capet et al. 2008; Dauhajre et al. 2019), sourcing realistic boundary conditions from global simulations that assimilate ever more complete observations of the ocean state, over spatial extents large enough to encompass larval dispersal. In addition, a dense network of Lagrangian drifting buoys provide a unique empirical insight into ocean dispersal (Álvarez-Noriega et al. 2020; Jönsson and Watson 2016; Aiken and Navarrete 2020).

But why should we include connectivity principles in spatial management and conservation? Is it not enough to secure governance in the exploitation of resources and that EBM destination zoning or MPA's are being accepted and complied by stakeholders? Metapopulation theory for spatially structured populations has shown

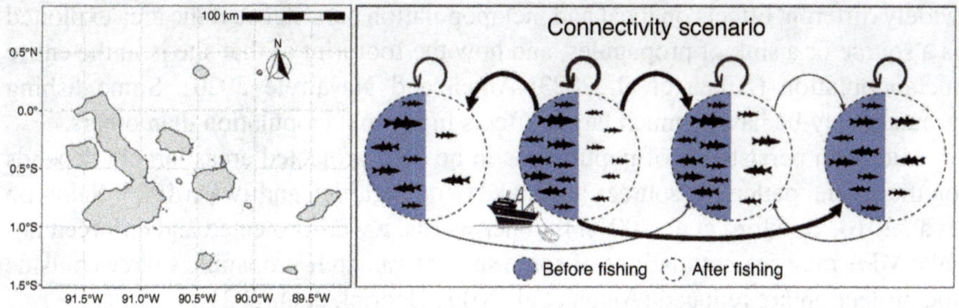

Fig. 18.2 Schematic representation of connectivity and spatial management that includes zones where fishing is allowed and zones where fishing is prohibited (no-take MPAs). The scheme could represent the Galapagos Marine Reserve Management Plan for the Galapagos Marine Reserve presented in the map. In the example, arrows and their thickness represent direction and strength of dispersal by propagules among the four different zones or local populations. Initially, before fishing (blue section of the local populations), fish biomass is high everywhere and especially in the zone where fishing is allowed. After fishing (white section of the local populations) and as expected, fish biomass, and fish size, are significantly reduced in the fished zone. But fish biomass is also reduced in protected areas (MPAs) because their productivity or population replenishment depends, to varying degrees, on the influx of new propagules from the fished population. In the example, fish biomass in the first zone on the left is drastically reduced because of the its dependence on influxes from the fished zone upstream, even though fish size in that MPA may not have been altered

the importance of the character of the dispersal process on population dynamics, coexistence, and persistence. The general principle being that mid-term persistence of spatially-structured populations, under natural conditions or when subjected to anthropically imposed mortality (e.g. fisheries, Fig. 18.1), depends on local dynamics and connectivity (Hastings 2014). Figure 18.2 illustrates a hypothetical, but realistic scenario of fishing and protected zones connected by dispersal. Four findings, well founded in theory suffice to demonstrate why basic principles of connectivity must be considered when planning destination zones, siting no-take MPAs, or assessing the performance of these spatial measures.

Local population productivity, *i.e.*, the replenishment of benthic populations within any EBM or MPA status, depends on the spatial structure of dispersal and emerging patterns of connectivity. Over most scales of zoning and no-take MPAs, complete retention and self-recruitment cannot be assumed (Lett et al. 2015) and therefore, 'effective reproduction', *i.e.* recruitment of dispersing propagules to local populations within the zone, depends from upstream populations. Using local knowledge of high historical productivity of exploited species at designated sites is an important piece of information, but it only indicates that propagules arrive at high rates at a given site, not where they come from (Burgess et al. 2014; Lett et al. 2015).

The resilience of populations to fishing mortality in fishing access areas depends on the patterns of connectivity and dependencies to other distant fishing grounds and to zones under fishing protection. In other words, the same fishing mortality exerted in a population, and even if the same biomass is being removed, can have widely different effects on local and metapopulation persistence if the area exploited is a source or a sink of propagules, and how the footprint of that site is in the entire metapopulation (Aiken et al. 2023; Aiken and Navarrete 2020). Some fishing grounds may be having much larger effects in the metapopulation than others.

Mid-term persistence of populations in no-take protected areas largely depends on the spatial pattern of sources and sinks (Roughgarden and Iwasa 1986; Salomon et al. 2010; Botsford et al. 2001). In other words, a well-protected and enforced no-take MPA may see populations of some species vanish if propagule sources outside the protection are reduced (Aiken et al. 2011; Dedrick et al. 2021).

Entire metapopulation persistence (extinction), especially under intensive fishing mortality, can be highly reliant on a few sites or localities that are still sustaining transient population growth (Aiken et al. 2023; Mari et al. 2017). Those sites can be very distant, beyond the areas our management plans attempt to protect (Dedrick et al. 2021). Protecting those sites, and not necessarily the apparent high-productivity areas within islands, must be a priority.

Thus, establishing the EBM zoning and siting no-take areas based solely on patterns of abundance, or habitat suitability, or tradeoffs among stakeholders' needs and goals, and ignoring recruit fluxes between populations, may result in less-than-expected benefits of the management plan, even if correctly enforced (Fig. 18.2). Efforts should thus be made to incorporate some of these basic principles when assessing the existing Galápagos Marine Reserve Management Plan (GMRMP) and when siting small no-take MPAs along the coast.

Why the Galápagos Islands Is an Ideal System and Why We Should also Look at Continental Populations?

As identified by theorists such as Ilka Hanski, islands provide ideal testing grounds for metapopulation theory, as a highly punctuated connectivity can accentuate processes that are difficult to observe when dispersal barriers are more diffuse. Although the shallow and intertidal habitats of the Galápagos are distant some 800 km from the south American west coast, some degree of connectivity between the two is likely for species with pelagic larval durations of weeks. Importantly, this connectivity is predominantly unidirectional. Figure 18.3 illustrates the probability of passive drifting from and to the Galápagos, derived from 45 years of data from the Global Drifter Program (Elipot et al. 2022) following the method of Aiken and Navarrete (2020). As such, the Galápagos Islands likely receive recruits from the mainland via the swift and constant south equatorial current, but the return journey is far less likely to occur (Fig. 18.3). This strongly biased dispersal can have profound consequences for persistence and coexistence in the Galápagos that must be considered within management strategies (Aiken and Navarrete 2011, 2014, 2020; Aiken et al. 2023). As a consequence of strongly biased westwards advection in the equatorial Pacific, a species with a planktonic larval dispersal time on the order of

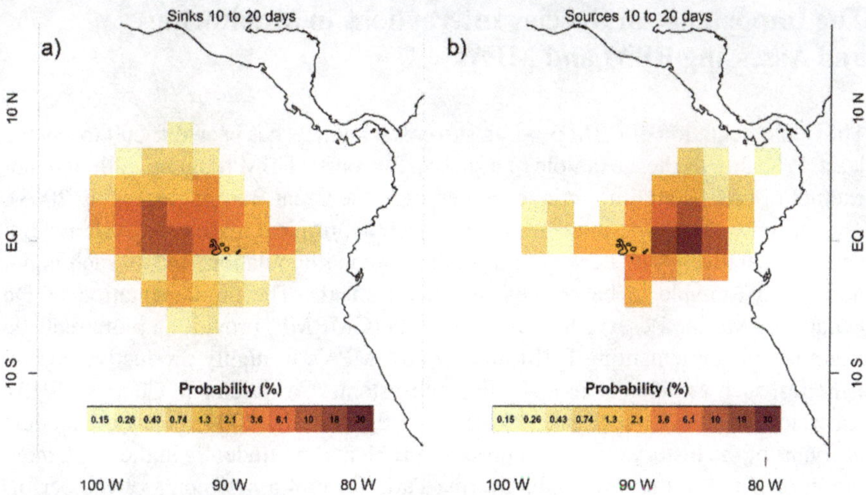

Fig. 18.3 Probability of drifting (**a**) from the Galapagos Islands and (**b**) to the Galapagos Islands, from the continent, withing the period of 10–20 days that simulates larval pelagic development times for many invertebrate and fish species. Probabilities are based on 45 years of data from the Global Drifter Program data, using the method presented in Aiken and Navarrete (2020)

multiple weeks may exhibit weak self-recruitment, and potentially depend upon subsidization by continental populations. This situation determines the metapopulation dynamics, with the result that large population fluctuations that are unrelated to local conditions can occur simply due to variations in larval supply from remote source populations. Clearly in such a case persistence of the species in the Galápagos depends, to some extent, on the level of protection afforded the continental source populations.

The reactive dynamics of long dispersers in the Galápagos has additional consequences for coexistence, *i.e.*, local diversity of species. Taking a pair of species as example, if one species were the dominant when competing for a local resource, but the subordinate had a shorter pelagic larval duration and hence more stable recruitment, "reactive coexistence" would be possible, whereby dispersal driven fluctuations in recruit supply are sufficient to allow persistence of the subordinate when the dominant is subject to biased dispersal. The subordinate survives by occupying a "dispersal niche" (Aiken and Navarrete 2014), being the inferior competitor in terms of using habitat resources, but using a dispersal strategy that is more robust. Again, co-oscillation in the populations of these two competitors, and even local extinction of one or the other, could be driven by variations in the recruit supply of the dominant alone, driven by either larval production rates in the continental source populations, the protected areas within the Galápagos archipelago, or an unfavorable ocean circulation.

The Importance of Species Interactions in Monitoring and Assessing EBM and MPAs

The implementation of EBM programs to costal habitats has been difficult to say the least. Probably the best example of a successful coastal EBM program, although not exempt of pitfalls and limitations, comes from the Great Barrier Reef (Day 2008). This program, however, manages an oligotrophic, tropical marine ecosystem adjacent to a relatively low density and wealthy human population, and as such is not necessarily suitable to be applied in other contexts. The implementation of the Galápagos Marine Reserve Management Plan (GMRMP) provides a more realistic example for implementing EBM and no-take MPA's in highly productive coastal fisheries, such as the ones in upwelling ecosystems. Castrejón and Charles (2013), Castrejón et al. (2014), Reck (2014) and Edgar et al. (2004) provide excellent accounts of the history, definition process, problems and tradeoffs in the implementation of EBM in the Galápagos Marine Park. Several assessments of the performance of the GMRMP have been conducted, mostly on the co-management system, structure of socio-economic systems, governance (Cáceres et al. 2022; Mestanza-Ramón et al. 2019; Heylings and Bravo 2007; Castrejón and Charles 2013) and, to a lesser extent, the biodiversity components (Edgar et al. 2008).

Adaptive monitoring is an essential step of any EBS and coastal spatial management program. Selected monitoring targets typically include the fished species subjected to the spatial management, integrated or composite community variables (e.g. species richness), as well as 'sentinel' species that may indicate or integrate somehow the state of the ecosystem (e.g. marine iguanas, sharks, (Vinueza et al. 2014; Castrejón and Charles 2013; Edgar et al. 2008). However, we argue that without the aid of a model that can help managers anticipate which species in the ecosystem will respond to management policy, monitoring programs will misrepresent or entirely miss the effects of the management plan and fail to attribute causes of biodiversity change to the implemented policy. Sentinel or abundant species may be useful to monitor for many reasons (*e.g.* indicators of climate change, pollution, etc.), but may not be the species that respond to a given fishery management policy. Monitoring all biodiversity of species that interact in a marine ecosystem and that transform biomass to finally sustain fisheries, is virtually impossible too. All standardized quantitative assessments of biodiversity (*e.g.* subtidal visual transects, quadrats, destructive samples, eDNA) are systematically biased towards or against small, rare, cryptic, large, highly mobile, infrequent, etc. species. Thus, efforts to assess the diversity of co-occurring species that can be affected by management are colossal and unsustainable over time.

Indeed, propagation of management-induced alterations through the ecological web cannot be assumed to be linear, proportional or even in the same direction to the magnitude of the alteration. The problem of propagation of species interactions was brightly captured in Yodzis's 'indeterminacy of ecological interactions' over 30 years ago (Yodzis 1988). Basically, alteration of one species biomass will have positive, negative or negligible effects on the abundance and persistence of other species that are not directly connected to the target one, potentially driving them to extinction or to a pest status (*e.g.* release from top-down control). Therefore, identifying which species in the community will be positively or negatively affected by removing biomass (fishing), is not possible without consideration of the type and intensity of species interactions.

Attributing Biodiversity Change to Management

For coastal regions with abundant ecological information, such as the Galápagos Marine Reserve (Riofrío-Lazo et al. 2021), models of intermediate to high complexity can be constructed to represent the ecological web with sufficient realism to simulate scenarios for different fisheries impacts (Ávila-Thieme et al. 2021; Riofrío-Lazo et al. 2021) and different levels of compliance to set policies (Navarrete et al. in press). Diverse multi-species or multi-component modeling strategies exists (Yodzis 2001; Pauly et al. 2000). Riofrío-Lazo et al. (2021) have implemented an Ecopath with Ecosim (EwE, Pauly et al. 2000) that could be the basis for the type of quantitative-qualitative modeling illustrated in Fig. 18.4. However, the allometric trophic web models (ATN), based on bioenergetic

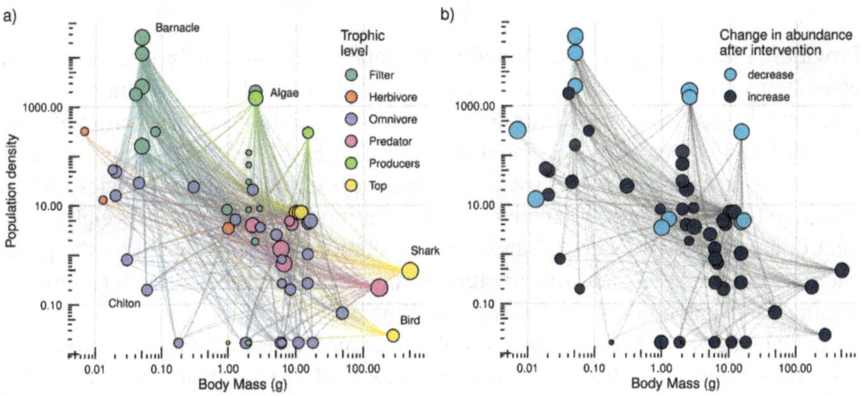

Fig. 18.4 (**a**) Schematic representation of an ecological web representing consumption type of interactions (food web) among 60 species (nodes) from all trophic levels, indicated by the different color nodes. Names of some nodes are indicated for reference. The links representing trophic interactions among nodes are taken from the rocky shore intertidal food web of central Chile (Kéfi et al. 2012) and therefore represent a realistically complex ecological system. The nodes and links are ordered in the Abundance (Population density) and Body Mass plane to show how small-bodied and moderate abundance species can participate in the fabric of nature and be affected by fisheries and management policies. Many of these species are not included in standardized monitoring programs. (**b**) Shows the change in abundance of species that takes place after a single species if fished in the system. Both, significantly positive (blue nodes) and significantly negative (black nodes) changes are observed with respect to the unfished system, and they appear in different sectors of the Abundance-Body Size spectrum. These nodes are 'interaction-indicator' species that should be included in spatial management and Ecosystem Based Management monitoring programs, which allows managers to attribute biodiversity changes to policy

equations for biomass transfer, present some advantages over EwW because part of the parameterization can be done from basic allometric principles (Martinez 2020; Brose 2010). The more mechanistic basis for this modeling framework is also more amenable to the inclusion of non-trophic interactions (Kéfi et al. 2012; Kefi et al. 2016), such as habitat provisioning that can be critically important for benthic coastal ecosystems.

Although quantitative results from multi-species models are highly dependent on parameters, it can be shown that qualitive outcomes to identify "interaction-indicator" species can be quite insensitive to a range of parameter values. Navarrete et al. (in press) showed that moderately complex models of species interactions can be used to link management policies and levels of compliance with set regulations, with biodiversity monitoring programs, through identifying which species most likely respond to the alteration of the biomass of fishery target species. Since the pathways of propagation of biomass and information are mostly defined by the topology of the interaction web, the identities of the species being affected by harvesting are relatively insensitive to parameter values (Fig. 18.4).

Concluding Remarks

Providing useful and effective guidelines to improve the sustainability of exploited ecosystems and conserving biodiversity, grounded on solid scientific findings, is a great challenge. Here, we illustrate how the consideration of basic principles of dispersal and connectivity in spatial planning and EBM, as well as multi-species models for assessing ecosystem-wide consequences of management policies, can be integrated in ecosystem and spatial management programs. This is especially relevant in areas where non-compliance with regulations is compounded by the gross shortage of resources and infrastructure of management agencies, such as most of South America and GMR productive upwelling ecosystems.

In Galápagos and Chile, high levels of non-compliance are rampant in most fisheries (Castrejón and Charles 2013; Fernández et al. 2020; Oyanedel et al. 2018, 2020). Fishers must effectively participate in the conservation and MPA siting process to reduce non-compliance. This requires maximizing the returns per unit area (stock fraction) that is set aside, and demonstrating that the oss of fishable biomass provides longer-term benefits. This can be aided using transient metapopulation theory and improved circulation models as illustrated here.

Decision makers are usually pushed to protect zones and siting no-take MPAs in areas that have less or no value for fishers and other stakeholders (Edgar et al. 2004) rather than areas where existing biological information would advise. They usually assess the effectiveness of a zoning system as if the sole creation of a protected area, anywhere in a metapopulation, was sufficient to improve fishery resilience and productivity. Many empirical studies suggest that the effectiveness of MPAs, even within their limits, can depend on several factors (Edgar et al. 2014). As discussed above, the effect on adjacent fishing areas and population persistence strictly depend on connectivity and this must be considered when assessing effectiveness of spatial management.

Precisely because a spatial management plan results from trade-offs among multiple needs and objectives (Douvere and Ehler 2009; Castrejón and Charles 2013), adaptive and carefully designed monitoring is essential for its success from the perspective of exploited marine populations. All-encompassing biodiversity monitoring is unfeasible in all real ecosystems and economically unsustainable over time. Multi-species dynamic models can be used to guide and focus monitoring programs to those 'interaction-indicator' species that most likely will respond to management policies and to lack of compliance with those regulations.

Acknowledgements This work and collaboration were made possible by ANID PIA/ BASAL FB0002 (CAPES), FB FB210021 (COPAS COASTAL), the Millennium Science Initiative Program NCN19_056 (SECOS), the Millennium Nucleus for Ecology and Conservation of Temperate Mesophotic Reefs (NUTME) and the European Union's Horizon 2020 research and innovation programme under the Marie Sklodowska-Curiegrant agreement N°896159 (INDECOSTAB). MIA thanks Fondecyt 3220110 and the Walton Family Foundation.

References

Aiken C, Navarrete SA (2011) Environmental fluctuations and asymmetrical -dispersal: generalized stability theory for studying Metapopulation persistence and marine protected areas. Mar Ecol Prog Ser 428:77–88

Aiken C, Navarrete SA, Castillo M, Castilla JC (2007) Along-shore larval dispersal kernels in a numerical ocean model of the Central Chilean Coast. Mar Ecol Prog Ser 339:13–24

Aiken CM, Navarrete SA (2014) Coexistence of competitors in marine meta-communities: environmental variability, edge effects and the dispersal niche. Ecology 95:2289–2302

Aiken CM, Navarrete SA (2020) Incorporating the connectivity timescale in metapopulation partitioning. Am Nat 196:145–156

Aiken CM, Navarrete SA, Pelegrí JL (2011) Potential changes in larval dispersal and alongshore connectivity on the Central Chilean Coast due to an altered wind climate. J Geophys Res 116. https://doi.org/10.1029/2011jg001731

Aiken CM, Navarrete SA, Jackson EL (2023) Reactive persistence, spatial management, and conservation of Metapopulations: an application to seagrass restoration. Ecol Appl 33(2):e2774. https://doi.org/10.1002/eap.2774

Álvarez-Noriega M, Burgess SC, Byers JE, Pringle JM, Wares JP, Marshall DJ (2020) Global biogeography of marine dispersal potential. Nat Ecol Evol 4:1196–1203

Ávila-Thieme MI, Corcoran D, Pérez-Matus A, Wieters EA, Navarrete SA, Marquet PA, Valdovinos FS (2021) Alteration of coastal productivity and artisanal fisheries interact to affect a marine food web. Sci Rep 11:1765

Botsford LW, Hastings A, Gaines MS (2001) Dependence of sustainability on the configuration of marine reserves and larval dispersal distance. Ecol Lett 4:144–150

Brose U (2010) Body-mass constraints on foraging behaviour determine population and food-web dynamics. Funct Ecol 24:28–34

Burgess SC, Nickols KJ, Griesemer CD, Barnett LAK, Dedrick AG, Satterthwaite EV, Yamane L, Morgan SG, White JW, Botsford LW (2014) Beyond connectivity: how empirical methods can quantify population persistence to improve marine protected-area design. Ecol Appl 24:257–270

Cáceres R, Pittman J, Castrejón M, Deadman P (2022) The evolution of polycentric governance in the Galapagos small-scale fishing sector. Environ Manag 70:254–272

Capet X, Mcwilliams JC, Molemaker MJ, Shchepetkin AF (2008) Mesoscale to submesoscale transition in the California current system. Part II: frontal processes. J Phys Oceanogr 38:44–64

Castrejón M, Charles A (2013) Improving fisheries co-management through ecosystem-based spatial management: the Galapagos marine reserve. Mar Policy 38:235–245

Castrejón M, Defeo O, Reck G, Charles A (2014) Fishery science in Galapagos: from a resource-focused to a social–ecological systems approach. In: Denkinger J, Vinueza L (eds) The Galapagos Marine Reserve: a dynamic soical-ecological system. Springer, New York

Charles A (2001) Sustaineable fishery systems. Blackwell Science, Oxford

Dauhajre DP, Mcwilliams JC, Renault L (2019) Nearshore Lagrangian connectivity: submesoscale influence and resolution sensitivity. J Geophys Res Oceans 124:5180–5204

Day JC (2008) The need and practice of monitoring, evaluating and adapting marine planning and management –lessons from the great barrier reef. Mar Policy 32:823–831

Dedrick AG, Catalano KA, Stuart MR, White JW, Montes HR, Pinsky ML (2021) Persistence of a reef fish metapopulation via network connectivity: theory and data. Ecol Lett 24:1121–1132

Douvere F, Ehler C (2009) Ecosystem-based marine spatial management: an evolving paradigm for the management of coastal and marine places. Ocean Yearbook Online 23:1–26

Edgar G, Banks S, Bensted-Smith R, Calvopiña M, Chiriboga A, Garske-Garcia L, Henderson S, Miller KA, Salazar S (2008) Conservation of threatened species in the Galapagos Marine Reserve through identification and protection of marine key biodiversity areas. Aquat Conserv Mar Freshwat Ecosyst 18:955–968

Edgar GJ, Bustamante RH, Fariña JM, Calvopiña M, Martínez C, Toral-Granda MV (2004) Bias in evaluating the effects of marine protected areas: the importance of baseline data for the Galapagos Marine Reserve. Environ Conserv 31:212–218

Edgar GJ, Stuart-Smith RD, Willis TJ, Kininmonth S, Baker SC, Banks S, Barrett NS, Becerro MA, Bernard ATF, Berkhout J, Buxton CD, Campbell SJ, Cooper AT, Davey M, Edgar SC, Forsterra G, Galvan DE, Irigoyen AJ, Kushner DJ, Moura R, Parnell PE, Shears NT, Soler G, Strain EMA, Thomson RJ (2014) Global conservation outcomes depend on marine protected areas with five key features. Nature 506:216

Elipot S, Sykulski A, Lumpkin R et al (2022) A dataset of hourly sea surface temperature from drifting buoys. Sci Data 9:567. https://doi.org/10.1038/s41597-022-01670-2

FAO (2020) The state of world fisheries and aquaculture 2020. Sustainability in Action, Rome

Farrell BF, Ioannou PJ (1996) Generalized stability theory. Part I: autonomous operators. J Atmos Sci 53:2025–2040

Fernández M, Kriegl M, Garmendia V, Aguilar A, Subida MD (2020) Evidence of illegal catch in the benthic artisanal fisheries of Central Chile: patterns across species and management regimes. Lat Am J Aquat Res 48:287–303

Gaines SD, Lester SE, Grorud-Colvert K, Costello C, Pollnac R (2010) Evolving science of marine reserves: new developments and emerging research frontiers. Proc Natl Acad Sci USA 107:18251–18255

Hastings A (2014) Persistence and management of spatially distributed populations. Popul Ecol 56:21–26

Heylings P, Bravo M (2007) Evaluating governance: a process for understanding how co-management is functioning, and why, in the Galapagos Marine Reserve. Ocean Coast Manag 50:174–208

IFOP (2014) Review of biological reference points in domestic fisheries. Report of the third international workshop held in Viña Del Mar by IFP

Jönsson BF, Watson JR (2016) The timescales of global surface-ocean connectivity. Nat Commun 7:11239

Kéfi S, Berlow EL, Wieters EA, Navarrete SA, Petchey OL, Wood SA, Boit A, Joppa LN, Lafferty KD, Williams RJ, Martinez ND, Menge BA, Blanchette CA, Iles AC, Brose U (2012) More than a meal... integrating non-feeding interactions into food webs. Ecol Lett 15:291–300

Kefi S, Miele V, Wieters EA, Navarrete SA, Berlow EL (2016) How structured is the entangled bank? The surprisingly simple organization of multiplex ecological networks leads to increased persistence and resilience. PLoS Biol 14. https://doi.org/10.1371/Journal.Pbio.1002527

Lett C, Nguyen-Huu T, Cuif M, Saenz-Agudelo P, Kaplan DM (2015) Linking local retention, self-recruitment, and persistence in marine Metapopulations. Ecology 96:2236–2244

Mari L, Casagrandi R, Rinaldo A, Gatto M (2017) A generalized definition of reactivity for ecological systems and the problem of transient species dynamics. Methods Ecol Evol 8:1574–1584

Martinez ND (2020) Allometric trophic networks from individuals to socio-ecosystems: consumer–resource theory of the ecological elephant in the room. Front Ecol Evol 8

Mestanza-Ramón C, Capa MS, Saavedra HF, Paredes JR (2019) Integrated coastal zone management in continental Ecuador and Galapagos Islands: challenges and opportunities in a changing tourism and economic context. Sustainability 11:6386

Murawski S, Methot R, Tromble G (2007) Biodiversity loss in the ocean: how bad is it? Science 316:1281–1284

Navarrete S, Ávila-Thieme MI, Valencia D, Génin A, Gelcich S (in press) Monitoring the fabric of nature: using allometric trophic network models and observations to assess policy effects on biodiversity. Proc R Soc B

Naylor RL, Kishore A, Sumaila UR, Issifu I, Hunter BP, Belton B, Bush SR, Cao L, Gelcich S, Gephart JA, Golden CD, Jonell M, Koehn JZ, Little DC, Thilsted SH, Tigchelaar M, Crona B (2021) Blue food demand across geographic and temporal scales. Nat Commun 12:5413

Oyanedel R, Gelcich S, Milner-Gulland EJ (2020) Motivations for (non-)compliance with conservation rules by small-scale resource users. Conserv Lett 13:E12725

Oyanedel R, Keim A, Castilla JC, Gelcich S (2018) Illegal fishing and territorial user rights in Chile. Conserv Biol 32:619–627

Pauly D, Christensen V, Walters C (2000) Ecopath, ecosim, and ecospace as tools for evaluating ecosystem impact of fisheries. ICES J Mar Sci 57:697–706

Reck G (2014) Development of the Galápagos Marine Reserve. In: Denkinger J, Vinueza L (eds) The Galapagos marine reserve: a dynamic soical-ecological system. Springer, New York

Riofrío-Lazo M, Reck G, Páez-Rosas D, Zetina-Rejón MJ, Del Monte-Luna P, Reyes H, Murillo-Posada JC, Hernández-Padilla JC, Arreguín-Sánchez F (2021) Food web modeling of the southeastern Galapagos Shelf Ecosystem. Ecol Indic 132:108270

Roughgarden J, Iwasa Y (1986) Dynamics of a metapopulation with space-limited subpopulations. Theor Popul Biol 29:235–261

Salas S, Chuenpagdee R, Charles A, Seijo JC (eds) (2011) Coastal fisheries of Latin America and the Caribbean. FAO, Rome

Salomon Y, Connolly SR, Bode L (2010) Effects of asymmetric dispersal on the coexistence of competing species. Ecol Lett 13:432–441

Travis J, Coleman FC, Auster PJ, Cury PM, Estes JA, Orensanz J, Peterson CH, Power ME, Steneck RS, Wootton JT (2014) Integrating the invisible fabric of nature into fisheries management (Vol 111, Pg 581, 2013). Proc Natl Acad Sci USA 111:4644–4644

Vinueza L, Post A, Guarderas P, Smith F, Idrovo F (2014) Ecosystem-based management for rocky shores of the Galapagos Islands. In: Denkinger J, Vinueza L (eds) The Galapagos Marine Reserve: a dynamic soical-ecological system. Springer, New York

Walsh SJ, Mena CF (2016) Interactions of social, terrestrial, and marine sub-systems in the Galapagos Islands, Ecuador. Proc Natl Acad Sci USA 113:14536–14543

Williams PD, Hastings A (2013) Stochastic dispersal and population persistence in marine organisms. Am Nat 182:271–282

Yodzis P (1988) The indeterminacy of ecological interactions as perceived through perturbation experiments. Ecology 69:508–515

Yodzis P (2001) Must top predators be culled for the sake of fisheries? Trends Ecol Evol 16:78–84

Sergio A. Navarrete is a Marine Biologist with over 30 years of experience in coastal marine ecosystems. His work centers at the interface of ecology and physical oceanography, including experimental ecology, modeling, and theoretical approaches to address questions of sustainability and long-term transformations of the complex and heavily exploited marine communities. Sergio obtained his PhD from Oregon State University and currently teaches ecology and statistics at Pontificia Universidad Católica de Chile. He is the Director of the Coastal Marine Research Station at Las Cruces and in charge of its no-take Marine Protected Area, the oldest one in South America.

Christopher M. Aiken works as a Research Fellow at CQUniveristy's Coastal Marine Ecosystem Research Centre in Gladstone, Australia. He has worked on a range of topics from ocean physics, climate dynamics, and marine ecology. Together with Professor Navarrete, he has published a series of papers that embed system reactivity within metapopulation theory.

M. Isidora Ávila-Thieme holds a PhD in Ecology from the Pontificia Universidad Católica de Chile. Her research focuses on the study of ecological networks to understand anthropogenic impacts on natural ecosystems. She uses a combination of theoretical, computational, and empirical approaches to produce useful scientific information for the sustainable management of natural resources. As a postdoc at the Coastal Social-Ecological Millennium Institute and the Advanced Conservation Strategies, she is assessing the impacts of noncompliance in kelp fisheries from a social-ecological perspective. Isidora was selected as one of the 25 women in science in Latin America by 3M.

Daniel Valencia is a Ph.D. candidate in Ecology from the Pontificia Universidad Católica de Chile. He integrates empirical and modeling approaches to understand the effects of trophic and non-trophic interactions on the population dynamics of interacting species, and how sensitive those systems are to the variation in regional-scale processes, such as temperature or dispersal, in marine environments.

Alexandre Génin is a post-doctoral community ecologist who obtained his PhD from Montpellier University, France. His research blends theoretical approaches with field observations to understand the impact of anthropogenic perturbations on complex ecological systems. He is currently employed as a Marie-Skłodowska Curie fellow to understand better the emergence of abrupt transitions in the coral reefs of Easter Island.

Stefan Gelcich is a Marine Biologist and associate professor, based at the Universidad Catolica de Chile, he studies the linkages and feedbacks between social and ecological systems. He is particularly interested in advancing science which contributes to the sustainable management of coastal oceans. He has been part of multiple interdisciplinary projects and centers over the past 20 years where he has led research focused on human behavior, ecosystem services and subtidal ecology. Current project themes include marine conservation and management, adaptive capacity of coastal industries and illegal fishing, among others. Stefan is the current director of the research center "Instituto Milenio en Socio-ecologia Costera".

Chapter 19
Levels of Upwelling are Important to Consider for Conservation

Michael J. Kingsford, Margarita Brandt, and Juan Manuel Alava-Jurado

Introduction

Conservation and sustainable practices ultimately require an understanding of societal needs and the processes that influence ecosystems (Cinner and Aswani 2007). The acquisition of knowledge, therefore, is critical to build robust practices. The Galápagos National Park (GNP) is no exception, particularly given its unique equatorial location where rocky reefs and associated fauna and flora are bathed in cool upwelled waters and warm tropical currents.

The Galápagos Islands are influenced by three dominant current systems, firstly, cool north-flowing currents influenced by the Humbolt Current and related Perú coastal current. Secondly, an equatorial counter current, the Cromwell Current, that generates east-flowing waters that are topographically forced to upwell on the eastern side of the archipelago. Thirdly, the west flowing Panamá current transports warm water from Central America (Liu et al. 2014).

The basaltic reefs of the Galápagos are subjected to upwelling of intensities that vary greatly in time and space. El Nino Southern Oscillation (ENSO) characterises temporal variation from above average levels of upwelling (La Nina) to El Nino where the archipelago is bathed in warm waters (Sachs and Ladd 2010; Tarakanov and Borisova 2013; Hartten and Gage 2000) caused by a deepening thermocline and

M. J. Kingsford (✉)
Marine Biology and Aquaculture, James Cook University, Townsville, QLD, Australia

Centre of Excellence for Coral Reefs Studies, Townsville, QLD, Australia
e-mail: michael.kingsford@jcu.edu.au

M. Brandt
College of Science, Biology and Environment, Universidad de San Francisco, Quito, Ecuador

J. M. Alava-Jurado
Marine Biology and Aquaculture, James Cook University, Townsville, QLD, Australia

S. J. Walsh et al. (eds.), *Island Ecosystems*, Social and Ecological Interactions in the Galapagos Islands, https://doi.org/10.1007/978-3-031-28089-4_19

a weakening Cromwell current Firing et al. (1983). During these times the nutrients essential for a rich growth of algae on reefs are constrained to depths below 30–60 m (Barber and Chavez 1986) and, therefore, reef assemblages would be affected. The changing oceanographic conditions according to ENSO can also affect patterns of connectivity among islands (van Sebille et al. 2019).

The impact that El Nino conditions have on the marine fauna is well documented with major impacts on plankton (Ochoa and Gomez 1987), benthic algal cover, and populations of pelagic fishes (Barber and Chavez 1986), penguins and megafauna that include marine iguanas and sea lions (Trillmich and Limberger 1985; Glynn 1988; Steinfartz et al. 2007; Vargus et al. 2006). Further, the resistance and resilience (*sensu* Brierley and Kingsford (2009) of organisms such as corals, to these perturbations can be poor with slow recovery rates (Glynn et al. 2015, 2018).

There is great spatial variation in upwelling, with areas of high upwelling found on the western side of Isabella and conversely low upwelling around the islands of Pinta and Marchena. The level of upwelling can also vary within an island, for example Floreana, Santa Cruz and San Cristobal (NOAA). It has been demonstrated experimentally that variation in upwelling within and among islands has can affect the dynamics of reef-based assemblages. Witman et al. (2010) concluded that upwelling influenced recruitment rates of barnacles and, predation on them by fishes and invertebrates. It is likely, therefore, that other members of reef assemblages and, the very nature of habitats that are major drivers in the distribution of taxa will vary.

Fishes are important contributors to reef assemblages and is well known that the abundance and diversity of species vary with habitat type that includes biotic components and abiotic characteristics such as rock and dead corals that determine topographic complexity (Hall and Kingsford 2021). On tropical reefs habitat-forming taxa such as hard corals and soft corals are major drivers (e.g., Jones et al. 2004; Epstein and Kingsford 2019) while in temperate waters algal cover and the presence of habitat determining grazers (including urchins and gastropods, Fletcher 1987; Andrew 1993) greatly influence the responses of fishes and habitat associations (Curley et al. 2002; Kingsford and Carlson 2010). Given the patchiness of bottom-up drivers these habitats could vary at spatial scales that equate with spatial and temporal variation in upwelling.

Reef fishes not only contribute to the food web of reefs (Okey et al. 2004), but they are critical to fisheries in the Galápagos where, for example, the families Carangidae, Haemulidae, Serranidae, Labridae, Lutjanidae and Scorpaenidae are targeted (e.g., Schiller et al. 2015; Zimmerhackel et al. 2015). Patterns of reef fish abundance have been related to levels of impact by humans, differences among broad habitats/environments such as mangroves, substrata including rocky reefs and artificial structures (Riofiro-Lazo et al. 2022). Inferences have also been made from a study at two islands Floreana and Santa Cruz that differences in temperature and nutrient concentrations most likely drive differences in community structure (Bruneel et al. 2021). There have, however, been no studies at larger spatial scales that have incorporated depth and reported levels of upwelling in the design.

Fundamental to conservation practices in the Galápagos is a marine zoning plan that provides different levels of protection throughout the Archipelago. As stated by

Moity (2018), the level of protection from the zoning of marine parks is under constant review as we learn more and therefore the sophistication and effectiveness of conservation practices can be improved. An understanding of the processes affecting reef associate organisms can assist managers to conserve and respond to environmental perturbations that are increasing in a changing world.

Understanding energetic contributions to the health of reefs is beneficial. Upwelling can influence the level of 'pelagic subsidy' of reefs from pelagic and reef associated planktivores and, this may be location dependent. For example, Docmac et al. (2017) used stable isotopes to conclude that 70–95% of fish production on the coast of Chile was from pelagic food sources. This contrasts with a more traditional view of food chains where producers provide for herbivorous consumers that in-turn are prey for predators. An upwelling driven archipelago begs the question, is there an algal subsidy that promotes bottom-up tropic links? We predicted that number of herbivores would be highest where the availability of algae was greatest (i.e., in areas of high upwelling).

Our objective was to determine the abundance and biomass of reef associated fishes at sites that encompassed different between levels of upwelling. Specifically, our aims were to (1) identify broad habitat characteristics in areas subjected to low, medium, and high upwelling. This was partly based on existing knowledge (e.g., Witman et al. 2010; NOAA 2015), and new data on thermal depth profiles by site; (2) determine how numbers of herbivorous fishes varied by level of upwelling and depth. What is the herbivore subsidy based on spatial variation in abundance? (3) discuss how our findings can assist with conservation and to make predictions on how reef processes may vary in space and time.

Material and Methods

Sampling Design and Analyses

Because upwelling affects algal growth through nutrient supply – we expected there would be a response from herbivorous fishes. Specifically, we predicted that abundance and biomass of herbivorous fishes would be greatest at sites with high upwelling. Our classification of levels of upwelling was based on NOAA imagery of chlorophyll signals around the Galápagos archipelago, historical data from temperature loggers (Witman et al. 2010). Furthermore, temperature profiles were obtained in waters adjacent to each site using a RDRconcerto CTD; the instrument was lowered at ≤ 1 m s^{-1}, the data were collected in July 2019.

We made qualitative assessment of habitats and estimated the abundance and biomass of herbivorous fishes at sites categorised as having a high, medium or low exposure to upwelling (Table 19.1). Two replicate sites were selected for each level of upwelling. Because the abundance patterns of fishes often vary with depth (Russ 1984; Kingsford et al. 1989), sampling was stratified as shallow, mid, deep. Actual

Table 19.1 Habitat characteristics of reef profiles that were sampled with different levels of upwelling

Habitat characteristics	Level of upwelling		
	High	Medium	Low
Algae	Macroalgae (*Spatoglossum* sp *Sargassum* spp.), and tufting algae (e.g. *Ulva* sp. *Hypnea* sp. *Padina* sp.) very abundant to abundant, filamentous algae in damselfish territories.	Macro and tufting algae, were often abundant in shallow water (e.g. *Sargassum* spp. *Ulva* sp.). Green and red filamentous algae rare to abundant on dead coral matrix and base rock in urchin grazed barrens.	Macroalgae rare to absent even in shallow waters, dead coral matrix with green and red filamentous algae.
Hard corals	Absent to rare	Rare (e.g. *Pocillopora* spp.)	*Pocillopora* spp. *Pavona* spp. scattered in individual colonies, especially in shallow water.
Sea urchins	Small groups of *Eucidaris galapagensis*	Large urchin grazed barrens with abundant, especially *Eucidaris galapagensis* and *Lytechinus semituberculatus,* mid to deep strata.	Abundant, especially *Diadema mexicanus*
Vertical relief (m) Mean(range) SE	1.8 (0.7–4) 0.21	1.3 (0.3–2.5) 0.18	1.2 (0.3–3) 0.21
Sites Latitude and Longitude	Cabo Douglas 0°15.815S 91°37.035 E Punta Espinosa 0°15.777S 91°26.656 E	Cerro Brujo 0 °45.254S 89°27.626E Tijeretas North 0°53.048S 89°36.323E	Pinta 0°32.478S 90°43.749E Marchana 0°23.148S 90°27.718E

Descriptions are presented based on the abundance of algae, hard corals, and 'habitat determining' sea urchins. A measure of habitat complexity is also provided (Vertical relief), as m above the substratum, estimated in each transect (n = 18 transects); mean, range and standard error shown. Abundance categories: "Very abundant" – high (>75%) percentage benthic cover and > 1 individual per m^2 of discrete organisms; "Abundant" – relatively high (>50%) percentage benthic cover and > 0.1 individuals per m^2 of discrete organisms; "Rare" – absent in some transects; "Absent" – not seen on the dive

depth depended on the extent and depth of the reef profile, generally to sand or flat substratum at the base of reefs. Shallow was 1.5–4.9 m, mid 5–11.9 m and deep 12–20 m. Estimates of fish abundance and total length were obtained using two sizes of belt transects within each depth stratum. Large and highly mobile fishes belonging to the families Acanthuridae, Girellidae, Scaridae and large *Microspathodon* spp (Pomcentridae) were counted in 50*10 m transects as the diver swam out a tape (n = 3 transects). Small, and often cryptic, herbivores belonging to the families Pomacentridae and Blenniidae were counted in 30*5 m transects (n = 5 transects). Divers searched over and under boulders when using this method of sampling. Because topographic complexity affect patterns of reef fish abundance a measure of complexity was obtained at the beginning of each transect and estimated as maximum height of boulders above the substratum.

The metric of herbivore fish biomass was important to determine the likely input of food to higher levels of the food chain, especially piscivores. Sizes of fish were allocated to size categories and an estimate of body weight was determined from a Length Weight Relationships (LWR) using the median size where L = total length in mm.

$$\text{Weight} = (aL^b)/10^4 \tag{19.1}$$

Where a = parameter describing body shape and proportion
b = isometric growth in body proportions

We followed the recommendations of Froese et al. (2014). Because the body shape of fishes varied (e.g., a blenny versus an acanthurid) values of a and b were optimised by species. Based on samples of fish we had collected and weighed we predicted weights based on formula (19.1). The variance between predicted and actual weights was then minimized by altering a and b using the EXCEL function Solver; b varied between 2.9 and 3.05 and a was unchanged at 0.2. For two species we had large samples sizes of individuals to compare measured weight versus predicted over a wide size range of fish. The estimates of weight based on TL were highly accurate for *Stegastes beebei*, n = 144, $r^2 = 0.975$ and *Stegastes arcifrons* n = 84, $r^2 = 0.983$. For all fish taxa visual estimates of total length were made by divers who had calibrated their estimates based on the sizes of fishes that were collected by spear (for a separate study) and published size maxima from Humann and Deloach (1993), Robertson and Allen (2016) and Fishbase (2022).

Data on abundance were analysed with a partially hierarchical Analysis of Variance (ANOVA). We followed the procedures of Underwood (1997). Data were tested for heterogeneity using a Cochran's test. If the test was significant data were transformed with Ln(x+1) transformations. The factors Upwelling (treatments low, medium, high) and Depth (shallow, mid, deep) were treated as a fixed factors and the site nested in the U*D interaction was treated as a random factor.

Results

Benthic Cover and Water Temperature

There were great differences in benthic cover between sites at different levels of upwelling (Fig. 19.1, Table 19.1). High upwelling sites were characterised by abundant macroalgae (e.g., *Sargassum* and *Spatoglossum*), fleshy green algae (e.g.,

Fig. 19.1 Images of habitats that were representative of the sites sampled and categorised as low, medium and high upwelling habitats

Ulva) and tufting reds in shallow water. At mid and deep strata of the reef fleshy and tufting algae in particular were abundant. Invertebrate grazers (e.g. the pencil urchins *Eucidaris galapagensis*) were observed, but were generally observed in small patches around cracks and crevices. At medium levels of upwelling macroalgae and fleshy algae was abundant in the shallows, and urchin grazed barrens we extensive at mid and deep strata. At sites characterised by low levels of upwelling macroalgae and fleshy algae was rare or absent on a rocky matrix, or sometimes dead coral, in the shallows. At greater depths the substratum was generally barren with a cover of short filamentous algae. The rocks at all sites and levels of upwelling were basalt with a vertical relief of 0.3–4 m above the base substratum. Vertical relief varied little with depth, 76–78% of transects at each depth stratum had rocks > 1 m high.

Temperature profiles to over 100 m at Cabo Douglas (Isla Fernandina) and Kicker Rock (San Cristobal) showed strong vertical stratification. Temperatures were over 21 °C at the surface to 15.5 °C at a thermocline at depths between 60 and 70 m (Fig. 19.2).

Over the depth range that fishes were counted there were clear differences between sites with high and low levels of upwelling. Differences in temperature of two to three degrees were common below about 8 m at these sites. Sites characterised as 'Mid' upwelling showed overlap in temperature with high and low upwelling sites.

Abundance and Biomass of Herbivorous Fishes

The most abundant large herbivore was the acanthurid *Prionurus laticavius* (Fig. 19.3). Other large herbivores that were detected included scarids and girellids. *P. laticavius* were most abundant in areas of low upwelling and lowest abundances were found in high upwelling treatments and, this resulted in a significant Upwelling effect (Table 19.2). Further, greatest abundances were always found in shallow water when compared to other depths. There were also significant differences between sites within the three levels of upwelling. Standard errors were indicative of aggregation behaviour, where individuals numbering in the hundreds were found in groups, particularly in shallow water.

Relatively low numbers of scarids were detected, but patterns of abundance at each depth varied significantly by level of upwelling resulting in an interaction between upwelling and depth (Table 19.2, Fig. 19.3). On average, a higher number of scarids were detected at low upwelling sites when compared to high upwelling. Although there was a trend for high numbers of scarids in shallow water at high and low levels of upwelling, this was not found in the medium treatment for upwelling. Of the 104 scarids that were counted, *Scarus ghobban* may up 91% of the total, with *Scarus rubroviolacaeus* second (7%) and *Scarus perrico* with 2%. Scarids were most commonly observed foraging at shallow to mid depths. *Girella freminvilli* were relatively rare, but when they were seen they were in large aggregations and greatest numbers of fish were seen at one site with high levels of upwelling.

Small herbivorous blennies, *Ophioblennius steindachneri* were observed in highest numbers in shallow water at all sites and levels of upwelling. These consistent patterns resulted in significant difference among depths (Fig. 19.4, Table 19.2). Although there was a trend for *O. steindachneri* to be more abundant at sites with low upwelling, this was not significant. Significant differences were found between sites within levels of upwelling.

The congeneric pomacentrids *Stegastes arcifrons* and *S. beebei* were found in high abundance. There was a strong and significant trend for *S. arcifrons* to be found in very high abundance at sites with low levels of upwelling, differences were also

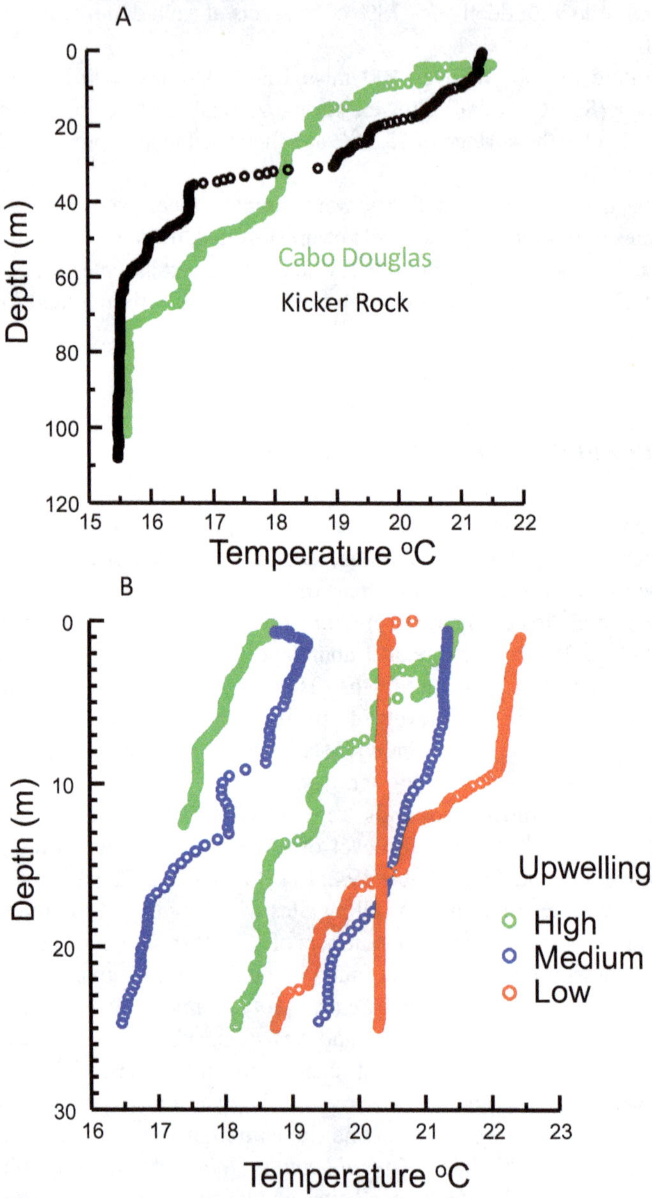

Fig. 19.2 Depth profiles of temperature – (**a**) Deep profiles in areas known for upwelling – Kicker Rock (San Cristobal) and Cabo Douglas (Fernandina); (**b**) depth profiles at sites with different exposure to nutrient rich upwelling, profiles represent the depth range over which fish were counted, (generally 20 m of less)

found between sites within levels of upwelling (Fig. 19.4, Table 19.2). In particular, at one low upwelling site *S. arcifrons* were abundant at all depths (Isla Marchena),

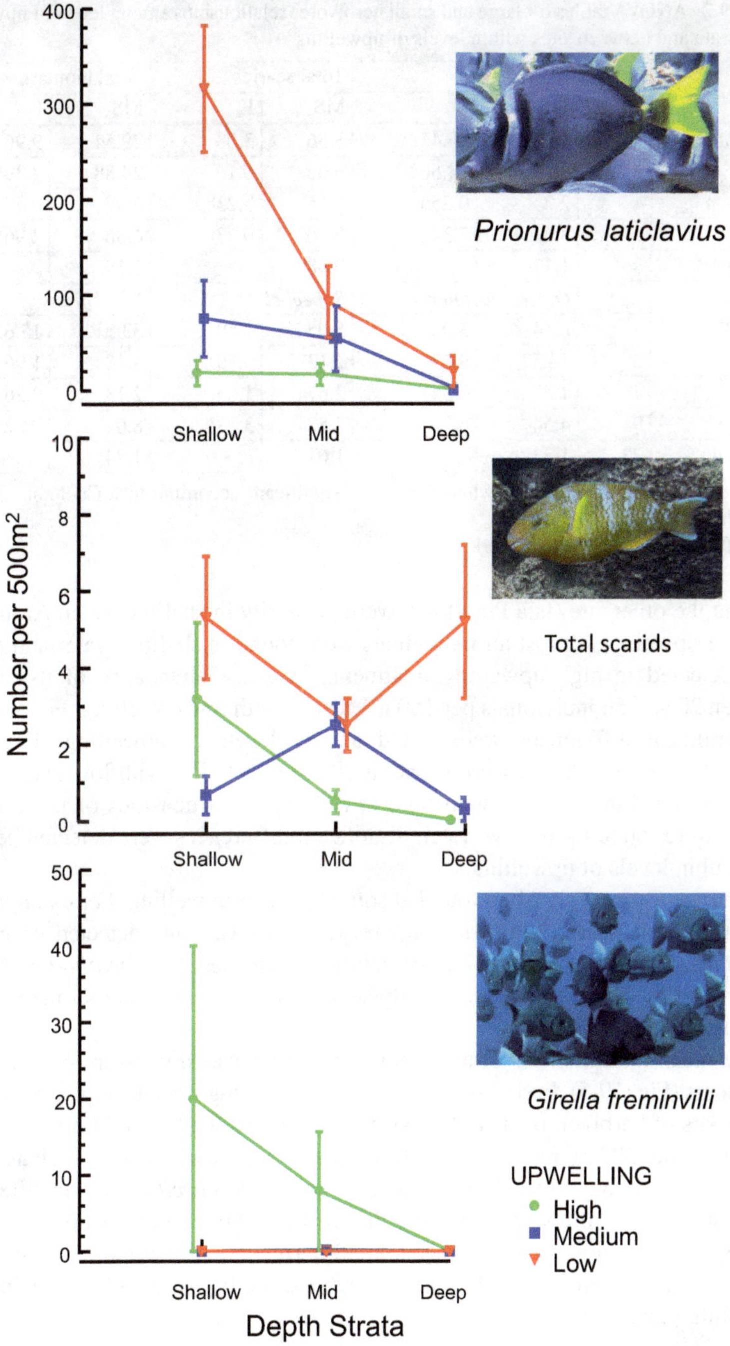

Fig. 19.3 Fish abundance large herbivores sampled in 50*10 m transects sampled shallow, mid and deep (n = 3 transects) at three levels of upwelling, ± 1SE

Table 19.2 ANOVA tables for large and small herbivores relationships among levels of upwelling, depth strata and between sites within levels of upwelling

		P. laticlavius		Total scarids		Total biomass	
Factor	df	MS	F	MS	F	MS	F
Upwelling	2	36.08	16.4*	3.56	3.08	29.54	9.96*
Depth	2	32.02	14.56*	0.02	0.16	24.88	8.39*
DxU	4	2.2	0.255	1.15	3.23*	2.97	0.47
Site (U)	9	8.62	7.34***	0.36	0.73	6.36	2.96*
Residual	36	1.17		0.48		2.14	
		O. steindachneri		*S. beebei*		*S. arcifrons*	
Upwelling	2	6.14	3.95	5.05	1.89	33.31	15.62*
Depth	2	14.53	9.27*	2.13	0.79	4.18	1.96
DxU	4	1.57	0.34	2.67	1.03	2.13	0.26
Site (U)	9	4.54	6.74***	2.57	3.18**	8.04	9.22***
Residual	72	1.34		1.61		1.74	

Data were log(x+1) transformed when they were significant according to a Cochran's Test for normality

$^*P < 0.05$, $^{**}P < 0.01$, $^{***}P < 0.001$

while at the other site (Isla Pinta) fish were primarily in shallow water. At medium levels of upwelling almost all individuals were found in shallow water and no fish were detected in high upwelling treatments. Average abundance of fish ranged between 20 and 50 individuals per 150 m^2 at sites with low upwelling. For *S. beebei* no significant differences were found among levels of upwelling (Fig. 19.4, Table 19.2). Although fish were found at all depths at sites with low and medium levels of upwelling, at sites with high upwelling mean densities of up to 82 per 150 m^2 were found in shallow water. Significant differences were detected between sites within levels of upwelling.

Some taxa were only often found at some levels of upwelling. For example. The territorial pomacentrid *Nexilosus latifrons* (n = 61) was only detected in areas of high upwelling. In contrast, two large damselfish species *Microspathodon dorsalis* (n = 8) and *M. bairdi* (n = 8) were only seen in the shallows water at medium and low levels of upwelling.

The mean total biomass of all herbivorous fishes was highest in low upwelling treatments (Fig. 19.5). In shallow water, at low upwelling sites, there was an average of 265 kgs of herbivorous fish of all sizes per 500 m^2; while at mid it was 125 and high 40 kg per 500 m^2 respectively. The biomass of large herbivores, such as *P. laticavius,* was 8.6× that of small herbivores (e.g., *Stegastes arcifrons*). The differences between size categories of herbivores dropped at higher levels of upwelling with large herbivores biomass 3.8× that of small herbivores at medium and 2× at high upwelling sites. Significant differences were detected between sites within levels of upwelling (Table 19.2).

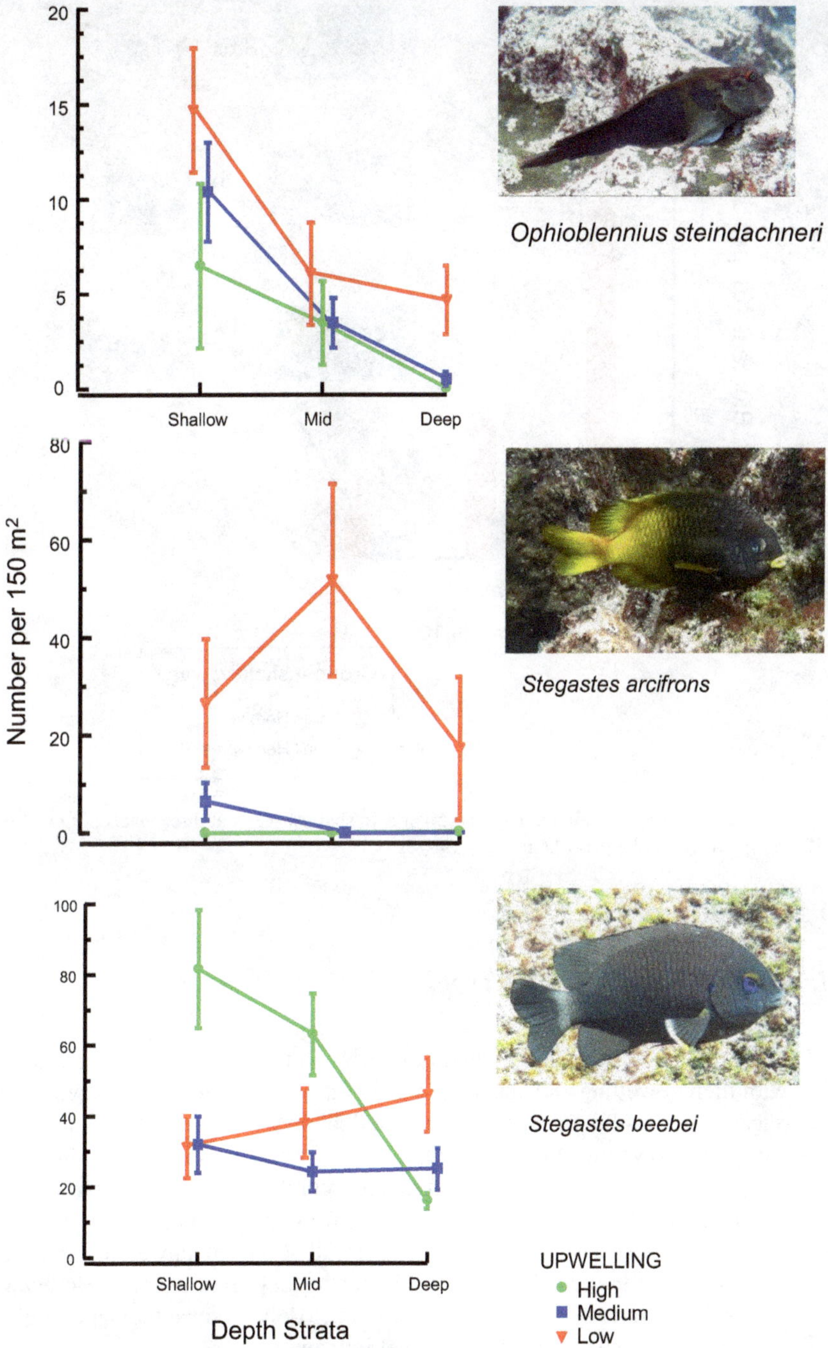

Fig. 19.4 Abundance of small herbivores sampled in 30*5 m transects, sampled in shallow, mid and deep (n = 5 transects) at three levels of upwelling, ± 1SE

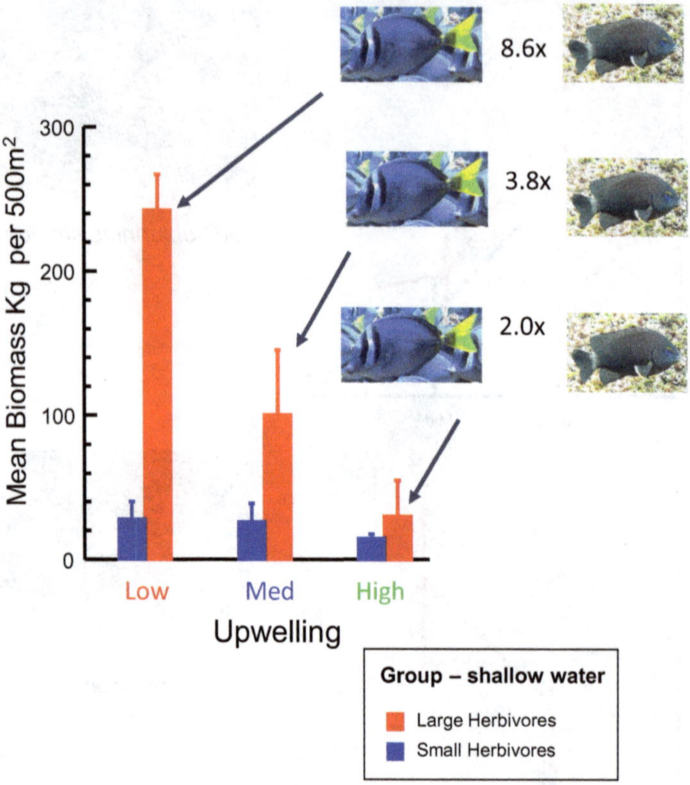

Fig. 19.5 Biomass of all small and large herbivores in shallow water at three levels of upwelling, ± 1SE all expressed as kg per 500 m^2

Discussion

Herbivore Abundance and Trophic Biology

We predicted that number and biomass of herbivorous fishes would be greatest at sites with high upwelling and abundant algae of a variety of forms. This prediction was rejected because greatest abundance and biomass of herbivores fishes was found at sites of low upwelling. Algae of different forms including macroalgae (i.e., *Sargassum* spp.), foliose (e.g., *Ulva*) tufting reds, coralline algae and patches of filamentous algae were abundant at sites of high upwelling. Although some small herbivores (e.g., *Stegastes beebie*) were found at all levels of upwelling and most depths, this was not the case for most herbivores be they large or small. Herbivores appeared to prefer habitats in shallow water with closely cropped filamentous algae and perhaps with a rich associated microbial in fauna.

Regardless of level of upwelling, herbivores were generally most abundant in very shallow water. From a conservation perspective, therefore, the algal resources and associated herbivores in 4 m or water or less are important for ecosystem health.

Furthermore, quantitative assessments of herbivore abundance need to include this important depth stratum.

The feeding preferences of different herbivore species gives insight to patterns of abundance by level of upwelling. The nutritional ecology of acanthurids and scarids varies from those that are browsers that feed on filamentous algae and processing food through acid digestion, to those that scrape or excavate calcareous surfaces to extract detritus, animal material and turfing algae (Choat et al. 2019). For the latter category a strong beak is beneficial, as found in scarids. In contrast, acanthurids generally have weak teeth which are not suited for scraping. The acanthurid *Prionurus laticavius* dominated patterns of abundance and in contributing to total fish biomass, particularly in shallow waters. In a detailed review of the ecological roles of herbivores Clements et al. (2016) concluded that prionurids were browsers, with carbohydrate fermentation, that is characteristic of herbivores. The blenny *P. laticavius* clearly preferred substrata that as devoid of macro algae and rich in cropped algae, particularly filamentous taxa. Algal gardens defended by pomacentrids and blennies were also common in shallow waters, also providing an additional algal resource (Hixon and Brostoff 1983) to roving prionurids that overwhelmed such territories *en-mass*.

Although other acanthurids have been recorded in the Galápagos archipelago (e.g., *Acanthurus xanthopterus* and *A. nigricans*), there were absent from counts. This suggest, therefore, that acanthurids as a 'functional group' (Welsh and Bellwood 2014) are primarily represented by one species, *P. laticavius*. This suggests a potential monospecific vulnerability to the composition and trophic importance of Galápagos fishes.

Scarids are more omnivorous (Clements et al. 2016) and this probably explains why *Scarus ghobban*, in particular, was found over all depths, although they were rare in deep strata at sites with medium and low upwelling. Girellids are common herbivores in temperate latitudes. They are omnivorous but generally consume over 80% read and green algae combined (Clements and Choat 1997). Although was made casual observations of girellids at sites with mid and low levels of upwelling, large groups of *Girella freminvilli* were only detected in counts at sites with high upwelling.

Small-bodied damselfishes were the dominant group of small herbivores; with the exception of *Microspathodon* spp. they less than 20 cm TL. Mostly common in the Galápagos (Robertson and Allen 2016), *Stegastes arcifrons* and *S. beebie* were often found in high abundance and, for *S. beebie* of they were abundant at all levels of upwelling and generally at all depths. Small herbivorous damselfish often defend an algal garden and this can have a positive effect on the availability algae to herbivores (Hixon and Brostoff 1983, 1996). Territorial damselfish often eat a high percentage of algal material but are omnivorous (Buckle and Booth 2009). We predict that the protein component of the diet is likely to be greater in *S. beebie* that are found in deep strata on a reef profile where algal input may be largely restricted to detached algal fragments. The blenny *Ophioblennius steindachneri* preferred shallow water and appeared sedentary in its behaviour. As for many of the other herbivorous fishes blennies were most abundant in area of low upwelling and were observed cropping filamentous algae.

Upwelling Driven Habitats, Algal Subsidy and Ecosystems Effects

The persistence of habitats – we identified three broad reef profiles as follows. (1) those that were dominated by macro and tufting algae over most of the reef profile with small patches of grazing invertebrates. (2) Those with macro and foliose algae in the shallows and extensive urchin grazed barrens with patches of filamentous and foliose algae such as *Ulva* and, (3) lastly habitats that were largely devoid of macro and foliose algae, but were rich in a closely cropped filamentous algae. Habitats 1–3 generally aligned with waters of different temperatures (Witman et al. 2010; our CTD results) and, reported levels of upwelling and related temperature profiles that manifest themselves as chlorophyll signals from high to very low (e.g., NOAA 2015; Forryan et al. 2021).

Algal biomass in the Galápagos will depend on species-specific responses to nutrient load, temperature and browsing pressure from invertebrates, fishes and vertebrate grazers such as marine iguanas and turtles. If algal growth rates are high then grazing pressure may not be great enough to curtail growth, this is especially true where some grasers can only manage algae of a certain length (e.g., limpets versus urchins, Fletcher 1987) or grazing halos result from the requirement of nearby shelter to reduce predation (Andrew 1993). The growth of filamentous algae is rapid in areas devoid of macro-algae and removal rates may be lower inside the defended territories of damselfishes (Russ 1987).

There are multiple ways that algal type and biomass could affect populations of herbivorous fishes (Fig. 19.6). Abundance of suitable algae may affect numerical processes (e.g., local population size) when combined with other factors such as the availability shelter and the likelihood of predation. Energetic processes include the availability of food, growth and condition, maturation, and fecundity (Jones and McCormick 2002). For example, the local population size and health *P. laticavius* appear heavily dependent on large areas of reef that are covered with short and easily accessible filamentous algae. In contrast, the territorial damselfish *Nexilosus latifrons* may depend on cool water, high upwelling and high growth rates of edible algae.

For a reef assemblage to function energy may be through top-down processes such as a pelagic subsidy or, a more traditional bottom-up set of trophic interactions the include nutrient input to the autotrophs, herbivory and different levels of predators that may include micro and macro predators and well as piscivores to top predators such as sharks. Pelagic subsidy from the consumption of imported plankton and baitfish is well acknowledged (e.g., Barber and Chavez 1986). For example, in temperate waters of Chile Docmac et al. (2017) concluded that 70–95% of fish production was from pelagic sources (Fig. 19.7).

Here we propose that many reef assemblages in the Galapagos have a major algal subsidy. This in turn creates a substantial herbivore subsidy that is available to predators and, conveniently herbivores come in a range of sizes. The biomass of herbivores at sites of low upwelling was an average of 0.265 tonnes per 500 m^2 and, the

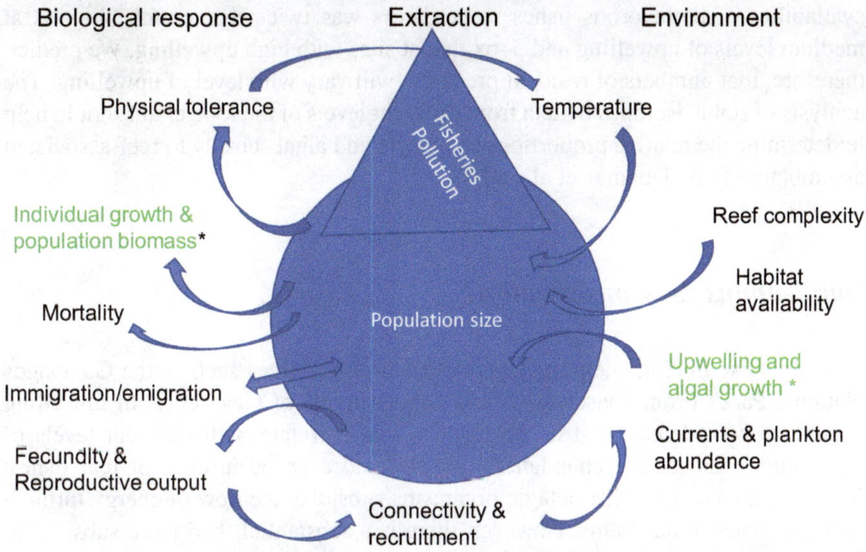

Fig. 19.6 Factors influencing the size of a fish population. Biological responses include natural population dynamics, energetic processes and physical tolerances; 'Extraction' equates with anthropogenic effects and, 'Environment' includes the physical environment in the water column and on reefs, as well as nutrient input that can facilitate a biological responses

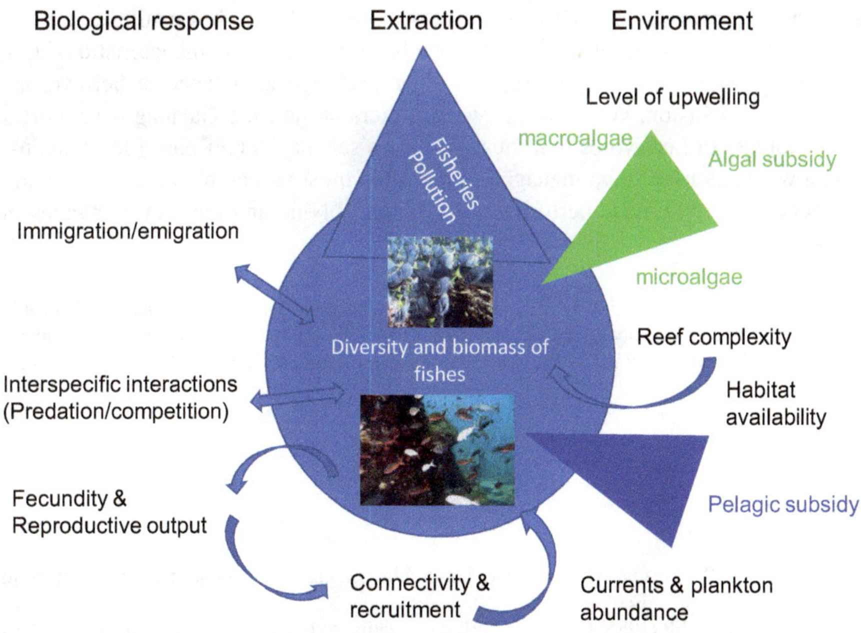

Fig. 19.7 Factors affecting an assemblage comprised of multiple species. Use of the terms: 'Biological response', 'Extraction' and 'Environment', as for Fig. 19.6. The green and blue arrows represent energetic input to the systems as algal and pelagic subsidies respectively

availability of herbivorous fishes to predators was twice that of the biomass at medium levels of upwelling and 5–6x that at sites with high upwelling. We predict, therefore, that numbers of resident predators will vary with level of upwelling. The analysis of stable isotopes in taxa from different levels of the food chain should help to determine the relative proportions of pelagic and algal subsidy to reef-associated assemblages (e.g., Docmac et al. (2017)).

Implications for Conservation?

How do these findings help managers of natural resources such as the Galápagos National Park? From conservation and zoning points of view levels of upwelling should be considered because the habitats that correlate with different levels of upwelling influence the abundance, biomass and even occurrence of reef fishes. There is an awareness that pelagic organisms subsidize the flow of energy through reef associated food chains. However, there is a substantial 'herbivore subsidy' to reefs and this is unexpectedly high where algal biomass is low. When the zoning plan of the Galápagos is revised a sensible consideration would be to identify and protect replicate habitats that are indicative areas of low and high upwelling. Further, make protected areas large enough to sustain high herbivores biomass. Prolonged El Nino conditions and related warm seas could change Galápagos habitats and food chain dynamics, as recognized by Barber and Chavez (1986) and other marine ecosystems (Wernberg et al. 2016). However, we predict that with warming seas there would be an expansion of habitats that we have identified as characteristic of low upwelling, with low standing crops of algae and high abundance of herbivorous fishes. In conclusion, very shallow warm waters around the Galapagos support a high biomass of herbivores that must provide a subsidy to reef-based food chains. Shallow waters need to be managed as are often most vulnerable to anthropogenic impacts and Pacific-wide perturbations such as El Nino and long-term changes in climate.

Acknowledgements We would like to thank our dive buddies, Mark O'Callaghan and multiple helpers from the GSC Laboratory. We also thank the referees for their constructive comment. Funding to MJK was provided by the Australian Research Council Centre of Excellence for Coral Reef Studies.

References

Andrew NL (1993) Spatial heterogeneity, sea urchin grazing, and habitat structure on reefs in temperate Australia. Ecology 74:292–302

Barber RT, Chavez FP (1986) Ocean variability in relation to living resources during the 1982-83 El-Nino. Nature 319:279–285

Brierley AS, Kingsford MJ (2009) Impacts of climate change on marine organisms and ecosystems. Curr Biol 19:R602–R614

Bruneel S, Van Echelpoel W, Ho L, Raat H, Schoeters A, de Troyer N, Sor R, Ponton-Cevallos J, Vandeputte R, van der Heyden C, de Saeyer N, Forio MAE, Bermudez R, Dominguez-Granda L, Luca S, Moens T, Goethals P (2021) Assessing the drivers behind the structure and diversity of fish assemblages associated with rocky shores in the Galapagos Archipelago. J Mar Sci Eng 9:18

Buckle EC, Booth DJ (2009) Ontogeny of space use and diet of two temperate damselfish species, Parma microlepis and Parma unifasciata. Mar Biol 156:1497–1505

Choat JH, Russell BC, Chin A (2019) The fish assemblages of the great barrier reef: their diversity and origin. In: Hutchings P, Kingsford M, Hoegh-Guldberg O (eds) The great barrier reef: biology, environment and management, 2nd edn. CSIRO Publishing, Collingwood

Cinner JE, Aswani S (2007) Integrating customary management into marine conservation. Biol Conserv 140:201–216

Clements KD, Choat JH (1997) Comparison of herbivory in the closely-related marine fish genera *Girella* and *Kyphosus*. Mar Biol 127:579–586

Clements KD, German DP, Piché J, Tribollet A, Choat JH (2016) Integrating ecological roles and trophic diversification on coral reefs: multiple lines of evidence identify parrotfishes as microphages. Biol J Linn Soc 120:729–751

Curley B, Kingsford M, Gillanders B (2002) Spatial and habitat related patterns of abundance of temperate reef fishes on the central coast of New South Wales: implications for the design of marine protected areas. Mar Freshw Res 53:1197–1210

Docmac F, Araya M, Hinojosa IA, Dorador C, Harrod C (2017) Habitat coupling writ large: pelagic-derived materials fuel benthivorous macroalgal reef fishes in an upwelling zone. Ecology 98:2267–2272

Epstein HE, Kingsford MJ (2019) Are soft coral habitats unfavourable? A closer look at the association between reef fishes and their habitat. Environ Biol Fishes 102:479–497

Firing E, Lukas R, Sadler J, Wyrtki K (1983) Equatorial undercurrent disappears during 1982-1983 El-Nino. Science 222:1121–1123

FISHBASE (2022). https://www.fishbase.se/search.php [Online]. [Accessed]

Fletcher WJ (1987) Interactions among subtidal australian sea urchins, gastropods, and algae: effects of experimental removals. Ecol Monogr 57:89–109

Forryan A, Naveira Garabato AC, Clément V, Nurser AJG, Hearn AR (2021) Galápagos upwelling driven by localized wind–front interactions. Sci Rep 11:1277

Froese R, Thorson JT, Reyes RB (2014) A Bayesian approach for estimating length-weight relationships in fishes. J Appl Ichthyol 30:78–85

Glynn W (1988) El Nino-Southern Oscillation 1982-1983: nearshore population, and ecosystem reponses. Annu Rev Ecol Syst 19:309–346

Glynn PW, Riegl B, Purkis S, Kerr JM, Smith TB (2015) Coral reef recovery in the Galapagos Islands: the northernmost islands (Darwin and Wenman). Coral Reefs 34:421–436

Glynn PW, Feingold JS, Baker A, Banks S, Baums IB, Cole J, Colgan MW, Fong P, Glynn PJ, Keith I, Manzello D, Riegl B, Ruttenberg BI, Smith TB, Vera-Zambrano M (2018) State of corals and coral reefs of the Galapagos Islands (Ecuador): past, present and future. Mar Pollut Bull 133:717–733

Hall A, Kingsford MJ (2021) Habitat type and complexity drive fish assemblages in a tropical seascape. J Fish Biol. https://doi.org/10.1111/jfb.14843

Hartten LM, Gage KS (2000) ENSO's impact on the annual cycle: the view from Galapagos. Geophys Res Lett 27:385–388

Hixon MA, Brostoff WN (1983) Damselfish as keystone species in reverse: intermediate disturbance and diversity of reef algae. Science 220:511–513

Hixon MA, Brostoff WN (1996) Succession and herbivory: effects of differential fish grazing on Hawaiian coral-reef algae. Ecol Monogr 66:67–90

Humann P, Deloach N (1993) Reef fish identification Galapagos. Cotopixel Ediciones, Quito

Jones GP, McCormick M (2002) Numerical and energetic processes in the ecology or coral reef fishes. In: Sale P (ed) Coral reef fishes: dynamics and diversity in a complex ecosystem, 2nd edn. Academic Press, San Diego

Jones GP, Mccormick MI, Srinivasan M, Eagle JV (2004) Coral decline threatens fish biodiversity in Marine Reserves. PNAS 101:8251–8253

Kingsford MJ, Carlson IJ (2010) Patterns of distribution and movement of fishes, *Ophthalmolepis lineolatus* and *Hypoplectrodes maccullochi*, on temperate rocky reefs of south eastern Australia. Environ Biol Fishes 88:105–118

Kingsford MJ, Schiel DR, Battershill CN (1989) Distribution and abundance of fish in a rocky reef environment at the subantarctic Auckland Islands, New Zealand. Polar Biology 9:179–186

Liu Y, Xie L, Morrison JM, Kamykowski D, Sweet WV (2014) Ocean circulation and water mass characteristics around the Galápagos Archipelago simulated by a multiscale nested ocean circulation model. Int J Oceanogr 2014:198686

Moity N (2018) Evaluation of no-take zones in the Galapagos Marine Reserve, zoning plan 2000. Front Mar Sci 5

NOAA (2015) El Niño and the Galápagos [Online]. Available: https://www.climate.gov/news-features/blogs/enso/el-ni%C3%B1o-and-gal%C3%A1pagos. Accessed

Ochoa N, Gómez O (1987) Dinoflagellates as indicators of water masses during El Niño, 1982–1983. J Geophys Res Oceans 92:14355–14367

Okey TA, Banks S, Born AR, Bustamante RH, Calvopina M, Edgar GJ, Espinoza E, Farina JM, Garske LE, Reck GK, Salazar S, Shepherd S, Toral-Granda V, Wallem P (2004) A trophic model of a Galapagos subtidal rocky reef for evaluating fisheries and conservation strategies. Ecol Model 172:383–401

Riofrio-Lazo M, Zetina-Rejon MJ, Vaca-Pita L, Murillo-Posada JC, Paez-Rosas D (2022) Fish diversity patterns along coastal habitats of the southeastern Galapagos archipelago and their relationship with environmental variables. Sci Rep 12:13

Robertson DA, Allen GR (2016) Shore fishes of the tropical eastern pacific indentification Guide App. The Smithsonian Tropical Resarch Insitute (STRI)

Russ G (1984) Distribution and abundance of herbivorous grazing fishes in the central Great Barrier Reef. II. Patterns of zonation of mid-shelf and outershelf reefs. Mar Ecol Prog Ser 20:35–44

Russ GR (1987) Is the rate of removal of algae by grazers reduced inside territories of tropical damselfishes? J Exp Mar Biol Ecol 110:1–17

Sachs J, Ladd S (2010) Climate and oceanography of the Galapagos in the 21st century: expected changes and research needs. Galápagos Res 67

Schiller L, Alava JJ, Grove J, Reck G, Pauly D (2015) The demise of Darwin's fishes: evidence of fishing down and illegal shark finning in the Galapagos Islands. Aquat Conserv-Mar Freshw Ecosyst 25:431–446

Steinfartz S, Glaberman S, Lanterbecq D, Marquez C, Rassmann K, Caccone A (2007) Genetic impact of a severe El Nino Event on Galapagos Marine Iguanas (Amblyrhynchus cristatus). Plos One 2:9

Tarakanov AO, Borisova AV (2013) Galapagos indicator of El Niño using monthly SST from NASA Giovanni system. Environ Model Softw 50:12–15

Trillmich F, Limberger D (1985) Drastic effects of El Nino on Galapagos Pinnipeds. Oecologia 67:19–22

Underwood AJ (1997) Experiments in ecology: their logical design and interpretation using analysis of variance. Cambridge University Press, Cambridge

Van Sebille E, Delandmeter P, Schofield J, Hardesty BD, Jones J, Donnelly A (2019) Basin-scale sources and pathways of microplastic that ends up in the Galapagos Archipelago. Ocean Sci 15:1341–1349

Vargas FH, Harrison S, Rea S, Macdonald DW (2006) Biological effects of El Nino on the Galapagos penguin. Biological Conservation 127:107–114

Welsh JQ, Bellwood DR (2014) Herbivorous fishes, ecosystem function and mobile links on coral reefs. Coral Reefs 33:303–311

Wernberg T, Bennett S, Babcock RC, De Bettignies T, Cure K, Depczynski M, Dufois F, Fromont J, Fulton CJ, Hovey RK, Harvey ES, Holmes TH, Kendrick GA, Radford B, Santana-Garcon J, Saunders BJ, Smale DA, Thomsen MS, Tuckett CA, Tuya F, Vanderklift MA, Wilson S (2016) Climate-driven regime shift of a temperate marine ecosystem. Science 353:169–172

Witman JD, Brandt M, Smith F (2010) Coupling between subtidal prey and consumers along a mesoscale upwelling gradient in the Galapagos Islands. Ecol Monogr 80:153–177
Zimmerhackel JS, Schuhbauer AC, Usseglio P, Heel LC, Salinas-de-Leon P (2015) Catch, bycatch and discards of the Galapagos Marine Reserve small-scale handline fishery. Peerj 3:22

Michael J. Kingsford is a Distinguished Professor in the Marine Biology and Aquaculture group of the College of Science and Engineering at James Cook University, Australia. He has published extensively on the ecology of reef fishes, biological oceanography, climate change and jellyfishes. He has forty years of research experience of studying fishes in temperate and tropical regions of Australia and other parts of the world. Since 2015 he has been contributing to our knowledge of Galapagos fishes.

Margarita Brandt is a Professor at the Universidad de San Francisco Quito. She has published multiple papers on the influence of upwelling on marine assemblages at the Galapagos. Further, she is actively working on the affects of climate change on marine organisms.

Juan Manual Alava-Jurado is a Lecturer at the Universidad de San Francisco Quito and is undertaking his PhD on Galapagos fishes and the dynamics of reef-based food chains.

Chapter 20
Ten Years of Wildlife Health and Conservation in the Galapagos, 2013–2022

Gregory A. Lewbart, Juan Pablo Muñoz Pérez, Diego Páez-Rosas, Carlos Valle, Daniela Alarcón-Ruales, Maximilian Hirschfeld, Diane Deresienski, and Kenneth J. Lohmann

Introduction

In June of 2013, our team embarked on the first of what would be many projects investigating the health of Galápagos wildlife. With a permit from the Galápagos National Park (PNG) in progress, and the support of the Galápagos Science Center (GSC), three of us (Greg Lewbart, Max Hirschfeld, Ken Lohmann), along with PNG Ranger Juan Garcia and several GSC volunteers, initiated and completed a 2-day health assessment of 28 green turtles (*Chelonia mydas*) and a single hawksbill turtle (*Eretmochelys imbricata*). While waiting for our research permit and our

G. A. Lewbart (✉) · D. Deresienski
School of Aquatic, Wildlife, and Zoological Medicine, North Carolina State University, Raleigh, NC, USA
e-mail: galewbar@ncsu.edu

J. P. Muñoz Pérez
College of Biological and Environmental Sciences and Galapagos Science Center, Universidad San Francisco de Quito, Quito, Ecuador

School of Science, Technology & Engineering, University of the Sunshine Coast, Sippy Downs, Australia

D. Páez-Rosas
College of Biological & Environmental Sciences, Universidad San Francisco de Quito, Quito, Ecuador

C. Valle
College of Biological & Environmental Studies and USFQ-UNC Galapagos Science Center, Universidad San Francisco de Quito, Quito, Ecuador

D. Alarcón-Ruales
Galapagos Science Center & University of the Sunshine Coast, Sippy Downs, QLD, Australia

M. Hirschfeld
James Cook University, Townsville, Australia

K. J. Lohmann
Department of Biology, University of North Carolina at Chapel Hill, Chapel Hill, NC, USA

S. J. Walsh et al. (eds.), *Island Ecosystems*, Social and Ecological Interactions in the Galapagos Islands, https://doi.org/10.1007/978-3-031-28089-4_20

Table 20.1 Baseline health data has been established for the following species

Green sea turtles (*Chelonia mydas*)
Hawskbill Sea turtle (*Eretmochelys imbricata*)
Marine iguana (*Amblyrhynchus cristatus*)
San Cristóbal Giant tortoise (*Chelonidis chathamensis*)
Galápagos Sea lion (*Zalophus wollebaeki*)
San Cristóbal Lava lizard (*Microlophus bivittatus*)
Red-footed booby (*Sula sula*)
Blue-footed booby (*Sula nebouxii*)
Nazca booby (*Sula granti*)
Great frigatebird (*Fregata minor*)
Magnificent frigatebird (*Fregata magnifiscens*)
Swallowtail gull (*Creagrus furcatus*)
Galápagos shearwater (*Puffinus subalaris*)
Red-billed tropicbird (*Phaethon aethereus*)
Galápagos yellow land iguana (*Conolophus subcristatus*)
Santa Fe land iguana (*Conolophus pallidus*)
Pink land iguana (*Conolophus marthae*)
Hybrid iguana (*Amblyrhynchus cristatus* X *Conolophus subcristatus*)
Española lava lizard (*Microlophus delanonis*)
Black-lined salema (*Xenocys jessiae*)
Sally Lightfoot Crab (*Grapsus grapsus*)

Bold type indicates peer reviewed published results

aviso de campo (field permit) to be finalized, we had almost a week to explore San Cristóbal and think about future projects related to veterinary medicine and wild animal welfare. A literature search turned up hundreds of articles on wildlife evolution, natural history, ecology, genetics, invasive species eradication, anatomy, and physiology. The veterinary literature was limited to about five dozen publications, most focused on avian species. Many avian taxa were covered and included topics such as parasites, bacterial diseases, viral diseases, pollution, and baseline health assessments (Calle et al. 2017). Very few papers addressed health and diseases of reptiles, and we decided this would be a good area to focus.

In 2014, we added, Diego Páez-Rosas, Diane Deresienski, Juan Pablo Muñoz Pérez, and Daniela Alarcón-Ruales to our team and continued to utilize the help and support of the PNG, GSC, and Universidad San Francisco de Quito (USFQ) volunteers. By June of 2014, we had published our green sea turtle health assessment paper in *PLoS ONE* and started working on marine iguanas and Galápagos sea lions, while continuing with our sea turtle health assessment work. In 2016, Carlos Valle, USFQ, joined our group, and we began a long and productive collaboration working on seabird health in the archipelago. To date, we have published baseline health assessments of 15 species (Table 20.1) with six more manuscripts in review or preparation at the time of writing this paper (Arguedas-Porras et al. 2018; Colosimo et al. 2022; Lewbart et al. 2017a, b, 2018b, 2019; Muñoz-Pérez et al. 2017; Paéz-Rosas et al. 2016; Posner et al. 2020; Souza et al. 2021; Tucker-Retter et al. 2021; Valle et al. 2018, 2019a, b, 2020). Other projects relate to wildlife health,

Table 20.2 The following projects and areas of research are either ongoing or completed

Effects of El Niño on marine iguanas
Tick-borne diseases of marine iguanas
Eye parasites in sea lions
Defining the shrinking marine iguana phenomenon
Determining if gastroliths serve as "ballast" for diving marine iguanas
Effects of tourist activity on nesting seabirds
Validating infrared temperature guns (giant tortoises, sea turtles, iguanas)
Blood cross-matching (sea lions, giant tortoises, sea turtles)
Quantifying microplastics effects on Galápagos wildlife
Health assessment, anesthesia, euthanasia of Sally Lightfoot crabs
Population health of cetaceans

Bold type indicates peer reviewed published results

anatomy, physiology, behavior, and ecology and have been undertaken, and in some cases, published (Table 20.2) (Cerreta et al. 2018, 2019; Christman et al. 2023; Dass et al. 2021; Gregory et al. 2023, in press; Lewbart et al. 2017b, c, 2018a; Masterson et al. 2022; Muñoz-Pérez et al. 2023, in press; Phillips et al. 2018; Posner et al. 2020; Rhea et al. 2023; Swanepoel et al. 2022).

Relationships

As with most things in life, strong, trusting relationships provide the foundation for productivity and accomplishment. Perhaps in no place is this more accurate or important than the Galápagos. With over 97% of the land mass, and an even larger area representing the Galápagos Marine Reserve under the jurisdiction and protection by the PNG, establishing good relations and trust with this dedicated organization is paramount to working with wildlife in Galápagos. For foreign scientists, it is essential that they establish strong collaborations with Ecuadorian scientists based in Galápagos and on the mainland of Ecuador. The GSC provides the springboard for finding collaborators and GSC staff make it possible to facilitate these vital connections and build local capacity. To be effective and efficient, one must also garner the help and support of Galápagueños, such as, commercial fishers, boat captains, naturalist guides, taxi drivers, shop merchants, and hotel owners. All of these organizations and people need to be treated with respect and as important collaborators. Without transportation, food, and beds, science in the archipelago would quickly grind to a halt.

Ideas

For the inquisitive and open-mind, the Galápagos is full of ideas, challenges, and conundrums. After working out the logistics for doing veterinary and health related fieldwork in the Galápagos (on sea turtles), we quickly realized we could apply the same logistical paradigm to other species, both aquatic and terrestrial. For the baseline health assessments, it was simply a matter of collaborating with experts who had access to and knowledge about the taxa to be examined. Several projects resulted from questions posed by local scientists and even citizens based on things they had observed (e.g., sea lions with ocular lesions).

Permitting Process

Every year, usually in the fall, the PNG places a call for scientists to submit research proposals. There is a well-defined template for these proposals and there is a requirement that foreign researchers collaborate with an Ecuadorian scientist(s). Local research institutions like the Galápagos Science Center and the Charles Darwin Foundation have their own forms and requirements. Storing and exporting biological samples also require forms and in some cases *Convention on International Trade in Endangered Species* (CITES) paperwork.

Materials, Supplies, and Logistics

Over the years our team has worked to develop and refine our data sheets and checklists to make sure we have everything we need in the field. In many cases we are very remote and not connected by cell phone or internet. Sometimes, if we are traveling via a research vessel, we may be over a day's travel from our home base, the Galápagos Science Center on San Cristobal Island. This disconnectedness forced us to be self-sufficient with the appropriate backups in place (in some cases two or three of key pieces of equipment and supplies). Table 20.3 illustrates a typical field checklist and Fig. 20.1 a typical data sheet.

Animal Capture and Handling

Once all of the permits and other paperwork are collected, the equipment and supplies organized and packed, and the trip logistics worked out, the team is ready for animal capture, restraint, physical examination, and sample collection. The protocols and logistics for each species type is different.

Coastal Invertebrates Like Sally Lightfoot Crabs (Fig. 20.2) Generally, these animals are captured by hand and then placed in a secure container like a sturdy bucket or cooler. Measurements and samples can either be obtained in the field or back in the laboratory. Normally latex or nitrile gloves are suitable for safe handling.

Generally, fish intended for research sampling are captured with specialized nets and then either sampled in the field and released or euthanized and further sampled in the laboratory (Fig. 20.3).

Table 20.3 A typical checklist used before heading into the field to collect samples and data

Iguana fieldwork checklist
Tape measure
Digital scale
Weight bag
Weight rope
Leather gloves
Doppler and charger
Four boxes of glass slides
Two empty slide boxes (100 slides)
Doppler lube (three bottles)
Lactate meter (2) and batteries
Lactate strips (75)
Clipboard
Thermocouple
Sharpie pens
Eppendorf tubes (at least 200)
Box of alcohol swabs
Sharps container
Three orange supply boxes
Table (folding)
Sexing probes
Data sheets
Zinc oxide
PIT tags
PIT tag reader
9 volt batteries
Tarp
Ice packs
Small cooler
Electric centrifuge (for spinning Eppendorf tubes)
Alcohol
Nitrile gloves (multiple sizes)
3 AA batteries for scale
Clinical refractometer
First aid kit
iSTATS (2)
iSTAT cartridges (75)
3 mL syringes (with needles)
Heparin
Small centrifuge (battery operated)
Microcapillary tubes
Tonovet®

(continued)

Table 20.3 (continued)

Blood spot cards (75)
5 mm biopsy punches (10)
4-0 PDS suture (10 packs)
Sterile swabs for bacterial culture
SAF media for feces
Metal pan and aluminum foil microplastics fecal samples
Lidocaine gel
Lidocaine 2%
IR camera and charger
IR flashlight

Sea Turtles In some ways sea turtles are the most challenging animals to capture and restrain. In most instances we utilize the efforts of between two and five experienced skin divers (snorkelers) who hand capture the turtles and carefully either swim them to shore or place them on a research vessel. Whether aboard a boat, or more commonly, on a sandy beach, the turtles are processed (measurements, weights, physical examination) and samples (blood, skin biopsy, scute biopsy, feces, etc.). Every effort is made to minimize the time out of water and the tagged turtles are quickly released (Figs. 20.4, 20.5, and 20.6).

Giant Tortoises Nearly all of our giant tortoise work has been with animals under human care in either a breeding facility or on a large farm that tourists frequent. In most cases a park ranger or rangers will identify which tortoises are to be examined and sampled. Generally, we are doing this work as a pre-release health assessment. Capture is easy as is restraint in most cases. Tortoises over about 10 kg are generally placed in dorsal recumbency (on their back) for ease of sampling and handling (Fig. 20.7). In most cases we complete our physical examination and sample collection within about 20 minutes. Smaller tortoises can be restrained and examined while being hand-held (Fig. 20.8).

Iguanas We have worked with all four Galápagos iguanids (*Amblyrhynchus cristatus, Conolophus subcristatus, Conolophus pallidus, and Conolophus marthae*) (Figs. 20.9, 20.10, 20.11, 20.12, and 20.13) In all cases the animals are located then hand-captured, usually while wearing thick protective gloves (Figs. 20.10 and 20.14). We usually complete our physical examination and sample collection within 15 min (Figs. 20.15, 20.16, 20.17, 20.18, and 20.19).

Lava Lizards Our team has worked with two species of lava lizards, *Microlophus bivittatus* and *M. delanonis*. Most of the islands in the archipelago support lava lizards. These animals are small, almost always less than 100 g, and are normally caught using a monofilament slip knot snare at the end of a long retractable pole. They are easy to restrain with one hand and restraint and sampling can be accomplished in 10 minutes or less (Figs. 20.20 and 20.21).

Pink Iguana (Conolophus marthae)_Microplastic and Health Study_Field and Lab Data Sheet. **Location:**
Field Team: & Date

ID	Time of CAPTURE	Sex	Species	Recapture or New	Time of FECAL sample	Time of BLOOD sample	Heart Rate (/min)	Resp. Rate (/min)	Body Temp_IR Body or Thermo-couple cloaca	Weight (Kg)	Girth (cm)	Total Length (cm)	Snout to Vent Length (cm)	Tail Length (cm)	Photo with scale	BCS>1: skiny. 2: normal, 3: obese .	MM_ Mucous membrane

Lat: **Long:**
Air temp: **Air temp:**

ID	Hydration	Pain Score	Eye, Ear, Nose and Throat (EENT)	Integ,Skin	Coelom	Urogenital	MSK_ musculoskeletal	Neuro	Time of RELEASE	chem8_Na	chem8_K	chem8_Cl	chem8_iCa	chem8_tCO2	chem8_Gluc	chem8_BUN	chem8_CREAT

Fig. 20.1 Field data sheet that can be printed out for hand collection of data using a pencil or indelible ink pen

ID	chem8_HCT	chem8_HB	chem8_anion GAP	Lactate (mmol/L)	PCV (%) Manual Hematocrit	TP (Total Protein)	Blood Spot Card Check	White Cell Estimate	Heterophils (%)	Lymphocytes (%)	Monocytes (%)	Eosinophils (%)	Basophils (%)	Notes

Fig. 20.1 (continued)

Fig. 20.2 Adult Sally lightfoot crab (*Grapsus grapsus*) near a Nazca booby (*Sula granti*)

Fig. 20.3 Large schools of black-lined salema (*Xenocys jessiae)* are common in the Galápagos, especially close to shore and underwater structure

Fig. 20.4 Large green sea turtle (*Chelonia mydas*) being brought ashore for examination and sample collecting

Fig. 20.5 Green sea turtle (*Chelonia mydas*) on a portable table in the field

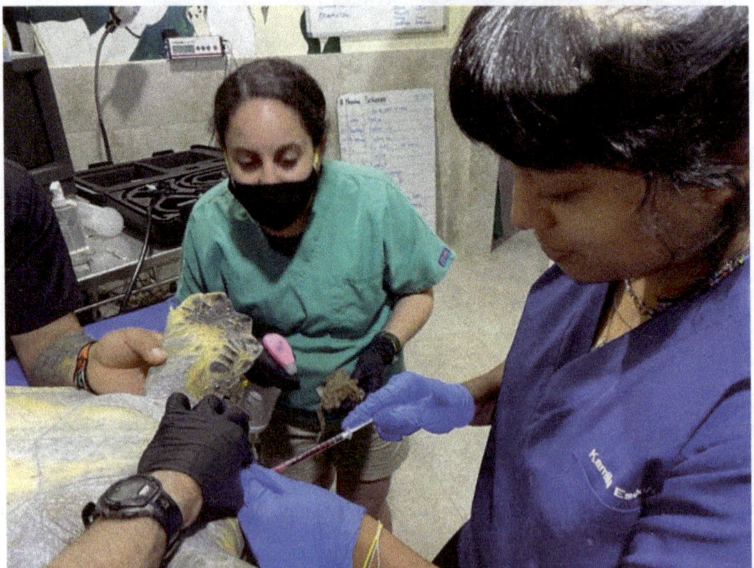

Fig. 20.6 This green turtle (*Chelonia mydas*) is having its blood collected from the dorsal tail vein

Fig. 20.7 This giant tortoise (*Chelonoidis chathamensis*) is being examined as part of an annual health assessment and pre-release program at a large breeding center. For short procedures tortoises do well on their carapace (back)

Fig. 20.8 This juvenile giant tortoise (*Chelonoidis chathamensis*) is being restrained for jugular vein blood collection. Blood from the jugular vein is ideal because there is little risk of lymphatic fluid contamination

Fig. 20.9 Large marine iguana (*Amblyrhynchus cristatus*) being radiographed without restraint

Fig. 20.10 Colorful marine iguana (*Amblyrhynchus cristatus*) from Floreana being manually restrained prior to physical examination

Fig. 20.11 Two marine iguanas (*Amblyrhynchus cristatus*) illustrating the sexual dimorphism that exists (male on left and female on right) from the island of Isla Lobos near San Cristóbal

Fig. 20.12 A large yellow land iguana (*Conolophus subcristatus*) on North Seymour Island in its burrow. Sometimes animals are captured in their burrows

Fig. 20.13 A large yellow land iguana (*Conolophus subcristatus*) in the wild with Daphne Major in the background

Fig. 20.14 A large Santa Fe land iguana (*Conolophus pallidus*) being manually restrained

Fig. 20.15 A large pink land iguana (*Conolophus marthae*) in the wild on Wolf Volcano, the only place in the world this species is found

Fig. 20.16 This hybrid iguana (*Amblyrhynchus cristatus* X *Conolophus subcristatus*) was captured and sampled on South Plaza Island

Fig. 20.17 This represents a typical field lab for processing land iguanas. This lab was set up on the island of Santa Fe

Fig. 20.18 Here a Santa Fe land iguana (*Conolophus pallidus*) is restrained for lateral tail vein blood collection

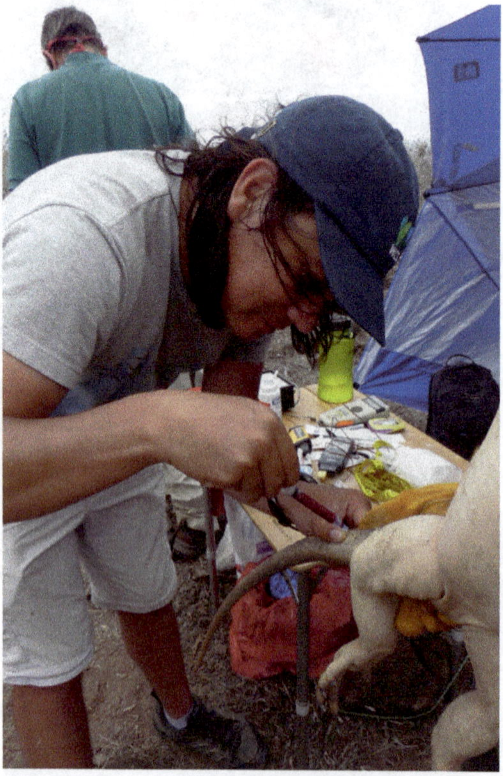

Fig. 20.19 In some cases, space on a research vessel can be used as a portable laboratory

Seabirds Seabirds are amazing creatures to work with. Our team has captured, measured, and sampled nine species of seabirds in the Galápagos since 2016. In most cases the birds are captured manually while on their nests. The only exceptions to this are the swallow tail gulls that are captured with hand-held nets and white-rumped petrels that are caught with mist nets. Two people are needed to safely restrain most species while a third person obtains measurements and blood samples (Figs. 20.22, 20.23, and 20.24).

Sea Lions Our team has been involved with a number of health-related projects of Galápagos sea lions (*Zalophus wollebaeki*). We generally work with juvenile or sub-adult animals weighing less than 60 kg. These animals are hand captured by PNG rangers using specialized, reinforced nets. Once captured they are manually restrained for measurements and sampling. We limit capture and handling time to 15 min (Figs. 20.25 and 20.26).

Cetaceans This is a relatively new area of research for our team and to date we have not handled any animals. Current efforts involve photographic data collection, drone imaging, and skin/blubber biopsies obtained with crossbow arrows. Data and samples have been obtained from both toothed (Odontocetes) and baleen (Mysticeti) whales (Figs. 20.27, 20.28, and 20.29).

Fig. 20.20 Lava lizards are dimorphic, with the females generally being more colorful. These two examples are from the island of Española. (**a**) Female *Microlophus delanonis*. (**b**) Male *Microlophus delanonis*

Sample Collecting, Storage, and Analysis

Our work requires a variety of sample types and sampling methods. These samples include hair, feathers, skin, blood, blubber, feces, oral swabs, nasal swabs, cloacal swabs, and ecto and endoparasites. In some cases specialized tools or instruments are required, and if the procedure involves entering the animals circulatory system or taking a biopsy, sterility and good hygiene will be a priority. In most situations we are analyzing samples and collecting data in the field using specialized, portable

Fig. 20.21 Galápagos National Park Veterinarian Andrea Loyola restrains a lava lizard (*Microlophus delanonis*) during a rainstorm

Fig. 20.22 A mature Nazca booby (*Sula granti*) is being restrained for blood collection from the ulnar vein

Fig. 20.23 Red-billed tropicbird (*Phaethon aethereus*) being manually restrained for its physical examination

equipment. Any unused field samples are usually stored on ice in a cooler until they can be appropriately analyzed or permanently stored in the Galápagos Science Center facilities. Depending on the sample and type of analysis, samples might be frozen, placed in specialized media, dried, fixed in formalin, fixed in ethanol, or simply kept refrigerated. Some samples require analysis on mainland Ecuador or in another country and these samples are stored and shipped when appropriate and after the proper permits are obtained.

Writing Up Results

Once the fieldwork is completed and the data collected and compiled it's time to write up the results. These are always collaborative efforts with one or two people taking the lead. Once a draft is ready it is shared with the team and everyone has their chance to review, edit, and revise the draft. Once everyone is satisfied with the manuscript and the journal of submission the paper is submitted and all authors are included with the subsequent correspondence.

Fig. 20.24 Some field sites are difficult to access and researchers need to be physically fit and nimble

Fig. 20.25 A recently captured juvenile sea lion (*Zalophus wollebaeki*) being carried to the field lab area

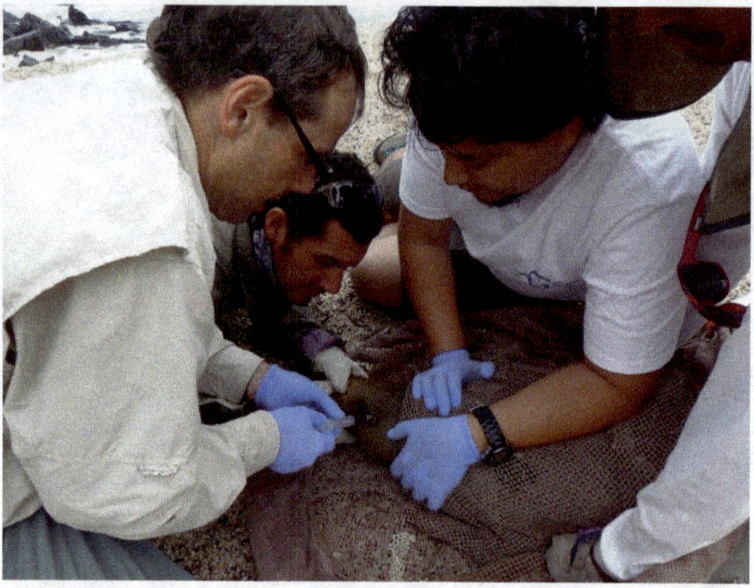

Fig. 20.26 A sea lion (*Zalophus wollebaeki*) being restrained for blood collection from the gluteal venous sinus

Fig.20.27 A veterinary student spotting dolphins and whales. A short-finned pilot whale (*Globicephala macrorhynchus*) is presenting its fluke in the background

Fig. 20.28 Here a researcher is preparing to dart a whale to collect a skin and blubber sample

Fig. 20.29 The researcher on the left is holding an arrow with a skin/blubber sample from a short-finned pilot whale (*Globicephala macrorhynchus*). This tissue can be used for genetics, heavy metal, infectious disease, and toxicology studies

Fig. 20.30 This figure lists the individuals, organizations, and institutions that have contributed to the efforts summarized in this book chapter. Without them this work would not be possible

Partners

Our team relies on a large number of individuals, institutions, and organizations to accomplish this work. Many individuals, named below and unnamed, contributed to and continue making contributions to these efforts (Fig. 20.30).

References

Arguedas-Porras R, Lewbart GA, Deresienski D, Lohmann K, Steinberg D, Muñoz-Pérez JP, Valle CA (2018) Haematology and biochemistry of the San Cristobal lava lizard (*Microlophus bivittatus*). Conserv Physiol 6(1). https://doi.org/10.1093/conphys/coy046

Calle P, Rodríguez J, Matamoros Y (eds) With the collaboration of: Cunningham A, Travis D, Rideout B, Jiménez G, Sutherland-Smith M, Ponder J, Nieto A, Huyvaert K, Dary L, Lewbart G, Arguedas R (2017) Workshop to develop a Wildlife Health Plan for the Galápagos Islands. © 2018 Wildlife Conservation Society

Cerreta AJ, Lewbart GA, Diaz R, Vasquez G, Muñoz-Pérez JP, Paéz-Rosas D (2018) *Chelonoidis chathamensis*. Methods of body temperature assessment. Herp Rev 49(4):696–697

Cerreta A, Vaden SL, Lewbart GA, Muñoz-Pérez JP, Páez-Rosas D (2019) Increased BUN and glucose in a group of San Cristóbal Galapagos tortoises (*Chelonoidis chathamensis*). Vet Rec Case Rep 7:3000699. https://doi.org/10.1136/vetreccr-2018-000699

Christman J, Lohmann K, Hirschfeld M, Muñoz-Pérez JP, Garcia J, Westermeyer H, Lewbart GA (2023) Intraocular pressure using rebound tonometry in the Galápagos marine iguana (Amblyrhynchus cristatus). Int J Vet Res 2(2):1–11

Colosimo J, Gentile G, Gerber G, Vera C, Sevilla C, Lewbart GA (2022) Conservation of a flagship species; health assessment of the pink land iguana, *Conolophus marthae*. PLoS One 17(3):e0257179. https://doi.org/10.1371/journal.pone.0257179

Dass K, Lewbart GA, Muñoz-Pérez JP, Yépez M, Loyola A, Chen E, Páez-Rosas D (2021) Evaluation of fatty acid whole blood concentrations in the San Cristóbal Galápagos tortoise (*Chelonoidis chathamensis*). PeerJ 9:e11582. https://doi.org/10.7717/peerj.11582

Gregory TM, Livingston I, Hawkins EC, Loyola A, Cave A, Vaden SL, Deresienski D, Breen M, Riofrio-Lazo M, Lewbart GA, Paez-Rosas D (Feb, 2023, in press) Dirofilaria immitis identified in Galapagos sea lions: A wildlife health and conservation concern. Accepted Journal of Wildlife Diseases

Lewbart GA, Ulloa C, Deresienski D, Regalado C, Muñoz-Pérez GJ, Hardesty BD, Valle CA (2017a) Health status of red-footed boobies (*Sula sula*) determined by hematology, biochemistry, blood gases, and physical examination. J Zoo Wild Med 48(4):1230–1233

Lewbart GA, Valle C, Muñoz-Perez J (2017b) *Microlophus indefatigabilis* diet. Herp Rev 48(4):851–852

Lewbart GA, Hirschfeld M, Ricuarte O, Lohmann. (2017c) *Amblyrhynchus cristatus* diet. Herp Rev 48(2):439

Lewbart GA, Cohen E, Hirschfeld M, Muñoz-Pérez JP, García J, Fu A, Chen EP, Lohmann KJ (2018a) Field-based radiographic imaging of marine megafauna: marine iguanas (*Amblyrhynchus cristatus)* as a case study. Front Mar Sci. https://doi.org/10.3389/fmars.2018.00040

Lewbart GA, Griffioen J, Savo A, Muñoz-Pérez JP, Ortega C, Loyola A, Roberts S, Schaaf G, Steinberg D, Osequeda SB, Levy MG, Páez-Rosas D (2018b) Biochemistry and hematology parameters of the San Cristóbal Galápagos tortoise. Conserv Physiol. https://doi.org/10.1093/conphys/coy004

Lewbart GA, Grijalva CJ, Calle PP, Ingerman K, Muñoz-Pérez JP, Quezada G, Vera CA, Gentile G, Valle CA (2019) Health assessment of *Conolophus subcristatus*, *Conolophus pallidus*, and *C. subcristatus* X *Amblyrhynchus cristatus* hybrid (Galápagos land iguanas). PLoS One 14(10):e0222884

Masterson CA, Locklear TR, Páez-Rosas D, Muñoz-Pérez JP, Ortega C, Lewbart GA, Vaden SL (2022) Intraocular pressure using rebound tonometry in the San Cristóbal Galápagos tortoise (*Chelonoidis chathamensis*). Int J Vet Res 2(1):13–18. (veterinary student as first and corresponding author)

Muñoz-Pérez JP, Lewbart GA, Hirschfeld M, Denkinger J, Castañeda JG, García J, Lohmann KJ (2017) Blood gases, biochemistry, and haematology of Galápagos hawksbill turtles (*Eretmochelys imbricata*). Conserv Physiol 5(1):cox028. https://doi.org/10.1093/conphys/cox28

Muñoz-Pérez JP, Lewbart GA, Alarcon-Ruales DE, Skehel A, Cobos E, Rivera R, Jaramillo A, Vivanco H, Zurita-Arthos L, Wallace BP, Valle CA, Townsend KA (Feb, 2023, in press) Galápagos and the plastic problem. Accepted, Frontiers in Sustainability

Paéz-Rosas D, Hirschfeld M, Deresienski D, Lewbart GA (2016) Health status of Galápagos Sea lions (*Zalophus wollebaeki*) on San Cristobal Island rookeries determined by hematology, biochemistry, blood gases, and physical examination. J Wildl Dis 52(1):100–105

Phillips BE, Páez-Rosas D, Flowers JR, Cullen JM, Law JM, Colitz C, Deresienski D, Lohmann KJ, Lewbart GA (2018) Evaluation of the ophthalmic disease and histopathologic effects due to the ocular trematode *Philophthalmus zalophi* on juvenile Galápagos Sea lions (Zalophus wollebaeki). J Zoo Wild Med 49(3):581–590

Posner LP, Applegate JA, Cannedy A, Deresienski D, Mozzachio K, Serrano M, Lewbart GA (2020) Total injectable anesthesia of dogs and cats for remote location veterinary sterilization clinic. BMC Vet Res 16:304. https://doi.org/10.1186/s12917-020-02525-x

Rhea S, Camacho BE, Amoriello CW, Correa M, Lewbart GA, Cruz M, Vélez A, Castillo P, Pairis-Garcia M (2023) Assessing animal welfare on small-scale multi-species farms located on Floreana Island, Galápagos Islands. Animals 13:686. https://doi.org/10.3390/ani13040686

Souza AR, Huffman AM, Muñoz-Pérez JP, Vaden S, Diaz R, Vasquez G, Donnelly KA, Páez-Rosas D, Lewbart GA (2021) Crossmatching the San Cristóbal Galápagos tortoise (*Chelonoidis chathamensis*). J Herpetol Med Surg 31(4):315–320

Swanepoel L, Loyola A, Flowers J, Lewbart GA, Garrett KB, Yabsley MJ (2022) Infection with a novel pentastome (*Raillietiella* sp.) in a juvenile endangered Galápagos tortoise (Chelonoidis niger). Veterinary Parasitology: Regional Studies and Reports. https://doi.org/10.1016/j.vprsr.2022.100782

Tucker-Retter E, Velsey-Gross Z, Deresienski D, Ulloa C, Muñoz-Pérez J-P, Skehel A, Passingham RK, Castaneda J, Lewbart GA, Valle CA (2021) Health status of Nazca boobies (*Sula granti*) on Daphne Major Island in the Galápagos determined by hematology, biochemistry, and physical examination. J Zoo Wildl Med 52(2):671–679

Valle CA, Ulloa C, Deresienski D, Regalado C, Muñoz-Pérez JP, Garcia J, Hardesty BD, Skehel A, Lewbart GA (2018) Health status of great frigatebirds (*Fregata minor*) determined by hematology, biochemistry, blood gases, and physical examination. Conserv Physiol. https://doi.org/10.1093/conphys/coy034

Valle CA, Ulloa C, Deresienski D, Regalado C, Muñoz-Pérez JP, Garcia J, Hardesty BD, Skehel A, Lewbart GA (2019a) Health status and baseline hematology, biochemistry and blood gas values of Galapagos shearwaters (*Puffinus subalaris*). J Zoo Wild Med 50(4):1026–1030

Valle CA, Grijalva CJ, Calle PP, Muñoz-Pérez JP, Quezada G, Vera CA, Lewbart GA (2019b) Methods of body temperature assessment in *Conolophus subcristatus, Conolophus pallidus* (Galápagos land iguanas), and *Amblyrhynchus cristatus* X *C. subcristatus* hybrid. PeerJ 7:e6291. https://doi.org/10.7717/peerj.629

Valle C, Ulloa C, Regalado C, Muño-Pérez JP, Garcia J, Hardesty BD, Skehel A, Deresienski D, Passingham RK, Lewbart GA (2020) Baseline hematology, biochemistry, blood gas values, and health status of the Galápagos swallow-tailed gull (*Creagrus furcatus*). Conserv Physiol. https://doi.org/10.1093/conphys/coaa064

Greg A. Lewbart is Professor of Aquatic, Wildlife, and Zoological Medicine and a diplomate of the American and European Colleges of Zoological Medicine. This marks his 31st year on the faculty of the NC State College of Veterinary Medicine where he teaches invertebrate, fish, amphibian, and reptile medicine. He has made 26 trips to the Galápagos doing conservation work on a wide variety of native wildlife since 2013.

Juan Pablo Muñoz Pérez has more than 20 years of experience diving and surfing in the Galápagos. For the past 10 years he has been an active researcher in the archipelago. Juan Pablo's research focuses on animal movement, ecology, and health, especially on sea turtles, marine iguanas, and other marine animals including whales and dolphins. Since 2014 he has been the regional pioneer and active scientist investigating plastic pollution in Galápagos with the primary objective of using science to find solutions. He is now completing a PhD at the University of The Sunshine Coast in Australia.

Diego Páez-Rosas is interested is in the ecological aspects of marine megafauna and the interface between trophic flexibility and environmental variability. He combines laboratory and fieldwork methods to track the foraging ecology of Galapagos marine predators and relate their feeding habits, home range and trophic behavior to their breeding success and energy budgets. This research also has ramifications in population dynamics, health status, and conservation management.

Carlos Valle is a professor of Ecology and Evolutionary at the Universidad San Francisco de Quito Campus Cumbayá and Campus Galápagos. Carlos earned his Ph.D. in Ecology and Evolutionary Biology at Princeton University in the Laboratory of Professors Peter and Rosemary Grant. He started researching in the Galapagos when he was still a university student more than 30 years ago. His main area of research is the behavioral ecology of seabirds and lava lizards. In recent years he has also projected his research in trophic ecology using the stable isotope technique in collaboration with various colleagues.

Daniela Alarcon-Ruales is a Research Scientist and Emblematic Projects Coordinator at the Universidad San Francisco de Quito and the Galápagos Science Center. She is also a Ph.D. Candidate in Environment and Sustainability at the University of the Sunshine Coast, Australia.

Maximilian Hirschfeld is a marine ecologist with over 10 years of experience living and working in the Galápagos. His expertise ranges from shark and ray ecology and evolution to wildlife health in isolated oceanic archipelagos. Since joining the Galápagos Science Center in 2012, he has studied sea lions, sea turtles, giant tortoises, marine iguanas, sharks and rays across the archipelago. He holds a master's degree from the Universidad San Francisco Quito and a PhD from James Cook University, where his thesis work focused on the endemic Galápagos bullhead shark.

Diane Deresienski is board certified by the American Board of Veterinary Practitioners (ABVP) in canine and feline medicine and has been practicing veterinary medicine for 34 years. She is a retired professor from the North Carolina State Universtiy College of Veterinary Medicine where she remains an adjunct faculty member. Diane has has made 15 trips to the Galápagos where her efforts focus on native avian and reptile health. She is currently a consultant to the Dean of the Universidad San Francisco Quito College of Veterinary Medicine.

Kenneth J. Lohmann is the Charles P. Postelle Distinguished Professor of Biology at the University of North Carolina at Chapel Hill. His research focuses on the behavior, sensory biology, and conservation of marine animals. He served as President for the International Sea Turtle Society in 2018–2019 and as a member of the Advisory Board for the Galapagos Science Center from 2012–2018.

Chapter 21
Challenges in the Application of the Ecosystem Approach to Fisheries Management in the Galapagos Islands

Marjorie Riofrío-Lazo, Manuel J. Zetina-Rejón, Gunter Reck, Diego Páez-Rosas, and Francisco Arreguín-Sánchez

Introduction

The ecosystems that support fishing and other economic activities are subject to significant alterations to their functioning, resilience, and the goods and services they can provide (Garcia et al. 2003). During the last few decades, most efforts to regulate fisheries and the conservation of the seas and oceans have had limited success in preventing the ongoing problems of overfishing, habitat degradation, and irreversible loss of marine biodiversity (Pacoureau et al. 2021). Therefore, there is a great concern about the state of fishing resources, as many are being overexploited (Cury and Pauly 2020; FAO 2020). Overfishing transforms an initially stable, mature, and efficient ecosystem into an immature and disturbed one (Pauly et al. 1998). By directing fishing toward resources of high economic value and reducing their abundance, food webs and flows of biomass and energy in the ecosystem are altered, affecting its organization and functioning (Scheffer et al. 2001, 2005; Bascompte et al. 2005; Lotze et al. 2011; Salcido-Guevara et al. 2012).

M. Riofrío-Lazo (✉) · D. Páez-Rosas
Universidad San Francisco de Quito USFQ, Colegio de Ciencias Biológicas y Ambientales COCIBA, Quito, Ecuador

USFQ & UNC-Chapel Hill, Galapagos Science Center GSC,
Isla San Cristóbal, Islas Galápagos, Ecuador
e-mail: mriofriol@usfq.edu.ec

M. J. Zetina-Rejón · F. Arreguín-Sánchez
Instituto Politécnico Nacional IPN, Centro Interdisciplinario de Ciencias Marinas CICIMAR, La Paz, Baja California Sur, Mexico

G. Reck
Universidad San Francisco de Quito USFQ, Colegio de Ciencias Biológicas y Ambientales COCIBA, Quito, Ecuador

S. J. Walsh et al. (eds.), *Island Ecosystems*, Social and Ecological Interactions in the Galapagos Islands, https://doi.org/10.1007/978-3-031-28089-4_21

Fishing alters the habitat, affecting the species' abundance, production capacity and spatial distribution (Rijnsdorp et al. 2018; Porobic et al. 2019; Takeshige et al. 2021). In addition, illegal, unreported, and unregulated fishing contributes to underestimating the total catch, significantly depletes fish populations, and poses a significant threat to marine biodiversity (Murawski 2000; Pitcher et al. 2002; Coll et al. 2008; Agnew et al. 2009). Including non-target species (Zhou et al. 2011; Ragheb et al. 2019), threatened species (Fariñas-Franco et al. 2018; Martins-Vieira et al. 2020), or species considered key to the ecosystem sustainability (Coll et al. 2016; Riofrío-Lazo et al. 2021).

As implemented in the 1940s, fisheries management is strongly based on ecosystem theory but focuses mainly on fishing activity and resource-oriented management (FAO 2003). The key problem with this mono-specific conventional management approach is that it focuses on the single target species production, does not consider the fishery impact on non-target species and marine habitats, and neglects human factors (social, economic, cultural, and institutional) that affect fisheries management (FAO 2009).

The adoption of a more holistic, Ecosystem-Based Management (EBM) approach, also called the Ecosystem Approach to Fisheries (EAF), was encouraged by recognizing the significant direct and collateral impacts that fisheries have on marine ecosystems (De Young et al. 2008; Long et al. 2015). The FAO (2003) defines EAF as "attempt to balance the various societal objectives, taking into account the knowledge and uncertainties of the biotic, abiotic and human components of the ecosystem and their interactions, and applying an integrated approach to fisheries within the ecosystem ecologically significant limits". It is considered a systematic and interdisciplinary approach to fisheries management in a defined geographic area that seeks the sustainability of the ecosystem and contributes to its resilience (Long et al. 2015; Bastardie et al. 2021).

To achieve sustainable development in fisheries, the EAF proposes principles and standards applicable to the conservation, management, and development of fisheries without contradicting or replacing conventional fisheries management. Instead, it seeks to improve its application and reinforce its ecological relevance (FAO 2003). The concept and principles of EAF have been adopted in the management systems of different nations, although they are not fully understood or operational (Bastardie et al. 2021). For example, in the Galapagos Islands, Ecuador, it was adopted in 1998 from the declaration of the Organic Law of the Special Regime of Galapagos (LOREG); however, some commercial species continue without recovering their maximum historical levels of abundance, and the management objectives have not been achieved; moreover, some fisheries remain unsustainable (Castrejón and Charles 2013). In this chapter, we provide an overview of the EAF principles and characteristics and discuss the challenges of its implementation in the marine resource management system of the Galapagos Islands.

Principles and Characteristics of the EAF

The political guidelines contained in the Code of Conduct for Responsible Fisheries (FAO 1995) constitute the fundamental principles on which the EAF is based and are (FAO 2009):

1. That governance must ensure the human population's well-being and the ecosystem's proper functioning.
2. Fisheries management must limit its adverse effects on the ecosystem.
3. Maintaining ecological relationships between fishery resources and associated and dependent species is necessary.
4. Management strategies must be compatible throughout the area where the fishery is developed.
5. The precautionary approach must be applied since the knowledge of the ecosystems is limited.

The EAF contemplates and integrates other existing management approaches for fisheries and management of marine and coastal resources (Fig. 21.1). A fundamental element of EAF is Co-management, which ensures that all stakeholders participate in the decision-making process (FAO 2009). The management actions to achieve the objectives of the EAF are the Coastal Zone Integrated Management and the Marine Space Management. Marine Protected Areas are a tool that, together with other strategies, make it possible to achieve effectiveness in fisheries management and biodiversity conservation (Staples et al. 2014).

Ecosystem-based spatial management effectively applies EAF in marine and coastal environments (Day 2008; Douvere 2008). Based on a strategic and

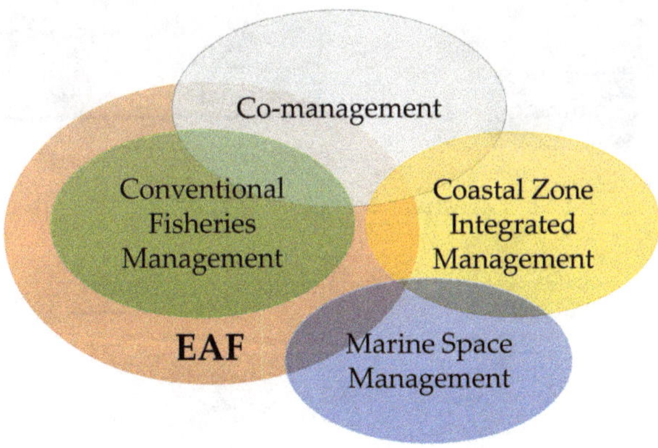

Fig. 21.1 Integration of management approaches in the EAF. Taken and modified from Staples et al. (2014)

comprehensive plan, it seeks to manage current and potential conflicts of use, reduce the accumulated effects of human activities, and optimize the sustainable development of socioeconomic activities while protecting marine areas (Douvere and Ehler 2009). A successful example of the implementation of this approach is the Great Barrier Reef Marine Park in Australia, where zoning has been used as a management strategy for decades (Arkema et al. 2006; Day 2008; Douvere 2008).

The main characteristics of the EAF are (Dimech et al. 2015):

(i) It is participatory at all levels of planning and execution where the principle of equity is met.
(ii) It is comprehensive, as it includes all the components of the fishing system: the ecological, socioeconomic, and political sectors, as well as external factors that may affect management.
(iii) It favors using the best available knowledge, scientific criteria, and local knowledge.
(iv) Use incentives as complementary management measures.
(v) Promotes the adoption of an adaptive management system and management risk assessment.

For the implementation of the EAF, the decision-makers must translate its principles and characteristics into fishery control measures through a fishery management plan (Fig. 21.2). The first step is planning with the definition of limits and scale of the EAF, stakeholders, and problems to be addressed. Objectives are then identified and prioritized; a cost-effective management plan is developed according to the high-priority goals. Next, management actions are executed that are monitored and evaluated to determine which ones can and should be excluded, changed, or added. The review of the management actions must be frequent and periodic to determine if an acceptable level of performance is being generated according to the objectives.

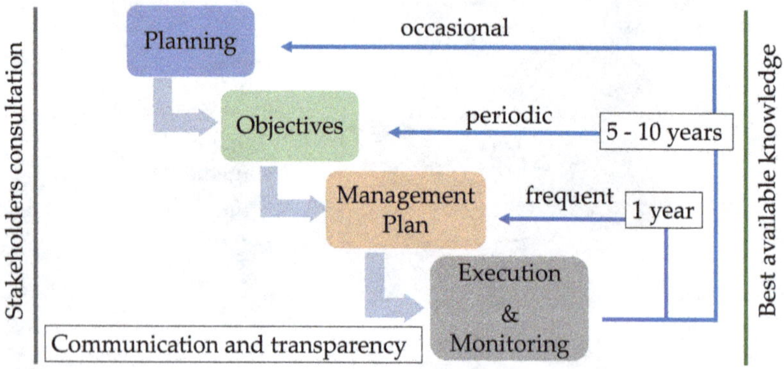

Fig. 21.2 Steps for implementing EAF principles in a management system

In addition, an occasional review of the entire management plan in a predetermined period should be carried out to assess whether it is relevant to the community's current conditions. Stakeholders must be consulted throughout the implementation process, and the best available knowledge must be used at each step. Communication and transparency are key aspects during execution, so knowledge about the fishery's evolution and the new management system must be available to all sectors involved (Ward et al. 2002; Garcia et al. 2003; FAO 2009).

The monitoring, follow-up, and evaluating of the performance of management actions is a critical step in adaptive management planning, a fundamental principle of EAF. It arises from the need to make decisions under conditions of high uncertainty (Holling 1978; Dobbs et al. 2011). Adaptive management implies a systematic process by which management policies and practices improve as our understanding of the socio-ecological system improves by analyzing its response to the management plan employed (Fig. 21.3).

Conceiving adaptability as a management strategy for a dynamic system is highly relevant. It implies modifying the conventional notion of management based on individual populations for a holistic notion of the ecosystem (Arreguín-Sánchez et al. 2015; Riofrío-Lazo 2018; Arreguín-Sánchez 2022). Under this perspective, "from the ecosystem to the fish stocks," the performance indicators to establish fishing limits for target species should express holistic attributes of the ecosystem related to the organization's maintenance, resilience, production capacity, among others (See Arreguín-Sánchez 2022). Subsequently, establish strategies for individual resources maintaining periodic monitoring of the ecosystem (Arreguín-Sánchez et al. 2021).

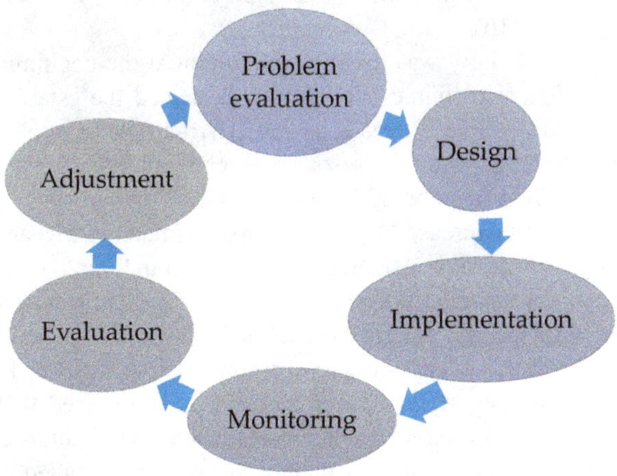

Fig. 21.3 Adaptive management cycle

Management of Fishery Resources in the Galapagos Archipelago

The Galapagos Islands are located 960 km from mainland Ecuador in the Eastern Tropical Pacific. This archipelago constitute an ecosystem with high conservation priority due to the high species diversity (~2900 species) and endemism (18%) that they present (Heylings et al. 2002). These characteristics promoted the creation in 1998 of the Galapagos Marine Reserve (GMR), which comprises an area of ~138,000 km^2, and the adoption of a Management Plan for Conservation and Sustainable Use of the GMR in 1999 for the regulation of fishing and tourist activities, and the protection of vulnerable species (Heylings et al. 2002).

The only type of fishing allowed in specific areas within the GMR is artisanal, according to the zoning scheme implemented in 2000 (Heylings et al. 2002). Currently, 1146 artisanal fishermen and 399 fishing vessels are registered in the Galapagos, divided into three types (pangas, fibers, and boats) according to their size and characteristics (Bucaram et al. 2013; Dirección del Parque Nacional Galápagos 2016). The main fisheries are sea cucumber (*Isostichopus fiscus*), spiny lobsters (*Panulirus gracilis* and *P. penicillatus*), and finfish. Other resources, slipper lobster (*Scyllarides astori*), octopus (*Octopus oculifer*), chitons (*Radsia goodallii*), and gastropods (Hexaplex sp., locally called churo) are harvested for local consumption on a smaller scale. The Fishing Calendar includes the objectives, action plan, and permanent and adaptive management measures for the fisheries that take place in the GMR (Dirección del Parque Nacional Galápagos 2016). Authorized fishing gear includes lobster snare, handline, beach seine, gillnet, and hook line trolling. The use of longlines and industrial fishing are not authorized, and shark fishing is prohibited within the GMR (Castrejón et al. 2014; Dirección del Parque Nacional Galápagos 2016).

The sea cucumber fishery was the most important at the beginning of the 90 s. However, the inefficient resource administration caused the establishment of six total closed periods over time (Reyes et al. 2013). After 6 years of closure, this fishery was reopened in 2022 upon the resource assessment consider it in recovering. The lobster fishery went from being industrial (in 1960) to artisanal (in 1984), and the percentage of catch decreased by 67% in approximately 25 years (Bustamante et al. 2000), for which it was considered overexploited in 2006 (Hearn et al. 2006). However, fishery performance indicators currently place the resource in recovery. The finfish is a multispecific fishery of demersal coastal and pelagic fish, and the oldest of all, with activity records in 1832 and commercialization in 1940, being the Galapagos sailfin grouper *Mycteroperca olfax*, locally referred to as Galapagos bacalao, the most important target species (Reck 1983). Also caught the Galapagos white-spotted sand bass (*Paralabrax albomaculatus*, locally called camotillo), the misty grouper (*Hyporthodus mystacinus*), the mottled scorpionfish (*Pontinus clemensi*), the wahoo (*Acanthocybium solandri*), the yellowfin tuna (*Thunnus albacares*), the longfin yellowtail (*Seriola rivoliana*) among others (Schiller et al. 2014).

Various studies conclude that some of the main commercial species (e.g., sea cucumber, spiny lobster, the Galapagos bacalao, and the camotillo grouper) are currently highly exploited (Usseglio et al. 2016; Buglass et al. 2018; Ramírez-González et al. 2020).

From Conventional Management to the Ecosystem Approach

The current management of fishery resources in the Galapagos has an ecosystemic connotation and occurred after the declaration of the LOREG in 1998. This transition emerged as a response to various problems that arose over time since the creation of the first signed management plan in 1974, and which intensified with the overexploitation of the most economically profitable resource, sea cucumber (De Miras et al. 1996; Zapata 2005; Castrejón et al. 2014).

There were no clear management guidelines, the fishing sector was disappointed at not being included in the planning process, and there were institutional conflicts because their functions overlapped. The lower catch per fishing effort caused social conflicts intensified by the presence of an Ecuadorian industrial fishing fleet that began exploiting tuna in the Galapagos in 1970. In addition, between the years 1980 and 2000, there was an accelerated increase in fishing efforts (90%), tourists (85%), and the population in general (Zapata 2005). The overexploitation of sea cucumbers revealed the implementation of control measures without scientific support and the fishery's lack of monitoring, control, and surveillance (De Miras et al. 1996).

With the entry into force of the LOREG, the GMR was created and established that the Directorate of the Galapagos National Park (Dirección Parque Nacional Galápagos, DPNG) is the authority to manage the GMR. A participatory or co-management system was established, consisting of a core group named "grupo núcleo" of the primary users of the marine reserve (fishermen, tour operators, naturalist guides, scientists, and the DPNG). This core group formed the Participatory Management Board (Junta de Manejo Participativo, JMP), which decided by consensus on implementing management actions. In addition, the Inter-institutional Management Authority (Autoridad Interinstitucional de Manejo, AIM) was also created, which ratified decisions made by the JMP by a majority vote (Zapata 2005). Among the actions implemented are the zoning of the GMR, the exclusion of industrial fishing within the reserve, the designation of use rights to local fishermen, and the first annual fishing calendar.

Thus, EAF was adopted in the Galapagos management system through co-management and ecosystem-based spatial management to reduce conflicts between users over incompatible demands related to marine space (e.g., tourism vs. fishing, large-scale vs. small-scale fishing). Diminish the impact of human activities on sensitive ecological areas (critical areas for the functioning of the system and conservation of threatened species), contribute to fisheries sustainability, and protect the marine area (Castrejón and Charles 2013).

After 17 years of implementation of co-management in the Galapagos, the objectives were not achieved, so a modification of the LOREG in 2015 reduced citizen participation to a non-binding advisory level on issues related to the administration of the GMR (Consejo de Gobierno del Régimen Especial de Galápagos 2016). As a result, the National Environmental Authority makes the final decision in coordination with the competent technical entities of the sector and the Government Council of the Special Regime of Galapagos (Consejo de Gobierno del Régimen Especial de Galápagos, CGREG). Subsequently, with the establishment in 2014 of the Management Plan for the Protected Areas of Galapagos for Good Living, the DPNG began a new zoning process for the terrestrial and marine systems of the Galapagos (Dirección Parque Nacional Galápagos 2014). This process increased to 33% the percentage of marine conservation areas with the establishment of the Darwin and Wolf Marine Sanctuary in 2016.

Next, the GMR management system and the level of compliance with the principles and characteristics of the EAF are analyzed to provide actions that benefit its application.

Participatory Management System

One of the characteristics of the EAF is to be participatory at all levels of planning and execution (Dimech et al. 2015). Although this was accomplished in the Galapagos with the creation of the JMP, it did not operate efficiently and the stated objectives were not achieved. To understand why co-management has not been fully effective as in other countries (Defeo et al. 2016; Pérez-Ramírez et al. 2012; De la Cruz-González et al. 2018; Gutiérrez et al. 2011), it is worth indicating the conditions under which it was adopted.

Co-management arose under a conflictive climate in the Galapagos; there was great concern about population growth, tourism development, and the management of marine resources. There was intense social pressure to open the sea cucumber fishery (Zapata 2005). The co-management did not know how to resolve conflicts due to the constant requirement of technical studies to define each topic discussed. There was a greater interest in fishing issues. Much time, money, and effort were invested in creating a management plan that did not provide planned and efficiently evaluated measures, which generated various conflicts. Fishermen were only involved in decision-making, and mistrust was generated in the JMP due to the asymmetrical power among the users (Zapata 2005).

Although the co-management system did not generate the expected results, such as the optimum use of fishing resources, nor did it ensure the sustainable development of the fishing sector, various achievements were attained (Heylings and Bravo 2007). Among them are the Management Plan of the GMR, the agreed provisional zoning scheme, the Fishing Calendar, and the approval and development of a fishing-tourism activity called "pesca artesanal vivencial" (Zapata 2006; Castrejón 2011).

Management Actions

The EAF takes into consideration the ecological, socioeconomic, and political sectors, as well as external factors; it favors the use of the best available knowledge and the precautionary approach and promotes the adoption of an adaptive management system that supports decision-making in conditions of high uncertainty (Dimech et al. 2015; Long et al. 2015). Until 2006 there were weaknesses in the legal framework of the Fisheries Management Plan of the GMR, and there was no fishing policy agreed upon at the provincial level. The lack of efficiency of the management actions is attributed to weak governance that made short-term decisions in the face of crises. Weak governance is considered one of the leading causes of overexploitation of benthic resources in the Galapagos, as well as excessive fishing capacity, i.e., too many boats and fishermen (Castrejón 2011).

The GMR Fishing Calendar 2002–2006, established in 2002, did not have clear operational objectives by fishery. No management objective was included that specified what was expected to be achieved from the management of each type of fishery from a biological, economic, social, or governance point of view, so decision-makers did not have a guide to assess the success of the applied strategies (Zapata 2005). Furthermore, the measures were not precautionary since an action measure was defined for a period of validity instead of a review period without knowledge of the resource's current state or the fishery's operational feasibility. Fisheries were not effectively monitored, results were not evaluated, and therefore adaptive management was not fulfilled (Castrejón 2011). Various management strategies were implemented, such as moratoria, allocation of use rights, reduction of the fishing season, and transferable individual catch quota (adopted only once in 2001). However, their late application and inefficient planning failed to prevent overcapitalization and overexploitation of sea cucumber and spiny lobster (Castrejón and Charles 2013).

In response to these difficulties, the "Capítulo pesca" was created in 2009, a new fisheries management plan with clear operational objectives, fishery performance indicators, and strategies that were reviewed and adapted according to the available knowledge about the resource and the fishery. (Comisión Técnica Pesquera de la Junta de Manejo Participativo 2009). Subsequently, the Fishing Calendar 2016–2021 was implemented, whose objective was to ensure the management and sustainable development of fisheries and the socioeconomic well-being of the fishing sector (Dirección del Parque Nacional Galápagos 2016). Currently, the DPNG, the Ministry of Production, Foreign Trade, Investments and Fisheries, the CGREG, the Artisanal Fishing Sector, and allied organizations analyze the technical-scientific information available on fishing resources for the construction of the Fishing Calendar 2022–2027.

Zoning

The zoning of the GMR is aimed at the persistence and prosperity of endemic or threatened species and the maintenance of processes and services in the ecosystem (Heylings et al. 2002). However, 22 years after its implementation, the objectives have not been achieved since there is overexploitation of the primary fishing resources (Usseglio et al. 2016; Buglass et al. 2018; Ramírez-González et al. 2020), and it does not provide sufficient protection to species threatened and key areas for the functioning of the ecosystem (Edgar et al. 2008, Moity 2018). Furthermore, there is non-compliance with the measures implemented as they are considered illegitimate (Viteri and Chávez 2007), and conflicts between the tourism and fishing sectors continue (Castrejón and Charles 2013).

The lack of legitimacy and reliability in the 2000 zoning scheme exists because its design did not have a comprehensive long-term approach, but rather an attempt was made to minimize the short-term impacts of zoning on the activities of interest in each sector. It was a sociopolitical process where each sector proposed, defended, and negotiated their zoning proposals according to their interests (Bustamante et al. 2005; Viteri and Chávez 2007). The consensus was achieved through incentives and pressure. For example, commercial diving and sport fishing licenses were allocated to fishermen who decided to change their fishing activity and become tour operators. Legal sea cucumber fishing motivated fishermen from mainland Ecuador to migrate to the Galapagos to obtain fishing licenses and alternative benefits (Heylings et al. 2002). As a result, the fishing sector increased by 55% from 1999 to 2000 and intensified the "race for the fish." The application of inadequate incentives did not allow the sea cucumber and spiny lobster fisheries to be conserved (Castrejón 2011). The population decrease of these resources affected the economic income of fishermen and increased illegal fishing and conflicts over access to resources among users. The fishermen looked for work alternatives as tour operators and in experiential artisanal fishing, and the catches diversified to other species (e.g., large pelagic fish) of the white finfish fishery (Castrejón 2011).

The zoning design was done without solid scientific knowledge and with an incorrect use designation (Bustamante et al. 2005). For example, the spatially heterogeneous distribution patterns of sea cucumber and lobster were not considered (Hearn et al. 2006; Toral et al. 2005). Furthermore, the physical delimitation of the management zones is considered inadequate as it is not based on a geographic coordinate system, which is relevant for fisheries that occur at night (Castrejón and Charles 2013). There was not enough information on coastal marine biodiversity; therefore, the designation of non-use areas was based on expert judgment due to limited scientific knowledge and the areas traditionally used for tourism (Bustamante et al. 2005). Thus, approximately 71% of the key biodiversity sites are protected from fishing in the GMR (Edgar et al. 2008). Regarding the 2016 zoning, the inadequate inclusion of social, economic, and political parameters continued in the planning of this process, the lack of consensus and effective communication with the affected sectors, and the feeling of illegitimacy on the part of the fishing sector (Burbano et al. 2020).

How to Benefit the Application of the EAF in the GMR Management System?

There is extensive information concerning the fisheries in the GMR, and work is being carried out that includes social and economic aspects (Castrejón et al. 2014); however, these must be integrated in an understandable way for decision-makers. Even though studies have shown the effects that fishing generates on the structure and functioning of marine ecosystems in the Galapagos (Okey et al. 2004; Ruiz et al. 2016; Riofrío-Lazo et al. 2021), the fishing system management follows the conventional target resource-based approach to fisheries and not the ecosystem-based perspective. The management measures used for each fishery are evaluated based on the resource state and the socioeconomic situation of the fishery without considering the impacts of fisheries on the structure, functioning, organization, and resilience of the ecosystem. Therefore, to benefit the application of the EAF in the GMR, we recommend:

- Widely recognize the usefulness of the EAF for achieving resource sustainability and ecosystem conservation by the authority responsible for management. Thus, the EAF principles will be incorporated within a political framework of appropriate laws and practices to control human activities.
- Involves stakeholders from the beginning of the planning process to establish intersectoral operational agreements and form a multidisciplinary team. In this manner, the credibility of the management system efficiency will be improved among the users of the GMR.
- Generate a long-term commitment. With the application of the EAF, the benefits are achieved in the long term, so the Ecuadorian government must commit to support and finance the strategies implemented. These must be designed with a comprehensive long-term approach that reflects the interest of the authorities and users.
- Establish clear fisheries management strategies for their implementation and the monitoring and evaluating of their performance.
- Incorporate in the fisheries management plan information on species affected by incidental catch (cetaceans, birds, turtles, among others) and simultaneously apply mitigation measures for their conservation and maintenance.
- Use adequate incentives for the fishing sector. These should target active full-time fishermen to reduce economic dependence on fishing and, consequently, the fishing effort.
- Use scientific information and local fisheries knowledge so that management decisions are preventive. Before its adoption, the conditions of the socio-ecological system should be evaluated and implemented appropriately to avoid poor performance of the fishery.
- Develop a comprehensive analysis of the ecosystem. The scientific sector should be encouraged to develop studies that assess the impacts of fisheries on the structure and functioning of the ecosystem and the impacts of environmental variability on fisheries and the ecosystem.

- Constantly monitor the abundance and distribution of biological resources and the ecosystem attributes. Thus, changes associated with fishing activity or environmental variability can be determined.
- Continuous evaluation of the zoning scheme at least every 5 years. So, a better delimitation of non-use or conservation zones, and extractive use zones can be done according to the best available knowledge.
- Fisheries management under a holistic approach. In this perspective, the aim is first to maintain the sustainability of the ecosystem and then to establish specific management actions for individual resources. A periodic ecosystem monitoring system must be maintained to evaluate the implemented decisions and make the necessary adjustments, i.e., adopt the concept of adaptability in management.

Acknowledgments We acknowledge the financial support received from the Secretaría de Educación Superior, Ciencia, Tecnología e Innovación SENESCYT (Ecuador); Universidad San Francisco de Quito, and the Galapagos Science Center. We also thank the Dirección Parque Nacional Galápagos for the research permit PC-64-22 and technical support.

References

Agnew DJ, Pearce J, Pramod G et al (2009) Estimating the worldwide extent of illegal fishing. PLoS One 4:e4570. https://doi.org/10.1371/journal.pone.0004570.g001

Arkema KK, Abramson SC, Dewsbury BM (2006) Marine ecosystem-based management: from characterization to implementation. Front Ecol Environ 4:525–532

Arreguín-Sánchez F (2022) Conventional fisheries management and the need for an ecosystem approach. In: Arreguín-Sánchez F (ed) Holistic approach to ecosystem-based fisheries management. Springer, Cham, pp 17–29. https://doi.org/10.1007/978-3-030-96847-2

Arreguín-Sánchez F, del Monte-Luna P, Zetina-Rejón M (2015) Climate change effects on aquatic ecosystems and the challenge for fishery management: pink shrimp of the southern Gulf of Mexico. Fisheries 40(1):15–19. https://doi.org/10.1080/03632415.2015.988075

Arreguín-Sánchez F, Del Monte-Luna P, Albañez-Lucero MO et al (2021) How much biomass must remain in the sea after fishing to preserve ecosystem functioning? The case of the sardine fishery in the Gulf of California, Mexico. In: Ortiz M, Jordán F (eds) Marine coastal ecosystems modelling and conservation. Springer, Cham, pp 143–161

Bascompte J, Melián CJ, Sala E (2005) Interaction strength combinations and the overfishing of a marine food web. Proc Natl Acad Sci USA 102:5443–5447

Bastardie F, Brown EJ, Andonegi E et al (2021) A review characterizing 25 ecosystem challenges to be addressed by an ecosystem approach to fisheries management in Europe. Front Mar Sci 7:629186. https://doi.org/10.3389/fmars.2020.629186

Bucaram SJ, White JW, Sanchirico JN et al (2013) Behavior of the Galapagos fishing fleet and its consequences for the design of spatial management alternatives for the red spiny lobster fishery. Ocean Coastal Manag 78:88–100

Buglass S, Reyes H, Ramírez-González J et al (2018) Evaluating the effectiveness of coastal no-take zones of the Galapagos Marine Reserve for the red spiny lobster, *Panulirus penicillatus*. Mar Policy 88:204–212

Burbano DV, Meredith TC, Mulrennan ME (2020) Exclusionary decision-making processes in marine governance: the rezoning plan for the protected areas of the 'iconic' Galapagos Islands. Ecuador Ocean Coastal Manag 185:105066

Bustamante R, Reck G, Ruttenberg B et al (2000) The Galapagos spiny lobster fishery. In: Phillips B, Kittaka J (eds) Spiny lobsters: fisheries and culture. Fishing News Book, Oxford, pp 210–222
Bustamante R, Orensanz JM, Parma AM et al (2005) Scientific support for fisheries management in the Galapagos Marine Reserve: towards the development of robust conservation management strategies. Narrative report of the first year. Pew Fellows Program in Marine Conservation, Santa Cruz, Galapagos
Castrejón M (2011) Co-manejo pesquero en la Reserva Marina de Galápagos: tendencias, retos y perspectivas de cambio. México: Fundación Charles Darwin/Kanankil/Plaza Valdés, Mexico
Castrejón M, Charles A (2013) Improving fisheries co-management through ecosystem-based spatial management: the Galapagos Marine Reserve. Mar Policy 38:235–245
Castrejón M, Defeo O, Reck G et al (2014) Fishery science in Galapagos: from a resource-focused to a social–ecological systems approach. In: Denkinger J, Vinueza L (eds) The Galapagos marine reserve, social and ecological interactions in the Galapagos Islands. Springer, New York, pp 159–185
Coll M, Libralato S, Tudela S et al (2008) Ecosystem overfishing in the ocean. PLoS One 3:e3881
Coll M, Shannon LJ, Kleisner KM et al (2016) Ecological indicators to capture the effects of fishing on biodiversity and conservation status of marine ecosystems. Ecol Indic 60:947–962
Comisión Técnica Pesquera de la Junta de Manejo Participativo (2009) Capítulo Pesca del Plan de Manejo de la Reserva Marina de Galápagos. Galapagos Islands
Consejo de Gobierno del Régimen Especial de Galápagos (2016) Plan de Desarrollo Sustentable y Ordenamiento Territorial del Régimen Especial de Galápagos. –Plan Galápagos. Puerto Baquerizo Moreno, Galapagos
Cury P, Pauly D (2020) Global marine fisheries: avoiding further collapses. In: Henry C, Rockström J, Stern N (eds) Standing up for sustainable world. Edward Elgar Publishing, Cheltenham, pp 384–395
Day J (2008) The need and practice of monitoring, evaluating and adapting marine planning and management—lessons from the great barrier reef. Mar Policy 32:823–831
De la Cruz-González FJ, Patiño-Valencia JL, Luna-Raya MC et al (2018) Self-empowerment and successful co-management in an artisanal fishing community: Santa Cruz de Miramar, Mexico. Ocean Coastal Manag 154:96–102
De Miras C, Andrade M, Carranza C (1996) Evaluación socioeconómica de la pesca experimental de pepino de mar en Galápagos. Fundación Charles Darwin y ORSTOM, Santa Cruz, Galapagos
De Young C, Charles A, Hjort A (2008) Human dimensions of the ecosystem approach to fisheries: an overview of context, tools and methods. Food and Agriculture Organization of the United Nations, FAO, Rome
Defeo O, Castrejón M, Pérez-Castañeda R et al (2016) Co-management in Latin American small-scale shellfisheries: assessment from long-term case studies. Fish Fish 17:176–192
Dımech M, Barros P, Bianchi G (2015) What is the ecosystem approach to fisheries management?. J Acad Documents Fish Aquacult 1(3):137–143. Available via DergiPark. https://dergipark.org.tr/en/pub/jadfa/issue/9804/121517. Accessed 1 Sep 2022
Dirección del Parque Nacional Galápagos (2014) Plan de Manejo de las Áreas Protegidas de Galápagos para el Buen Vivir. Puerto Ayora, Galapagos
Dirección del Parque Nacional Galápagos (2016) Estudio Técnico de Calendario Pesquero 2016–2021. Comisión Técnica Pesquera y Sector Pesquero Artesanal de Galápagos. Puerto Ayora, Galapagos
Dobbs K, Day J, Skeat H et al (2011) Developing a long-term outlook for the great barrier reef, Australia: a framework for adaptive management reporting underpinning an ecosystem-based management approach. Mar Policy 35:233–240
Douvere F (2008) The importance of marine spatial planning in advancing ecosystem-based sea use management. Mar Policy 32:762–771
Douvere F, Ehler C (2009) Ecosystem-based marine spatial management: an evolving paradigm for the management of coastal and marine places. Ocean Yearbook 23:1–26

Edgar GJ, Banks S, Bensted-Smith R et al (2008) Conservation of threatened species in the Galapagos marine reserve through identification and protection of marine key biodiversity areas. Aquat Conserv Mar Freshwat Ecosyst 18(6):955–968

FAO (1995) Code of conduct for responsible fisheries. FAO, Rome

FAO (2003) The ecosystem approach to marine capture fisheries. FAO technical guidelines for responsible fisheries, No. 4, Suppl. 2. Rome, p 112

FAO (2009) Fisheries management. 2. The ecosystem approach to fisheries. 2.2 Human dimensions of the ecosystem approach to fisheries. FAO technical guidelines for responsible fisheries. No. 4, Suppl. 2, Add. 2. Rome, p 88

FAO (2020) El estado mundial de la pesca y la acuicultura 2020. La sostenibilidad en acción. FAO, Roma. https://doi.org/10.4060/ca9229es

Fariñas-Franco JM, Allcock AL, Roberts D (2018) Protection alone may not promote natural recovery of biogenic habitats of high biodiversity damaged by mobile fishing gears. Mar Environ Res 135:18–28

Garcia SM, Zerbi A, Aliaume C et al (2003) The ecosystem approach to fisheries. Issues, terminology, principles, institutional foundations, implementation and outlook. FAO fisheries technical paper, No. 443. Rome

Gutiérrez NL, Hilborn R, Defeo O (2011) Leadership, social capital and incentives promote successful fisheries. Nature 470:386–389. https://doi.org/10.1038/nature09689

Hearn A, Murillo JC, Nicolaides F et al (2006) Evaluación de la pesquería de langosta espinosa (*Panulirus penicillatus* y *P. gracilis*) en la Reserva Marina de Galápagos 2005. In: Hearn A (ed) Evaluación de las pesquerías en la Reserva Marina de Galápagos, Informe Compendio 2005. Quitor, Fundación Charles Darwin, pp 46–116

Heylings P, Bravo M (2007) Evaluating governance: a process for understanding how co-management is functioning, and why, in the Galapagos Marine Reserve. Ocean Coastal Manag 50:174–208. https://doi.org/10.1016/j.ocecoaman.2006.09.003

Heylings P, Bensted-Smith R, Altamirano M (2002) Zonificación e historia de la reserva marina de galápagos. In: Danulat E, Edgar GJ (eds) Reserva Marina de Galápagos. Línea Base de la Biodiversidad. Fundación Charles Darwin y Servicio Parque Nacional Galápagos, Santa Cruz, pp 10–21

Holling CS (ed) (1978) Adaptive environmental assessment and management. No. 333.7 A221. Wiley international series on applied systems analysis, Chichester

Long RD, Charles A, Stephenson RL (2015) Key principles of marine ecosystem-based management. Mar Policy 57:53–60. https://doi.org/10.1016/j.marpol.2015.01.013

Lotze HK, Coll M, Dunne JA (2011) Historical changes in marine resources, food-web structure and ecosystem functioning in the Adriatic Sea, Mediterranean. Ecosystems 14:198–222

Martins-Vieira ML, Alexandre-de-Lima CL, Botelho-de-Souza JR et al (2020) Effects of beach seine fishing on the biodiversity of seagrass fish assemblages. Reg Stud Mar Sci 40:101527

Moity N (2018) Evaluation of no-take zones in the Galapagos marine reserve, zoning plan 2000. Frontiers 5:244. https://doi.org/10.3389/fmars.2018.00244

Murawski SA (2000) Definitions of overfishing from an ecosystem perspective. ICES J Mar Sci 57:649–658

Okey TA, Banks S, Born AF et al (2004) A trophic model of a Galápagos subtidal rocky reef for evaluating fisheries and conservation strategies. Ecol Model 172:383–401

Pacoureau N, Rigby CL, Kyne PM et al (2021) Half a century of global decline in oceanic sharks and rays. Nature 589:567–571. https://doi.org/10.1038/s41586-020-03173-9

Pauly D, Christensen V, Dalsgaard J et al (1998) Fishing down marine food webs. Science 279:860–863

Pérez-Ramírez M, Ponce-Díaz G, Lluch-Cota S (2012) The role of MSC certification in the empowerment of fishing cooperatives in Mexico: the case of red rock lobster co-managed fishery. Ocean Coastal Manag 63:24–29. https://doi.org/10.1016/j.ocecoaman.2012.03.009

Pitcher TJ, Watson R, Forrest R et al (2002) Estimating illegal and unreported catches from marine ecosystems: a basis for change. Fish Fish 3:317–339

Porobic J, Fulton EA, Parada C et al (2019) The impact of fishing on a highly vulnerable ecosystem, the case of Juan Fernández ridge ecosystem. PLoS One 14(2):e0212485

Ragheb E, Akel EHK, Rizkalla SI (2019) Analyses of the non-target catch from the Egyptian Mediterranean trawlers, off Port Said. Egypt J Aquat Res 45:239–246

Ramírez-González J, Moity N, Andrade-Vera S et al (2020) Overexploitation and more than a decade of failed management leads to no recovery of the Galapagos Sea cucumber fishery. Front Mar Sci 7:554314

Reck G (1983) The coastal fisheries in the Galapagos Islands, Ecuador. Description and consequences for management in the context of marine environmental protection and regional development. Dissertation, The Christian-Albrecht University of Kiel

Reyes H, Ramírez J, Schuhbauer A (2013) Evaluación de la pesquería de pepino de mar en la Reserva Marina de Galápagos. In: DPNG, GCREG, FCD y GC (eds) Informe Galápagos 2011–2012. Puerto Ayora, Galapagos, pp 144–149

Rijnsdorp AD, Bolam SG, Garcia C et al (2018) Estimating sensitivity of seabed habitats to disturbance by bottom trawling based on the longevity of benthic fauna. Ecol Appl 28(5):1302–1312

Riofrío-Lazo M (2018). Manejo basado en el ecosistema de la región sureste del archipiélago de Galápagos: Sostenibilidad y adaptabilidad. Dissertation, Instituto Politécnico Nacional

Riofrío-Lazo M, Reck G, Páez-Rosas D et al (2021) Food web modeling of the southeastern Galapagos shelf ecosystem. Ecol Indic 132:108270. https://doi.org/10.1016/j.ecolind.2021.108270

Ruiz DJ, Banks S, Wolff M (2016) Elucidating fishing effects in a large-predator dominated system: the case of Darwin and Wolf Islands (Galapagos). J Sea Res 107:1–11. https://doi.org/10.1016/j.seares.2015.11.001

Salcido-Guevara LA, Del Monte-Luna P, Arreguín-Sánchez F et al (2012) Potential ecosystem level effects of a shrimp trawling fishery in La Paz Bay, Mexico. Open Jf Mar Sci 2:85–89

Scheffer M, Carpenter S, Foley JA et al (2001) Catastrophic shifts in ecosystems. Nature 413:591–596

Scheffer M, Carpenter S, de Young B (2005) Cascading effects of overfishing marine systems. Trends Ecol Evol 20:579–581

Schiller L, Alava JJ, Grove J et al (2014) The demise of Darwin's fishes: evidence of fishing down and illegal shark finning in the Galapagos Islands. Aquat Conserv Mar Freshwat Ecosyst 25(3):431–446

Staples D, Brainard R, Capezzuoli S et al (2014) Essential EAFM. Ecosystem approach to fisheries management training course. Volume 1 – for trainees. FAO Regional Office for Asia and the Pacific, RAP Publication, Bangkok

Takeshige A, Miyamoto M, Narimatsu Y et al (2021) Evaluation of impacts of bottom fishing on demersal habitats: a case study off the Pacific coast of North-Eastern Japan. Fish Res 238:105916

Toral V, Murillo JC, Piu M et al (2006) La pesquería de pepino de mar (*Isostichopus fuscus*) en la Reserva Marina de Galápagos en el año 2005. In: Hearn A (ed) Evaluación de las pesquerías en la Reserva Marina de Galápagos, Informe Compendio 2005. Fundación Charles Darwin, Quito, pp 6–45

Usseglio P, Friedlander AM, Koike H et al (2016) So long and thanks for all the fish: overexploitation of the regionally endemic Galapagos Grouper *Mycteroperca olfax* (Jenyns, 1840). PLoS One 11(10):e0165167. https://doi.org/10.1371/journal.pone.0165167

Viteri C, Chávez C (2007) Legitimacy, local participation and compliance in the Galapagos marine reserve. Ocean Coastal Manag 50:253–274

Ward T, Tarte D, Hegerl E et al (2002) Policy proposals and operational guidance for ecosystem based management of marine capture fisheries. WWF Australia, Sydney

Zapata C (2005) Entre el conflicto y la colaboración: El manejo participativo en la Reserva Marina de Galápagos. Sistematización, evaluación y factores de éxito del modelo participativo. Fundación para el Desarrollo Alternativo Responsable para Galápagos (FUNDAR Galápagos), Puerto Ayora, Galapagos

Zapata F (2006) La pesca artesanal vivencial en Galápagos. Sistematización del proceso. Fundación Futuro Latinoamericano y Agencia de los Estados Unidos para el Desarrollo Internacional (USAID), Quito
Zhou S, Smith ADM, Fuller M (2011) Quantitative ecological risk assessment for fishing effects on diverse data-poor non-target species in a multi-sector and multi-gear fishery. Fish Res 112:168–178

Marjorie Riofrío-Lazo focuses her research the ecology modelling of populations and marine ecosystem dynamics. She focuses on generating information that allows exploring effects of environmental variability and designing management strategies related to conservation, fishing activity and climate change. Her research also addresses a variety of issues of management concern including the functional role of species at ecosystems and the interaction between introduced and endemic fauna on Galapagos Islands.

Manuel J. Zetina-Rejón is a Doctor of Marine Sciences from the Instituto Politécnico Nacional (Mexico). For more than two decades, he has focused on the dynamics of interrelationships between marine species and fisheries dynamics through network modeling. Using this experience and encouraging his graduate students, in recent years, he has focused on social-ecological vulnerability and resilience analysis.

Gunter Reck is Co-Director of the USFQ Institute for Applied Ecology. He is interested in marine-coastal conservation and the Galapagos Islands, the development of the Visitor Management System in protected areas (APs), and supporting and sustaining a greater use of recreational, environmental services that can be provided in the APs. He conducts research in various fields, including marine conservation and tourism.

Diego Páez-Rosas is interested in marine megafauna's ecological aspects and the interface between trophic flexibility and environmental variability. He combines laboratory and fieldwork methods to track the foraging ecology of Galapagos marine predators and relate their feeding habits, home range, and trophic behavior to their breeding success and energy budgets. His researches also have ramifications in population dynamics, health status, and conservation management.

Francisco Arreguín-Sánchez is a Full Professor at the IPN. He focuses on the Dynamics of Ecosystems and Exploited Populations, Climate Change, and Fisheries Management. He is an Emeritus Member of the National System of Researchers and the Mexican Academy of Sciences and collaborates as a consultant with international agencies. He has more than 300 scientific publications, has directed more than 60 postgraduate theses, and has taught more than 130 courses.

Chapter 22
Cetaceans of the Galapagos Archipelago: Species in Constant Change and the Importance of a Standardized and Long-Term Citizen Science Program

Daniela Alarcón-Ruales, Judith Denkinger, Leo Zurita-Arthos, Salome Herrera, Santiago Díaz-Pazmiño, Eduardo Espinoza, Juan Pablo Muñoz Pérez, Bonnie J. Holmes, and Kathy A. Townsend

Introduction

Cetaceans are widely distributed globally and established in a range of different habitats, including, estuarine and coastal environments, offshore oceanic waters, and deep seas (Jefferson et al. 2015; Plagányi and Butterworth 2009). These animals play key roles in the function of the marine environment, are sentinels on the health of the ecosystems, and provide multiple benefits to humans (Bowen et al. 1992;

D. Alarcón-Ruales (✉)
School of Science, Technology & Engineering, University of the Sunshine Coast, Sippy Downs, QLD, Australia

Colegio de Ciencias Biológicas y Ambientales, Universidad San Francisco de Quito, Puerto Baquerizo Moreno, Galápagos, Ecuador

Galápagos Science Center GSC, Puerto Baquerizo Moreno, Galápagos, Ecuador
e-mail: dealarcon@usfq.edu.ec

J. Denkinger · L. Zurita-Arthos
Colegio de Ciencias Biológicas y Ambientales, Universidad San Francisco de Quito, Puerto Baquerizo Moreno, Galápagos, Ecuador

Galápagos Science Center GSC, Puerto Baquerizo Moreno, Galápagos, Ecuador

S. Herrera
Colegio de Ciencias Biológicas y Ambientales, Universidad San Francisco de Quito, Puerto Baquerizo Moreno, Galápagos, Ecuador

S. Díaz-Pazmiño
Animal Behavior Lab, University of Sydney, Sydney, NSW, Australia

E. Espinoza
Dirección del Parque Nacional Galápagos, Puerto Ayora, Galápagos, Ecuador

S. J. Walsh et al. (eds.), *Island Ecosystems*, Social and Ecological Interactions in the Galapagos Islands, https://doi.org/10.1007/978-3-031-28089-4_22

Kiszka et al. 2021; Schaeffer et al. 2008). Nevertheless, knowledge and proper data on some cetacean species are still remarkably poor.

Understanding of diversity, demography, and population structure information requires the use of a variety of methodologies. Using sporadic sightings and citizen science reports can provide unique insights, and the reports may be used as a first step in assessing demographic data, thus aiding in the construction of baseline species list. These data could then be used to construct measures of population status and seasonal variation in abundance (Evans and Hammond 2004). In addition, sightings reports may significantly contribute to the understanding of cetacean species worldwide. This is especially true in developing countries where substantial funding is generally difficult to obtain, such as, in Galapagos islands (Denkinger et al. 2013; Garcia-Cegarra et al. 2021; Lodi and Tardin 2018; Mwango'mbe et al. 2021).

The bathymetric characteristics including coastal, oceanic, and deep waters around the Galapagos Islands provide ideal environmental conditions for exceptional biodiversity for both resident and migratory cetacean species (Day 1994; Palacios and Salazar 2002). Large-scale monitoring in the area undertaken by various studies on cetaceans within the Galapagos also indicates there is a greater diversity of cetaceans in relation to other areas of the Eastern Tropical Pacific (ETP), which makes it a global hotspot for marine mammal diversity (Alava 2009; Denkinger et al. 2013; Ferguson et al. 2005; Kaschner et al. 2011; Smith and Whitehead 1999). Recorded observations show that the densest populations of cetaceans are near the coasts of the larger islands or in high-productivity zones, especially in the west and southwest areas of the Galapagos (Denkinger et al. 2013). Bottlenose dolphins (*Tursiops truncatus*) and Bryde's whales (*Balaenoptera edeni*) are the most commonly sighted species near coastal areas around the islands, while the common dolphin (*Delphinus delphis*) is the most abundant species in the region overall (Merlen 1995; Palacios 1999; Smith and Whitehead 1999; Wade and Gerrodette 1993). Other species such as humpback whales (*Megaptera novaeangliae*) or fin whales (*Balaenoptera physalus*) show a seasonal incidence, with higher numbers of sightings during the colder months of June to November (Félix et al. 2020; Palacios and Salazar 2002). Conversely, blue whales (*Balaenoptera musculus*), killer whales (*Orcinus orca*), and sperm whales (*Physeter macrocephalus*), among other species, maintain a year-round presence around the islands (Denkinger et al. 2020; Merlen 1999; Reilly and Thayer 1990; Whitehead 1999).

Due to a greater diversity of cetacean species in the ETP, the region was ideal for whale hunting during the nineteenth and twentieth centuries, contributing to the depletion of some populations (Clapham et al. 1999). A moratorium banned whaling in 1961, however, population numbers for many species, such as the blue whale, were already heavily affected. During the same year, small cetacean bycatch

J. P. Muñoz Pérez
College of Biological and Environmental Sciences and Galapagos Science Center, Universidad San Francisco de Quito, Quito, Ecuador

School of Science, Technology & Engineering, University of the Sunshine Coast, Sippy Downs, Australia

B. J. Holmes · K. A. Townsend
School of Science, Technology & Engineering, University of the Sunshine Coast, Sippy Downs, QLD, Australia

increased due to the development of a new tuna purse-seine fishing technique in the ETP. Several hundred thousand dolphins died in tuna purse-seine nets each year, reducing the populations of dolphins of some species like spotted and spinner dolphins (*Stenela* sp.) in the ETP by 70–80% (Edwards 2007; Wade 1994).

As a result of these anthropogenic impacts on many cetacean species in the ETP, the 200 nautical miles surrounding the Galapagos archipelago and continental waters were declared a marine mammal sanctuary in 1991 (Evans 1991). This meant that hunting for cetacean species was banned. In 1998, the Galapagos Marine Reserve (GMR) was further established, encompassing an additional 40 nautical miles around the islands. When first established, the GMR was one of the biggest marine reserves in the world (Anderson et al. 2003) and provided relief from fishing pressure around the islands. These reserves have proven to successfully protect resident coastal species, however, migratory cetaceans are still subject to fishing pressure outside of these zones. In sight of this, a marine migratory corridor for the ETP was created in 2022, using data collected from the migratory pathways of various species including sharks and turtles. The corridor connects the Galapagos-Cocos ridge on the Ecuadorian side, while expanding the protected area in some strategic regions around the GMR by 30,000 km^2 (Alava et al. 2022; Reck 2014). The main objective of these conservation areas is to protect marine environments as well as the migratory animals within the ETP. Currently, there is strong scientific support and interest to create marine corridors protecting large baleen species, identifying the main threats to these animals and possible mitigation strategies (Johnson et al. 2022). The most common threats for cetaceans currently include maritime traffic (Arbelo et al. 2013; Peltier et al. 2019), pollution (Lusher et al. 2018; Nabi et al. 2018; Tanabe et al. 1983), habitat loss (Laidre et al. 2015), climate variability and climate change (Albouy et al. 2020; Learmonth et al. 2006), which is expected to include greater and more unpredictable severity of El Niño-La Niña Southern Oscillation events and changes in upwelling, nutrients and food availability (Kislik et al. 2017; Sachs and Ladd 2010).

When the Galapagos Marine Reserve was created in 1998, a baseline document about cetacean species in the Galapagos was produced by Palacios and Salazar (2002). This report used data from occasional sightings and research efforts around the islands, collected over a 30-year period, from 1973 to 2000. They reported a diversity of at least 26 different species, which had previously been reported by Day (1994). Then, Denkinger et al. (2013) presented a data analysis over an 18-year period, 1993 to 2010, consisting of reports from naturalistic guides in the Galapagos waters. These results show a variation of species over the years and a clear impact of oceanographic conditions – decrease of species and sightings during El Niño events with warmer conditions and low productivity, followed by an increase during colder conditions of La Niña. The study also shows a shift and reduction of species diversity during the last 10 years of the analysis. Furthermore, Denkinger et al. (2013) and Denkinger et al. (2020) reported possible resident patterns for some species such as orcas, bryde whales and bottlenose dolphins based on these reports. However, specific data on population numbers, individual identification registers, genetic and spatial connectivity are limited to just four species – sperm whales, blue whales, humpback whales and orcas- found at the marine reserve (Cantor et al. 2016; Denkinger et al. 2020; Eguiguren et al. 2021; Félix et al. 2020; LeDuc et al. 2017).

During the past decade, tourism in the Galapagos has continuously grown with the exception of pandemic years (Alava et al. 2022). Similarly, citizen effort and data collected has increased in a parallel manner. Consequently, this project aims to update information on species diversity and temporality over the last decade. With the aim to analyze species diversity, temporality and map the distribution patterns of the most sighted cetaceans in Galapagos using citizen science and scientific cruises that report occasional encounters of species around the Galapagos collected by the CGP from the last 12 years (2010–2022) and review the significance of citizen science contributions.

Methods

We will be concentrating on three questions:

1. What species of cetaceans have been reported in Galapagos in the last 12 years?
2. Is there a seasonality for the species reported in the archipelago?
3. What is the relevance of citizen science contribution to the study of cetaceans in the Galapagos?

Study Area

Composed of a group of 13 main islands and more than 100 islets and rocks, the Galapagos Archipelago is located in the Eastern Tropical Pacific (ETP) region, in the equatorial sector of the Pacific Ocean, approximately 1000-km west of the coast of Ecuador, South America, between 01 40' North Latitude – 01 25′ South Latitude and −89 15′ West Longitude −92 00′ West Longitude.

The set of islands within the Galapagos archipelago represent the tops of volcanoes that constitute a relatively shallow shelf (<200 m), but are surrounded by deep water – +1000 – 4000 m (Bustamante et al. 2000; Snell et al. 1996). The region's characteristics include shallower thermocline and productive waters (Ballance et al. 2006). The equatorial and Costa Rica Dome upwellings are among the most productive parts of the pelagic ETP 12/21/22 1:09:00 PM (Love et al. 1972) delimited by (Wyrtki 1966), within 25′ North Latitude, 10′ South Latitude, the American continents and 13O′ West Longitude (Fig. 22.1).

Oceanographic studies show that despite the Galapagos having been described as a high-nutrient, low chlorophyll (HNLC) area (Palacios 2003), the Galapagos supports a heterogeneous environment represented by high biological productivity (primary and secondary productivity). This is due to upwelling by oceanographic and physical conditions, feasible for the establishment of typical communities of the tropics and equatorial water masses of both coastal and pelagic habitats (Alava 2009). However, these environments are affected by cyclical oceanographic

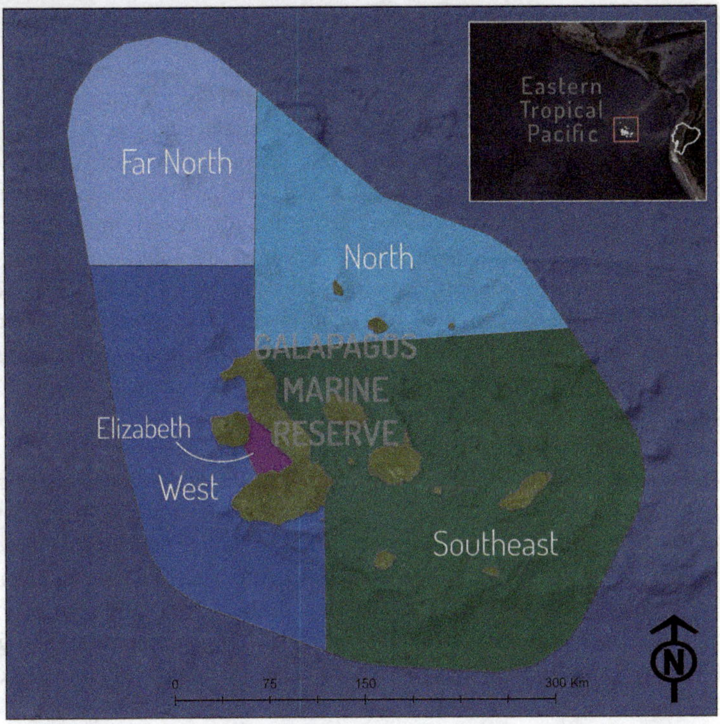

Fig. 22.1 The Galapagos Marine Reserve, within the five bioregions: Far North, North, Southeast, West, Elizabeth. Plus, information on location on the Eastern Tropical Pacific (ETP)

changing conditions such as El Niño and La Niña (Sachs and Ladd 2010). Productivity tends to be higher in the west and southwest region of the archipelago. This is a result of the island mass effect (increased chlorophyll), the proximity to the Equatorial Front, the active volcanic activity and the congregation of main oceanic currents, which show a marked seasonality in terms of their intensity and direction (Kislik et al. 2017; Palacios 2003).

There are various currents dominating nutrient flow and temperature in the Archipelago. The South Equatorial Current, Humboldt Current, that dominates in the Garua season (from May to November), carries cold and productive waters to the ETP. Meanwhile, the Panama Current, an extension of the north-equatorial counter-current carries warm waters in the wet season (December–June) (Banks 2002). From the west, the Equatorial undercurrent (EUC) or Cromwell Current that moves west to east flows deep along the equator and surges as it strikes the Galapagos platform, supplying phytoplankton to the region. This brings nutrients such as iron, nitrogen, phosphorus, and silicate providing upwelling conditions associated with high productivity throughout the year (Dickson 2006; Pennington et al. 2006). This current has been attributed to the high productivity observed in the Elizabeth and West bioregions, in conjunction with localized topographic upwelling, and iron provided by coastal sediments (Alava 2009).

Occasional Sightings Contributions

Reports on cetacean encounters have been collected by various sources of the Cetacea Galapagos Program (CGP). This includes citizen science, as well as contributions made by trained nature guides from the Galapagos National Park, who navigate year-round on the waters close to the coastal area but also traverse between the main islands. Additional data includes contributions by researchers searching for cetaceans on the islands, participation by fishers working with the program, inputs from the local community – notably, students and other researchers not working on cetaceans and information collected on social media. The latter were reported on Instagram and *YouTube* and were made by nature guides, tourists, local agencies, and the community. Sightings were also reported from users and followers of the Cetacea Galapagos Program Instagram account.

For all data, species identifications were always confirmed by the authors and users were contacted for more detailed information. Data from occasional sightings over a 12-year period – 2010 to 2022 – were analyzed to search for trends on presence and distribution of the selected species and to update species diversity information presented by Denkinger et al. (2013). All entries used for the analysis were characterized by information about the species, date of observation, georeferenced location, and island. Number of individuals, behavior, observer, or photographic register was collected in most of the cases. Data locations were extracted and mapped using the ArcGIS program.

The data were further analyzed to summarize and describe information on citizen science contributions, species diversity, and seasonality for the most common species, which were classified as those that accumulate more than 20 sightings. In addition, data for average annual temperature were obtained from NOAA to assess whether changes in the temperature conditions affected species diversity and abundance.

Furthermore, we compared our data to baseline analysis of Cetaceans in the Galapagos by Palacios and Salazar (2002) and the species diversity analysis by Denkinger et al. (2013) that uses similar methodologies in reporting sightings information on the Galapagos.

Results

total of 2560 sightings were collected by 268 different observers, with an average of n = 199 (+/- 112) observations per year. From those sightings 73% (n = 1869) were obtained by citizen science reports from different sectors including nature guides, tourists, fishers, students, and the local community. Of all reports, 29% (n = 547) were reports from the social media platforms Instagram and YouTube (Fig. 22.2). For all observations the 22 top observers are responsible for 50% (n = 1282) of all the sightings.

In 2020 only 60 observations were recorded due to the decrease of tourism and the limitations placed on activities caused by the COVID-19 pandemic. Most sightings were reported in 2011 with 450 observations, due to a research effort of seven consecutive months of observations in the west area of the Galapagos in the Bolivar Channel, with 95% (n = 430) of sightings collected by the author during that year (Fig. 22.3).

Observations were more abundant during the cold season (n = 1475) from June to November and less sightings (n = 1080) were made during the warm season months from December to May. Similarly, more frequent encounters (higher number of sightings) were reported during La Niña events, conversely there is a decrease in sightings in warmer conditions during El Niño (Fig. 22.3).

The total observations made between 2010 and 2022 in the Galapagos Marine Reserve, include 19 different species: six mysticetes and 13 odontocetes in five different families (Fig. 22.4). The number of recorded species and sightings fluctuate over the years, with an average of n = 9 species per year for cetacean diversity, with n = 13 species as maximum in 2019 and n = 6 species minimum in 2011 (Fig. 22.3).

For odontocete species, the bottlenose dolphin (n = 754) is the most sighted cetacean, followed by killer whales (n = 479) and the common dolphin (n = 221) (Figs. 22.5–22.6). In the case of mysticetes the most common species are Bryde's whale (n = 565), followed by the humpback (n = 185) and blue whale (n = 125) (Figs. 22.5–22.6). Three other species have been frequently sighted with between 20 and 100 sightings: the pilot whale (n = 53), sperm whale (n = 47) and false killer whale (n = 23). In the case of Cuvier's beaked whales, a total of ten (n = 10) sightings were recorded, six (n = 6) for the striped dolphin and the fin whale. Seven species obtain less than five sightings during the 12-year period, such as the sei whale with just four sightings (n = 4), Risso's dolphin, minke whale and pantropical spotted dolphin (n = 3), dwarf sperm whale (n = 2), spinner dolphin and melon headed whale just one (n = 1) sighting over the period.

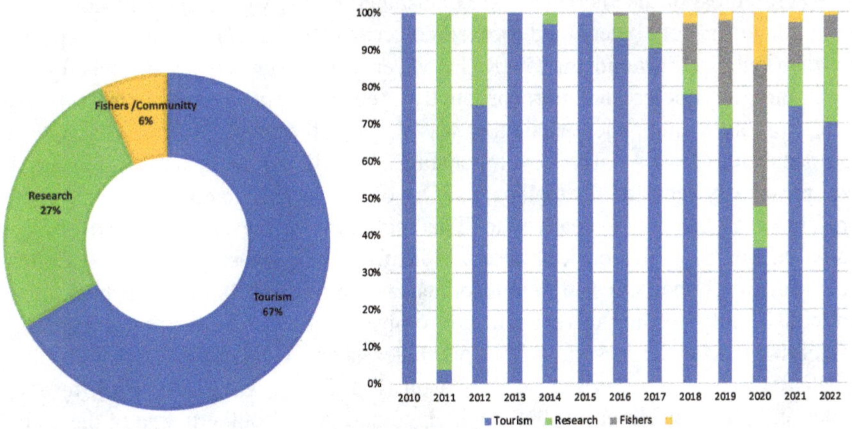

Fig. 22.2 Input types from the users of the Galapagos Marine Reserve registering opportunistic cetacean sightings in Galapagos

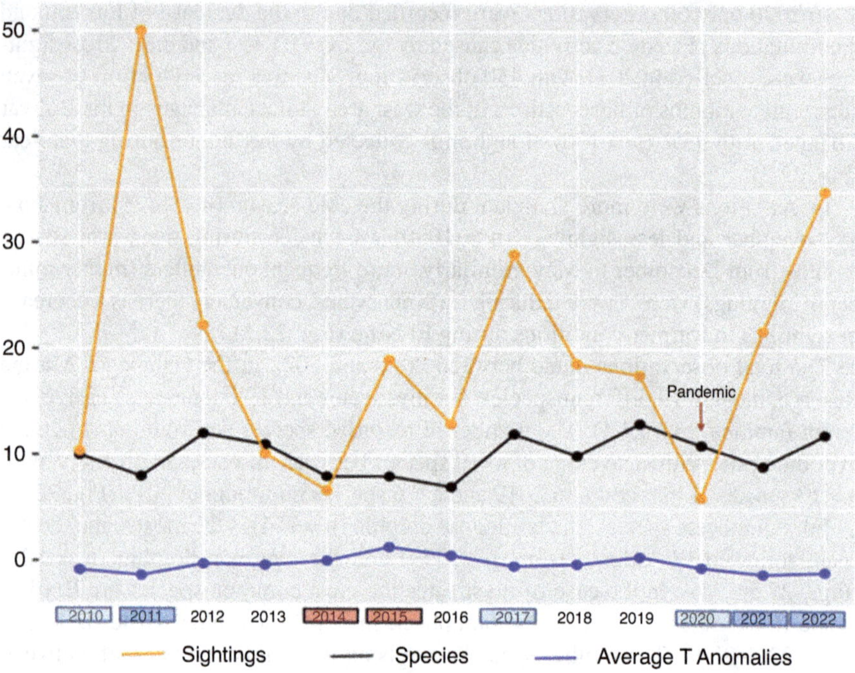

Fig. 22.3 Number of sightings expressed in 10^1 and species per year in the GMR during the 2010–2022 period, represented in blue boxes colder years register in 2010, 2011, 2017, 2020, 2021, 2022 and in red boxes warm years in 2014 and 2015. Average annual temperature anomalies in degrees adapted from http://www.cpc.ncep.noaa.gov

Species Distribution and Presence

In general, most of the encounters are distributed on the west and south side of the archipelago for both baleen and toothed cetaceans (Fig. 22.7). Odontocete species are distributed all around the GMR. However, for baleen whales there is a clear preference for cooler and more productive seasons and environments, with the exception of the humpback whale that was sighted all around the archipelago close to coastal areas from April to November (Fig. 22.8). This is consistent with information previously reported for Galápagos. Out of the 19 registered species, eight are present in the Galapagos year-round. This includes two mysticeti species: Bryde's and blue whales, with a peak of observations from July to October, and six odontocete species: the bottlenose dolphins, orcas, and common dolphins, which are present consistently during the year with more observations in the warm months from December to May. Observations of Pilot whales, sperm whales, and Cuvier's beaked whales are also higher in the warm season (Fig. 22.7), with distribution in deep water and productive areas of the archipelago and in the southern area of the archipelago for sperm whales (Fig. 22.8).

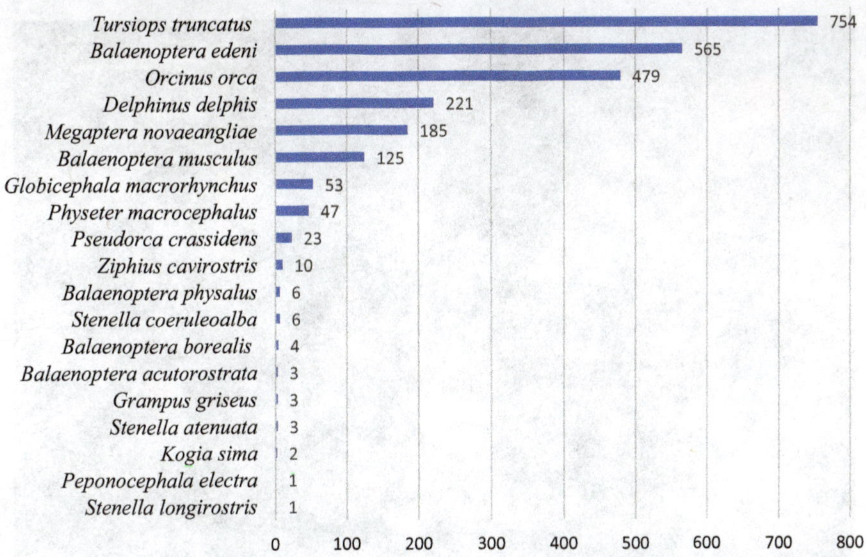

Fig. 22.4 Cumulative sighting records for cetacean species in the Galapagos Marine Reserve from 2010 to 2022

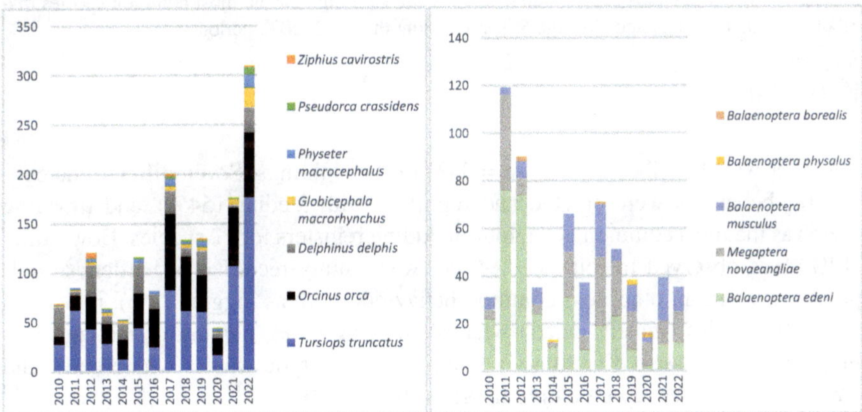

Figs. 22.5–22.6 Sighting frequencies of the most common odontocetes and mysticetes sightings from 2010 to 2022

When comparing our data with historical data presented by Palacios and Salazar (2002) and Denkinger et al. (2013), we can observe that bottlenose dolphins were the most common sighted species in the three periods in coastal areas, while the reported abundance of other species has shifted over the years (Fig. 22.9).

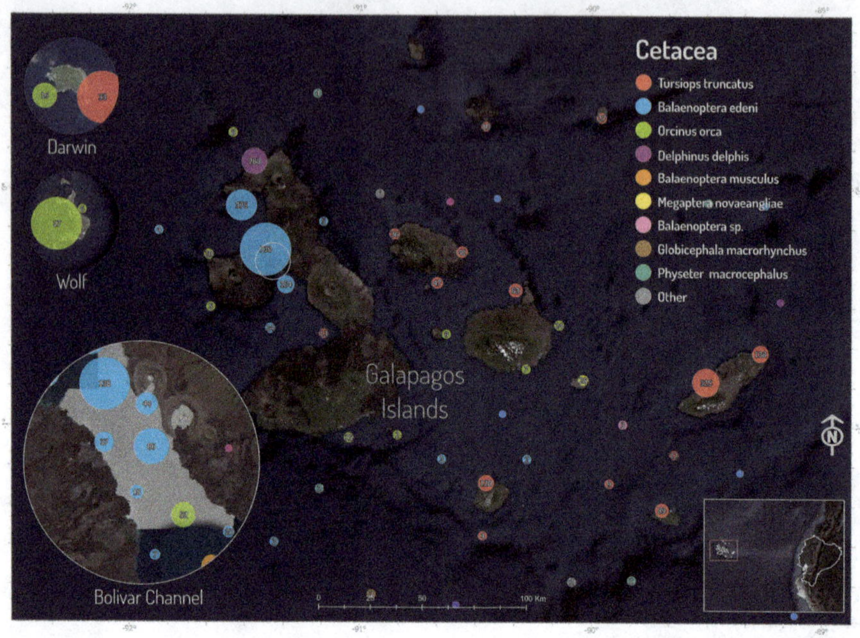

Fig. 22.7 Map of the clustered cumulative cetacean sightings for the most common species distributed inside the Galapagos Marine Reserve during the 2012–2022 period

Behavior

Behavior was described in 49% (n = 543) of the sightings. Seven different categories for behaviors were created and reported, with feeding (54%), and traveling (27%) as the most common behaviour including registers for 12 species. Bow riding (12%) was observed in four species. but was mainly recorded in bottlenose dolphins. In addition, other less common behaviors, such as logging (2%), registered for eight species (mostly baleen whales), playing (3%), nursing or mating (1%) involve less than 4 species (mainly humpback and bryde whales with a calf), and harassing (1%) was reported only in orcas (Fig. 22.10).

Tourist Swimming with Cetaceans in Galapagos.

As part of the observations, it was annotated every time that users were swimming with cetaceans, with a total of the 2.5% (n = 59) of the observations involving people swimming with whales or dolphins. These interactions were recorded for six species, and were mostly recorded on social media, in 39 cases. The frequency of this type of interaction was lower between 2012 and 2018, however, it increased exponentially in 2019, 2021 and 2022, with more than 15 cases every year (Fig. 22.11).

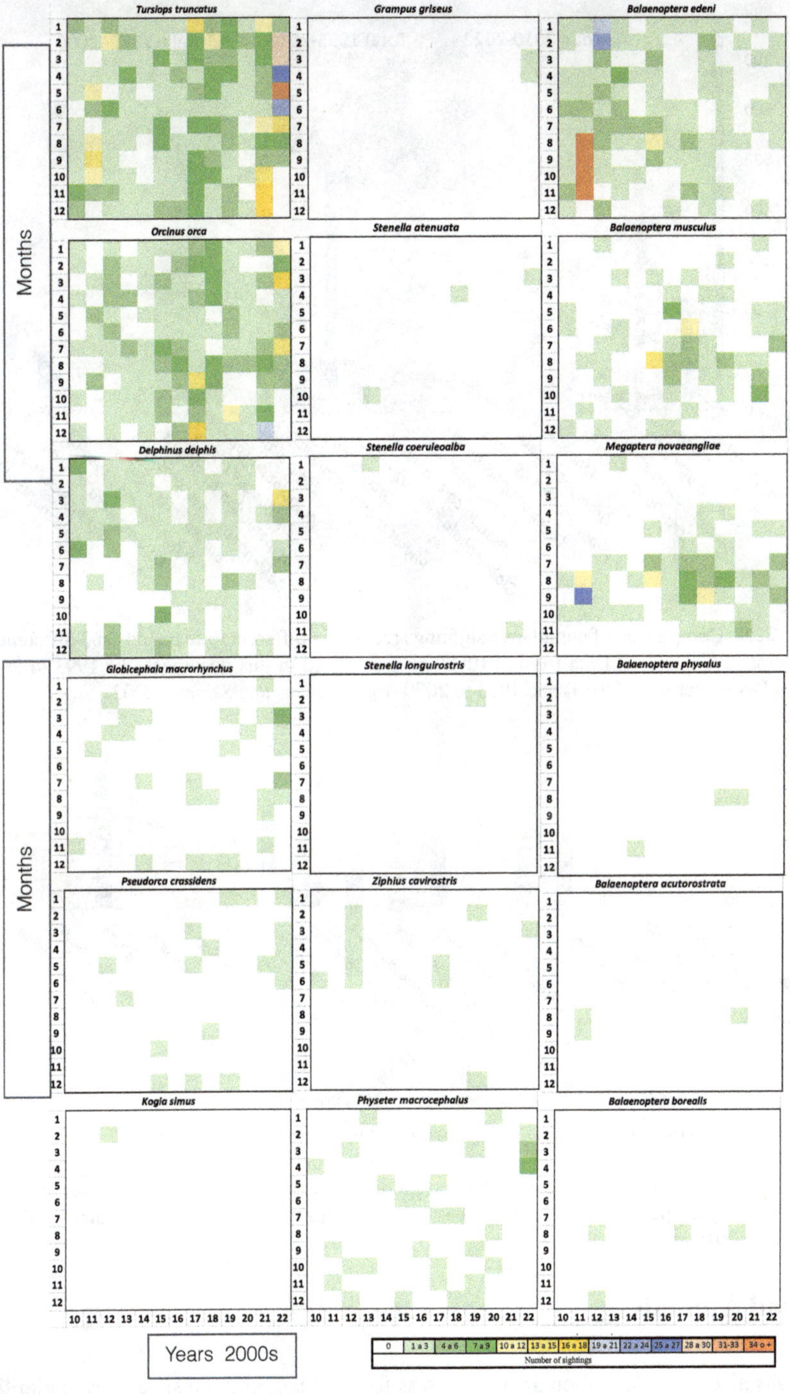

Fig. 22.8 Monthly sum of the number of sightings of cetacean species reported between 2010 and 2020

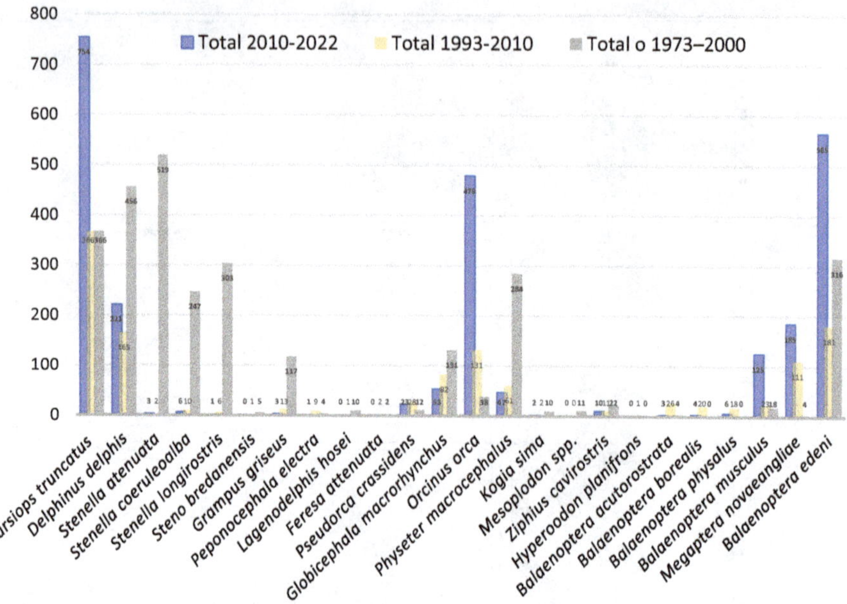

Fig. 22.9 Comparison of cumulative sightings records in different time periods and different ecosystems in Galápagos. Data from 2010–2020 is analyzed in this document, the 1993 to 2010 is from Denkinger et al. (2013) and 1973 to 2000 from Palacios and Salazar (2002)

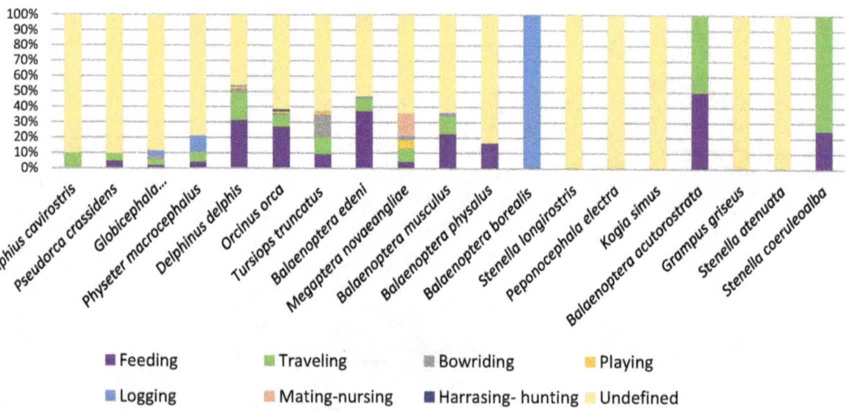

Fig. 22.10 Description of the (n = 543) sightings reporting percentages of the seven different categories per species for behaviour that has been described under the cumulative sightings between 2010 and 2022

Dead Animals

It was also recorded when an animal was found dead, and the species was identified when possible. Over the 2010–2022 period, we have six reports on four different

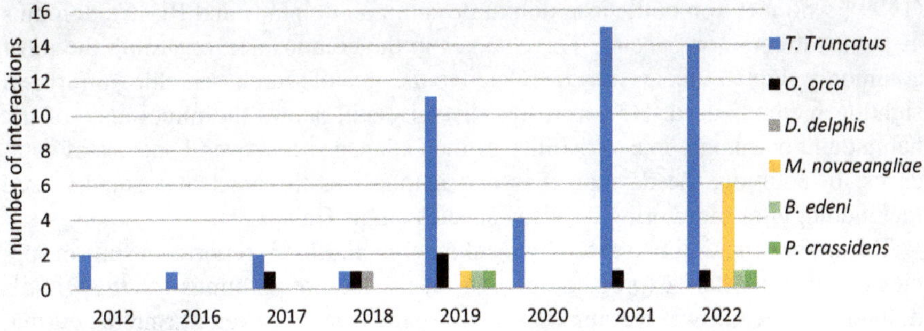

Fig. 22.11 Description per species involved on the (n = 63) observations including people swimming with whales or dolphins during the years between 2010 and 2022, reported by naturalistic guides or posted on social media platforms

species, including humpback whale (n = 3), blue whale (n = 1), pilot whale (n = 1) and an unidentified species of dolphin. For three of the cases, it was reported that the animals presented entanglements with fishing gear, however, due to the lack of a necropsy being performed, it was not possible to conclude if that was the primary cause of death.

Discussion

This analysis highlights the importance to standardize citizen science contributions, especially in remote areas such as the Galapagos, where access is limited, and meteorological conditions and logistics make it difficult to access all bioregions on the archipelago. This coincides with previous studies that have published information using trained user's contributions like Palacios and Salazar (2002), Denkinger et al. (2013), Garcia-Cegarra et al. (2021). Therefore, it is important to highlight the role and importance of citizen reports and social media, which can complement or compensate for the lack of research effort, providing valuable information to study cetacean species (Jarić et al. 2020; Morais et al. 2021).

We successfully collected information on 19 species of the 26 registered for the region, according to Palacios (2003). These number of species could be attributed to the focus of research efforts that center themselves on coastal areas. It is important to mention that there is a lack of an actual identification guide, which means that uncommon species could be underestimated.

The species composition indicates that the bottlenose dolphin, orca and common dolphins are the most frequently sighted toothed cetaceans within the GMR, followed by the short- finned pilot whale and false killer whale. These results are different to those by Palacios and Salazar (2002), who reported striped dolphins as one of the most common species, followed by the Risso's dolphin and the short-finned pilot whale. Our latest results also differ from reports by Smith and Whitehead

(1999), who mention bottlenose dolphins, common dolphins and Risso's dolphins as the most common species. The changes in the composition regarding the most commonly sighted species may be related to the area of search, searching effort and sighting methods used. However, this diversity still shows the importance of the habitat and oceanographic variability of the Galapagos for several species of cetaceans. In addition, the data justifies and emphasizes the need of a standardized methodology for monitoring cetacean species on the Galapagos.

Recent information presented by Denkinger et al. (2013) reported a shift in species composition and expressed concern on the lack of sightings for the Risso's dolphins. Indeed, it was a common species in the past, however, during the current study only three Risso's sightings were reported. These sightings were far from the coast, which confirms the concern about the status of the species in Galapagos waters. On the other hand, blue whale and orca sighting frequency increased when compared to other studies. Before 2010, blue and killer whales were considered a rare species but are now sighted year-round with peaks during colder months (Denkinger et al. 2020; Denkinger et al. in press). To obtain more and better information regarding cetacean species in the Galapagos, it is imperative to encourage and train citizen scientists to collect standardized, good quality data.

Knowledge on connectivity with other regions is limited for all species of cetacean in Galapagos. Regarding blue whales, Torres-Florez et al. (2015) reported the first evidence of migratory movements between Galapagos and Chile. This evidence comes from data collected in 1998 by a multiyear series of marine mammal line-transect surveys in the Eastern Tropical Pacific, by the Southwest Fisheries Science Center (SWFSC) and information collected by Blue Whale Centre/Universidad Austral de Chile between January and April of 2002–2013. Based on photo identification techniques and molecular markers they identified a genetic and visual match between a blue whale sampled in southern Chile and the Eastern Tropical Pacific close to the Galapagos. Similarly, a genetic review with limited information of the population structure of blue whales in the Eastern Pacific also shows connectivity by the individuals using the Galapagos with the populations on ESP and ENP (Leduc et al. 2017). Additionally, information using spatial analysis by Hucke-Gaete et al. (2018) who fitted 10 Chilean Northern Patagonia (CNP) blue whales with satellite transmitters, reports the first record of two complete migratory paths between CNP and Galapagos. Moreover, Denkinger et al. (in press) describe an individual blue whale photographed in the Galapagos and then in the Costa Rica Dome and then again in the Galapagos archipelago showing connectivity between these regions.

In the case of orcas, two instances exist of an individual identified with photo ID and then reported with a match between regions in the ETP. The first case, described by Guerrero-Ruiz et al. (2005), was of a match between a photographed adult male in Peruvian waters that when compared with a Mexican photo ID catalogue found a match with an animal previously photographed two times in the Mexican Pacific. Subsequently Pacheco et al. (2019) reported a sighting of an adult male orca in northern Peru, who has also been observed at several locations throughout coastal, island, and offshore archipelago areas of the Eastern Tropical Pacific including the Galapagos. Regarding other species, connectivity with other regions has not been described at the archipelago, little is known about these populations on the islands.

In addition, social media gives us an indicator of the importance of outreach. It is necessary to inform the local communities on how they can participate and contribute to scientific discoveries and conservation. In Galapagos, regulations prohibit the users like tourist to swim with cetaceans. However, animals occasionally approach and interact with people who are already in the water. In 1991, the Marine Mammal Sanctuary was created by the Ecuadorian Environmental Government Regulatory Agency. The country adopted the Action Plan for the Conservation of Marine Mammals of the Southeast Pacific, committing to conserve all species and to promote the care for animals and security of the people practicing activities with cetaceans. This agreement includes technical standards for observation, to protect and guarantee the integrity of the vessel occupants, and minimize disturbance to cetaceans. Therefore, ship captains and observers must follow specific rules including prohibiting swimming with or harassing animals, as well as upholding minimum distances between vessels and animals. It is important to highlight these regulations with visitors and to create awareness on what to do if a cetacean is encountered when practicing aquatic activities.

All cetacean populations are protected against whaling, and population numbers may be increasing for some species. However, global threats such as climate variability, pollutants, maritime traffic and commercial fishing continue to endanger cetaceans. This is particularly problematic when data is unavailable to assess how many animals are affected each year (Alava et al. 2021; Bedriñana-Romano et al. 2021; Muñoz-Arnanz et al. 2019). To mitigate this, more information is needed, for instance, tissue sampling could be used to test molecular and chemical biomarkers, which will help us understand/establish population structure and contaminants for free ranging individuals. On the other hand, other methodologies - such as stable isotopes or acoustic studies - have also been tested to better understand the problems, with the idea that marine mammals can be used as indicators of oceanic health (Fossi et al. 2020; Schirinzi et al. 2020).

Conclusion

Records of species diversity and distribution fluctuate across years and seasons, with higher numbers when cold and productive conditions are present. The non-scientific sector, such as, tourism, fisheries, or the community as a citizen science platform, is key to provide long term data on cetaceans in the region and changes over time. The importance of citizen science was highlighted during the COVID-19 pandemic when records of sightings dropped. Training and an identification guide are fundamental tools to guarantee correct species identification. Additionally, a photographic register for individual identification is still required.

The most common region to sight cetaceans is around Isabela Island and the west area of the archipelago. However, the eastern southern islands (San Cristobal and Española) in the Galapagos show an increasing number of sightings, possibly due to more developed tourism in this region compared with the previous analysis by Denkinger et al. (2013). There is still limited information on population structure and origin for cetacean species in Galapagos. Dolphins (*Delphinus delphis* and *Tursiops truncatus*) are sighted in the highest numbers and have on average the highest number of animals per sighting. Future research on cetaceans must include molecular marker analysis, diet, and acoustic and spatial data to better understand populations of this species in the Galapagos, which in turn will enhance respective conservation efforts.

Acknowledgments We appreciate all the users that shared cetacean sightings to the project, especially to Manolo Yepez, Ana Eguiguren, Dr. Hal Whitehead and his lab at the Department of Biology, Dalhousie University. We are also grateful to Sofia Green, Daniel Herrera, Martin Cox, Paulo Tobar, Caro Pesantez and all the great naturalistic guides from Lindblad expeditions, with special recognition of Lynn Fowler and Cindy Manning, for their long trajectory contributing with cetacean science in the Galapagos. In addition, we thank the Galapagos Science Center for all the support and all the staff working there, especially the Founding Directors, Carlos Mena and Steve Walsh, also to the Universidad San Francisco de Quito and the Geocentro USFQ. We want to thank Dr. Hector Guzmán and the Smithsonian Tropical Research Institute in Panama, Jen Jones and the Galapagos Conservation Trust, and the Decanato de Investigación and the Galapagos POA grants. We also recognize the contribution of Alysa Valentine and Emily Christiansen for the organization of the database. The Galapagos National Park for all the support since the beginning off the project. The present project has been carried out under the research permit PC 54-22.

References

Alava JJ (2009) Productividad y flujo de carbón en los ecosistemas marinos de la Reserva Marina de Galápagos basado en abundancia de cetáceos e índices tróficos. Revista de Biología Marina y Oceanografía 44(1): 109–122. ISSN 0717-3326. https://dialnet.unirioja.es/servlet/oaiart?codigo=2969977

Alava JJ, Rosero RP, Alarcón D, Eguiguren A, Denkinger J, Haase B, Merlen G, Tirira DG (2021) Ballena azul (Balaenoptera musculus). In: Libro Rojo de los mamíferos del Ecuador, Versión 2021.1. (3a edición). Asociación Ecuatoriana de Mastozoología, Fundación Mamíferos y Conservación, Pontificia Universidad Católica del Ecuador y Ministerio del Ambiente, Agua y Transición Ecológica del Ecuador. https://bioweb.bio/faunaweb/mamiferoslibrorojo/FichaEspecie/Balaenoptera%20musculus/%5BToda%20la%20poblaci%C3%B3n%20del%20Ecuador%5D

Alava JJ, McMullen K, Jones J, Barragán-Paladines MJ, Hobbs C, Tirapé A, Calle P, Alarcón D, Muñoz-Pérez JP, Muñoz-Abril L, Townsend KA, Denkinger J, Uyaguari M, Domínguez GA, Espinoza E, Reyes H, Piedrahita P, Fair P, Galloway T et al (2022) Multiple anthropogenic stressors in the Galápagos Islands' complex social–ecological system: Interactions of marine pollution, fishing pressure, and climate change with management recommendations, Integr Environ Assess Manag:ieam.4661. https://doi.org/10.1002/ieam.4661

Albouy C, Delattre V, Donati G, Frölicher TL, Albouy-Boyer S, Rufino M, Pellissier L, Mouillot D, Leprieur F (2020) Global vulnerability of marine mammals to global warming. Sci Rep 10(1):Article 1. https://doi.org/10.1038/s41598-019-57280-3

Anderson DJ, Huyvaert KP, Wood DR, Gillikin CL, Frost BJ, Mouritsen H (2003) At-sea distribution of waved albatrosses and the Galápagos Marine Reserve. Biol Conserv 110(3):Article 3. https://doi.org/10.1016/S0006-3207(02)00238-0

Arbelo M, Espinosa de los Monteros A, Herráez P, Andrada M, Sierra E, Rodríguez F, Jepson P, Fernández A (2013) Pathology and causes of death of stranded cetaceans in the Canary Islands (1999-2005). Dis Aquat Org 103(2):Article 2. https://doi.org/10.3354/dao02558

Ballance LT, Pitman RL, Fiedler PC (2006) Oceanographic influences on seabirds and cetaceans of the eastern tropical Pacific: a review. Prog Oceanogr 69(2–4):Article 2–4. https://doi.org/10.1016/j.pocean.2006.03.013

Banks S (2002) Ambiente Físico. In: En: Reserva Marina de Galápagos. Línea Base de la Biodiversidad. Fundación Charles Darwin/Servicio Parque Nacional Galápagos, pp 22–37

Bedriñana-Romano L, Hucke-Gaete R, Viddi FA, Johnson D, Zerbini AN, Morales J, Mate B, Palacios DM (2021) Defining priority areas for blue whale conservation and investigating overlap with vessel traffic in Chilean Patagonia, using a fast-fitting movement model. Sci Rep 11(1):Article 1. https://doi.org/10.1038/s41598-021-82220-5

Bowen BW, Meylan AB, Ross JP, Limpus CJ, Balazs GH, Avise JC (1992) Global population structure and natural history of the green turtle (*Chelonia mydas*) in terms of matriarchal phylogeny. Evolution 46(4):Article 4. https://doi.org/10.1111/j.1558-5646.1992.tb00605.x

Bustamante RH, Reck GK, Ruttenberg BI, Polovina J (2000) The Galápagos spiny lobster fishery. In: Fishing news books: Oxford, pp 210–220

Cantor M, Whitehead H, Gero S, Rendell L (2016) Data from: Cultural turnover among Galápagos sperm whales. http://hdl.handle.net/10255/dryad.113412

Clapham PJ, Young SB, Brownell RL (1999) Baleen whales: conservation issues and the status of the most endangered populations. Mamm Rev 29(1):Article 1. https://doi.org/10.1046/j.1365-2907.1999.00035.x

Day D (1994) List of cetaceans seen in Galápagos. https://aquadocs.org/bitstream/handle/1834/23783/NG_53_1994_Day_List_of_cetaceans.pdf?sequence=1

Denkinger J, Oña J, Alarcón D, Merlen G, Salazar S, Palacios DM (2013) From whaling to whale watching: Cetacean presence and species diversity in the Galapagos Marine Reserve. In: Walsh SJ, Mena CF (eds) Science and conservation in the Galapagos Islands, vol 1. Springer, New York, pp 217–235. https://doi.org/10.1007/978-1-4614-5794-7_13

Denkinger J, Alarcon D, Espinosa B, Fowler L, Manning C, Oña J, Palacios DM (2020) Social structure of killer whales (*Orcinus orca*) in a variable low-latitude environment, the Galápagos Archipelago. Mar Mamm Sci 36(3):Article 3. https://doi.org/10.1111/mms.12672

Denkinger J, Douglas AB, Douglas B, Sears R, Narvaez M, Alarcon-Ruales D (in press) Year-round presence of northern and southern hemisphere Blue Whales (*Balaenoptera musculus*) at the Galapagos Archipelago. JNL387.

Dickson T (2006) A comparison of concentrations of macronutrients and chlorophyll a in high-and low-chlorophyll concentration areas around the Galápagos Islands

Edwards EF (2007) Fishery effects on dolphins targeted by tuna purse-seiners in the eastern tropical Pacific Ocean. Int J Comp Psychol 20(2):Article 2. https://doi.org/10.46867/IJCP.2007.20.02.05

Eguiguren A, Pirotta E, Boerder K, Cantor M, Merlen G, Whitehead H (2021) Historical and contemporary habitat use of sperm whales around the Galápagos Archipelago: implications for conservation. Aquat Conserv Mar Freshw Ecosyst 31(6):Article 6. https://doi.org/10.1002/aqc.3496

Evans (1991) Ecuadorian whale sanctuary is created providing protection for the Galapagos Islands. Ambio 20(2):Article 2

Evans PGH, Hammond PS (2004) Monitoring cetaceans in European waters. Mamm Rev 34(1–2):Article 1–2. https://doi.org/10.1046/j.0305-1838.2003.00027.x

Félix F, Palacios DM, Salazar SK, Caballero S, Haase B, Falconí J (2020) The 2005 Galápagos humpback whale expedition: a first attempt to assess and characterise the population in the Archipelago. J Cetacean Res Manag:291–299

Ferguson MC, Barlow J, Reilly SB, Gerrodette T (2005) Predicting Cuvier's (Ziphius cavirostris) and Mesoplodon beaked whale population density from habitat characteristics in the eastern tropical Pacific Ocean (No. 3). 7(3):Article 3

Fossi MC, Baini M, Simmonds MP (2020) Cetaceans as ocean health indicators of marine litter impact at global scale. Front Environ Sci 8:586627. https://doi.org/10.3389/fenvs.2020.586627

Garcia-Cegarra AM, Toro F, Gonzalez-Borasca V (2021) Citizen science as a tool to assess cetacean diversity in the Atacama Desert coast. Ocean Coast Manag 213:105858. https://doi.org/10.1016/j.ocecoaman.2021.105858

Guerrero-Ruiz M, García-Godos I, Urban R (2005) Photographic match of a killer whale (Orcinus orca) between Peruvian and Mexican waters. Aquat Mamm 31(4):438–441

Hucke-Gaete R, Bedrinana-Romano L, Viddi FA, Ruiz JE, Torres-Florez JP, Zerbini AN (2018) From Chilean Patagonia to Galapagos, Ecuador: novel insights on blue whale migratory pathways along the Eastern South Pacific. Peer J, 6, e4695

Jarić I, Roll U, Arlinghaus R, Belmaker J, Chen Y, China V, Douda K, Essl F, Jähnig SC, Jeschke JM (2020) Expanding conservation culturomics and iEcology from terrestrial to aquatic realms. PLoS Biol 18(10):e3000935

Jefferson TA, Webber MA, Pitman RL (2015) Marine mammals of the world: a comprehensive guide to their identification, 2nd edn. Elsevier/Academic Press. https://books.google.com.ec/books?hl=en&lr=&id=_IY2n07t7QYC&oi=fnd&pg=PP1&dq=%E2%97%8F%09Jefferson,+T.+A.,+Webber,+M.+A.,+y+Pitman,+R.+L.+(2015).+Marine+mammals+of+the+world:+a+comprehensive+guide+to+their+identification.+2nd+ed.+Elsevier+Inc.,+Academic+Press,+San+Diego.+California&ots=cCbMCs2NnE&sig=4n4sQK_uzPVdTbV2lJW2b5zUWPI&redir_esc=y%23v=onepage&q&f=false

Johnson C, Reisinger R, Palacios D, Friedlaender A, Zerbini A, Willson A, Lancaster M, Battle J, Graham A, Cosandey-Godin A, Jacob T, Felix F, Grilly E, Shahid U, Houtman N, Alberini A, Montecinos Y, Najera E, Kelez S (2022) Protecting blue corridors, challenges and solutions for migratory whales navigating international and National Seas. WWF, Oregon State University, University of California, Santa Cruz, WWF International. https://wwfeu.awsassets.panda.org/downloads/wwf_blue_corridors_report_feb2022.pdf

Kaschner K, Tittensor DP, Ready J, Gerrodette T, Worm B (2011) Current and future patterns of global marine mammal biodiversity. PloS One. https://doi.org/10.1371/journal.pone.0019653

Kislik E, Saltos GM, Torres G, Borbor-Córdova M (2017) Biological hotspots in the Galápagos Islands: exploring seasonal trends of ocean climate drivers to monitor algal blooms ... https://doi.org/10.5281/zenodo.1133040

Kiszka JJ, Caputo M, Méndez-Fernandez P, Fielding R (2021) Feeding ecology of elusive Caribbean killer whales inferred from Bayesian stable isotope mixing models and whalers' ecological knowledge. Front Mar Sci *8*:648421. https://doi.org/10.3389/fmars.2021.648421

Laidre KL, Stern H, Kovacs KM, Lowry L, Moore SE, Regehr EV, Ferguson SH, Wiig Ø, Boveng P, Angliss RP, Born EW, Litovka D, Quakenbush L, Lydersen C, Vongraven D, Ugarte F (2015) Arctic marine mammal population status, sea ice habitat loss, and conservation recommendations for the 21st century: Arctic marine mammal conservation. Conserv Biol 29(3):Article 3. https://doi.org/10.1111/cobi.12474

Learmonth JA, MacLeod CD, Santos MB, Pierce GJ, Crick HQP, Robinson RA (2006) Potential effects of climate change on marine mammals. In: Oceanography and marine biology, vol. 44, p 431. https://books.google.com.ec/books?hl=en&lr=&id=R7-TfdYeLEgC&oi=fnd&pg=PA431&dq=%E2%97%8F%09Learmonth,+J.+A.,+MacLeod,+C.+D.,+Santos,+M.+B.,+Pierce,+G.+J.,+Crick,+H.+Q.+P.,+%26+Robinson,+R.+A.+(2006).+Potential+effects+of+climate+change+on+marine+mammals.+Oceanography+and+Marine+Biology,+44,+431&ots=dw_n66biGm&sig=fYXx-tume7icaBQCvQ1ff8OxC-E&redir_esc=y#v=onepage&q&f=false

Leduc RG, Archer FI, Lang AR, Martien KK, Hancock-Hanser B, Torres-Florez JP, Hucke-Gaete R, Rosenbaum HC, van Waerebeek K, Brownell RL, Taylor BL (2017) Genetic variation in blue whales in the eastern pacific: Implication for taxonomy and use of common wintering grounds. Mol Ecol 26(3):Article 3. https://doi.org/10.1111/mec.13940

Lodi L, Tardin R (2018) Citizen science contributes to the understanding of the occurrence and distribution of cetaceans in southeastern Brazil – a case study. Ocean Coast Manag 158:45–55. https://doi.org/10.1016/j.ocecoaman.2018.03.029
Love C, Taft B, Laurs R (1972) EASTROPAC Atlas, vol. 2. National Marine Fisheries Service, Circular, 330, 7
Lusher AL, Hernandez-Milian G, Berrow S, Rogan E, O'Connor I (2018) Incidence of marine debris in cetaceans stranded and bycaught in Ireland: recent findings and a review of historical knowledge. Environl Pollut 232:467–476. https://doi.org/10.1016/j.envpol.2017.09.070
Merlen G (1995) Guía de campo de los mamíferos marinos de Galápagos. Fundación Charles Darwin para las Islas Galápagos
Merlen G (1999) The Orca in Galapagos: 135 sightings, p 7
Morais P, Afonso L, Dias E (2021) Harnessing the power of social media to obtain biodiversity data about cetaceans in a poorly monitored area. Front Mar Sci 8:765228. https://doi.org/10.3389/fmars.2021.765228
Muñoz-Arnanz J, Chirife AD, Galletti Vernazzani B, Cabrera E, Sironi M, Millán J, Attard CRM, Jiménez B (2019) First assessment of persistent organic pollutant contamination in blubber of Chilean blue whales from Isla de Chiloé, southern Chile. Sci Total Environ 650:1521–1528. https://doi.org/10.1016/j.scitotenv.2018.09.070
Mwango'mbe MG, Spilsbury J, Trott S, Nyunja J, Wambiji N, Collins T, Gomes I, Pérez-Jorge S (2021) Cetacean Research and Citizen Science in Kenya. Frontn Mar Sci 8:642399. https://doi.org/10.3389/fmars.2021.642399
Nabi G, McLaughlin RW, Hao Y, Wang K, Zeng X, Khan S, Wang D (2018) The possible effects of anthropogenic acoustic pollution on marine mammals' reproduction: an emerging threat to animal extinction. Environ Sci Pollut Res 25(20):Article 20. https://doi.org/10.1007/s11356-018-2208-7
Pacheco AS, Castro C, Carnero-Huaman R, Villagra D, Pinilla S, Denkinger J, Palacios-Alfaro JD, Sánchez-Godinez C, González-Ruelas R, Silva S, Alcorta B, Urbán RJ (2019) Sightings of an adult male killer whale match humpback whale breeding seasons in both hemispheres in the eastern tropical Pacific. Aquat Mamm 45(3):Article 3. https://doi.org/10.1578/AM.45.3.2019.320
Palacios DM (1999) Blue whale (Balaenoptera musculus) occurrence off the Galapagos islands, 1978-1995. J Cetacean Re. Manag 1:41–51
Palacios D (2003) Oceanographic conditions around the Galápagos Archipelago and their influence on cetacean community structure (Thesis Disertation). ProQuest. https://www.proquest.com/openview/6b8754a8969c51dafcbcd81cdec3ffbe/1?pq-origsite=gscholar&cbl=18750&diss=y
Palacios D, Salazar S (2002) Cetáceos. In: Reserva Marina de Galápagos, Línea Base de la Biodiversidad, pp 291–304
Peltier H, Beaufils A, Cesarini C, Dabin W, Dars C, Demaret F, Dhermain F, Doremus G, Labach H, Van Canneyt O, Spitz J (2019) Monitoring of marine mammal strandings along French coasts reveals the importance of ship strikes on large Cetaceans: a challenge for the european marine strategy framework directive. Front Mar Sci 6:486. https://doi.org/10.3389/fmars.2019.00486
Pennington JT, Mahoney KL, Kuwahara VS, Kolber DD, Calienes R, Chavez FP (2006) Primary production in the eastern tropical Pacific: a review. Prog Oceanogr 69(2–4):285–317
Plagányi ÉE, Butterworth DS (2009) Competition with fisheries. In: Encyclopedia of marine mammals. Elsevier, pp 269–275. https://doi.org/10.1016/B978-0-12-373553-9.00065-1
Reck G (2014) Development of the Galápagos Marine Reserve. In: Denkinger J, Vinueza L (eds) The Galapagos Marine Reserve. Springer, pp 139–158. https://doi.org/10.1007/978-3-319-02769-2_7
Reilly SB, Thayer VG (1990) Blue whale (Balaenoptera musculus)distribution in the eastern tropical Pacific. Mar Mamm Sci 6(4):Article 4. https://doi.org/10.1111/j.1748-7692.1990.tb00357.x
Romano L, Viddi FA, Ruiz JE, Torres-Florez JP, & Zerbini AN (2018) From Chilean Patagonia to Galapagos, Ecuador: novel insights on blue whale migratory pathways along the Eastern South Pacific. PeerJ, 6, e4695

Sachs JP, Ladd SN (2010) Climate and oceanography of the Galapagos in the 21st century: expected changes and research needs. Galapagos Res 67:50–54
Schaeffer BA, Morrison JM, Kamykowski D, Feldman GC, Xie L, Liu Y, Sweet W, McCulloch A, Banks S (2008) Phytoplankton biomass distribution and identification of productive habitats within the Galapagos Marine Reserve by MODIS, a surface acquisition system, and in-situ measurements. Remote Sens Environ 112(6):Article 6. https://doi.org/10.1016/j.rse.2008.03.005
Schirinzi GF, Pedà C, Battaglia P, Laface F, Galli M, Baini M, Consoli P, Scotti G, Esposito V, Faggio C, Farré M, Barceló D, Fossi MC, Andaloro F, Romeo T (2020) A new digestion approach for the extraction of microplastics from gastrointestinal tracts (GITs) of the common dolphinfish (Coryphaena hippurus) from the western Mediterranean Sea. J Hazard Mater 397:122794. https://doi.org/10.1016/j.jhazmat.2020.122794
Smith SD, Whitehead H (1999) Distrution of dolphins in galapagos waters. Mar Mamm Sci 15(2):Article 2. https://doi.org/10.1111/j.1748-7692.1999.tb00819.x
Snell HM, Stone PA, Snell HL (1996) A summary of geographical characteristics of the Galapagos Islands. J Biogeogr 23(5):Article 5. https://doi.org/10.1111/j.1365-2699.1996.tb00022.x
Tanabe S, Mori T, Tatsukawa R, Miyazaki N (1983) Global pollution of marine mammals by PCBs, DDTs and HCHs (BHCs). Chemosphere 12(9–10):Article 9–10. https://doi.org/10.1016/0045-6535(83)90132-7
Torres-Florez JP, Olson PA, Bedriñana-Romano L, Rosenbaum H, Ruiz J, LeDuc R, Hucke-Gaete R (2015) First documented migratory destination for eastern South Pacific blue whales. Marine Mammal Science, 31(4):1580–1586
Wade PR (1994). Abundance and population dynamics of two eastern Pacific dolphins, Stenella attenuata and Stenella longirostris orientalis. Doctoral thesis, University of California. https://www.proquest.com/openview/893d96213e575f59532ecca0e17175f9/1?pq-origsite=gscholar&cbl=18750&diss=y
Wade PR, Gerrodette T (1993) Estimates of cetacean abundance and distribution in the eastern tropical Pacific. Report of the International Whaling Commission vol. 43, pp 477–493. https://swfsc-publications.fisheries.noaa.gov/publications/CR/1993/9392.PDF
Whitehead H (1999) Variation in the visually observable behavior of groups of galapagos sperm whales1. Mar Mamm Sci 15(4):Article 4. https://doi.org/10.1111/j.1748-7692.1999.tb00884.x
Wyrtki K (1966) Oceanography of the eastern equatorial Pacific Ocean
https://bioweb.bio/faunaweb/mamiferoslibrorojo/ListaEspeciesPorFamilia/145
https://psl.noaa.gov/enso/mei/
https://www.turismo.gob.ec/wp-content/uploads/2016/04/NORMATIVA-QUE-REGULA-OBSERVACION-DE-BALLENAS-Y-DELFINES-EN-ECUADOR.pdf.

Daniela Alarcón-Ruales is a Research Scientist and Emblematic Projects Coordinator at the Universidad San Francisco de Quito and the Galápagos Science Center. She is also a PhD Candidate in Environment and Sustainability at the University of the Sunshine Coast, Australia.

Judith Denkinger is a professor of Ecology and Marine Ecology at the college of biology and environmental sciences at the University of San Francisco de Quito. She has been researching the behavior, demography, and conservation of marine mammals since 1993 with research trips to the Atlantic Arctic and the North Sea. In Ecuador, she is coordinating research projects on humpback whales since 1996 and in the Galapagos Archipelago since 2009.

Leo Zurita-Arthos is a professor of Ecology and coordinator of the GEOcentro at the college of biology and environmental sciences at the University of San Francisco de Quito. His research is mainly focused on the mapping and modeling of ecosystem services at a regional and global level, applying and developing relevant geotechnologies.

Salome Herrera is an undergraduate biology student at San Francisco de Quito University with a concentration in marine ecology.

Santiago Díaz-Pazmiño is a Marine Biologist, whose research focuses on collective animal behavior, namely, social behavior, learning, and communication across the animal kingdom. Currently working as a team member of the Cetacea Galapagos Program and Ph.D. applicant at the University of Sydney, Australia.

Eduardo Espinoza is a marine biologist from the University of Guayaquil who graduated in 1998 and has worked in the Galapagos islands since 1991, he started as a scientific assistant in several research programs, then as a research associate in the fisheries monitoring program at the Charles Darwin's foundation since 1998, after which undertook a Master program in the Tokyo University of Marine technology in Japan Starting from 2005, Eduardo has connected with the Galapagos National Park Service (GNPS) as project manager for the Marine Monitoring Program, and later on became responsible for the Department of Marine Resources, at the moment he is a specialist of the department of Monitoring of the Marine Ecosystems in the GNPS.

Juan Pablo Muñoz Pérez, with more than 20 years of experience in diving and surfing in the Galápagos. Then ten years doing research. Juan Pablo's research focuses on animal movement and ecology, especially sea turtles and other marine animals such as whales and dolphins. He also understands animal health. Since 2014 he has been the pioneer and very active in investigating plastic pollution in the Galápagos with the primary objective of using science to find solutions. Area of study he is now completing a Ph.D. at the University of The Sunshine Coast in Australia.

Dr. Bonnie J. Holmes' research focuses on the biology and ecology of sharks, fish, and cetaceans. She specializes in revealing biological links to animal movements to answer complex questions about how marine species use their habitats in time and space. The outputs of her work have resulted in changes to sustainability initiatives and species management across the broader Indo-West Pacific region.

Assoc Prof Kathy A. Townsend is a marine scientist, whose research focuses on marine conservation and human impacts on the marine environment. Dr. Townsend is an invited member of the United Nations Marine Litter and Plastic Reduction Program, and GCT's Plastic Free Galapagos. She supervises multiple Ph.D. students working out of the Galapagos Science Centre. Her research programs draw their strength from a multidisciplinary approach and focus on global-scale issues such as the impacts of marine debris and marine pollution. Her research programs also contribute to a greater understanding of the conservation status of marine species identified by the IUCN Red List for threatened species.

Chapter 23
Establishing Standardized Health Baselines for Green Turtle Populations

Caitlin E. Smith, Ben L. Gilby, Juan Pablo Muñoz Pérez, Jason P. van de Merwe, and Kathy A. Townsend

Introduction

Marine turtles are subject to anthropogenic threats such as negative fisheries interactions (Riskas and Tiwari 2013; Riskas et al. 2016), climate change (Smith et al. 2021a), marine debris (Smith et al. 2021b; Schuyler et al. 2012; Schuyler et al. 2016), commercial harvesting (LaCasella et al. 2021), boat strike (Denkinger et al. 2013), invasive species predation (Welicky et al. 2012) and contaminants (Finlayson et al. 2016; Barraza et al. 2021). Exposure to these threats has led to all seven species of marine turtles facing population declines. As a consequence, marine turtles are regularly used as environmental proxies due to their longevity, high site fidelity and habitat use in areas of high anthropogenic stress (Flint et al. 2010). Green turtles, *Chelonia mydas*, are the most abundant marine turtle in Australia, and are listed as 'vulnerable' at a federal level (EPBC Act 2000) and 'endangered' under the International Union for Conservation of Nature (IUCN) Red List of Threatened Species (Seminoff 2004). Thus, there is an increasing need to understand the drivers

C. E. Smith (✉) · B. L. Gilby · K. A. Townsend
School of Science, Technology and Engineering, The University of the Sunshine Coast, Sunshine Coast, QLD, Australia
e-mail: caitlin.smith@research.usc.edu.au

J. P. Muñoz Pérez
College of Biological and Environmental Sciences and Galapagos Science Center, Universidad San Francisco de Quito, Quito, Ecuador

School of Science, Technology & Engineering, University of the Sunshine Coast, Sippy Downs, Australia

J. P. van de Merwe
Coastal and Marine Research Centre, Griffith University, Gold Coast, QLD, Australia

School of Environment and Science, Griffith University, Gold Coast, QLD, Australia

S. J. Walsh et al. (eds.), *Island Ecosystems*, Social and Ecological Interactions in the Galapagos Islands, https://doi.org/10.1007/978-3-031-28089-4_23

of population decline within this species (Flint et al. 2010). Several studies have linked environmental impacts with negative changes in marine turtle body condition and blood biochemistry (Aguirre and Balazs 2000; Ley-Quiñónez et al. 2017). Therefore establishing baselines, or reference intervals, for a suite of health parameters is a robust method of studying population health over multiple spatial and temporal scales (Mcfadden et al. 2014).

Reference or baseline blood parameters are required to enable future assessments of changes in health at a population level (Lewbart et al. 2014). When establishing health baselines, screening tests and examinations should follow the same basic assessments used in domestic veterinary medicine, such as, haematology, biochemistry, blood gas, histology and toxicology to allow for accurate analysis and diagnoses (Wyneken et al. 2014). These datasets can be a useful tool in animal health management (Lewbart et al. 2014), and can be used to explore health differences over environmental gradients, assisting with the rehabilitation of sick and injured turtles, and assessing the impacts of various human induced threats. Reference intervals for free-living healthy populations have been published for many chelonian species at various foraging and nesting locations, for which all parameters differ spatially (Labrada-Martagón et al. 2010; Harris et al. 2011; Ley-Quiñónez et al. 2017; Muñoz-Pérez et al. 2017; Samsol et al. 2020). However, very few studies have established robust baseline health parameters in free-living foraging populations of green turtles, which are often used as a proxy for environmental health (Flint et al. 2010; Lewbart et al. 2014; Mcfadden et al. 2014; Flint et al. 2019). Blood can be obtained from the jugular veins of marine turtles in a minimally invasive way, and used to determine biochemical, blood gas and haematological parameters that are useful in animal health management (Lewbart et al. 2014). Green turtles are a migratory species, however, sub-adult and juvenile turtles are resident foragers and have not yet entered the adult migratory stage of their life cycles, making them excellent indicators of site-specific variation (Leusch et al. 2021). For this reason, sub-adults and juveniles are often sampled for health and toxicological studies.

Detecting variation in marine turtle health as a result of environmental change is essential in understanding drivers of population decline. In a study conducted by Anderson et al. (2011), the health of Kemp's ridley turtles was compared between healthy free-living baselines and individuals after a cold-stunning event. Turtles that had been exposed to the cold stunning event exhibited hypoglycaemia, hypocalcaemia, and elevations in uric acid and blood urea nitrogen (BUN) levels. Similarly, Aguirre and Balazs (2000) explored the variation in baseline levels of blood biochemistry of healthy juvenile green turtles to those that had been exposed to fibropapillomatosis (FP). This study indicated that a number of turtles with severe FP were azotaemic, hypoproteinamic, hypoalbuminaemic, hypoferraemic, and had low cholesterol, indicating chronic stress as a result of the infection. These studies highlight the importance of obtaining and maintaining baseline health data across populations to explore the effects of environmental variations and disease prevalence.

Visual assessments of individual body condition can be biased by the individual observer and have high variation between populations and studies. Therefore, weight to length ratios or residual analyses have been used in many studies to remove observer bias to allow for quantitative and comparative analysis (Peig and Green 2009). Scaled Mass Index (SMI) analysis has been used as a non-biased analyses to determine individual and mean population body condition in the northern Great Barrier Reef (nGBR) genetic stock of green turtles (Bell et al. 2019), granting more accurate quantitative analysis. This in turn, allows researchers to make global comparisons of populations and compare environmental and anthropogenic drivers that may be influencing negative health consequences, i.e., land-based pollutants, marine debris, fishing pressures, lack of food availability.

The aim of this study was to establish standardized health baselines for two sub-adult foraging green turtle populations in eastern Australia through standard health assessment methods, external examinations and blood sampling. It was hypothesised that health parameter ranges in blood gas, biochemistry and haematology would vary between the two sites due to varying environmental conditions, food availability, and proximity to land based pollutants. This study is a part of a larger project exploring the impacts of marine debris ingestion and land-based contaminant on green and hawksbill sea turtles. Health parameter ranges established during this study will be used to determine health variation between sites that are exposed to different marine debris and chemical profiles.

Methods

Study Sites

Lady Elliot and Heron Island are two islands located in the Capricorn Bunker group on the southern Great Barrier Reef in Queensland, Australia (Fig. 23.1). Resident foraging green turtles on these reef systems are part of the southern Great Barrier Reef (sGBR) genetic stock. Both islands are tourist hotspots with eco-resorts. These two sites are located approximately 110 km apart and were chosen due to their differing proximities to anthropogenic pressures. Heron Island is located off the coast of Gladstone and Lady Elliot Island is located approximately 120 km northeast of Hervey Bay and is therefore slightly more remote than Heron Island (Fig. 23.1).

Turtle Collection

Fifteen sub-adult (65–85 cm) and juvenile (< 65 cm) green turtles were collected from both Lady Elliot Island and Heron Island (total 30 individuals), during February and April 2022, respectively (Chaloupka and Limpus 2005). The sex of the sampled turtles could not be determined as they were all in the sub-adult and juvenile size class for green turtles, where external indicators of sex (e.g., tail length)

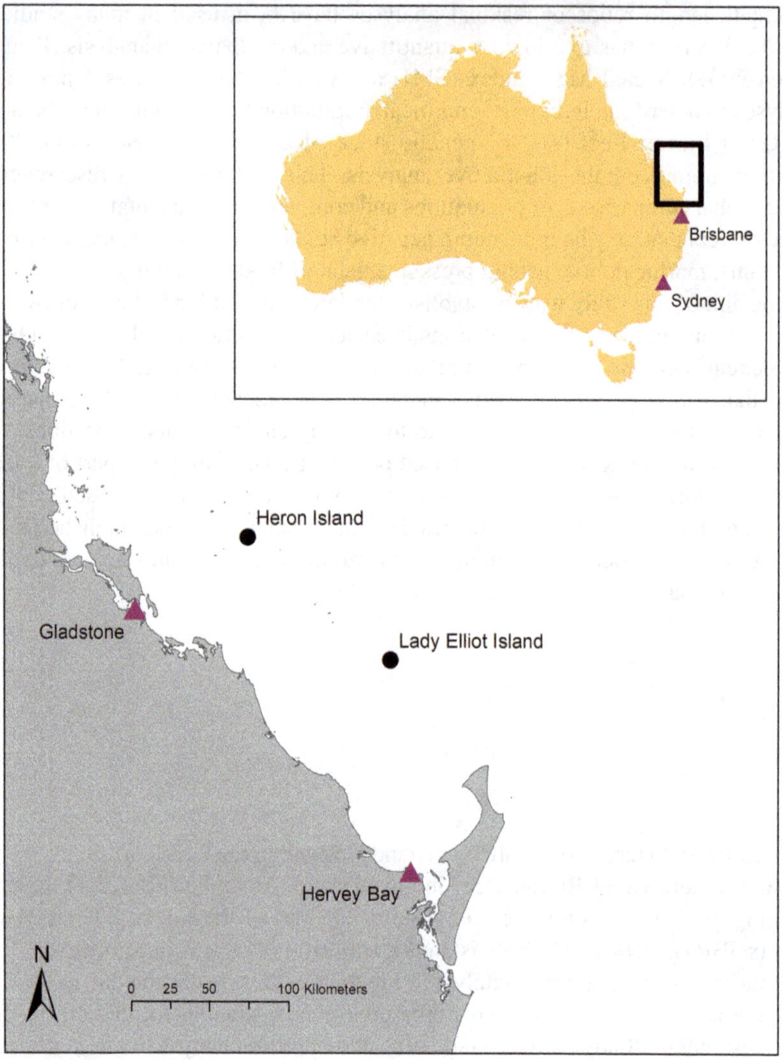

Fig. 23.1 Lady Elliot and Heron Island (black circles) on the Great Barrier Reef are two island sites in which sub-adult and juvenile green turtles were sampled to establish health parameter baselines. Major city centres are noted by purple triangles

cannot be used. Turtle collection was conducted by hand capture at both low and high tide. At high tide, turtles were collected by snorkel and brought back to the beach for sample collection. On Lady Elliot Island, green turtles remain on the reef flat in small shallow pool at low tide. In this case, researchers hand captured individuals in knee deep water and walked them to the beach for sampling. All animals were held for a maximum of 1 hour. Post-sampling, all individuals were released back into the water as close to the collection site as possible.

Flipper Tagging

Unique titanium tags, supplied by the Department of Environment and Science, Queensland Government, were applied to each individual on the trailing edge on the left and right front flippers, and registered in the Queensland Turtle Tagging Database. If turtles are re-captured in subsequent years, temporal variations can be examined.

External Measurements and Observations

Weight (kg) and curved carapace length (cm) were measured for each individual to the nearest gram and millimetre. An external examination, assessing the general health condition of each individual was also carried out, and epibiota such as barnacle numbers and algal growth (% cover) was recorded. Individuals were visually assessed for injuries such as tissue, bone or carapace loss/injury as well as any signs of disease, or dehydration such as fibrous growths and sunken eyes and carapace. Plastron concavity or malnourishment severity was also noted by sinking between the plastron ridges.

Blood Collection, Storage and Analysis

Blood was collected via the dorsal cervical sinus located on the neck of the turtle (Owens and Ruiz 1980) and was taken using either a 21- or 23- gauge Terumo™ needle (dependent on the individual's size) and a 20 mL Terumo™ syringe. Total collection amount was a maximum of 20 mL and did not exceed 2 mL/kg. Blood was aliquoted for various analyses (Fig. 23.2).

Immediately post-collection, approximately 0.01 mL of whole blood was added to a CG8+ iSTAT™ cartridge, which were stored between +2 °C and + 8 °C until the day of use, then kept at room temperature until the time of use. An iSTAT Portable Clinical Analyser (Heska Corporation, Fort Collins, Colorado, USA) was used to obtain in-field results. To avoid overheating while in the field, cartridges and the analyser unit was stored in an insulated box with ice packs to maintain temperature below 25 °C. The CG8+ cartridge provided values for the following parameters: sodium (Na; mmol/L), potassium (K; mmol/L), ionised calcium (iCa; mmol/L), glucose (Glu; mmol/L), haemoglobin (Hgb; g/dL), pH, partial pressure of carbon dioxide (pCO_2; mmHg), bicarbonate (HCO_3^-; mmol/L), total carbon dioxide (tCO_2; mmol/L), base excess in the extracellular fluid compartment (BE), and saturated oxygen percentage (sO_2; %). iSTAT values are calculated at a temperature of 37 °C, therefore parameters that are dependent on body temperature (HCO_3, pH, pO_2, pCO_2, iCa), were corrected for green turtle body temperatures. Individual body

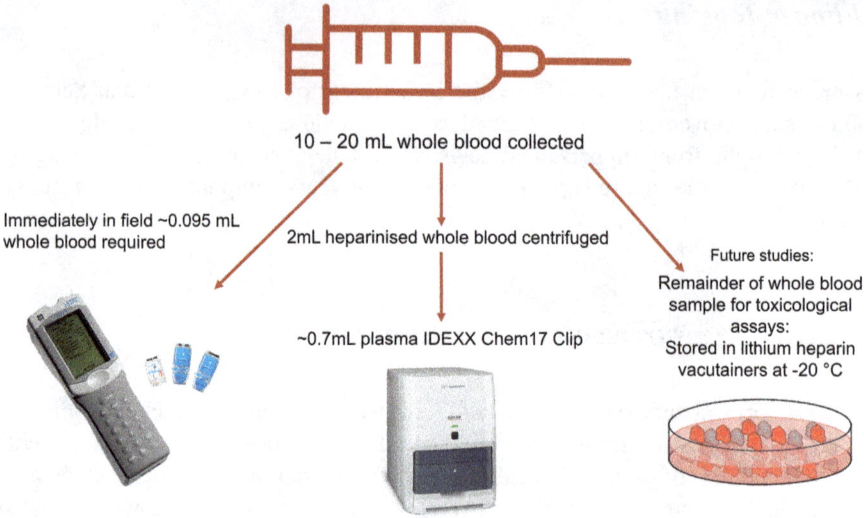

Fig. 23.2 Flowchart of blood collection with subsequent analysis and storage

temperature was taken using an infrared thermometer held approximately 5 cm from the neck of the individual. The temperature-dependent parameters were then manually adjusted for each sampled turtle, using the following calculations (Stabenau and Heming 1993; Anderson et al. 2011):

$$pH_M = 0.168 + 0.005\left(\Delta T_{25-Ti}\right) + pH_I \left[at\,temperatures < 25°C\right]$$

$$iCa_M = iCa_I \left(1 + 0.53[\,pH_I + pH_M\right)$$

$$pO_{2-M} = pO_{2I}\left(10^{-0.0058\Delta T}\right)$$

$$pCO_{2-M} = pCO_{2I}\left(10^{-0.019\Delta T}\right)$$

$$HCO_{3-M} = \alpha CO_2 \times pCO_{2-M} \times 10^{\left(pH_M - pK_a\right)}$$

Following the iSTAT analysis, the remaining blood was transferred into 10 mL lithium heparin BD™ vacutainers and stored in a cooler until the blood could be centrifuged and frozen for storage. Haematocrit % or packed cell volume (PCV) was measured using heparinised whole blood, and transferred by capillary action into microhematocrit tubes, and centrifuged at 12,000 rpm for 3 mins (JorVet ZipCombo). The percentage of red blood cells was calculated using a microhematocrit reader. Heparinised blood (2 mL) was also transferred into Eppendorf™ tubes and centrifuged for 3 mins at 12,000 rpm. The supernatant plasma was then pipetted into a separate Eppendorf™ tube, and along with the remaining red blood cells, were frozen at −20 °C and archived.

Laboratory Analysis

An IDEXX Catalyst One Laboratory Analyser™ was used to establish ranges of blood biochemistry (creatinine, urea, urea/creatinine, cholesterol, amylase and lipase). Analysis was carried out at SeaLife Aquarium Mooloolaba, a facility which regularly rehabilitates sick and injured marine turtles. 'IDEXX Chem 17'™ clips were used to calculate values of biochemical parameters.

Scaled Mass Index

Scaled mass index (SMI) of each turtle was calculated following Peig and Green (2009) and Bell et al. (2019):

$$\textit{Scaled mass index}\left(SMI\right) = W_i \left[\frac{L_O}{L_i}\right]^{b_{SMA}}$$

Using a log transformed standardised major axis regression (SMA) of weight (kg) versus curved carapace length (cm), the slope of the linear regression provides the allometric scaling exponent (b_{SMA}). L_i and W_i are the curved carapace length (CCL; cm) and weight (kg) for each individual, respectively, and L_O is the CCL to which the index is standardised. This study used the overall curved carapace length mean of both sites of 55.5 cm, similarly to Bell et al. (2019). This method was used to standardise the body condition of each individual for the purpose of population comparison.

Statistical Analysis

Due to the non-normal distribution of most measured blood and external variables and the small sample sizes for each site, a Kruskal-Wallis test in the R statistical framework (RCoreTeam 2022) was used to compare all measured variables between sites. A standard alpha level of $p = 0.05$ was used.

Results

Demography and External Indicators

All turtles sampled at both sites appeared clinically healthy, as they had no barnacle overgrowth (<10 barnacles in total) and no external indication of poor health. Internal body temperatures at both foraging populations, ranged from 23.5 °C to 25.6 °C with a mean of 24.2 ± 0.17 °C. The mean curved carapace length (CCL) of the Lady Elliot and Heron Island populations were 54.96 ± 1.56 cm and 56.19 ± 2.56 cm, respectively. Therefore, there was no difference in the size of sampled individuals between sites (chi-squared = 0.723, p = 0.395, df = 1). However, the maximum CCL sampled on Heron (76.2 cm) was slightly higher than Lady Elliot (64.7 cm).

Scaled Mass Index

The SMI of the two populations did not differ significantly between Lady Elliot Island (21.73 ± 4.18) and Heron Island (29.23 ± 6.90) (chi-squared = 0.723, p = 0.395, df = 1). However, due to the sampling of three larger sub-adults at Heron Island, the SMI range was larger in this population, with a maximum SMI of 94.84, compared with a maximum of 55.24 on Lady Elliot, due to the outliers (Fig. 23.3).

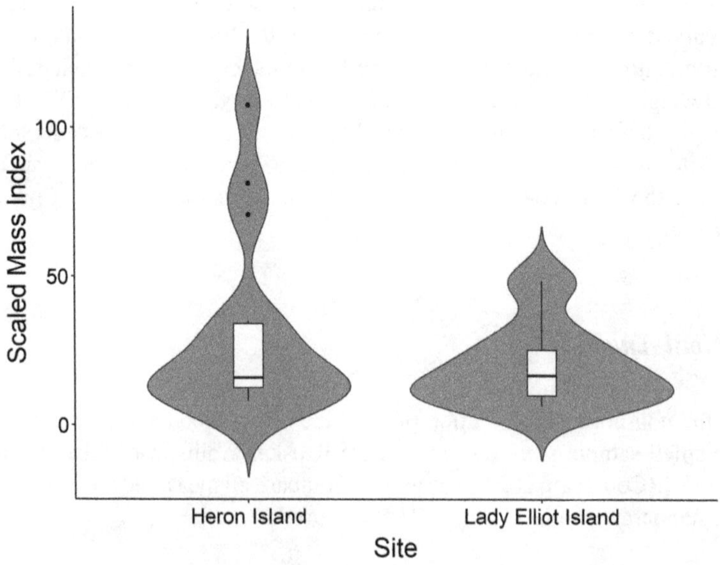

Fig. 23.3 Violin plot of scaled mass index at Lady Elliot Island (mean = 21.73 ± 4.18) and Heron Island (mean = 29.23 ± 6.90)

Values calculated at each site suggest that the body condition of all turtles sampled was considered 'good', and is similar to values published by Bell et al. (2019) for the nGBR genetic stock.

Health Parameters

Minimum/maximum ranges and means were calculated for 18 health parameters at both study sites. Ranges for tCO_2 (chi-squared = 8.47, p = 0.003, df = 1), HCO_3 (chi-squared = 10.351, p = 0.001, df = 1), BE (chi-squared = 8.648, p = 0.003, df = 1), cholesterol (chi-squared = 6.792, p = 0.009, df = 1; Fig. 23.4) and amylase (chi-squared = 12.397, p = 0.0004, df = 1; Fig. 23.5) were significantly higher at Lady Elliot Island (Table 23.1).

Discussion

On a global scale, marine turtle health assessments can be a key indicator of the sustainability of island ecosystems. Data from these assessments can guide government and management bodies in developing more stringent legislation for

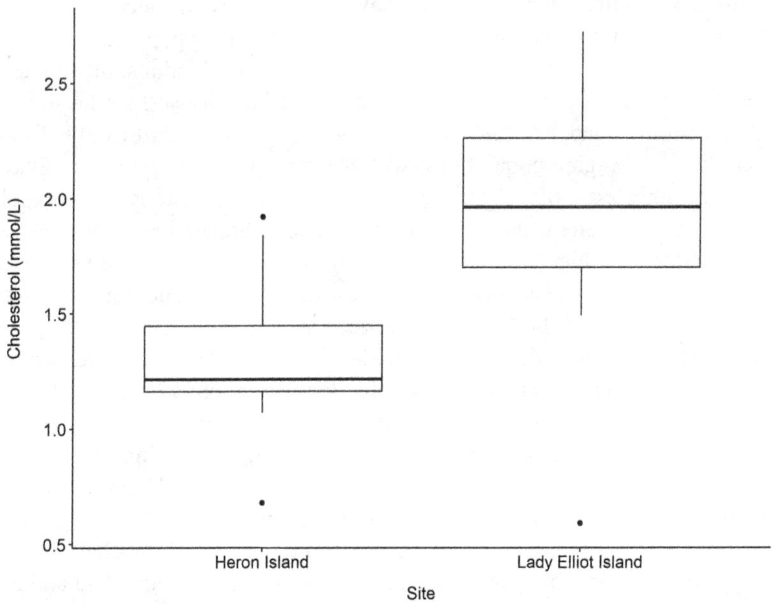

Fig. 23.4 The range of cholesterol (mmol/L) values for individuals on Heron Island (mean = 1.31 ± 0.11) and Lady Elliot Island (mean = 1.90 ± 0.17)

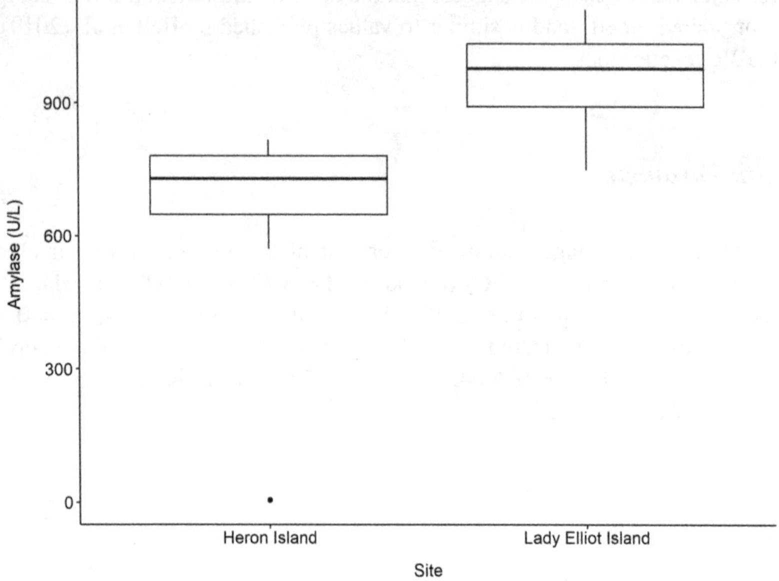

Fig. 23.5 Amylase (U/L) values for green turtles on Heron Island (mean = 654.2 ± 76.29) and Lady Elliot Island (mean = 962 ± 32.48)

environmental protection. Sub-adult and juvenile green turtles are ideal sentinels for assessing site specific impacts, as they have high fidelity to their foraging area. Outside of nesting and breeding seasons, green turtle foraging populations on Heron and Lady Elliot Islands are predominantly juveniles and sub-adults, thereby allowing for assessments of populations-wide impacts of key stressors on turtle health. The demographics and external characteristics were not significantly different between the two foraging populations, with the mean CCL of sampled individuals at each site within 2 cm of one another. Using external characteristics, researchers can standardize measurements so that datasets can be compared between sites, studies, and datasets. By standardizing body condition, researchers can eliminate subjectivity and observer bias, giving a more quantitative value for comparative analysis. While the SMI did not differ between sites in this study, the sample size used (n = 15 per site) may decrease the model's accuracy. Therefore, increasing the sample size and size range at these sites will allow for a more robust model, and the potential to detect SMI changes in the future. Other ecological studies assessing population health, in both terrestrial and marine settings, have singularly relied on body condition as a measure of health (Burgess et al. 2013; Noren et al. 2015). While this is useful in various settings, it is not an absolute measure of health, and therefore a more in-depth assessment of health (i.e., blood gas, biochemistry, haematology) is necessary to make detailed inferences regarding individual and population health status.

Table 23.1 Blood gas, haematological and biochemical health parameters as well as external measurements recorded from sub-adult and juvenile green turtles on Heron and Lady Elliot Islands

	Lady Elliot Island			Heron Island		
Parameter	Min.	Mean ± SE	Max.	Min.	Mean ± SE	Max.
CCL (cm)	46.70	54.96 ± 1.56	64.70	44.90	56.19 ± 2.56	76.20
Scaled mass index	5.90	21.73 ± 4.18	55.24	8.32	29.23 ± 6.90	94.84
Sodium (mmol/L)	145.0	150.7 ± 0.88	158.0	144.0	152.3 ± 0.97	159.0
Potassium (mmol/L)	3.80	5.14 ± 0.37	8.60	4.0	4.5 ± 0.08	5.1
Ionised calcium (mmol/L)	0.89	1.36 ± 0.06	1.83	1.0	1.2 ± 0.02	1.3
Glucose (mmol/L)	3.10	3.73 ± 0.10	4.40	3.2	3.9 ± 0.09	4.6
Hematocrit (% RBC)	8.0	25.4 ± 1.69	32.0	9.5	24.8 ± 1.74	31.0
Haemoglobin (g/dL)	5.80	7.92 ± 0.31	9.50	5.4	7.2 ± 0.26	8.5
pH	7.021	7.179 ± 0.020*	7.356	7.030	7.230 ± 0.030*	7.510
pCO_2 (mmHg)	63.22	89.82 ± 4.43	109.57	36.91	59.10 ± 3.31	75.55
TCO_2 (mmol/L) *	36.0	40.5 ± 1.04*	47.0	28.0	34.8 ± 1.27*	42.0
HCO_3 (mmol/L) *	43.29	50.73 ± 1.22*	59.32	31.75	41.39 ± 1.71*	49.38
BE (mmol/L) *	4.0	9.25 ± 1.05*	15.0	−7.0	2.1 ± 1.56*	9.0
sO_2 (%)	52.0	74.8 ± 3.15	90.0	52	72 ± 2.66	82
Creatinine (μmol/L)	12.0	50.7 ± 16.82	172.0	9.0	76.1 ± 24.27	266.0
Urea (mmol/L)	2.0	5.0 ± 1.71	21.7	2.2	2.8 ± 0.20	4.1
BUN/creatinine (mmol/L)	7.0	31.8 ± 7.16	75.0	4.0	13.25 ± 4.67	45.0
Cholesterol (mmol/L) *	0.59	1.90 ± 0.17*	2.72	0.68	1.31 ± 0.11*	1.92
Amylase (U/L)*	749	962 ± 32.48*	1111	5.0	654.2 ± 76.29*	817.0
Lipase (U/L)	10.0	416.5 ± 84.42	920.0	224.0	557.0 ± 89.07	1167.0

Asterisk represents significance between sites

This study has established ranges for several blood gas, biochemical and haematological parameters, that will be used to assess changes between foraging populations. This data enables researchers to make inferences on the health of various bodily functions and systems (i.e., gut health, liver, kidney function, circulatory system) at a site/population level. For example, amylase is a gut enzyme that is responsible for breaking down starch, glucose, maltose and maltriose into simple sugars (Espinoza-Romo et al. 2018). Increased amylase values can be attributed to inflammation of the pancreas and can be linked to fatigue, and lower body condition scores (Anderson et al. 2013). The Lady Elliot Island population indicated higher amylase values, in addition to increased base excess, cholesterol, bicarbonate and total carbon dioxide. However, both amylase ranges fall within other published juvenile green turtle reference intervals, suggesting that the differences in amylase values between the two populations could be due to diet variation (McFadden et al. 2014). Aguirre and Balazs (2000) observed two free-living "healthy" populations of juvenile green turtles at two different island sites, similar to this study, and reported significantly different baseline ranges at a site level for various blood parameters. This validates that "healthy" ranges can drastically change between foraging populations, highlighting the importance of establishing baseline parameters for various size classes, species and foraging sites. In addition, Stewart et al. (2016) observed marked differences between the health of green turtles in Barbados in relation to their exposure to tourism attractions. Both Lady Elliot Island and Heron Island foraging turtles experience comparable levels of tourist activity, therefore tourism exposure is unlikely a factor in the significant differences observed between baseline ranges at the site level.

For example, the Galapagos Islands presents a unique comparative opportunity as it is one of the most remote archipelagos on Earth (Rivas-Torres et al. 2018). Comparing sGBR green turtle populations to the Galapagos green turtle population allows researchers to evaluate ideal wild population quality and health as the Galapagos represents a population that is fairly removed from anthropogenic stressors. Lewbart et al. (2014) conducted green turtle health assessments on the Island of San Cristobal and established ranges for several blood gases and electrolytes. When comparing standardized health parameters between the sGBR and Galapagos, all ranges vary between Australia and the Galapagos. Sodium, potassium and ionised calcium baseline levels are slightly lower in the Galapagos compared to the sGBR, whereas haemoglobin and pH levels are slightly higher in the Galapagos; and PCV and bicarbonate ranges are very similar. The body condition and the external assessments of the Galapagos and sGBR populations suggest that both are clinically "healthy", which further adds evidence that blood parameter ranges can vary significantly between genetic populations while not negatively effecting overall health.

To make any definitive inferences on the health of the two populations, data on the different types and severity of anthropogenic and environmental stressors is required. Alternatively, some researchers have suggested that health assessments are not robust, and invite confirmation bias (Roman et al. 2021). However, health baseline datasets are very rarely used in isolation to inform population health and are useful in providing insight into future research needs. Future research in this region will confirm any negative health changes by exploring contaminant accumulation (organic and inorganic), quantifying plastic ingestion, and validate data using *in vitro* bioassays that explore changes in cellular function (i.e., cell death, oxidative stress and genotoxicity).

In conjunction with external observations and standardized body condition scoring, these baseline blood biochemistry ranges provide researchers with the data to make educated inferences regarding individual and population health variation. Many studies have reiterated that blood reference values should be established for green turtles at the population level and by geographic area, whilst considering age class, disease status, seasonal variability and sex (Aguirre and Balazs 2000). The haemolysis of samples also has the potential to change the reliability of various biochemical ranges, therefore it is recommended that in future when replicating this study's methodology that the upmost care is taken in the drawing and storage of blood. As this study has planned annual resampling to account for temporal variation, using a blood smear method to look at white blood counts would strengthen haematological findings. In addition to contaminant accumulation and plastic quantification, parasite invasion may also be an indicator of population health. An investigation into parasite load could therefore be useful in disentangling the causes of health variation over time and space (Corner et al. 2022).

Acknowledgments The authors would like to acknowledge SeaLife Mooloolaba, in particular, Brittany Attwood for facilitating laboratory analysis; and would like to extend their gratitude to Sonnie Flores, Dr. Christine Dudgeon and Dr. Asia Haines for their assistance during field work and data collection.

Ethics Statement Data collection at Heron Island and Lady Elliot Island was performed under a Scientific Purposes Permit (SPP18-001035), Marine Parks Permit (MPP18-001041-2) and Great Barrier Reef Marine Park Permit (G18/40908.1) issued by the Queensland Department of Environment and Science (DES) and the Great Barrier Reef Marine Park Authority (GBRMPA). Animal ethics was approved by the University of the Sunshine Coast Animal Ethics Committee, approval number ANS2076.

References

Aguirre AA, Balazs G (2000) Blood biochemistry values of green turtles, Chelonia mydas, with and without fibropapillomatosis. Comp Haematol Int 10:132–137

Anderson ET, Harms CA, Stringer EM, Cluse WM (2011) Evaluation of hematology and serum biochemistry of cold-stunned green sea turtles (Chelonia mydas) in North Carolina, USA. J Zoo Wildlife Med 42:247–255

Anderson ET, Socha VL, Gardner J, Byrd L, Manire CA (2013) Tissue enzyme activities in the loggerhead sea turtle (Caretta caretta). J Zoo Wildl Med 44:62–69

Barraza AD, Finlayson KA, Leusch FDL, van de Merwe JP (2021) Systematic review of reptile reproductive toxicology to inform future research directions on endangered or threatened species, such as sea turtles. Environ Pollut 286:117470

Bell IP, Meager J, VAN DE Merwe JP, Madden HOF, C. A. (2019) Green turtle (Chelonia mydas) population demographics at three chemically distinct foraging areas in the northern Great Barrier Reef. Sci Total Environ 652:1040–1050

Burgess EA, Brown JL, Lanyon JM (2013) Sex, scarring, and stress: understanding seasonal costs in a cryptic marine mammal. Conserv Physiol 1

Chaloupka M, Limpus C (2005) Estimates of sex- and age-class-specific survival probabilities for a southern great barrier reef green sea turtle population. Mar Biol 146:1251–1261

Corner RD, Cribb TH, Cutmore SC (2022) Vermetid gastropods as key intermediate hosts for a lineage of marine turtle blood flukes (Digenea: Spirorchiidae), with evidence of transmission at a turtle rookery. Int J Parasitol 52:225–241

Denkinger J, Parra M, Muñoz JP, Carrasco C, Murillo JC, Espinosa E, Rubianes F, Koch V (2013) Are boat strikes a threat to sea turtles in the Galapagos marine reserve? Ocean Coast Manag 80:29–35

Espinoza-Romo B, Sainz-Hernández J, Ley-Quiñónez C, Hart C, Leal-Moreno R, Aguirre A, Zavala-Norzagaray A (2018) Blood biochemistry of olive ridley (Lepidochelys olivacea) sea turtles foraging in northern Sinaloa, Mexico. PLoS One 13:e0199825

Finlayson KA, Leusch FDL, VAN DE Merwe JP (2016) The current state and future directions of marine turtle toxicology research. Environ Int 94:113–123

Flint M, Morton JM, Limpus CJ, Patterson-Kane JC, Murray PJ, Mills PC (2010) Development and application of biochemical and haematological reference intervals to identify unhealthy green sea turtles (Chelonia mydas). Vet J 185:299–304

Flint M, Brand A-F, Bell IP, Hof CAM (2019) Monitoring the health of green turtles in northern Queensland post catastrophic events. Sci Total Environ 660:586–592

Harris HS, Benson SR, Gilardi KV, Poppenga RH, Work TM, Dutton PH, Mazet JA (2011) Comparative health assessment of western Pacific leatherback turtles (Dermochelys coriacea) foraging off the coast of California, 2005–2007. J Wildl Dis 47:321–337

Labrada-Martagón V, Méndez-Rodríguez LC, Gardner SC, Cruz-Escalona VH, Zenteno-Savín T (2010) Health indices of the green turtle (Chelonia mydas) along the Pacific coast of Baja California Sur, Mexico II Body condition index. J Chelonian Conserv Biol 9:173–183

Lacasella, E. L., Jensen, M. P., Madden HOF, C. A., BELL, I. P., FREY, A. & Dutton, P. H. 2021. Mitochondrial DNA profiling to combat the illegal trade in tortoiseshell products. Front Mar Sci, 7, 595853

Leusch FDL, Hollert H, Holmes G (2021) Editorial – virtual special issue (VSI) green turtles as silent sentinels of pollution in the great barrier reef – Rivers to reef to turtles project. Sci Total Environ 757:144188

Lewbart GA, Hirschfeld M, Denkinger J, Vasco K, Guevara N, García J, Muñoz J, Lohmann KJ (2014) Blood gases, biochemistry, and hematology of Galapagos green turtles (Chelonia mydas). PLoS One 9:e96487

Ley-Quiñónez CP, Rossi-Lafferriere NA, Espinoza-Carreon TL, Hart CE, Peckham SH, Aguirre AA, Zavala-Norzagaray AA (2017) Associations between trace elements and clinical health parameters in the North Pacific loggerhead sea turtle (Caretta caretta) from Baja California Sur, Mexico. J Environ Sci Pollut Res 24:9530–9537

Mcfadden KW, Gómez A, Sterling EJ, Naro-Maciel E (2014) Potential impacts of historical disturbance on green turtle health in the unique & protected marine ecosystem of Palmyra atoll (Central Pacific). Mar Pollut Bull 89:160–167

Muñoz-Pérez JP, Lewbart GA, Hirschfeld M, Alarcón-Ruales D, Denkinger J, Castañeda JG, García J, Lohmann KJ (2017) Blood gases, biochemistry and haematology of Galápagos hawksbill turtles (Eretmochelys imbricata). Conserv Physiol 5

Noren SR, Udevitz MS, Triggs L, Paschke J, Oland L, Jay CV (2015) Identifying a reliable blubber measurement site to assess body condition in a marine mammal with topographically variable blubber, the Pacific walrus. Mar Mamm Sci 31:658–676

Owens DW, Ruiz GJ (1980) New methods of obtaining blood and cerebrospinal fluid from marine turtles. Herpetologica:17–20

Peig J, Green AJ (2009) New perspectives for estimating body condition from mass/length data: the scaled mass index as an alternative method. Oikos 118:1883–1891

RCORETEAM (2022) R: a language and environment for statistical computing. R Foundation for Statistical Computing, Vienna

Riskas KA, Tiwari M (2013) An overview of fisheries and sea turtle bycatch along the Atlantic coast of Africa. Munibe 1:1–82

Riskas KA, Fuentes MM, Hamann M (2016) Justifying the need for collaborative management of fisheries bycatch: a lesson from marine turtles in Australia. Biol Conserv 196:40–47

Rivas-Torres GF, Benítez FL, Rueda D, Sevilla C, Mena CF (2018) A methodology for mapping native and invasive vegetation coverage in archipelagos:an example from the Galápagos Islands. Progr Phys Geogr Earth Environ 42:83–111

Roman L, Gilardi K, Lowenstine L, Hardesty BD, Wilcox C (2021) The need for attention to confirmation bias and confounding in the field of plastic pollution and wildlife impacts: comment on "clinical pathology of plastic ingestion in marine birds and relationships with blood chemistry". Environ Sci Technol 55:801–804

Samsol S, Abd Wahid ME, Li T-H, Rusli MUU (2020) Hematology, blood gases and biochemistry profiles of wild-nesting sea turtles in Terengganu, Malaysia. Malays Appl Biol 49:25–31

Schuyler Q, Hardesty BD, Wilcox C, Townsend K (2012) To eat or not to eat? Debris selectivity by marine turtles. PLoS One 7:e40884

Schuyler QA, Wilcox C, Townsend KA, Wedemeyer-Strombel KR, Balazs G, VAN Sebille E, Hardesty BD (2016) Risk analysis reveals global hotspots for marine debris ingestion by sea turtles. Glob Chang Biol 22:567–576

Seminoff J (2004) Chelonia mydas. The IUCN red list of threatened species 2004: supplementary material

Smith CE, Booth DT, Crosby A, Miller JD, Staines MN, Versace H, Madden-Hof CA (2021a) Trialling seawater irrigation to combat the high nest temperature feminisation of green turtle Chelonia mydas hatchlings. Mar Ecol Prog Ser 667:177–190

Smith CE, Walker J, Donnan K, Riskas K, Madden HOF, C. A. (2021b) Determining the associated risk of beach-washed logs and their origin at Milman Island, a nesting ground for the endangered hawksbill turtle (Eretmochelys imbricata). Mar Freshw Res. https://doi.org/10.1071/MF20286

Stabenau EK, Heming TA (1993) Determination of the constants of the Henderson-Hasselbalch equation,(alpha) CO2 and pKa, in sea turtle plasma. J Exp Biol 180:311–314

Stewart K, Norton T, Mohammed H, Browne D, Clements K, Thomas K, Yaw T, Horrocks J (2016) Effects of "swim with the turtles" tourist attractions on GREEN sea turtle (CHELONIA MYDAS) health in Barbados, West Indies. J Wildl Dis 52:S104–S117

Welicky RL, Wyneken J, Noonburg EG (2012) A retrospective analysis of sea turtle nest depredation patterns. J Wildl Manag 76:278–284

Wyneken J, Lohmann KJ, Musick JA (2014) The biology of sea turtles: volume III. Mar Turt Newsl 141:18–19

Caitlin Elizabeth Smith is a PhD candidate working to assess the toxicity of marine debris and its effects on marine turtle health and establishing health profiles for multiple turtle populations. She has been a research assistant at the Australian Institute of Marine Science, exploring the effects of climate change on coral and sponge physiology, and the Australian National University, looking at microplastic pollution. Caitlin's most recent position was as a turtle ecologist for WWF-AUS researching strategies to help protect turtles against climate change.

Ben L. Gilby is an award-winning researcher with expertise in coastal ecology, coastal management and ecological monitoring. His research works to optimise the ways we manage, conserve, and restore ecosystems for greatest ecological, social and economic outcomes. He has over 10 years of professional scientific experience across industry and universities, and has published over 80 peer reviewed articles.

Juan Pablo Muñoz Pérez, Co-Guest Editor, has more than 20-years of experience in diving and surfing in the Galápagos and 10-years conducting research. Juan Pablo's research focuses on animal movement and ecology, especially sea turtles and other marine animals such as whales and dolphins. Since 2014 he has been a pioneer in investigating plastic pollution in Galápagos with the primary objective of using science to find solutions. Currently, he is completing his PhD at the University of the Sunshine Coast.

Jason P. van de Merwe is a marine ecologist and ecotoxicologist, with expertise in marine turtle ecology, the fate and effects of chemical contaminants in the aquatic environment and human impacts in the coastal zone. Jason's research is currently focused on ethical alternatives to assessing the accumulation and effects of chemical contaminants in the environment, particularly in marine wildlife species such as marine turtles, dugong, whales and dolphins.

Kathy A. Townsend Associate Professor Kathy Townsend is a marine scientist, whose research focuses on marine conservation and human impacts on the marine environment. Dr Townsend is an invited member of the United Nations Marine Litter and Plastic Reduction Program, and GCT's Plastic Free Galapagos. She supervises multiple PhD students who are working out of the Galapagos Science Centre. Her research programs draw their strength from a multidisciplinary approach, and focus on global scale issues such as the impacts of marine debris and marine pollution. Her research programs also contribute to a greater understanding of the conservation status of marine species identified by the IUCN Red List for threatened species.

Part VII
Island Ecosystems: Interdisciplinary Science for Sustainability

Chapter 24
An Agent-Based Model of Household Livelihood Strategies in the Galapagos Islands: Impact of Jobs in Fishing, Fishing Restrictions, and Fishing Deregulation on Household Employment Decisions

Stephen J. Walsh and Carlos F. Mena

Introduction

In the Galapagos Islands of Ecuador employment opportunities have historically revolved around jobs in fisheries, agriculture, and tourism. Today, tourism is the dominant employment sector for residents living in the Galapagos Islands as over 80-percent of residents are associated with tourism, often as their primary household livelihood alternative. Since the 1970s, tourism has exploded in the Galapagos Islands providing jobs, but also provoking national and international concerns for the environment and island sustainability more generally (Villacis and Carrillo 2013). With high flows of national and international tourists traveling to the Galapagos as well as a significant increase in population immigration, primarily Ecuadorians traveling to the Galapagos from the continent to work in the burgeoning tourism industry, the expanding human dimension has brought satisfaction for those employed in tourism and concern for those engaged in conservation (Walsh and Mena 2016). As such, accelerated population migration has hastened the introduction of invasive species, degraded ecosystems goods and services, expanded the consumption of local resources, and challenged local communities to provide basic services to support the expanding human dimension and at the same time to protect the environment (Epler 2007; Taylor et al. 2008). In 2019, approximately 35,000

S. J. Walsh (✉)
Department of Geography, University of North Carolina at Chapel Hill, Chapel Hill, NC, USA
e-mail: swalsh@email.unc.edu

C. F. Mena
College of Biological & Environmental Studies, Universidad San Francisco de Quito, Quito, Ecuador
e-mail: cmena@usfq.edu.ec

S. J. Walsh et al. (eds.), *Island Ecosystems*, Social and Ecological Interactions in the Galapagos Islands, https://doi.org/10.1007/978-3-031-28089-4_24

residents populated the Galapagos Islands and 275,000 tourists visited the Galapagos as boat-based and/or land-based tourists. With nearly 150 visitation sites identified throughout the Galapagos Islands for tourism, and a population spread primarily over four islands within the archipelago, tourists and residents are consuming the amenity and other natural resources of the Galapagos Islands, thereby, threatening the sustainability of its island ecosystems (Kerr 2005; Honey 2008; Johannes de Haan et al. 2019).

The number of hotel rooms in the Galapagos now exceeds the number of boat berths. In 2001, 30-percent of Ecuadorian and 86-percent of foreign tourists stayed on boats compared with 9-percent and 66-percent in 2010, respectively (World Wildlife Fund 2012). With communities rapidly expanding and boat- and land-based tourism exhibiting differential demands on the environment as well as on the community infrastructure, there is clear evidence that the urban infrastructure is not keeping pace with the consumptive demands of visitors and residents. Land-based tourism impacts resource consumption patterns in significant ways, particularly, through a demand for services supplied by local communities as well as housing, water, and power by tourists and the imported labor force. By comparison, boat-based tourism has less of an impact on local communities, but negative impacts on the environment continue to occur, for instance, through oil spills during transport and off-loading activities, dumping of refuse at sea as well as propeller strikes of sea-turtles and sea-lions. Island hopping on private speed boats moves people and products between islands, day trips move tourists to alternate islands for recreation opportunities, and cargo ships continually transit between the Ecuadorian mainland and the Galapagos Islands delivering food and household and community products for consumption and sale. Fuel is also delivered to the islands for the generation of power, mainly from diesel generators that supply power to the electric grid as well as fuel for boats, cars, and trucks as well as cooking fuel for local use. In short, the human dimension in the Galapagos is significant and expanding its intensity and geographic reach. The direct influence of people in the Galapagos is most clearly seen in the numbers of people arriving each day into the Galapagos for recreational services, and indirectly through the consumptive demands of visitors and resident of products imported into the islands. The rate of increase in the annual number of tourists visiting the Galapagos and the corresponding demand for additional residents to work in the tourism industry challenges human-environment interactions in sustainable ways.

The above has highlighted "normal" conditions in the Galapagos Islands, but, as a consequence of COVID-19, life in the Galapagos was anything but normal. With nearly 80% of Galapagos households involved in tourism, the pandemic shocked the socio-economic life of the Islands through international travel bans, suspension of fights from the Ecuadorian mainland to the Galapagos Islands, closure of the Galapagos National Park and Galapagos Marine Reserve, and the collapse of the tourism industry in the Galapagos, thereby, shutting hotels, restaurants, and support services for tourism and inducing considerable uncertainty among households for their livelihoods and the future of the islands. When Pizzitutti et al. (2014 and 2017) and Walsh and Mena (2016) developed spatial simulations models for the Galapagos that examined human-environment interactions, with a focus on the impact of tourism on economic development and resource conservation, the authors never imagined such a sudden and complete cessation of tourism in the Galapagos. Instead they modelled low growth, high growth, and collapse, but not at the unprecedented rate and magnitude of what has been seen in the Galapagos as a consequence of COVID-19. Even with the summer 2020 re-opening of the Galapagos National Park and Galapagos Marine Reserve, resumption of flights to the Galapagos Islands, and a government declared "COVID-19 Free" for the Galapagos," hesitant international travelers, national travel restrictions in their places of origin, and travel regulations imposed by the Galapagos Islands and the Ecuadorian national government all point towards a delayed economic recovery in the Galapagos Islands and the long-term challenges of life in the Galapagos without, yet, a return to a robust tourism industry.

While the direct and indirect effects of COVID-19 are not modeled here, the changes in job preferences, desire to transition out of tourism to fisheries and/or farming, and the availability of jobs in these economic sectors suggest that the transition from tourism is now more urgent, and more essential, and more uncertain. The globalization of the Galapagos Islands has introduced more opportunities to local households, but also more vulnerabilities, particularly, related to the pandemic, but also to economic uncertainties linked to social-ecological shocks through exogenous forces of change.

To assess human-environment interactions in the Galapagos islands and job diversification strategies of residents, the *Galapagos Fishers-Agent Based Model* (***GF-ABM***) is used to examine the decision-making strategies of local fisher households to engage in tourism, fisheries, and/or government as a consequence of household demographics and the "pushes" and "pulls" of fishers into the tourism industry, the largest and most lucrative employment sector in the Galapagos (Gonzalez et al. 2008; Malanson and Walsh 2015). While not all fishers wish to leave fisheries part-time or completely, several factors are considered through the creation of scenarios that are modeled within the ***GF-ABM***. As described by Walsh et al. (2019), the ***GF-ABM*** contains a demographic element that models basic demographic change at the household level. The model also contains an employment management component in which fisher agents select jobs among employment sectors – fisheries, tourism, and government (Hearn 2008). The tourism and government sectors have three tiers that require increasing agent skills. Fishers make their employment decisions based on their preferences to remain in fishing, the availability of jobs in three employment sectors, and their personal and professional qualifications that facilitate their movement among the employment sectors. Households contain members that are non-fisher agents, and fishers belong to households. Income and expenses are calculated for bother fishers and household agents (Hearn et al. 2005; Engie and Quiroga 2014). We model hypothetical scenarios of tourism development through the ***GF-ABM*** by examining how each scenario differs from base (default) model outcomes in the behavior of fishers to their demographic characteristics, personal preferences, job availability in specified employment sectors in the Galapagos, and government regulation of the fishing industry (Cinner et al. 2009).

The four scenarios tested against base or default parameters include the following: (1) Increased Percentage of Jobs in Fishing, (2) Decreased Percentage of Jobs in Fishing, (3) Imposition of Fishing Restrictions, and (4) Deregulation of the Fishing Industry. The context for the selection of the four tested scenarios relates to the possible increased employment in fisheries as a consequence of high-valued catches for export, such as lobster, sea cucumbers, and grouper. The model also includes an attraction of the tourism industry to fishers for full- or part-time employment in temporary or permanent jobs. We also model the imposition of increased fishing regulations by the Galapagos National Park through the setting of catch quotas as well as an annual determination of whether a fishing season will be allowed based on the perceived health and vitality of the fishing industry. We also examine the possible deregulation of the fishing industry through the relaxation of the fishing regulations, such as, the annual opening of the fishing season, expanded catch quotas as well as a reduction in the cost of obtaining a sailing certification for employment on-board tourist vessels as sailors, cooks, deck hands, and other boat services (Walsh et al. 2019).

Study Area

The Galapagos Islands are an oceanic, island archipelago comprised of 11 large islands and 100 s of small islands and islets located in the Eastern Pacific Ocean approximately 1000-km off the coast of the Ecuadorian mainland (Fig. 24.1). Bifurcated by the equator, the Galapagos Islands' geographic position has created unique marine and terrestrial environments that welcome warm- and cold-water species, native and endemic flora and fauna, and stunning volcanic landscapes. As a National Park, Marine Reserve, and a UNESCO World Heritage Site, the islands are renowned for their iconic species, including, giant tortoises, marine iguanas, hammerhead sharks, flightless cormorants, and Darwin Finches.

With an international reputation for endemism and biodiversity, the Galapagos Islands have also generated a rich history of human exploration, settlement, and exploitation. Historically plundered for its marine resources, such as, whales, fur seals, and giant tortoises, the contemporary exploitation has transitioned to the consumption of amenity resources, illegal harvesting of shark fins, harvesting of sea cucumbers and lobsters often beyond sanctioned limits imposed by the Galapagos National Park, and the direct and indirect consequences of the expanding human dimension seen through the importation of food, fuel, and consumer products for home and commercial sale (Watkins and Cruz 2007). In 1990, the Galapagos Islands recorded 10,000 residents and 40,000 tourists, and in 2019, the number of residents increased to 35,000 and the number of tourists expanded to 275,000 (The Nature Conservancy 2017). The associated demand for community services has exposed several limitations of the existing community infrastructure, fueled by residents and tourists, which has resulted in the degradation of the existing water supplies and distribution systems as well as health and education systems within the Islands. While considerable progress is being made to support the rapidly increasing human

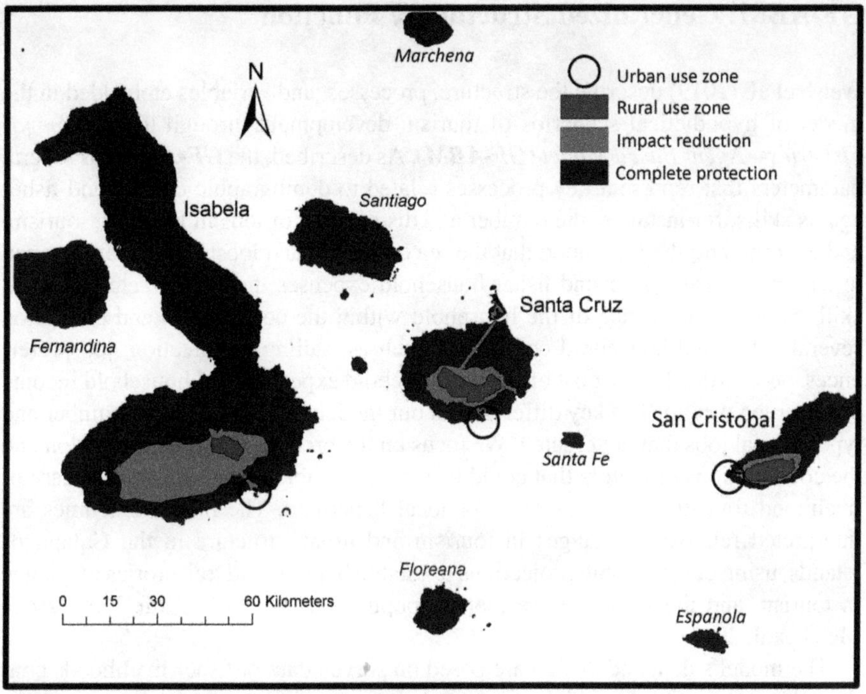

Fig. 24.1 Main islands of the Galapagos Archipelago of Ecuador; circles indicate urban use zones and the primary coastal communities of Puerto Villamil, Puerto Ayora, and Puerto Baquerizo Moreno on Isabela, Santa Cruz, and San Cristobal Islands, respectively

dimension in the islands, more still needs to be achieved to address the needed social services as well as the household necessities of life in the Galapagos Islands.

In addition, the Galapagos National Park continues to suffer from inconsistent leadership, replacing Directors with considerable frequency. Residents continue to return to the mainland for health care despite a relatively new hospital on San Cristobal Island and an existing hospital on Santa Cruz Island and clinics on Isabela and Floreana Islands. Households in the Galapagos continue their employment in traditional sectors of agriculture and fisheries, but tourism continues to employ a growing share of households, now exceeding 80% of all households in the Galapagos (Epler 2007; Taylor et al. 2008; Walsh and Mena 2016). Employment is increasing in the government sector, including, jobs within the Galapagos National Park as well as community government and the Government Council, the provincial authority in the Galapagos. For fisher households, opportunities exist for employment in the tourism industry, particularly, working on-board cruise boats (Pollnac et al. 2001). Demographic characteristics, employment preferences, job availability, government restrictions, and household responses to exogenous shocks, such as, El Nino events, when fish stocks are negatively affected, or economic downturns, periodically alter job opportunities in both fisheries and tourism (Mahon et al. 2008, Rozzi et al. 2010, Schuhbauer and Koch 2013).

GF-ABM: Generalized Structure & Function

Walsh et al. (2019) describe the structure, processes, and variables embedded in the model of hypothetical scenarios of tourism development through the *Galapagos Fisheries – Agent Based Model (**GF-ABM**)*. As described, the ***GF-ABM*** has several parameters that represent key processes related to demographic change and fisher agents skills, for instance, the number and distribution of jobs in fisheries, tourism, and government; the likelihood that the sea cucumber and lobster fisheries are open in any given year; fisher and fisher household expenses; fisher characteristics and skill levels; the influence of the household within the community; and checks on several fisher and household conditions, such as, sailing certification, job preferences, new births/deaths, cost of living, household expenses, and household income and accrued wealth. The key difference in our modeled scenarios is the number and type of local jobs that are created. We focus on the process of local job creation and the combinations of factors that could be most amenable to maximizing the ease of livelihood transitions into tourism for local fishermen. The model outcomes are interpreted relative to changes in tourism and urban structure in the Galapagos Islands, using demographic projections to establish trends and trajectories of change in tourism and the associated residential population (World Wildlife Fund 2003, Mena et al. 2011).

The model's data and design are based on survey data of fisher livelihoods conducted by Engie (2015), which recorded the career trajectories, household capitals, social responsibilities, and job preferences of former and active fishers (N = 166). The ***GF-ABM*** examines the transition of people away from fishing and into tourism. It does so by looking at three central elements of job transitions. First, it models major "pushes" and "pulls" affecting local Galapagos residents when making job decisions (Miller et al. 2010, Pizzitutti et al. 2017). Galapagos fishermen have long been compelled to consider job diversification to confront the socioeconomic uncertainty and changes in fish concentrations due to fluctuating marine productivity in the highly dynamic waters of the archipelago, that are strongly affected by ENSO (El Nino – Southern Oscillation) events. Policy uncertainty is also high given potential change in fishing quotas, seasonal closures, and license requirements managed by the Galapagos National Park. Secondly, the model touches upon the individual preferences of the job seekers themselves. Not all fishers wish to transition to non-fishing jobs, nor are people equally equipped for this transition. Finally, it recognizes that not all jobs are created equally within Galapagos economic sectors, by assigning levels to tourism and government jobs that correspond to increasingly higher skill requirements and salaries. We create three tiers of levels in each sector, with Level 1 being the least skilled and Level 3 the most skilled.

Fishers are agents, each modelled individually, are organized into households that can also contain non-fishing members. This structure replicates the reality of the varying social supports of fishers, which affect their income needs, job preferences, and available financial and social capital. The income and expenses calculated for each agent is affected by the structure of their household. Households are also agents and comprise a collection of individuals that collectively consume or contribute to annual income and overall wealth. The ***GF-ABM*** contains parameters relating to (1) demographic change, (2) livelihood skills, (3) available jobs on the islands, (4) the opening of sea cucumber and lobster fisheries, (5) household income, and (6) household expenses. We have translated the household survey of fishers conducted by Engie (2015) into a mathematical calculus of decision-making and probabilities of job switching based on certain conditions (Wilen et al. 2000). For modeling purposes, real fisher behavior is stylized into chains of decisions, all represented as probabilities.

GF-ABM: Hypothetical Scenarios of Change

The four scenarios tested against base or default parameters through use of the ***GF-ABM*** include the following: (1) Increased Percentage of jobs in Fishing, (2) Decreased Percentage of Jobs in Fishing, (3) Imposition of Fishing Restrictions, and (4) Deregulation of the Fishing Industry. The hypothetical values for each variable in the respective scenarios are compared to the default values used in the model. To assess the "Increased Fishing Scenario," the percentage of jobs in fishing is increased to 80% (versus 60% set as the default), percentage of jobs in tourism is decreased to 10% (versus 20% set as the default), and percentage of jobs in government is decreased to 10% (versus 20% set as the default). For the "Decrease Fishing Scenario" the percentage of jobs in fishing is decreased to 10% (versus 60% set as the default), percentage of jobs in tourism is increased to 55% (versus 20% set as the default), and percentage of jobs in government is increased to 35% (versus 20% set as the default). For the "Imposition of Fishing Regulations Scenario," the percentage of jobs in fishing is decreased to 10% (versus the 60% set as the default), percentage of jobs in tourism is increased to 35% (versus the 20% set as the default), percentage of jobs in government is increased to 55% (versus the 20% set as the default), probability of a sea cucumber season is decreased to 15% (versus the 33% set as the default), and the probability of a lobster season is decreased to 15% (versus the 85% set as the default). To assess the "Deregulation Scenario," the percentage of jobs in fishing is decreased to 10% (versus the 60% set as the default), percentage of jobs in tourism is decreased to 45% (versus the 60% set as the default), percentage of jobs in tourism is increased to 35% (versus the 20% set as the default), percentage of jobs in government is decreased to 10% (versus 20% set as the default), probability of a sea cucumber season is increased to 90% (versus the 33% set as the default), probability of a lobster season is increased to 90% (versus the 85% set as the default), and the cost of obtaining a sailing certification is decreased to $250 (versus the $1000 set as the default).

Analysis & Discussion

The four scenarios that are tested against default parameters through use of the ***GF-ABM*** include the following: (1) Increased Percentage of jobs in Fishing, (2) Decreased Percentage of Jobs in Fishing, (3) Imposition of Fishing Restrictions, and (4) Deregulation of the Fishing Industry. The model defaults also are presented in all graphs as a point of reference. For each scenario, plots are generated for selected household variables, based on 100 model runs per scenario. Each graph contains one boxplot per scenario for every model tick, i.e., one of 10-years of the model simulation. The boxplots are overlaid with lines showing the mean values for each scenario output. The presence of a **red** asterisk (*) indicates that the output from that scenario for that model tick was significantly different (t-test at the 0.05 level) than the output from the default scenario (See Figs. 24.2, 24.3, 24.4, 24.5, 24.6, 24.7, 24.8 and 24.9). Boxplots are a standardized way of displaying the distribution of data on a five-number summary – minimum, first quartile (Q1 – 25th percentile), median (Q2 – 50th percentile), third quartile (Q3 – 75th percentile), and maximum. Box plots may also have lines or points extending from the boxes indicating variability or spread outside the upper and lower quartiles.

In Figs. 24.2, 24.3, 24.4, 24.5, 24.6, 24.7, 24.8 and 24.9, the first bar of the graphics represents "default conditions" used to construct the ***GF-ABM***, the second bar represents "increased fishing jobs," the third bar represents "decreased fishing jobs," the fourth bar represents "fishing restrictions," and the fifth bar of the graphic represents "fishing deregulation." **Red** asterisks (*) indicate significantly different t-test values at the 0.05 level for model outcomes versus default values for each scenario. Figure 24.2 shows the influence of the household variables across all four scenarios and the increasing slope of the trend line indicates the mean values of each scenario across the 10-year annual time steps. No scenarios were significantly different (t-test at the 0.05 level) compared to the default values for each model tick.

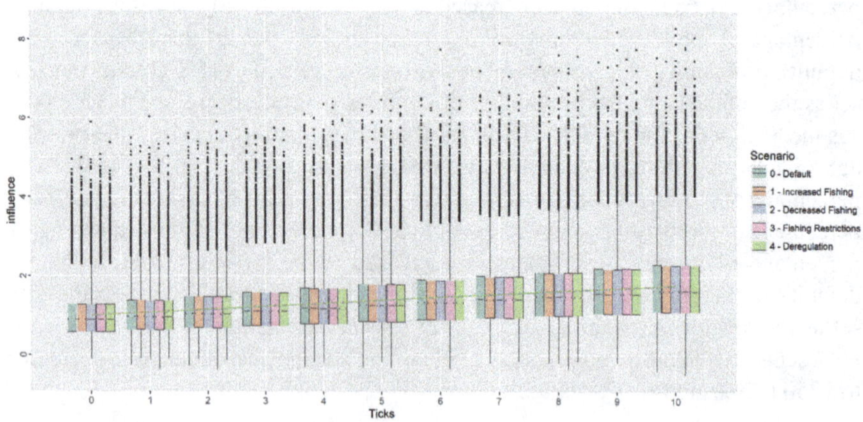

Fig. 24.2 Household influence across all scenarios

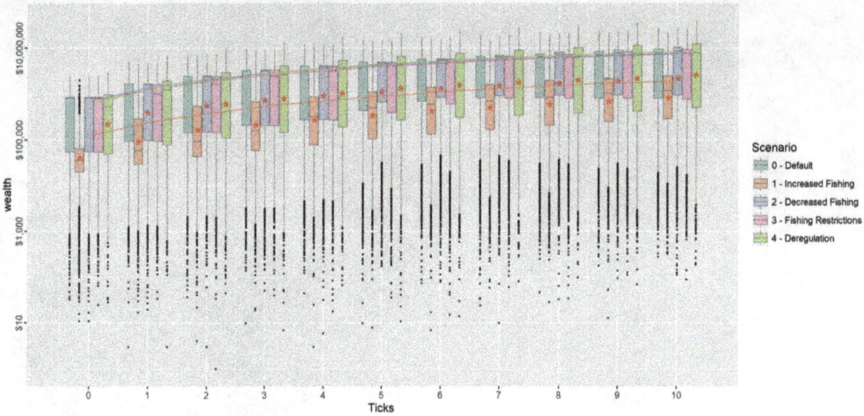

Fig. 24.3 Household wealth across all scenarios

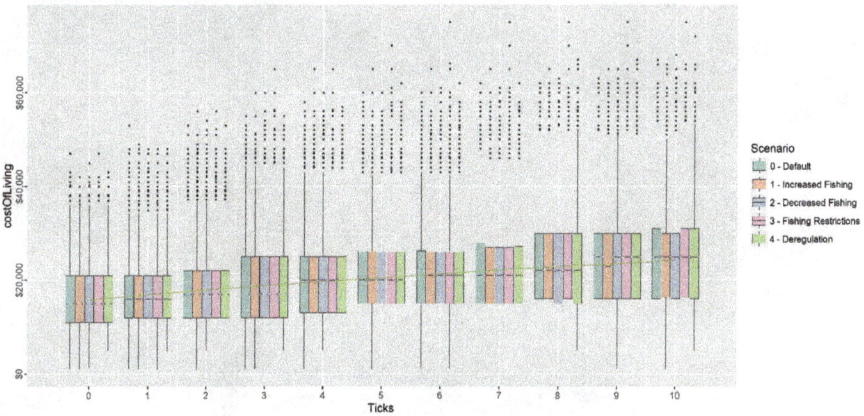

Fig. 24.4 Household cost of living across all scenarios

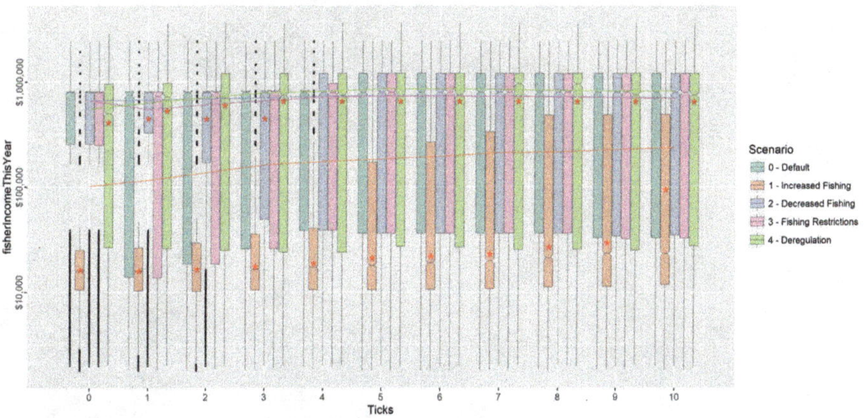

Fig. 24.5 Household income of fishers across all scenarios

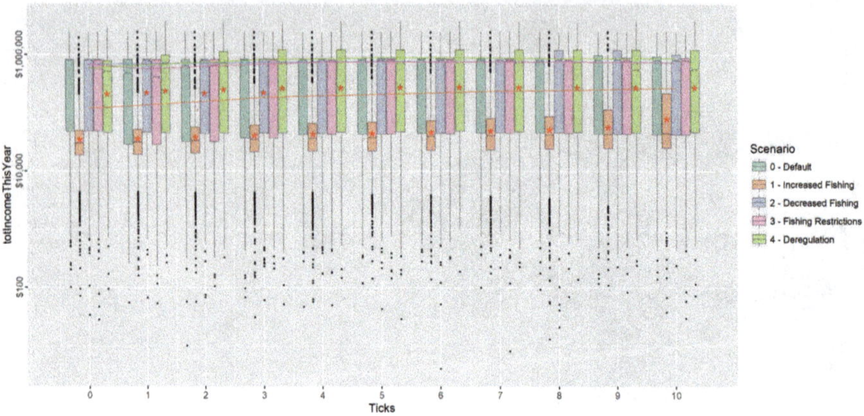

Fig. 24.6 Total household income across all scenarios

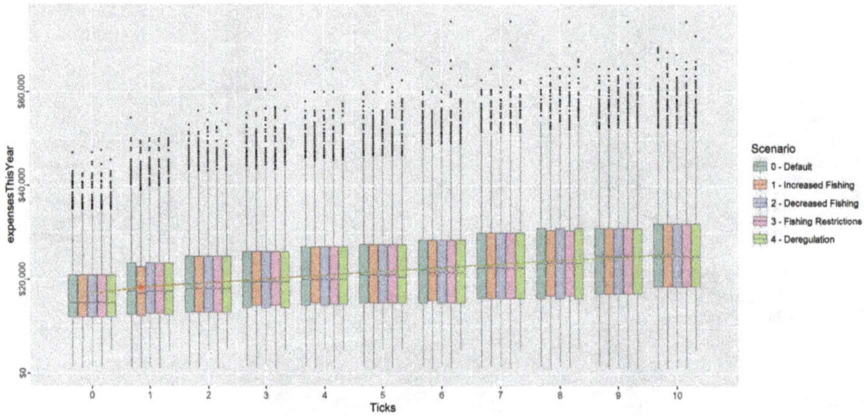

Fig. 24.7 Total household expenses across all scenarios

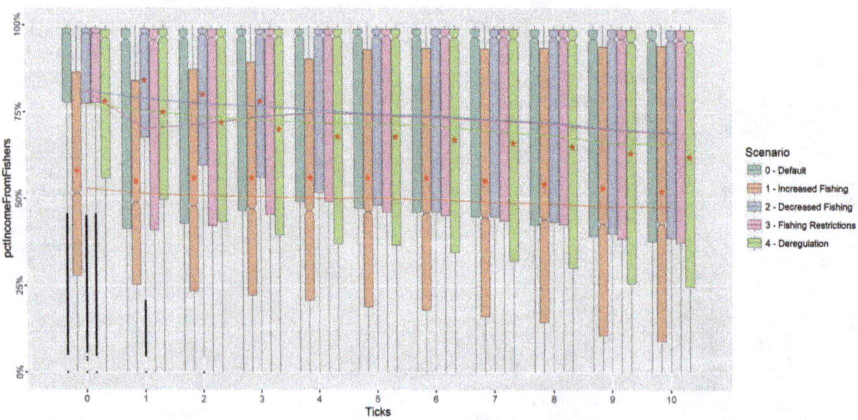

Fig. 24.8 Percentage of household income from fishers across all scenarios

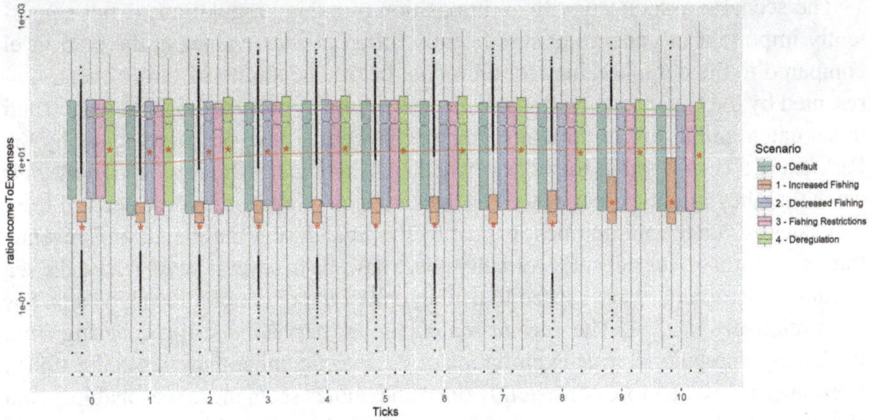

Fig. 24.9 Income-expense ratio across all scenarios

Figure 24.3 shows household wealth across all scenarios. In nearly all situations, the scenarios that indicate an increase in fishing jobs, a decrease in fishing jobs, and deregulation of the fishing industry are significantly different (t-test at the 0.05 level) compared to the default values at each model tick. Figure 24.4 shows the household cost of living across all scenarios. The scenarios related to an increase in fishing jobs and deregulation of the fishing industry are significantly different (t-test at the 0.05 level) compared to the default values at each model tick. Fig. 5 shows the income from fisher households across all scenarios. In nearly all situations, the scenarios that indicate an increase in fishing jobs, a decrease in fishing jobs, and deregulation of the fishing industry are significantly different (t-test at the 0.05 level) compared to the default values at each model tick. Figure 24.6 shows the total household income across all scenarios. The scenarios related to an increase in fishing jobs and deregulation of the fishing industry are significantly different (t-test at the 0.05 level) compared to the default values at each model tick. The scenario related to a decrease in fishing jobs is significantly different (t-test at the 0.05 level) compared to the default values at the initial three model ticks. Figure 24.7 shows the total household expenses across all scenarios. No scenarios were significantly different (t-test at the 0.05 level) compared to the default values for each model tick. Figure 24.8 shows the percentage of household income from fishers across all scenarios. The scenarios related to an increase in fishing jobs and deregulation of the fishing industry are significantly different (t-test at the 0.05 level) compared to the default values at each model tick. The scenario related to a decrease in fishing jobs are significantly different (t-test at the 0.05 level) compared to the default values at the initial three model ticks. Fig. 9 shows the income-expense ratio of households across all scenarios. The scenarios related to an increase in fishing jobs and deregulation of the fishing industry are significantly different (t-test at the 0.05 level) compared to the default values at each model tick. The scenario related to a decrease in fishing jobs is significantly different (t-test at the 0.05 level) compared to the default values at the initial three model ticks.

The scenario associated with an imposition of fishing regulations is not consistently important in generating significantly different t-test values at the 0.05 level compared to the default values for all model ticks. The ability of government, represented by the Galapagos National Park, to institute or prevent a fishing season and to set catch quotas are important regulatory powers. These decisions made by the Park exert a considerable impact on whether fishers can fish and the size of their catch that they might expect as well as the anticipated level of income derived from fish sales to local communities, export to the Ecuadorian mainland, and revenue through international sales. Historically, there have been several years where the sea cucumber season has been closed to fishing and/or the quota of allowable catch has been relatively modest. The current sea cucumber harvest is still recovering from the largely unregulated over-exploitation of the sea cucumber fisheries in the 1990s. Lobster seasons are more consistently open and quotas seem more reasonable, given stated available stocks by Park managers and catch expectations of fishers. Even with fewer jobs in fishing and additional jobs in tourism and government, represented in the scenario, the ***GF-ABM*** runs did not show significant differences compared to default values. In short, the default values and the model values were sufficiently similar, thereby, reducing the differences in t-test values at the 0.05 level across model ticks.

For the scenario that emphasizes an increase percentage of jobs in fishing (60–80%), decrease percentage of jobs in tourism (20–10%), and decrease percentage of jobs in government (20–10%), significant differences from default values across nearly all model ticks were observed. Registered fishers are either full-time or part-time, or they have transitioned to alternate employment, generally, in tourism. For part-time fishers, their transition back to full-time fishing is relatively easy to achieve, while a transition of former fishers who now have part- or full-time jobs in tourism is more difficult to achieve when conditions and opportunities warrant. There are opportunity costs associated with job transitions from tourism to fisheries as well as operational issues to confront, particularly, if they do not own their own boat, hence, making arrangements for a short-term return to fishing as a hired fisher is more difficult to achieve when many are seeking the same job transition. Family and friendship ties are an important social network for finding jobs in fisheries, whereas those who own boats and hire them out to other licensed fishers could easily be readied for fishing when circumstances dictate financial opportunities.

For the scenario associated with a decrease percentage of jobs in fishing (60–10%), an increase percentage of jobs in tourism (20–55%), and an increase percentage of jobs in government (20–35%), significant differences from default values across nearly all model ticks were observed. These factors are important motivations for fishers to seek other forms of employment, but most fishers wish to fish even when difficult circumstances occur. Many fishers are relatively unequipped to seek other forms of employment in tourism due to limited education levels and poor English language skills. For those that are able to move from fishing into tourism, jobs are relatively plentiful, as fishers are well equipped for work on the cruise boats as sailors and deck hands. But with only approximately 75 licensed tour boats operating in the Galapagos, there are employment limits for fishers even during periods of high demand by tourists.

Deregulation of the fishing industry can occur through several mechanisms. In this scenario, the total jobs modifier was increased (1.5–2.0), percentage of jobs in fishing decreased (60–45%), percentage of jobs in tourism increased (20–45%), percentage of jobs in government decreased (20–10%), probability of a sea cucumber season increased (33–90%), probability of a lobster season increased (85–90%), and the cost of obtaining a sailing certification decreased ($1000–$250). Relaxing government regulations related to the fishing industry would be welcome news for fishers. They have long complained of over regulation by the Galapagos National Park that prevents fishers from consistently obtaining a fair wage for their efforts. Fishers have lamented the closure of fishing seasons, often for sea cucumbers, but for lobster as well. In addition, the setting of fishing quotas has appeared to be made without scientific rigor or reliable population counts for various fishing zones. While many fishers feel compelled to move into non-fishing jobs, permanently or temporarily, a relaxation of regulations would generate a change in household livelihood strategies that would more explicitly value fishing as a viable alternative.

Conclusions

Population migration is the primary driving force that is shaping the Galapagos Islands of Ecuador. Whether through the temporary migration of tourists or the semi-permanent or permanent migration of population, primarily from the Ecuadorian mainland, relocating in the Galapagos for jobs in the burgeoning tourism industry, the movement of people and products to the islands has considerable implications for the present as well as the future. With a strong correlation between the number of tourists and the number of residents, rapid increases in national and international tourism places stress on community infrastructure as migration brings new workers to the Galapagos, often through social networks. Unfortunately, development and smart growth lags considerably behind the rate and number of migrants entering the islands, producing a situation in which community services and social programs are generally deficient, falling well below what is expected and demanded by visitors and residents alike (Johannes de Haan et al. 2019). Over 80% of the residential population are involved in the tourism industry, and as more tourists arrive in these "Enchanted Islands," greater pressure is exerted on the touristic visitation sites distributed throughout the archipelago (Pizzitutti et al. 2017). Few high-quality visitation sites are still available to accommodate the expanding tourism industry, so challenges persist in the management of the Galapagos National Park and Marine Reserve for tourism and for jobs by residents in the tourism industry, centered in communities having limited development infrastructure to accommodate social and needs and programs, such as, water quality, healthcare, and education. At the same time, the complex interactions between people and environment are critical to the future of the Galapagos Islands (Walsh and Mena 2016). Maintaining high quality ecosystems and their goods and services, iconic species and landscapes, terrestrial and marine touristic visitation sites, and a sustainable ecological system that balances the needs of the human dimension are vital, but elusive to the local population, conservationists, and managers within the Galapagos Islands.

Aside from the current implications of the pandemic on the Galapagos, the tourism industry has become the leading economic sector in the Galapagos, completely over-shadowing employment and revenues related to fisheries and agriculture. Like other places around the globe, the near complete reliance on a single economic sector can induce a socio-economic vulnerability when conditions that support the industry are curtailed or suspended through endogenous or exogenous forces of change. Equilibrium conditions are dynamic, shocks to systems can change future trajectories, and feedback mechanisms that reorganize systems can surprise and frustrate. The use of spatial simulations is a useful approach for considering scenarios of change and the primary factors that influence system dynamics, encourage adaptive behaviors of people and programs, and facilitate the exploration of alternate futures.

The *Galapagos Fisheries-Agent Based Model* (***GF-ABM***) has proved to be an effective tool to assess "what if" scenarios of change in the Galapagos Islands, and their implications on household livelihood alternatives achieved through job diversification strategies, primarily, in tourism and government employment as well as the customary employment sectors in fisheries. The scenarios tested explores household adaptation to typical forces and factors of change that represent key social-ecological dynamics in the Galapagos Islands. The intention of the modeling is to assess what really matters in the behavior of fisheries as they seize job opportunities in tourism and government when jobs in fishing increased, jobs in fishing decreased, fishing restrictions increased, and fishing regulations decreased. Household plots are presented for each of the four scenarios and the default or base-line conditions for each tick of the model are noted relative to modelled scenario outcomes. Results indicate the relevance and important of ***GF-ABM*** for assessing household alternatives and behavioral shifts by fishers to changing employment conditions and exogenous factors of change. The model outcomes reaffirm the importance of endogenous and exogeneous forces of change in the Galapagos Islands that influence fisher behaviors and their adaptation in the face of opportunities and shocks to base-line systems. Government regulations, market conditions, ENSO-events, and labor alternatives in non-fishing enterprises are associated with livelihood decisions made by households and alternative job diversification strategies made in the wake of uncertainty.

Acknowledgements This work would not have been possible without the assistance and cooperation of Brian Frizzelle, Carolina Population Center, University of North Carolina at Chapel Hill and Kim Engie, Department of Geography, University of North Carolina at Chapel Hill.

References

Cinner JE, Daw T, McClanahan TR (2009) Socioeconomic factors that affect artisanal fishers' readiness to exit a declining fishery. Conserv Biol 23(1):124–130

Engie K (2015) Adaptation and shifting livelihoods of small-scale fishers in the Galapagos Marine Reserve, Ecuador. PhD dissertation (Department of Geography, University of North Carolina at Chapel Hill, Chapel Hill, NC)

Engie K, Quiroga D (2014) The emergence of recreational fishing in the Galapagos Marine Reserve: adaptation and complexities. In: Denkinger J, Vinueza L (eds) The galapagos marine reserve: a dynamic social-ecological system, Galapagos Good Series, "Social and Ecological Interactions in the Galapagos Islands," Walsh, S.J., Mena, C.F., Editors, pp 203–226

Epler B (2007) Tourism, the economy, population growth, and conservation in Galapagos. Charles Darwin Foundation, Puerto Ayora

Gonzalez JA, Montes C, Rodríguez J, Tapia W (2008) Rethinking the Galapagos Islands as a complex social-ecological system: implications for conservation and management. Ecol Soc 13(2) dic.dlib.indiana.edu/dic/handle/10535/2456

Hearn A (2008) The rocky path to sustainable fisheries management and conservation in the Galapagos marine reserve. Ocean Coastal Manag 51(8–9):567–574

Hearn A, Martinez P, Toral-Granda VM, Murillo JC, Polovina J (2005) Population dynamics of the exploited sea cucumber Isostichopus fuscus in the western Galapagos Islands, Ecuador. Fish Oceanogr 14(5):377–385

Honey M (2008) Ecotourism and sustainable development: who owns paradise? Island Press

Johannes de Haan F, Quiroga D, Walsh SJ, Bettencourt L (2019) Scales and transformative change – transitions in the galapagos. In: Urban Galapagos – transitions to sustainability in complex adaptive systems (Kvan T, Karakiewicz JA, Guest eds), "Social and ecological interactions in the Galapagos Islands" (Walsh SJ, Mena CF, Series eds). Springer, pp 43–58

Kerr SA (2005) What is small Island sustainable development about? Ocean Coastal Manag 48(7–8):503–524

Mahon R, McConney P, Roy RN (2008) Governing fisheries as complex adaptive systems. Mar Policy 32(1):104–112

Malanson GP, Walsh SJ (2015) ABM: individuals interacting in space (invited commentary). J Appl Geogr 56:95–98

Mena CF, Walsh SJ, Frizzelle BG, Malanson GP (2011) Land use change of household farms in the Ecuadorian Amazon: design and implementation of an agent based model. J Appl Geogr 31(1):210–222

Miller BW, Breckheimer I, McCleary AL, Guzman-Ramirez L, Caplow SC, Walsh SJ (2010) Using stylized agent-based models for population-environment research: a case from the Galapagos Islands. Popul Environ 31(6):401–426

Pizzitutti F, Mena CF, Walsh SJ (2014) Modeling tourism in the Galapagos Islands: an agent-based model approach. J Artif Soc Soc Simul 17(1):1–25

Pizzitutti F, Walsh SJ, Rindfuss RR, Reck G, Quiroga D, Tippett R, Mena CF (2017) Scenario planning for tourism management: a participatory and system dynamics model applied to the Galapagos Islands of Ecuador. J Sustain Tour 25(8):1117–1137

Pollnac RB, Pomeroy RS, Harkes IHT (2001) Fishery policy and job satisfaction in three Southeast Asian fisheries. Ocean Coast Manag 44(7–8):531–544

Rozzi R, Massardo F, Cruz F, Grenier C, Muñoz A, Mueller E (2010) Galapagos and Cape Horn: ecotourism or greenwashing in two emblematic Latin American archipelagoes? Environ Philos 7(2):1–32

Schuhbauer A, Koch V (2013) Assessment of recreational fishery in the Galapagos marine reserve: failures and opportunities. Fishery Res 144:103–110

Taylor JE, Hardner J, Stewart M (2008) Ecotourism and economic growth in the Galapagos: an Island economy-wide analysis. Environ Dev Econ 1:139–162

The Nature Conservancy (TNC) (2017) What is Ecotourism? Eco-tips and travel. Retrieved from https://www.nature.org/greenliving/what-is-ecotourism.xml

Villacis B, Carrillo D (2013) The socioeconomic paradox of Galapagos. In: Walsh SJ, Mena CF (eds) Science and conservation in the Galapagos Islands: frameworks & perspectives, Social and Ecological Interactions in the Galapagos Islands (Walsh SJ, Mena CF, Series Editors). Galapagos Islands, Ecuador, pp 69–85

Walsh SJ, Mena CF (2016) Interactions of social, terrestrial, and marine sub-systems in the Galapagos Islands, Ecuador. Proc Natl Acad Sci 113(51):14536–14543

Walsh SJ, Engie K, Page PH, Frizzelle BG (2019) Demographics of change: modeling the transition of fishers to tourism in the Galapagos Islands. In Urban Galapagos – transitions to sustainability in complex adaptive systems (Kvan T, Karakiewicz JA, Guest eds). "Social and Ecological Interactions in the Galapagos Islands" (Walsh SJ, Mena CF, Series Editors), Springer, pp 61–83.

Watkins G, Cruz F (2007) Galapagos at risk: a socioeconomic analysis of the situation in the archipelago. Charles Darwin Foundation, Puerto Ayora

Wilen JE, Stewart M, Layton DF (2000) Economic analysis of the Galapagos marine reserve resource management plan, Final report. University of California, Davis

World Wildlife Fund. (2003). Migration and environment in the Galapagos Islands, Quito, Ecuador

World Wildlife Fund. (2012). Carbon footprint of tourist activity on Galapagos Islands, Santa Cruz Island, Galapagos, Ecuador

Stephen J. Walsh, Co-Guest Editor, is the Founding Emeritus Director of the UNC Center for Galapagos Studies and the UNC-USFQ Galapagos Science Center. He is also an Emeritus Distinguished Professor of Geography and is a Reserch Professor in the UNC Department of Geography. His research focuses on understanding the multi-scale and multi-thematic drivers of land use/land cover change in vulnerable and fragile ecosystems through remote sensing, spatial analysis, and spatial simulation model of "what if" scenarios of change.

Carlos F. Mena, Co-Guest Editor, Co-Director of the Galapagos Science Center, Director of the Institute of Geography, and Professor in the College of Biological & Environmental Sciences, Universidad San Francisco de Quito, Ecuador. His research assesses the importance of land use/ land cover change on social and terrestrial systems in the Galapagos Islands as well as in the Ecuadorian Amazon.

Chapter 25
The Role for Scientific Collections and Public Museums in Island Conservation

John Dumbacher and Jaime A. Chaves

Introduction

Conservation and land management require deep knowledge of species' natural history, including information about species ranges, historical habitat associations, climate impacts, species interactions, and other key data. Natural history information contains clues about such things as the causes of species declines, shifting of species ranges, and the sources or timing of invasive species. In addition to basic knowledge to effectively carry out conservation action, the broader community needs to understand the situation, embrace the challenges, and help craft and enact solutions. Thus, natural history museums can provide key information about natural history, they are repositories for documenting the actions of today, and they provide vital connections between science and the broader community.

Natural history museums are scientific institutions that collect and preserve specimens of natural history objects. These objects document a variety of natural history phenomena, but central among these are specimens made from living species, which form the most complete, permanent, and tangible record of life on earth. Biological collections form a sort of reference library for living species, and they document not

J. Dumbacher (✉)
Department of Ornithology and Mammalogy, California Academy of Sciences, San Francisco, CA, USA
e-mail: JDumbacher@calacademy.org

J. A. Chaves
Department of Biology, San Francisco State University, San Francisco, CA, USA

Department of Ornithology & Mammalogy, California Academy of Sciences, San Francisco, CA, USA

S. J. Walsh et al. (eds.), *Island Ecosystems*, Social and Ecological Interactions in the Galapagos Islands, https://doi.org/10.1007/978-3-031-28089-4_25

only the individuals and species that are preserved, but also the times and places that they lived. Traditionally, specimens have provided the scientific foundation for stable taxonomy and the naming of species, but they are also used for many other types of biodiversity inquiry - including genetics, systematics, population studies, biogeography, ecology, zoonoses and disease, invasive species, biology of extinctions, habitat associations and niches, toxicology, etc. (Suarez and Tsutsui 2004; Wandeler et al. 2007). These studies provide critical data for conservation management, sustainable planning, and addressing current and future issues like climate change, species range and population changes, and environmental threats (Winker and Zink 2005; Winker 2004).

In addition to the collections themselves, museums have missions that include scientific research programs, education at a variety of levels, exhibits, collaborative engagement, outreach, and a breadth of public programs, and in many cases, direct conservation action or science activism. These each offer opportunities to leverage the collections to increase the impact of these activities, and to work independently to advance science and public education goals.

In some ways, museums have never been so public and accessible. The internet has made it possible to share collections data remotely, but also to share derivative information, such as digitized specimen images, CT scans, and genetic sequences. These data can be accessed on museum websites, but also in aggregated data sites such as GBIF (the Global Biodiversity Information Facility, https://www.gbif.org/), GenBank (https://www.ncbi.nlm.nih.gov/genbank/), MorphoSource (https://www.morphosource.org/), and others. Furthermore, the uses for museum specimens in research have expanded significantly since their original collection. Many of the research and education tools available now were not even imagined when specimens were first collected.

Despite this, the average person seems to understand less today about a museum's collection or how it is used. In the past, museums were a primary source of information for exotic or foreign species or for close-up viewing of organisms that were typically difficult for the average person to see. Today, video, other publishing methods, and even inexpensive travel have largely overshadowed the museum's impact in these traditional realms. Although many people continue to visit pubic exhibits at museums, fewer are aware of the breadth and depth of collections held at museums or of the expanse of their work (Hein 2000; Bradley et al. 2014). Museum educators are often reluctant to use real specimens in public programs or to share the numbers of specimens stored behind the scenes - especially of vertebrate animals - as there are public misperceptions that specimens were collected frivolously, that animals suffered needlessly, or that specimens were collected primarily to entertain visitors (Johnson 2018; Filardi 2015). Often even new museum board members do not understand the value, utility, importance, and irreplaceability of collections, and have suggested that we dispose of items that are not used often or recently. And as universities turn toward higher-cost research and away from traditional museum education and research, universities are divesting of their museums, and so fewer science students and young science professionals learn of the value and relevance of collections to their own work.

Here, we review what is the core work of museums, and how the typical vision and mission of public museums can be leveraged for island conservation, and we explore how museums could adapt to have greater impact in the future. We focus our examples on the Galápagos Islands, due it its rich history with scientific collections and natural history studies, and because its current conservation challenges are common among many islands. In addition to science, we also discuss the role of public museums in being a resource for local communities, to help engage people of all ages, and be a gathering place for education and important discussions of policy regarding natural history, sustainable management, and other nature-related themes. And we highlight the kinds of work that museums can and should do in service of island sustainability—a timely topic as our home institution, the California Academy of Sciences, embarks on a new commitment to work on islands. The purpose of this paper is to remind those interested in conservation and sustainability about the key work of museums and how we can leverage collaborative partnerships to support museum collections, improve conservation outcomes, and to also challenge museum professionals to engage creatively in island conservation.

The Collections

The Physical Specimen

The *voucher* or physical specimen is the sample taken from an organism to document what it is, what it looks like, and usually those key components used to identify it and study it. Examples of typical specimens vary broadly among taxa (Fig. 25.1). For plants it might be a whole small herbaceous plant pressed and dried, or for larger plants it may be a small branch with dried leaves and flowers. For marine invertebrates, fish, or herps it might be the entire animal fixed in formalin and stored in alcohol. For arthropods, these are often pinned and dried whole insects, or whole arachnids stored in ethanol. For birds or mammals, it may be an empty skin stuffed with cotton to look somewhat like the animal that it came from. But it could also be a bird egg, a skull or partial skeleton, a blood or tissue sample, a piece of baleen from a whale, a fossil, or any other whole or partial organism that is preserved for study. The goal has been to collect as much material from the original specimen as possible, focusing on the materials that are believed to serve science the most, while keeping the costs for specimen preparation, curation, and storage as modest as possible. As scientific uses for specimens have expanded, modern museums seek to preserve more material from each specimen taken, including frozen tissues, DNA samples, and try to preserve specimens in ways that reduce damage or contamination during preservation so that there is more material available for future study.

Fig. 25.1 Typical museum collection specimens. Clockwise from top left: botanical herbarium sheets; pinned insect tray; alcohol preserved fish and herps; avian museum skins; anthropology artifacts; frozen tissues

The Specimen Data and Metadata

In addition to specimens, researchers record a variety of primary data or metadata about each specimen in the collection catalog, and these data describe the specimen and the circumstances of the collection. Data fields include the date and locality of the collection, the person who made the collection, various measurements that might not be obtainable from the prepared specimen, and often much other information. These data fields are standardized (Wieczorek et al. 2012), which also allow data from many collections to be aggregated and served online on data portals such as Vertnet or GBIF (Constable et al. 2010; Robertson et al. 2014) (see the example in Table 25.1 and Fig. 25.2). Often biologists take copious field notes containing additional information about the expedition that contextualizes the collection, and even fieldnotes have been standardized to reliably capture the most important data,

Table 25.1 Specimens of *Geospiza* (Galápagos Finches) from museums worldwide found in a Global Biodiversity Information Facility (GBIF) search

Museum	Total Number of specimens	Earliest	Latest
Sam Noble Oklahoma Museum of Natural History	11	1962	1962
The University Museum of Zoology, Cambridge	16	1891	1897
Cornell University Museum of Vertebrates	18	1860	1897
University of Michigan Museum of Zoology	21	1891	1981
University of Oslo	25	1925	1925
Leibniz Institute for the Analysis of Biodiversity Change	31	1962	1963
Denver Museum of Nature & Science	38	1960	1960
Royal Ontario Museum	42	1890	1973
Naturalis Biodiversity Center, Leiden	50	1835	1966
Carnegie Museums	60	1936	1941
Senckenberg. Collection Aves	87	1891	1957
Natural History Museum of Los Angeles	88	1932	1957
Academy of Natural Sciences, Philadelphia	119	dates not listed	
Natural History Museum (London)	132	1835	1902
Museum of Comparative Zoology, Harvard University	143	1700	1933
Charles Darwin Research Station	201	1962	2021
Field Museum of Natural History	212	1891	1974
UC Berkeley Museum of Vertebrate Zoology	493	1818	1964
National Museum of Natural History, Smithsonian	1325	1868	1980
American Museum of Natural History	1357	1891	1935
California Academy of Sciences	3570	1898	1978

The search returned 8128 finch specimens, including 15 institutions that held fewer than 10 specimens total (not shown in table). Note that only the Charles Darwin Research Station has specimens collected after the 1980s

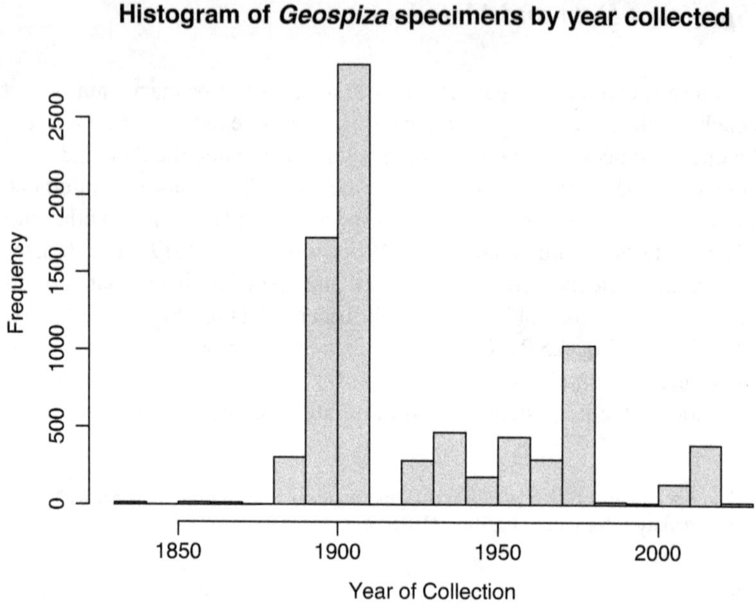

Fig. 25.2 Histogram of *Geospiza* specimen records from Global Diversity Information Facility (GBIF) search, by year collected. Note that the earliest collections are those made by Charles Darwin, and the most recent specimens are at the Charles Darwin Research Station

especially for vertebrates (Herman 1986). Serving these data online facilitates researchers anywhere in the world who do not require the actual specimen, however having a voucher allows researchers to verify the authenticity of the data, check the identity of the specimen in question, and allows further investigation.

In addition, scientists derive other data from collections, and these should be included or linked to the actual collections. Derivative data include genetic sequence data, photos and 3-D images, CT-scans, stable isotope analyses, morphological measurements, and even scientific findings or reports based upon collections (Fig. 25.3). Ideally these should be integrated with the primary catalog data and be at least as accessible as the specimens themselves.

These practices allow collections to be easily searchable, cross-referenced, linked to other data, and used for a variety of purposes. Thus, the specimen is not just a sample of a particular species, but it is also a sample taken from that *time period* and that *place*, and contains information about local environments, environmental contaminants, diet, parasites and pathogens, reproductive phenology, genetics and many other phenomena. Today, many researchers who visit museums to study the specimens are extracting data and doing projects that could not have been imagined by the original collectors.

Fig. 25.3 Derivative data and digitized products, all of which are data rich and can be served online to users anywhere in the world. Example include (**a**) data from specimen tags, (**b**) high resolution images, (**c**) computed tomography scans, (**d**) DNA sequence data

Core Functions: Building, Maintaining, and Using the Collections

The core work of collections professionals is to *build, maintain, and use* the collections. *Building* the collections involves adding new specimens and data. Collections were traditionally built with a particular focus and goal and often to create scientific collections for taxonomy and systematics. This is still done for many taxa (e.g., arthropods, marine invertebrates, some plant groups) or for localities that are poorly known, but as there are many other uses of specimens, collections can and should continue to grow even when taxonomy is considered "finished" (Winker 2004; Winker et al. 1991; Patterson 2002; Remsen 1995). Even when there is concern for declining populations or ethical reasons for not killing certain animals, collecting can usually be done humanely with little to no impact at the population level (Winker 2000; Winker et al. 2010; Patterson 2002). Furthermore, today's scientific collectors must obtain multiple permits and pass institutional ethics committees that help to ensure appropriate and sustainable collecting (Rocha et al. 2014). Building collections also involves anticipating what researchers may need in the future, and making sure that needed materials and data are preserved (Winker 2000). Without doubt, the greatest importance of these collections may be the unpredictable research value these will have in the near or long-term future to help solve problems that we cannot yet imagine (Remsen 1995).

Maintaining the collections is the essential work of ushering the collections safely into the future and is a field unto itself supported by scientific societies (e.g., the Society for the Preservation of Natural History Collections). Curators and collection managers are responsible for keeping specimens and data safe from loss or degradation (e.g., insects, fungus, and bacterial pests, theft, light or temperature damage, damaging chemicals, etc.), but their work also involves day-to-day record keeping, refreshing preservative chemicals, maintaining appropriate air-handling systems, excluding pests, and many other tasks. In addition, museum staff oversee loans and access to the collections, and like librarians, they make sure that records are kept of specimen uses, that they are used appropriately and returned in good condition, and that specimens are reshelved in their proper place.

Finally, museums must *use* the collections for the good of science and society. This means trying to make sure that they have maximum value and utility to society in every way. This is not meant just for in-house researchers, but the specimens should be shared with anyone with a bona fide need to use them. This involves traditional uses, but we must always be innovating new ways to leverage and use the data present in collections (Webster 2018; Winker 2004; Winker et al. 1991). In addition to scientific research, museum collections are used for teaching, exhibits, science communication, various public programs - including in the production of field guides, educational materials, videos, and podcasts. Collections and their data are used by various government agencies to assess management needs and to assist in law enforcement. Given our current biodiversity crisis, we have a responsibility to leverage the collections for management and sustainability, and to assist researchers who need access to specimens.

The Expanded Mission of Museums

Because of the many uses of the collections and the public interest in museums, the mission of museums has grown to include many activities and programs that use or benefit from the collections, including public exhibits, education, public programs, citizen science, conservation, and others (Hein 2000; Dillenburg 2011; Weil 2007). Even during these polarizing times, science museums have enjoyed strong reputations as trusted institutions and trusted sources of scientific information, partly because they are considered fact-based and non-partisan (Bradley et al. 2014, American Alliance of Museums 2021), and thus it is as important as ever to be creative and active in the larger public learning environment (Sutton et al. 2017; Wood 2018; Flowers 2022).

Public exhibits are well-known museum programs. Although exhibits formerly focused on showing and interpreting specimens, museum exhibits have grown to communicate and experience science more broadly. Exhibits provide excellent opportunities to share a more intimate view of the organisms, to illustrate important aspects of their natural history, or showcase other work that is being done by scientists (Fig. 25.4).

In addition to exhibits, museums often offer lectures, classes, teacher or educator training, design modules for augmenting classroom education, support citizen or community science, etc. It is critical for museums to assess and work with their audience and be creative and collaborative to design public programs that will fulfill the unique needs of their communities (Flowers 2022). In Galápagos, for example, there are a number of engaged and interested constituents, including the Galápagos National Park, ecotour guides, tourism business partners, local schools, engaged agricultural and fishing industries, and the many other people that engage with these primary industries. They are all affected by nature, by the history of the islands, by the species of plants and animals that live there, and for most of them, their health and the health of their businesses are directly tied to a healthy ecosystem. Nearly everyone seems interested in the science that is being done, what it means, and how it affects them. It is critical to have institutions where people can gather reliable first-hand information or even participate in the science that affects them.

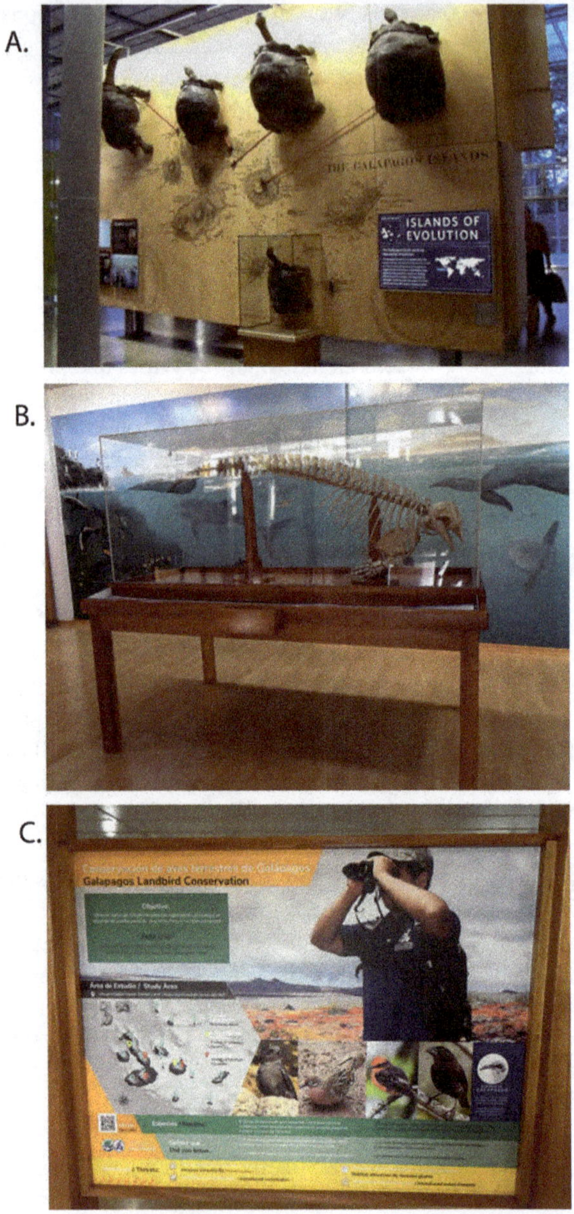

Fig. 25.4 Museum exhibits can use specimens or be purely interpretive. (**a**) Tortoises showing variation among islands, (**b**) an articulated skeleton of Dwarf Sperm Whale at the Charles Darwin Foundation, (**c**) an interpretive placard at CDF highlighting research done by their researchers

How Museums and Collections Support Island Conservation

The Traditional Work of Museums: Taxonomy and Systematics

Accurate species taxonomies are critical for conservation work. Conservation actions are often designed to prevent species extinction or use species or other taxonomic designations (e.g., subspecies, varieties) in legislation (Mace 2004; Garnett and Christidis 2017). Accurate species-level taxonomies are necessary to identify what species are present in an area, which species are under threat, or which areas have high species richness. The IUCN red list relies upon taxonomic hypotheses to prioritize species to protect, identify threats to species, and recommend conservation actions (Vogel Ely et al. 2017). Without taxonomies that are based in science, broadly accepted, and well supported, it can be impossible to properly prioritize populations or areas that need the most help or attention, and it may be impossible to legally defend particular actions without accepted taxonomies (Garnett and Christidis 2017; Garnett et al. 2020).

The classic work of museums has been to describe and classify species, document their geographic ranges, and study their relationships. This requires sizeable collections to understand the variation present at many levels among individuals and parse it into its various causes, including variation by sex, age, condition, geography, as well as the intraspecific variation among individuals, and then to sort these into species. For example, the first scientific collections were made in the Galápagos by Charles Darwin during the voyage of the HMS Beagle (Darwin 1839), and the first published accounts of the birds were published in collaboration with John Gould and his wife Elizabeth (Gould and Darwin 1839). Significant later collections of birds and other taxa include those made by Harvard biologist Louis Agassiz, by Georg Baur, by Rollo Beck and others for Walter Rothschild in England (who later also purchased many of Baur's collections), and Snodgrass and Heller for Stanford University. The largest collection of specimens to date was made by the California Academy of Sciences (James 2018, 2020) in a multi-disciplinary expedition that spent over a year in the Galápagos Islands during 1905–1906. Each of these collections resulted in new descriptions of species, species' ranges, and treatises on Galápagos birds (Ridgway 1896; Rothschild 1899; Rothschild 1903; Salvin 1876; Gifford 1913, 1919; Loomis 1918). With each collection of new specimens, a more definitive morphological taxonomy of bird species (Swarth 1929; Swarth 1931) was made possible, especially for the Galápagos finches which had proved difficult to convincingly parse into species (Dumbacher and West 2010). Adequate collections provided the foundation for these avian taxonomies, as is true for other taxa as well (e.g., tortoises (Bell and Scheinberg 2021, Van Denburgh 1914), lizards (Slevin 1935; Van Denburgh 1912a, b; Arteaga et al. 2019)).

In addition to understanding taxonomy and systematics, morphological work with specimens has also provided important insights into the ecology, adaptations, and evolution of species. Work on Galápagos finches has shown that bill size and shape has evolved in response to local diet and food availability (Bowman 1961) and in response to the presence of other competitors (Lack 1947). These and many other studies have become classic works in ecology and evolution.

Insights from Museum DNA

Most museum specimens were collected prior to our understanding of molecular genetics or even the discovery of DNA. Thus, the DNA that is still present in museum specimens tends to be degraded into short fragments and sometimes cross-linked or bound to proteins due to reactions with preservative chemicals. In recent decades, however, researchers have found ways to extract DNA from museum specimens that can be used in a variety of genetic research (Bi et al. 2013; Billerman and Walsh 2019; Tsai et al. 2020; Yeates et al. 2016; Wandeler et al. 2007). For example, DNA from old museum skins has allowed researchers to include members of extinct populations in studies of population genetics and phylogenies. This allowed Hoeck and colleagues to show that populations of Floreana Mockingbirds (*Mimus trifasciatus*) on both Champion and Gardner Islands harbor unique alleles that were formerly present on Floreana, and thus that both populations should be used to repopulate Floreana where they have been extinct since the late 1800's (Hoeck et al. 2010a, b). Work done on Galápagos Rail (*Laterallus spilonota*) museum specimens in combination with modern samples to show that they are most closely related to mainland Black Rails (*L. jamaicensis*) that dispersed to the Galápagos a little over one million years ago (Chaves et al. 2020). Likewise, in a range-wide phylogeny of Vermilion Flycatchers (*Pyrocephalus rubinus*), Carmi et al. (2016) included Galápagos subspecies from museum skins collected in 1899 and 1905–06. They showed that Galápagos birds are distinct species from mainland forms and showed that there is a deep split between San Cristobal Island populations and all other Galápagos Vermilion Flycatchers. These changes have been accepted by the ornithological community and have brought renewed and much-needed conservation focus to the Galápagos Vermilion Flycatchers (*Pyrocephalus nanus*) as well as to the San Cristóbal population (now known as *P. dubius*) which is on the brink of extinction if it is not already extinct (Jiménez-Uzcátegui et al. 2019). Similarly, a recent study of giant tortoise populations found that several specimens found in a cave on San Cristóbal Island were significantly different at both mitochondrial and nuclear genes (Jensen et al. 2022) and likely represent a previously undescribed and now-extinct taxon.

DNA from specimens collected in different years can be compared to document genetic changes in populations over time, especially to explore the loss of genetic diversity or changes due to introgression from other species. For example, Lawson et al. used both modern samples and historical museum specimens to show that Mangrove Finches (*Camarhynchus heliobates*) have been losing genetic diversity over 100 years of sampling, and have picked up diversity from nearby populations of Woodpecker Finches (*Camarhynchus pallidus*) (Lawson and Petren 2017).

These genetic studies are critical for island conservation, especially for management that moves individuals among populations, reintroduces species, or attempts larger rewilding efforts on islands where species have been lost (Perino et al. 2019; Taylor et al. 2017). Efforts to restore island ecosystems require an understanding of what species and populations were present, which are the closest living relatives, which adaptations might be needed, and therefore which populations to reintroduce.

Documenting and Studying Parasites and Wildlife Diseases

From the point of view of symbionts, including parasites and pathogens, the host is habitat and part of a larger ecosystem (Lutz et al. 2018). During the process of preparing a specimen, researchers examine the specimen closely, and with many taxa, even dissect the specimen. Often symbionts are discovered and can be collected and preserved, including mites, lice, ticks, botflies, or internal parasites like nematodes. Routine (e.g., frozen tissues) and special preparations (e.g., blood smears) can be made to document and preserve these. Even without special preparations, researchers have developed ways to extract or document diseases in older specimens. These have proven critical in discovering active infections that may affect the fitness of host species.

To help populations that are affected by disease, researchers need to identify the disease, accurately assess the threat, understand the source of the disease, which species are affected, or how long the pathogen has been present in the population. Museum specimens may have diagnostic lesions, pathogen DNA, or other characteristics that allow researchers to document the disease in former populations. For example, Avian pox (*Avipoxvirus*) has been identified as a serious threat in island bird systems and infects multiple Galápagos passerines. Patty Parker and colleagues were able to examine over 3600 bird specimens collected from the 1890's through the early 1900's (Parker et al. 2011). They found pox-like lesions that tested positive for avian pox in both histopathology exams as well as with genetic testing. Because Avian pox is a DNA virus, samples taken from these 100-year-old lesions contained both bird and pox DNA, and their PCR analyses were able to identify the strain of canarypox that was present. And because so many specimens were available from multiple islands, they were able to determine that the virus was introduced sometime in the mid to late 1890's, and at that time, it was mostly present on islands populated by people (Parker et al. 2011). Likewise, other researchers have successfully used museum specimens to document the presence of the "sin nombre" hantavirus in *Peromyscus* mice specimens, multiple chytrid fungus pathogens in amphibians, and papillomavirus in birds, and even avian malaria from museum tissues (Schmitt et al. 2019; Lutz et al. 2016; Fecchio et al. 2019; Pérez-Tris et al. 2011; Talley et al. 2015).

Mapping and Documenting Invasive Species

Islands are highly susceptible to invasive species, and most islands have complex histories of intentional and unintentional species introductions - many of which pose primary conservation threats. In Galápagos, there are over 450 introduced insect species (~23% of the total number of known species)(Causton et al. 2006), over 750 alien vascular plant species (Guézou et al. 2010), and 44 species of vertebrates (Phillips et al. 2012). Having good early collections is important for understanding when and how many of these species arrived in the Galápagos (Peck et al. 1998). Because many species are continuing to arrive in Galápagos (Toral-Granda et al. 2017), ongoing surveys and collecting are critical for establishing modern baselines of species presence as well as to monitor for new invasions. The introduction of the parasitic vampire fly, *Philornis downsi*, is an excellent example. The fly was first recognized in 1998 on Santa Cruz as a nest parasite on multiple species of Darwin's Finches (Fessl et al. 2001). Subsequent work has documented the destructive impact of the flies on nesting birds (Fessl et al. 2006, 2012; Dudaniec and Kleindorfer 2006). To determine the approximate timing that *Philornis* was introduced, researchers looked through museum collections of flies as well as collections of birds, looking for the characteristic scarring of the upper bill and nasal openings caused by the feeding larvae (Fessl et al. 2017). By looking through insect collections held in various museums, they found that the earliest known *Philornis* specimens were collected in Galápagos in 1964 on Santa Cruz Island from both the highlands and lowlands (Causton et al. 2006; Fessl et al. 2017), suggesting a recent introduction prior to or around the early 1960's. Kleindorfer and Sulloway examined museum specimens of finches, and they found no evidence of *Philornis*-induced damage to the nares or bill prior to 1962, suggesting again that the introduction likely occurred around the early 1960's (Kleindorfer and Sulloway 2016).

Environmental Contaminants

Although we know of no studies in Galápagos examining environmental toxins or contaminants from museum collections, specimens have been used elsewhere to test for mercury, heavy metals, and other persistent environmental contaminants, and even to study impacts of various toxins on animals. For example, studies of DDT in bird specimens were able to document both the presence of DDT in tissues, and the impact of eggshell thinning (Dosch 2007; Rocque and Winker 2005; Kiff 2005). As elevated marine mercury and marine plastics become more pervasive and their impacts on island biota are being hypothesized and tested, it is now possible to compare baseline levels of these contaminants from specimens collected long ago to those collected more recently, or around the time that these chemicals became more pervasive (Dahmardeh Behrooz and Poma 2021; Monteiro and Furness 1995). Similarly, the presence of soot particles in the feathers of museum specimens has documented exposure to contaminants before and after the passage of the Clean Air Act in the 1960s (Dubay and Fuldner 2017). These applications also require and justify continued collecting as pesticide use and other pollutants are continually changing.

Stable Isotopes, Diet, and Habitat Association

Common chemical elements (e.g., carbon, nitrogen) occur in nature with an array of different forms or isotopes that vary in molecular weight due to differing numbers of neutrons in the atom nucleus. The ratios of isotopes vary in predictable ways based upon habitat or biological processes that favor some isotopes over others. For example, the ratio of ^{15}N to ^{14}N increases with increasing trophic level, and the ratio of ^{13}C to ^{12}C increases as you move from marine to onshore environments. This has been exploited in a number of studies that have examined difference in diet among species, across time, or even to identify migratory routes in birds and whales (Hobson et al. 2014a, b; Boecklen et al. 2011). For example, researchers have measured stable isotopes from Galápagos tortoises collected during the 1800's to early 1900's to establish a baseline, and have shown that that modern tortoise diets have shifted over time from more grasses to more woody plants (Bell and Scheinberg 2021). Using carbon and nitrogen isotopes from Galápagos pinnipeds, researchers have documented differences in diet between Galápagos sea lions and Galápagos fur seals (Páez-Rosas et al. 2012), and shown that sea lions from different parts of the archipelago have colony-specific feeding areas and show fidelity to these areas throughout the year (Drago et al. 2016).

Collections of Fieldnotes and Photos: Other Museum Archives

Beyond the rich information that can be extracted from specimens, the museum archives stores fieldnotes, photos, letters, and other documents and memorabilia of biologists who made the collections and observed the species in their natural habitat. These documents contain details on behavior, habitat associations, the condition of the habitat, and various other ecological notes that are critical for assessing the ecology of species and understanding changes in the ecology over time. For example, Charles Darwin made notes on distributions of species that inspired some of his later classic works; Edward Winslow Gifford (California Academy of Sciences) observed and described tool use in the Woodpecker Finch in 1905–06 (Dumbacher and West 2010; Gifford 1906); and Bob Bowman (SFSU) took copious notes and field photographs regarding his collections and the state of the habitat during his work in the 1960's. Museum collectors (and sometimes seafarers) took notes on tortoise sources and their abundance, and sometimes other details reflecting the health of the populations and the health and condition of the individuals (Bell and

Scheinberg 2021). These records have been important for understanding the impacts of early seafarers, whalers, and pirates as well as local residents, and have helped develop an understanding of the historical condition of the islands, which is important for many restoration efforts.

Museums in the Galápagos

In the Galápagos, there are multiple active science centers, and at least one multidisciplinary science museum. The Charles Darwin Research Station, located in Puerto Ayora, Santa Cruz, currently hosts sizeable research collections of plants, insects and terrestrial invertebrates, marine invertebrates, and vertebrates that have been collected throughout the archipelago since their founding in 1959. The collections are digitized and data are available through their web datazone and via GBIF portals. In addition, they have an excellent library, archive, research laboratories, as well as sizeable research staff and a public exhibit space that is heavily visited by tourists. They are involved with multiple community programs as well as collaborative partnerships with other local environmental institutions.

Similarly, the Galápagos Science Center (GSC) was founded jointly by the Universidad San Francisco de Quito (USFQ) and the University of North Carolina, and is located in Puerto Baquerizo Moreno, San Cristóbal. It is a scientific research center that provides training at the university and other levels, conducts scientific research, and involves local students and other participants in these programs. Affiliated GSC researchers collect scientific specimens and they host a variety of programs for the community. Most specimens collected correspond to tissue samples (e.g., blood, tissue) or environmental samples stored in freezers as sources of genetic data for current and future studies. A number of specimens reside in the USFQ campus museum in Quito, and a growing number of samples are stored at the GSC. The GSC is also developing a biobank for the systematic collection and storage of samples and their associated data in accordance with international biobanking standards.

Despite the tradition of museum collections on the Galápagos, new approaches need to be incorporated so that the whole community supports the role of these institutions. Although several outstanding efforts have been made by these science centers to connect their mission to the community, work needs to be done to shift their local public image. Many local Galápagos residents are confused or misinformed about the work of museums and science centers so that many consider museums just cabinets of little-used artifacts useful only for school and tourist visits. By increasing exposure and access to positive specimen-based experiences for locals, it may be possible to demystify these collections and the science that is done in these institutions.

Future Recommendations

To continue to have positive impacts for science and sustainability, museums must continue to evolve and adapt to the latest challenges, research needs, and public requirements for information. For museums to remain relevant and continue to be a valuable community partner, we make some of the following recommendations.

1. Collect and preserve specimens today. It is important for our generation to preserve a record of what is here now for both today's and for future research. For many taxa such as plants, insects, soil, etc., we can preserve complete specimens without significant impacts on most populations. For taxa that are sensitive or rare, we can collect just DNA or tissue samples as a part of broader scientific research. We can also preserve salvage - specimens that were not killed for the collections but that died other ways, for example road kill, window kill, animals euthanized in wildlife hospitals, fisheries bycatch, or organisms that washed up dead on beaches. There are typically many more salvage specimens available than museum staff can possibly prepare. In building modern collections, we should try to include as many materials as possible, including DNA and RNA samples, and whole specimens prepared different ways, in order to anticipate future research needs. The Galápagos National Park could permit and incentivize park rangers and local research staff to collect salvage specimens. Failing to do so, important biological information will be lost forever, including data that could help identify and mitigate the factors causing mortality.
2. Build, fund, and empower local museums, science institutions, and local staff. Local institutions have greater rapport and connection with local communities, and they can be a place for gathering, education, and other community work. As museums are usually trusted institutions, local museums and staff are even more trusted than those from outside. In this age of dis-information, local museums provide a place for communities to share first-hand science and to work on challenges such as climate change and sustainability. Training and employment of local community members is being done well by both the Charles Darwin Foundation and the Galápagos Science Center, which continue to train and hire local curators, scientists, staff, and field personnel. This additionally helps to diversify the workplace and empower local community members to protect their local environment.
3. Digitize collections and collection data to make the information accessible to as many people as possible (Page et al. 2015). This should be done for all collections, but it is especially important to disseminate and connect the legacy data that are dispersed worldwide in other museums.
4. Design projects that do fundamentally great work, but that involve the community on a variety of levels as well as expand the collections. The Galápagos Barcode Project is an excellent example of science that is studying all forms of life in the Galápagos, creating amazing biodiversity resources in terms of both voucher specimens and genetic data, and has hired many local people to

participate in a variety of aspects of the project - especially during the COVID pandemic when tourism was not viable (see Chap. 29).

5. Researchers can and should leave data and specimens from their work in museums to document and memorialize the work they have done. In the course of scientific work, researchers often collect valuable field notes and many specimens or samples. Most these specimens are not completely used up in the research, and all too often when researchers retire or move to other institutions, the unused specimens are discarded or forgotten, buried in boxes or in freezers. At the very least, these should be offered to museums that can accession them and make them available to others who could use them. But ideally, researchers should coordinate to have unused material or legacy fieldnotes moved to collections early in their work, when the contents and value of the collection is fresh in their minds and when they can coordinate with the museum to help accession and preserve the materials. Museums are often able to track and store the material more efficiently than individual researchers or labs, and researchers can always request or access the material again at a later date.
6. Expand local and online impacts through a variety of innovative programs. In a place like Galápagos, visiting scientists should be asked to give a public seminar or evening salon on their work, and museums or other science centers are great places to host these. Programs will have to fit the local needs and interests, but the Galápagos in particular should be an easy place to do this. By hosting a variety of facilities tours and participatory programs for local community members, museums will help build mutual trust with the community and help build connections with science and sustainability.

International museums, including those in Galápagos do not operate in a vacuum. Museums must continue to collaborate on new initiatives, developing new tools for museum specimens, new ideas for public engagement, and always strive to be relevant with new social and biological challenges. It is not enough to do our traditional work well, we must constantly find new ways to grow the collections and to leverage museum assets in order to address important challenges of our times.

Acknowledgements We would like to thank the organizers and staff of the World Summit on Island Sustainability, the Galápagos Science Center, the Center for Galápagos Studies, and Steve Walsh, Emeritus Founding Director of the GSC for organizing this conference and the resulting book.

Literature Cited

American Alliance of Museums (2021). Influence of Museums: Museums and Trust 2021 [Online]. Available: https://www.aam-us.org/2021/09/30/museums-and-trust-2021/. Accessed 31 Aug 2022

Arteaga A, Bustamante L, Vieira J, Tapia W, Guayasamin JM (2019) Reptiles of the Galápagos: life on the Enchanted Islands, Quito, Ecuador, Tropical Herping

Bell RC, Scheinberg LA (2021) The collectors: beginnings of scientific inquiry and the lasting impacts of living and museum collections. In: Gibbs JP, Cayot LJ, Tapia Aquilera W (eds) Galápagos Giant tortoises: conservation from genes to landscapes. Academic, Cambridge, MA

Bi K, Linderoth T, Vanderpool D, Good JM, Nielsen R, Moritz C (2013) Unlocking the vault: next-generation museum population genomics. Mol Ecol 22:6018–6032

Billerman SM, Walsh J (2019) Historical DNA as a tool to address key questions in avian biology and evolution: A review of methods, challenges, applications, and future directions. Mol Ecol Resour 19:1115–1130

Boecklen WJ, Yarnes CT, Cook BA, James AC (2011) On the use of stable isotopes in trophic ecology. Annu Rev Ecol Syst 42:411–440

Bowman RI (1961) Morphological differentiation and adaptation in the Galápagos finches. Brussells, Palais Des Academies

Bradley RD, Bradley LC, Garner HJ, Baker RJ (2014) Assessing the value of natural history collections and addressing issues regarding long-term growth and care. Bioscience 64:1150–1158

Carmi O, Witt CC, Jaramillo A, Dumbacher JP (2016) Phylogeography of the vermilion flycatcher species complex: multiple speciation events, shifts in migratory behavior, and an apparent extinction of a Galápagos-endemic bird species. Mol Phylogenet Evol 102:152–173

Causton CE, Peck SB, Sinclair BJ, Roque-Albelo L, Hodgson CJ, Landry B (2006) Alien insects: threats and implications for conservation of Galápagos Islands. Ann Entomol Soc Am 99:121–143

Chaves JA, Martinez-Torres PJ, Depino EA, Espinoza-Ulloa S, García-Loor J, Beichman AC, Stervander M (2020) Evolutionary history of the Galápagos rail revealed by ancient mitogenomes and modern samples. Diversity 12:425

Constable H, Guralnick R, Wieczorek J, Spencer C, Peterson AT, Bart H, Bates J, Cotter G, Hanken J, Moritz C, Simmons N, Trueb L (2010) VertNet: a new model for biodiversity data sharing. PLoS Biol 8:1–4

Dahmardeh Behrooz R, Poma G (2021) Evaluation of mercury contamination in Iranian wild cats through hair analysis. Biol Trace Elem Res 199:166–172

Darwin CR (1839) Journal of researches into the geology and natural history of the various countries visited by H.M.S. Beagle. Henry Colburn, London

Dillenburg E (2011) What, if anything, is a museum? Exhibitionist. Spring, pp 8–13

Dosch JJ (2007) Jerald J. Dosch on dead birds' Tales: museum specimen feathers as historical archives of environmental pollutants. Environ Hist 12:661–665

Drago M, Franco-Trecu V, Cardona L, Inchausti P, Tapia W, Páez-Rosas D (2016) Stable isotopes reveal long-term fidelity to foraging grounds in the Galápagos Sea lion (*Zalophus wollebaeki*). PLoS One 11:1–16

Dubay SG, Fuldner CC (2017) Bird specimens track 135 years of atmospheric black carbon and environmental policy. Proc Natl Acad Sci 114:11321–11326

Dudaniec RY, Kleindorfer S (2006) Effects of the parasitic flies of the genus *Philornis* (Diptera:Muscidae) on birds. Emu 106:13–20

Dumbacher JP, West B (2010) Collecting Galápagos and the Pacific: how Rollo Howard Beck shaped our understanding of evolution. Proc Calif Acad Sci 61:211–243

Fecchio A, Collins MD, Bell JA, García-Trejo EA, Sánchez-González LA, Dispoto JH, Rice NH, Weckstein JD (2019) Bird tissues from museum collections are reliable for assessing avian Haemosporidian diversity. J Parasitol 105:446–453

Fessl B, Couri MS, Tebbich S (2001) *Philornis downsi,* dodge and Aitken, new to the Galápagos Islands (Diptera, Muscidae). Studia Dipterologica 8:317–322

Fessl B, Sinclair BJ, Kleindorfer S (2006) The life-cycle of *Philornis downsi* (Diptera: Muscidae) parasitizing Darwin's finches and its impacts on nestling survival. Parasitology 133:739–747

Fessl B, Heimpel GE, Causton CE (2012) Invasion of an avian Nest parasite, *Philornis downsi*, to the Galápagos Islands: colonization history, adaptations to novel ecosystems, and conservation challenges

Fessl B, Anchundia D, Carrión J, Cimadom A, Cotin J, Cunninghame F, Dvorak M, Mosquera D, Nemeth E, Sevilla C, Tebbich S (2017) Galápagos landbirds (passerines, cuckoos, and doves): status, threats, and knowledge gaps. GNPD, GCREG, CDF, and GC, Puerto Ayora

Filardi CE (2015) Why I collected a Moustached Kingfisher [Online]. The Audubon Society: The Audubon Society. Available: https://www.audubon.org/news/why-i-collected-moustached-kingfisher. Accessed 5 Sept 2022

Flowers J (2022) We can't do it alone: reimagining the role of museums as partners for community change. J Educ Equit 18:65–72

Garnett ST, Christidis L (2017) Taxonomy anarchy hampers conservation. Nature 546:25–27

Garnett ST, Christidis L, Conix S, Costello MJ, Zachos FE, Bánki OS, Bao Y, Barik SK, Buckeridge JS, Hobern D, Lien A, Montgomery N, Nikolaeva S, Pyle RL, Thomson SA, Van Dijk PP, Whalen A, Zhang ZQ, Thiele KR (2020) Principles for creating a single authoritative list of the world's species. PLoS Biol 18:1–10

Gifford EW (1906) Galápagos expedition journal, 1905-06. California academy archives: digital internet archive and biodiversity heritage library (online)

Gifford EW (1913) The birds of Galápagos Islands, with observations on the birds of Cocos and Clipperton Islands (Columbiformes to Pelicaniformes). Proc Calif Acad Sci Fourth Ser 2:1–132

Gifford E (1919) Field notes on the land birds of the Galápagos Islands and of Cocos Island, Costa Rica. Proc Calif Acad Sci Fourth Ser 2:189–258

Gould J, Darwin C (1839) The zoology of the voyage of the H.M.S. Beagle, under the Command of Capt. Fitzroy, R.N., during the years 1832–1836. Part 3: Birds, with a notice of the habits and ranges, Issue 1–5. Smith, Elder, and Co., Cornhill, London

Guézou A, Trueman M, Buddenhagen CE, Chamorro S, Guerrero AM, Pozo P, Atkinson R (2010) An extensive alien plant inventory from the inhabited areas of Galápagos. PLoS One 5:1–8

Hein HS (2000) The Museum in transition. Smithsonian Institution Press, Washington, DC

Herman SG (1986) The naturalist's field journal. A manual of instruction based on a system established by Joseph Grinnell, Buteo Books. Vermillion

Hobson KA, van Wilgenburg SL, Faaborg J, Toms JD, Rengifo C, Sosa AL, Aubry Y, Brito Aguilar R (2014a) Connecting breeding and wintering grounds of Neotropical migrant songbirds using stable hydrogen isotopes: a call for an isotopic atlas of migratory connectivity. J Field Ornithol 85:237–257

Hobson KA, Wilgenburg SLV, Wesołowski T, Maziarz M, Bijlsma RG, Grendelmeier A, Mallord JW (2014b) A multi-isotope (δ2H, δ13C, δ15N) approach to establishing migratory connectivity in Palearctic-Afrotropical migrants: an example using Wood warblers *Phylloscopus sibilatrix*. Acta Ornithol 49:57–69

Hoeck PEA, Beaumont MA, James KE, Grant RB, Grant PR, Keller LF (2010a) Saving darwin's muse: evolutionary genetics for the recovery of the Floreana mockingbird. Biol Lett 6:212–215

Hoeck PEA, Bollmer JL, Parker PG, Keller LF (2010b) Differentiation with drift: a spatio-temporal genetic analysis of Galápagos mockingbird populations (Mimus spp.). Philos Trans R Soc B Biol Sci 365:1127–1138

James M (2018) Collecting evolution. Am Sci 106:368–375

James M (2020) Collecting evolution: the Galápagos expedition that vindicated Darwin Oxford University Press, Oxford

Jensen EL, Quinzin MC, Miller JM, Russello MA, Garrick RC, Edwards DL, Glaberman S, Chiari Y, Poulakakis N, Tapia W, Gibbs JP, Caccone A (2022) A new lineage of Galápagos giant tortoises identified from museum samples. Heredity 128:261–270

Jiménez-Uzcátegui G, Wiedenfeld D, Valle CA, Vargas H, Piedrahita P, Muñoz-Abril LJ, Alava JJ (2019) Threats and vision for the conservation of Galápagos birds. Open Ornithol J 12:1–15

Johnson KW (2018) The ornithologist the internet called a murderer. The New York Times, 15 June 2018

Kiff LF (2005) History, present status, and future prospects of avian eggshell collections in North America. Auk 122:994–999

Kleindorfer S, Sulloway FJ (2016) Naris deformation in Darwin's finches: experimental and historical evidence for a post-1960s arrival of the parasite *Philornis downsi*. Glob Ecol Conserv 7:122–131

Lack D (1947) Darwin's finches: an essay on the general biological theory of evolution. Harper Torchbooks, New York

Lawson LP, Petren K (2017) The adaptive genomic landscape of beak morphology in Darwin's finches. Mol Ecol 26:4978–4989

Loomis LM (1918) A review of the albatrosses, petrels, and diving petrels. Proc Calif Acad Sci Fourth Ser 2:1–187

Lutz HL, Marra NJ, Grewe F, Carlson JS, Palinauskas V, Valkiūnas G, Stanhope MJ (2016) Laser capture microdissection microscopy and genome sequencing of the avian malaria parasite, plasmodium relictum. Parasitol Res 115:4503–4510

Lutz HL, Tkach VV, Wechstein JD (2018) Methods for specimen-based studies of avian symbionts. In: Webster MS (ed) The extended specimen: emerging Frontiers in collections-based ornithological research. CRC Press; Taylor & Francis Group, Boca Raton

Mace GM (2004) The role of taxonomy in species conservation. Philos Trans R Soc B Biol Sci 359:711–719

Monteiro LR, Furness RW (1995) Seabirds as monitors of mercury in the marine environment. In: Porcella DB, Huckabee JW, Wheatley B (eds) Mercury as a global pollutant: proceedings of the third international conference held in whistler, British Columbia, July 10–14, 1994. Springer, Dordrecht

Páez-Rosas D, Aurioles-Gamboa D, Alava JJ, Palacios DM (2012) Stable isotopes indicate differing foraging strategies in two sympatric otariids of the Galápagos Islands. J Exp Mar Biol Ecol 424–425:44–52

Page LM, Macfadden BJ, Fortes JA, Soltis PS, Riccardi G (2015) Digitization of biodiversity collections reveals biggest data on biodiversity. Bioscience 65:841–842

Parker PG, Buckles EL, Farrington H, Petren K, Whiteman NK, Ricklefs RE, Bollmer JL, Jiménez-Uzcátegui G (2011) 110 years of Avipoxvirus in the Galápagos Islands. PLoS One 6:e15989

Patterson BD (2002) On the continuing need for scientific collecting of mammals. J Neotrop Mamm 9:253–262

Peck SB, Heraty J, Landry B, Sinclair BJ (1998) Introduced insect Fauna of an oceanic archipelago: the Galápagos Islands, Ecuador. Am Entomol 44:219–237

Pérez-Tris J, Williams RAJ, Abel-Fernández E, Barreiro J, Conesa JJ, Figuerola J, Martinez-Martínez M, Ramírez A, Benitez L (2011) A multiplex PCR for detection of poxvirus and papillomavirus in cutaneous warts from live birds and museum skins. Avian Dis 55:545–553

Perino A, Pereira HM, Navarro LM, Fernández N, Bullock JM, Ceauşu S, Cortés-Avizanda A, van Klink R, Kuemmerle T, Lomba A, Pe'er G, Plieninger T, Rey Benayas JM, Sandom CJ, Svenning J-C, Wheeler HC (2019) Rewilding complex ecosystems. Science 364:eaav5570

Phillips RB, Wiedenfeld DA, Snell HL (2012) Current status of alien vertebrates in the Galápagos Islands: invasion history, distribution, and potential impacts. Biol Invasions 14:461–480

Remsen JV (1995) The importance of continued collecting of bird specimens to ornithology and bird conservation. Bird Conserv Int 5:145–180

Ridgway R (1896) Birds of the Galápagos archipelago. Proc U S Nat Museum 19:459–670

Robertson T, Döring M, Guralnick R, Bloom D, Wieczorek J, BRAAK K, Otegui J, Russell L, Desmet P (2014) The GBIF integrated publishing toolkit: facilitating the efficient publishing of biodiversity data on the internet. PLoS ONE 9

Rocha LA, Aleixo A, Allen G, Almeda F, Baldwin CC, Barclay MVL, Bates JM, Bauer AM, Benzoni F, Berns CM, Berumen ML, Blackburn DC, Blum S, Bolaños FR, Bowie CK, Britz R, Brown RM, Cadena CD, Carpenter KMLC, Chakrabarty P, Chaves G, Choat JH, Clements KD, Collette BB, Collins A, Coyne J, Cracraft J, Daniel T, Carvalho MRD, Queiroz KD, Dario FD, Drewes R, Dumbacher JP, Engilis J, Erdmann MV, Eschmeyer W, Feldman CR, Fisher BL, Fjeldså J, Fritsch PW, Fuchs J, Getahun A, Gill A, Gomon M, Gosliner T, Graves GR, Griswold CE, Guralnick R, Hartel K, Helgen KM, Ho H, Iskandar DT, Iwamoto T, Jaafar Z, James HF,

Johnson D, Kavanaugh D, Knowlton N, Lacey E, Larson HK, Last P, Leis JM, Lessios H, Liebherr J, Lowman M, Mahler DL, Mamonekene V, Matsuura K, Mayer GC Jr, Mccosker J, Mcdiarmid RW, Mcguire J, Miller MJ, Mooi R, Mooi RD, Moritz C, Myers P, Nachman MW, Nussbaum RA, Foighil DO, Parenti LR, Parham JF, Paul E, Paulay G, Pérez-Emán J, Pérez-Matus A, Poe S, Pogonoski J, Rabosky DL, Randall JE, Reimer JD, Robertson DR, Rödel MO, Rodrigues MT, Roopnarine P, Rüber L, Ryan MJ, Sheldon F, et al (2014) Specimen collection: an essential tool. Science 344:814–815

Rocque DA, Winker K (2005) Use of bird collections in contaminant and stable-isotope studies. Auk 122:990–994

Rothschild W (1899) A review of the ornithology of the Galápagos Islands. With notes on the Webster-Harris expedition. Novitates Zoologicae 6:85–205

Rothschild W (1903) Notes on birds of the Galápagos and Cocos Islands. Bull Br Ornithol Club 13:6–7

Salvin O (1876) On the avifauna of the Galápagos archipelago. Proc Zool Soc London 9:447–510

Schmitt, C. J., Cook, J. A., Zamudio, K. R. & Edwards, S. V. 2019. Museum specimens of terrestrial vertebrates are sensitive indicators of environmental change in the Anthropocene. Philos Trans R Soc B Biol Sci: 374

Slevin JR (1935) An account of the reptiles inhabiting the Galápagos Islands. Bull New York Zool Soc 38:2–24

Suarez AV, Tsutsui ND (2004) The value of museum collections for research and society. Bioscience 54:66–74

Sutton SW, Wylie E, Economopoulos B, O'brien C, Shapiro S, Xu S (2017) Museums and the future of a healthy world: "just, verdant and peaceful". Curator 60:151–174

Swarth HS (1929) A new bird family (Geospizidae) from the Galápagos Islands. Proc Calif Acad Sci Fourth Ser 18:29–43

Swarth HS (1931) The avifauna of the Galápagos Islands. Occasional Papers Calif Acad Sci 18:1–299

Talley BL, Muletz CR, Vredenburg VT, Fleischer RC, Lips KR (2015) A century of *Batrachochytrium dendrobatidis* in Illinois amphibians (1888–1989). Biol Conserv 182:254–261

Taylor G, Canessa S, Clarke RH, Ingwersen D, Armstrong DP, Seddon PJ, Ewen JG (2017) Is reintroduction biology an effective applied science? Trends Ecol Evol 32:873–880

Toral-Granda MV, Causton CE, Jager H, Trueman M, Izurieta JC, Araujo E, Cruz M, Zander KK, Izurieta A, Garnett ST (2017) Alien species pathways to the Galápagos Islands, Ecuador. PLoS One 12:1–21

Tsai WLE, Schedl ME, Maley JM, Mccormack JE (2020) More than skin and bones: comparing extraction methods and alternative sources of DNA from avian museum specimens. Mol Ecol Resour 20:1220–1227

van Denburgh J (1912a) The geckos of the Galápagos archipelago. Proc CAS Fourth Ser 1:405–430

van Denburgh J (1912b) The snakes of the Galápagos islands. Proc CAS Fourth Ser 1:323–374, pls. 22–30

van Denburgh J (1914) The gigantic land tortoises of the Galápagos archipelago. Proc CAS Fourth Ser 2:203–374, pls. 12–124

Vogel Ely C, Bordignon SADL, Trevisan R, Boldrini II (2017) Implications of poor taxonomy in conservation. J Nat Conserv 36:10–13

Wandeler P, Hoeck PEA, Keller LF (2007) Back to the future: museum specimens in population genetics. Trends Ecol Evol 22:634–642

Webster MS (ed) (2018) The extended specimen: emerging frontiers in collections-based ornithological research

Weil SE (2007) In: Sandell R, Janes RR (eds) From being about something to being for somebody: the ongoing transformation of the American museum. Routledge, London/New York

Wieczorek J, Bloom D, Guralnick R, Blum S, Döring M (2012) Darwin Core: an evolving community-developed biodiversity data standard. PLoS One 7:29715–29715

Winker K (2000) Obtaining, preserving, and preparing bird specimens. J Field Ornithol 71:250–297

Winker K (2004) Natural history museums in a postbiodiversity era. Bioscience 54:455–459
Winker K, Zink RM (2005) Bird collections: development and use of a scientific resource. Auk 122:966–971
Winker K, Fall BA, Klicka JT, Parmelee DF, Tordoff HB (1991) The importance of avian collections and the need for continued collecting. Loon 63:238–246
Winker K, Michael Reed J, Escalante P, Askins RA, Cicero C, Hough GE, Bates J (2010) The importance, effects, and ethics of bird collecting. Auk 127:690–695
Wood CM (2018) Visitor trust when museums are not neutral. Master of Arts, University of Washington
Yeates DK, Zwick A, Mikheyev AS (2016) Museums are biobanks: unlocking the genetic potential of the three billion specimens in the world's biological collections. Curr Opin Insect Sci 18:83–88

John Dumbacher is Curator of Ornithology and Mammalogy at the California Academy of Sciences and is an ecologist and molecular evolutionary biologist. John has done work in New Guinea, Africa, Himalayas, Galápagos, and California on evolution of birds and mammals and on chemical defense in birds.

Jaime A. Chaves is an Ecuadorian native Evolutionary Biologist with over a decade dedicated to scientific research on the Galápagos. From finches, to Giant Daises, from microbes to fur seals, Jaime's phylogenetic approaches have shed light into the evolution of endemic species. Jaime received his undergraduate degree at the Pontificia Universidad Catolica del Ecuador, a Master's on Conservation and Ecology at San Francisco State University (SFSU), California, and his PhD at UCLA in the Department of Ecology and Evolutionary Biology. Three year postdoctoral fellow at UNLV and UM (Miami) led him to teach Evolution as a professor at USFQ. He is currently an Assistant Professor at SFSU, Research Associate at the California Academy of Sciences, Research Associate at the GSC, and an Adjunct Professor at UNC.

Chapter 26
The Museum Effect: Platforms for Advocacy and Sustainability in Insular Environments

Eric Dorfman and Javan Sutton

Introduction

While the theme of these proceedings is island sustainability, in this paper, we broaden the focus to include insular environments generally. This is because, at least from the perspective of museums, the features of islands that make them interesting biologically – being bounded in space and time (Brown 1978) – are true for other environments and events as well. For instance, urban islands (e.g., New York's Central Park) display some of the same characteristics as oceanic islands (Davis and Glick 1978) and extinct species, for instance, are bounded in time (Simberloff 1986). These boundaries provide context to museums' collections, and a fulcrum for research tracking evidence of spatial and temporal change.

Research at natural history museums is fundamental to their existence (Howarth 2018) and has inputs that vary greatly in spatial and temporal scales. At the North Carolina Museum of Natural Sciences (NCMNS), study systems can be microbial (Council et al. 2016), organismal (Archis et al. 2018), habitat-based (O'Shea 2018), or global (Kays et al. 2022). They can also range from milliseconds (Bertone et al. 2022) to millions of years (Zanno et al. 2019). Whether genomic, ecological, or taxonomic, these studies have a high degree of complexity. At the extreme ends of this spectrum, data can be complex and arcane. The challenge for museums is to assist guests in interpreting and identifying with the results from scientific studies in a memorable and engaging way without being overly simplistic.

The variety of a natural history museum's outputs goes a long way to achieving this. Scholarly papers from natural history museums are routine and the best ones rank

E. Dorfman (✉)
Marine Earth and Atmospheric Sciences, North Carolina State University, Raleigh, NC, USA
e-mail: ejdorfma@ncsu.edu

J. Sutton
North Carolina Museum of Natural Sciences, Raleigh, NC, USA

S. J. Walsh et al. (eds.), *Island Ecosystems*, Social and Ecological Interactions in the Galapagos Islands, https://doi.org/10.1007/978-3-031-28089-4_26

among the world's most prominent research. Whereas scholarly articles are written principally for academics, NCMNS uses an education team to unpack this information for a general audience. These teams interpret scientific research, from both within and outside the institution, using formal and informal teaching modes, long- and short-term exhibitions, podcasts (e.g., *Love Nature*; NCMNS 2022a), and demonstrations of science in action through visible laboratories. One example is the very popular "Window on Animal Health," in which live veterinary examinations and procedures including veterinary clinical research are undertaken in full view of the public. Staff veterinarians also give talks in front of the window, interacting with guests and explaining the work being done within the clinic. An important aspect of this experience is the opportunity for a two-way conversation between the guests and NCMNS staff. This creates a relationship not only between the practitioner and audience, but the conversation develops a sense of comfort with the scientific process as a whole.

In this paper, we consider the temporal and spatial scales of various types of outputs at NCMNS to elucidate personalization of otherwise arcane scientific data. We also use the development of a new gallery devoted to marine research at NCMNS to explore these considerations.

North Carolina Museum of Natural Sciences

Founded in 1879, NCMNS is the largest natural history museum in the Southeast United States as measured by visitor numbers. Pre-pandemic visitation was about 1.2 million annually. Its ~250 staff include 29 full-time permanent researchers. The NCMNS houses about 3,000 live animals and stewards about four million specimens on behalf of the State of North Carolina (NCMNS 2022b).

The mission of the NCMNS is "To illuminate the natural world and inspire its conservation" (NCMNS 2021). This statement unpacks into two basic elements that combine into a single perspective (Fig. 26.1). The first is "illumination," that concentrates on *effect* and principally left-brain activities like research, education

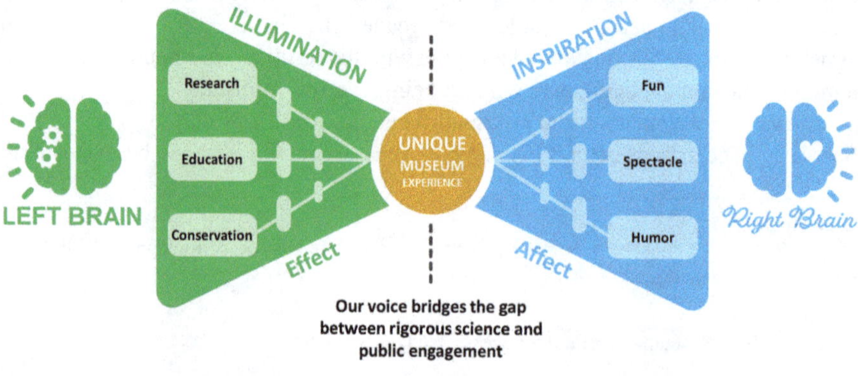

Fig. 26.1 A conceptual model of the North Carolina Museum of Natural Sciences' mission, joining the dual concepts of illumination (effect; left brain) and inspiration (affect; right brain)

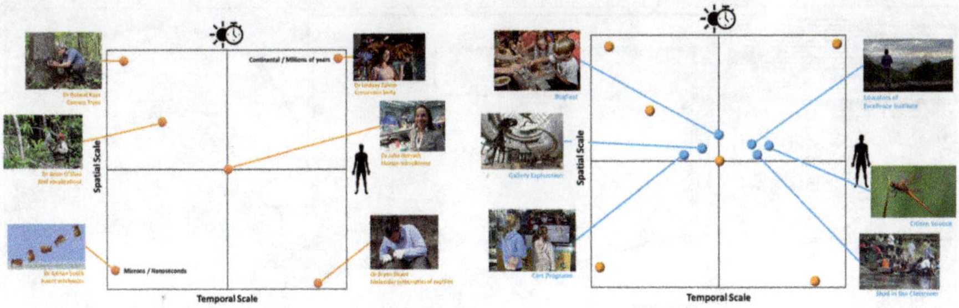

Fig. 26.2 (**a**) (left) Six examples of North Carolina Museum of Natural Sciences (NCMNS) research demonstrating the breadth of temporal and spatial scales. (**b**) (right) Six examples of NCMNS programs (blue) mapped onto the spatiotemporal graph in Figure Two (orange). Axes are approximately logarithmic; center lines represent 24 hours and two meters

programming, and conservation action. It is balanced by "inspiration," encompassing *affect* and such right-brain concepts as spectacle, fascination, and humor. Meeting in the middle, NCMNS bridges the gap between these two realms, blending the sciences and arts through work in science communication, STEAM[1] education, and research into environmental humanities.

Scales and Relevance

Six examples of research at NCMNS are mapped onto a graph (Fig. 26.2a), demonstrating the breadth of temporal and spatial scales. The graph is on a logarithmic scale, decreasing and increasing from the central midpoint, which represents one body-length in space and one day in time.

When overlaid onto the spatiotemporal map of research (Fig. 26.2b), these educational modes are closer to the center. We interpret this more centralized set of points as being more focused on the scale of direct human experience, making science more relevant and accessible to students and other guests who engage with it.

Marine Research Gallery

These issues play out tangibly when new galleries are developed or when existing ones are upgraded. The Exhibitions and Digital Media team at NCMNS are, at the time of writing, leading the concept development of a new gallery focused on marine research in the NCMNS' Nature Research Center. The "Marine Mysteries" gallery (working title; Fig. 26.3) is a 5400 square foot space that includes a

[1] Science, Technology, Engineering, the Arts, and Mathematics

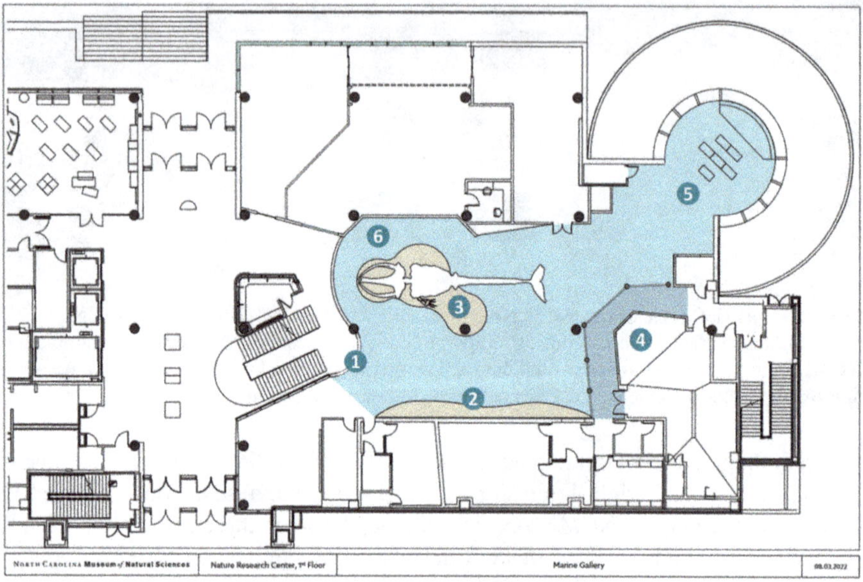

Fig. 26.3 Layout of the planned Marine Mysteries Gallery at the North Carolina Museum of Natural Sciences. Blue represents the gallery footprint and numbers represent segments described in the text

spherical, 1100 square foot theater that seats about 200. When installed, the guest experience will explore a variety of methods used to study ocean ecosystems.[2]

The gallery has six key elements, each with a different message and opportunities for engagement. Numbers of each section correspond with Fig. 26.3.

Portal

Key Message Guests are entering a place with compelling content about research to keep the planet healthy.

Experience Guests pass through a robustly-built gateway arch, decorated with marine motifs in aquas and contrasting colors. The main gallery title creates a sense of place, and wayfinding assists in navigating guests to other attractions that lie beyond. Lighting, visible from the entrance, reflects the marine environment and a sense of mystery.

Affective Goals Build excitement and shift guests' expectations and mindset on entering a well-defined new space. Pique their curiosity with what they can see beyond.

[2] These descriptions are current at the time of writing and may not reflect the final result.

Effective Goals (1) Enter this portal for the marine research gallery. (2) Understand that this is also the way to other experiences.

Living Shoreline

Key Messages (1) Introduction: many types of marine research are represented in this gallery; (2) Living shoreline design is more effective than standard break-walls and increases biodiversity; (3) Salt marshes and other wetland areas are natural filters for soil impurities and act as nurseries for fish and other wildlife; (4) Worldwide, 90% of wetlands have been destroyed in the last 300 years, most since 1900.

Experience As guests move through the portal they encounter an exhibition-level graphic that explains the key messages of the gallery. As they walk past, they see a model segment of coastline planted with artificial dune grasses at the end of which is a dead snag with a taxidermied double-crested cormorant *Nannopterum auritum*. In front of the planting, guests see a long, shallow tank with small inshore fish, (e.g., sand perch *Bairdiella chrysoura*) and Atlantic horseshoe crabs *Limulus polyphemus*, representing the biodiversity within a living shoreline. Guests inspect a part of this display that is removed, revealing the grid-like infrastructure used to create it. They read about living shorelines research as well as coastal ecology.

Affective Goals (1) Engage guests in the overall themes of the gallery; (2) Impress them with the innovative research being done on living shorelines; (3) Give a sense of the scope of biodiversity in coastal habitats; (4) Connect them to nature's ability to take care of us if we work with it.

Effective Goals (1) Orient guests to the organization of the gallery and its content; (2) Expose them to research happening in the intertidal; (3) Help them understand the importance of protecting our shorelines.

Open Ocean

Key Message With oceans covering 70% of the Earth's surface, research that helps to understand this environment and keep it healthy is critically important.

Experience Guests move across the hall to the 45-foot North Atlantic right whale *Eubalaena glacialis* and read about her untimely death when the pregnant female was struck by a boat. They will move around the island that holds the whale, reading and interacting with displays on sea turtle health, plankton and upwelling, marine mammals, and fisheries. At various points, electronic interactives unpack more in-depth stories. Guests view the plankton and upwelling content at the back of the island, closest to the deep-sea display (section "Deep Sea", below), tying together the two environments.

Affective Goals (1) Impress guests with the size of the whale up close and its vulnerability in a crowded ocean; (2) Fascinate them with the complexity and diversity of scientific research on the ocean; (3) Allow them to feel the importance of understanding how oceans influence global biological and physical patterns; (4) Give them hope, through initiatives like health support of sea turtles, that the tide can be turned on conservation of global oceans.

Effective Goals (1) Expose guests to research and research methods covering a diversity of topics in oceanography and marine biology; (2) Provide them with an understanding that the oceans are a large part of the climate change story.

Shallow Coast

Key Messages (1) 40% of people (3.1 billion) worldwide live within 100-km of the coast, so human lives are tangibly entwined with the ocean; (2) Research in this environment can shed light on topics as diverse as human vulnerability to sea level rise and the evolution of fishes.

Experience Looking across the gallery from the whale, guests see a house on stilts. Approaching, they see a 10,000 gallons ("Changing Oceans") aquarium. They walk up to it as if walking under a pier (Fig. 26.3). Once under it, guests see that it is lit from beneath, giving the impression of looking at it from underwater. They notice models of organisms encrusting the artificial pier pilings. At the aquarium, they see the same type of pilings in the water, home to real organisms that match those outside. They watch as a variety of small to medium coastal fish such as sergeant majors (*Abudefduf saxatilis*) and Spanish hogfish (*Bodianus rufus*) swim among the pilings.

Affective Goals (1) Excite guests by the unusual presentation of this environment; (2) Help them identify with the person living in the house above the water; (3) Intrigue them with information about the relationships among fish taxa.

Effective Goals (1) Give guests an understanding that sea level rise poses a special risk for people living on the coast; (2) Intrigue them through awareness that relationships among fish are complex and extend back millions of years.

Daily Planet Theater

On the outside of the building, the State Employees' Credit Union (SECU) Daily Planet Theater is a replica of the Earth. At 72-feet in diameter, it is the world's second largest globe. Inside, the theater features a 42-foot-tall three-quarter-spherical screen that is programmable by in-house technicians. A stage provides an additional focal point for guest speakers and three floors of audience space accommodates audiences of about 200 people. Content for this space is varied, relating to many of the museum's curatorial strengths, but new material will focus on the two major galleries near the theater: marine research and Cretaceous dinosaurs. Content clips range from 30 seconds to 5 minutes.

Key Messages (for the Marine Content) (1) Oceans are vast, beautiful, and varied; (2) Human impact is pervasive; (3) Marine research can be incredibly exciting.

Experience Visitors enter the theater from any point in this relatively open gallery, attracted by the content. They sit in one of the roughly 80 moveable chairs in the theater, unless in one of the upper floor galleries, which is standing-room only. They may also walk closer to the screen for a more immersive experience. Guests will view marine content showing schools of large, oceanic marine fish, whales and other iconic species in open water, as well as underwater footage of coral reefs, diving birds, and hydrothermal vents. Segments will also feature marine scientists in action, collecting samples from boats and remotely operated vehicles (ROVs).

Affective Goals (1) Guests are awed by the beauty and scale of the marine environment; (2) They are concerned about the threats posed by human activity; (3) feel committed to making a difference; (4) Young people, especially, are excited by the idea of a career in marine science.

Effective Goals (1) Guests slow down their trajectory through the Museum; (2) They acquire a greater depth of information than is available through signage or interactives.

Deep Sea

Key messages (1) The deep sea takes up most of the planet's surface, is the least known environment, and remains vulnerable to human disturbance; (2) Specialized equipment has been designed to study life in this realm of high pressure and low temperature; (3) Ecological and physical systems in the deep sea are major drivers for the rest of life on Earth.

Experience Guests entering the gallery from the portal see the deep-sea environment presented beyond the whale skeleton. Against the gallery wall, they see imagery depicting deep sea organisms, while a small model ROV hangs above, suspended from the ceiling. Guests read graphics that explain movements and connect this content to the plankton research featured nearby, underneath the whale skeleton. They look into a case in which a small number of deep-sea specimens are housed, evidence of these creatures' existence.

Affective Goals (1) Guests are fascinated by this seemingly alien environment; (2) They are impressed at the innovative technology used to explore this final frontier on Earth.

Effective Goals (1) Guests become aware of several new species of deep-sea organisms; (2) Other displays within NCMNS (e.g., hydrothermal vents shown one floor above) are reinforced.

Discussion

Exhibiting science is at the core of any natural history museum's mission because it explores and celebrates work undertaken by the institution. Over the long term, it builds a society that better understands science and, as a result, gains trust for it. From guests' perspective a successful exhibition is one that is engaging, providing information in a relatable way (Witcomb 2013). From a museum perspective, an effective display is one in which priority ideas are conveyed that make a long-term impact. A meeting of minds around these goals makes for an exhibition that is relevant both to visitors and the institution. Achieving mutual relevance is critical, because human activity is both the cause of current planetary environmental issues and, perhaps, has the power to reverse some of their effects. From a purely business standpoint, it is one of the main factors allowing any natural history museum to achieve its mission (Dorfman 2018).

The planned *Marine Mysteries* gallery at NCMNS aims to strike a balance between spectacle and content, delighting people while deepening their understanding of the world and, hopefully, creating life-long memories. The more relevant the presentation (Fig. 26.3), the better chance there is of making a long-term impact (Falk and Dierking 2000). Use of different modes of content assist learners who engage with different kinds of content, such as live animals, film, interactives, or static display. Importantly, each element supports the central idea and leaves no ambiguity as to the message.

Inclusion of a broad spectrum of research also gives NCMNS an opportunity to include facets of science that complete a scientific picture in areas where the institution does not currently conduct research. For instance, upwelling is a key component, touching not only trophic relationships but physical oceanography. By showcasing research in this area (e.g., Marchetti 2019), NCMNS helps visitors understand regimes of food availability as they wax and wane through cycles of La Niña and El Niño, respectively.

Programming within the space leverages the staff talents and infrastructure of NCMNS. By including content such as "Cart Programs" (Fig. 26.2b), in which educators bring relevant specimens, models and other materials into the space, NCMNS provides opportunities for displays to be interpreted in even more depth (Fig. 26.3). Such content will be added to the planned *Marine Mysteries* gallery at the other end of the building complex, as well as to informal tours and online presentations.

At the beginning of this chapter, we made the point that the concepts of island conservation apply broadly to work within the museum field. Natural habitats everywhere are becoming fragmented into islands through changing landscape use,

Fig. 26.4 Conceptual layout of the living shoreline display at the North Carolina Museum of Natural Sciences (number 2 in Fig. 26.3)

Fig. 26.5 A house on stilts over water (left) and the concept translated to the fit-out of the 10,000 gallon aquarium at the North Carolina Museum of Natural Sciences (right; number 3 in Figure Four). (Photo (left): Simon Berger. Accessed 3 Aug 2022, Wikimedia Commons)

climate-driven changes, invasive species and others. Loss of historically contiguous landscape has direct and complex impacts on biodiversity (reviewed by Fahrig 2003). The effects of the increasing numbers and decreasing sizes of these unintended "mainland islands" are perfectly suited to documentation through the collections of natural history museums. Data from these collections, coupled with relevance-enhancing displays and interpretation of the information, demonstrate the powerful platform for change represented by this field (Figs. 26.4 and 26.5).

Acknowledgements We thank the Galapagos Science Center, University of North Carolina Chapel Hill, and the Universidad San Francisco de Quito for inviting our participation in the *World Summit on Island Sustainability*. Bryan Stuart contributed to initial conceptualization. We are also grateful for feedback from Wendy Lovelady and Dan Dombrowski on drafts of this article.

References

Archis JN, Akcali C et al (2018) Is the future already here? The impact of climate change on the distribution of the eastern coral snake (*Micrurus fulvius*). PeerJ 6:e4647. https://doi.org/10.7717/peerj.4647

Bertone MA, Gibson JC et al (2022) A novel power-amplified jumping behavior in larval beetles (Coleoptera: Laemophloeidae). PLoS One 17(1):e0256509. https://doi.org/10.1371/journal.pone.0256509

Brown JH (1978) The theory of insular biogeography and the distribution of boreal birds and mammals. Great Basin Nat Mem 2:209–227

Council SE, Savage AM et al (2016) Diversity and evolution of the primate skin biome. Proc R Soc B 283:20152586. https://doi.org/10.1098/raspberry.2015.2586

Davis AM, Glick TF (1978) Urban ecosystems and island biogeography. Environmental Conservation 5(4):299–304

Dorfman EJ (2018) Natural history museums as enterprises of the future. In: Dorfman E (ed) The future of natural history museums. Routledge, New York, pp 200–2014

Fahrig L (2003) Effects of habitat fragmentation on biodiversity. Annu Rev Ecol Evol Syst 34:487–515

Falk JH, Dierking LD (2000) Learning from museums: visitor experiences and the making of meaning. Alta Mira Press, Oxford

Howarth F (2018) The future of research in natural history museums. In: Dorfman E (ed) The future of natural history museums. Routledge, New York, pp 65–81

Kays R, Davidson SC et al (2022) The Movebank system for studying global animal movement and demography. Methods Ecol Evol 13(2):419–431

Marchetti A (2019) A global perspective on iron and plankton through the Tara Oceans Lens. Global Biogeochemical Cycles 33:1–4. https://doi.org/10.1029/2019GB006181

NCMNS (2021) Strategic plan 2021–2023. North Carolina Museum of Natural Sciences. https://naturalsciences.org/interactive-docs/strategic-plan/. Accessed 14 Aug 2022

NCMNS (2022a) Love nature podcast website. https://love-nature.org/ Accessed 14 Aug 2022.

NCMNS (2022b) Annual report FY 2020–21. Museum and Friends of the Museum. North Carolina Museum of Natural Sciences. https://naturalsciences.org/interactive-docs/annual-report/2020-2021/. Accessed 14 Aug 2022

O'Shea, B.J. (2018). A rapid assessment of the forest avifauna of the Upper Berbice Region, Guyana. In: Alonso LE, J. Persaud et al (eds). 2018. Biodiversity assessment survey of the Upper Berbice Region, Guyana. BAT survey report no. 3. WWF-Guianas, Georgetown.

Simberloff D (1986) The proximate causes of extinction. In: Raup DM, Jablonski D (eds) Patterns and processes in the history of life, Dahlem workshop reports, vol 36. Springer, Berlin, Heidelberg. https://doi.org/10.1007/978-3-642-70831-2_14

Witcomb A (2013) Understanding the role of affect in producing a critical pedagogy for history museums. Mus Manag Curatorship 3:255–271

Zanno L, Tucker RT et al (2019) Diminutive fleet-footed tyrannosauroid narrows the 70-million-year gap in the North American fossil record. Commun Biol 2:64. https://doi.org/10.1038/s42003-019-0308-7

Eric Dorfman is a Research Professor at North Carolina State University in Marine, Earth, and Atmospheric Sciences, with a research concentration on tropical waterbirds. He is author of several popular books on New Zealand natural history and climate change, as well as scholarly papers on museum education, public programming, and estuarine ecology. His most recent book is *The Future of Natural History Museums* (Routledge 2018), and he is co-host of the popular podcast *Love Nature*.

Javan Sutton is the Director of Exhibits and Digital Media for the North Carolina Museum of Natural Sciences. He provides strategy, leadership, and creative direction for the Museum's exhibitions and digital productions. Javan previously has 10 years of experience designing exhibits and graphics for museums.

Chapter 27
Microgrids: An Opportunity for Sustainable Development on Islands

Noah Kittner

Introduction

Distributed energy resources are becoming more cost-competitive, particularly in island areas that have strict constraints on land resources. Importing energy in the form of diesel can be costly both from an economic and environmental perspective, and, therefore, islands have led the world in experimentation with distributed energy systems integration. This is most pronounced in the form of microgrids – distributed energy systems – that are flexible, controllable, and can disconnect from the traditional electric grid and operate autonomously. Microgrids provide added resilience features to traditional centralized power grid designs and may be able to provide flexibility for different geographies such as islands with unique features.

Microgrids have increased steadily in their deployment globally, becoming a multi-billion dollar industry. Furthermore, with rapid advances in distributed energy technologies such as solar PV cells, lithium-ion battery packs, and smart AC/DC inverters, microgrids are becoming a steady fixture of a more resilient and distributed power grid. Under climate change, with increasing attention toward extreme weather events, microgrids can often serve localized communities or neighborhoods that otherwise may be impacted by storms for weeks or months. For island communities, which are often remote, shortening the duration of outages becomes increasingly important from a livability perspective.

Perhaps the most interesting aspect of the microgrid's ability to transcend traditional centralized grids and become a modern-day fixture in power systems is that the pioneering microgrid design and features mostly occurred on island experiments. Microgrids are more likely found on physical terrestrial island nations

N. Kittner (✉)
Department of Environmental Sciences and Engineering, University of North Carolina at Chapel Hill, Chapel Hill, NC, USA
e-mail: kittner@email.unc.edu

S. J. Walsh et al. (eds.), *Island Ecosystems*, Social and Ecological Interactions in the Galapagos Islands, https://doi.org/10.1007/978-3-031-28089-4_27

because typically islands in the tropics have relied on diesel as a fuel source for power. On islands, microgrids have become testbeds to integrate higher shares of variable renewable energy options, such as solar photovoltaic electricity or wind power. New designs and technologies are often tested first in new arrangements compared to larger, centralized land-based electric grids.

Diesel-based microgrids and distributed energy resources have existed for a long time. Historically, importing diesel onto islands has been necessary for island energy use. Now with increased attention to global climate change and the opportunity to shift energy sources from fossil fuel and carbon-intensive options such as diesel to fuel-free power systems, distributed energy microgrids have become a viable alternative. The emergence of technologies such as lithium-ion battery storage and flow battery systems have created opportunities to replace diesel as a back-up source of power generation and enabled solar electricity to provide increasing levels of utilization in microgrids across the world.

Why Island Microgrids Are Testbed for Future Development

Island-based microgrids are opportunities to increase access to electricity for areas with underserved electricity needs. The systems are also ways to provide baseload and reliable electricity for regions that have consistently lacked reliable electricity. With smaller systems, controlling the variability of wind and solar electricity could be considered more manageable than in a larger, centralized grid. Diesel has served as a major back-up energy source for these microgrids during the past. Air quality and climate change concerns with diesel generation have renewed interest in displacing diesel as a backup generation source and finding zero-carbon and fuel-free electricity options on islands.

Mainland microgrids disconnect and connect to the main grid without problem. In effect, they may operate in island-mode, without regard to other physical connections. These microgrids provide support to the main grid as backup during natural disasters. Microgrids on islands can also become part of a larger grid and add resilience. For islands that may be vulnerable to hurricanes and other extreme weather events, added flexibility could allow a quicker restoration of power to parts of the island – to avoid lengthy outages and crises such as the disaster that unfolded following Hurricane Maria in Dominica, Saint Croix, and Puerto Rico in 2017. As islands typically are reliant on importing diesel fuel for electricity generation – there are opportunities – solar and wind-based systems may not need to compete directly with large thermal power plants such as natural gas turbines or coal. Instead, storage systems that complement the variability of wind and solar generation only need to outcompete diesel fuel. This rationale has led to many innovative projects on islands that are informing microgrid development projects that mesh with mainland activities.

New Technologies Emerging as Potential Drivers

Lithium-ion battery costs have plunged more than 90% from 1990 levels on a $/kWh levelized cost of storage basis – in some cases reaching below $100/kWh, a major accomplishment for grid reliability and battery adaptability (Kittner et al. 2017, 2020). Lithium-ion batteries have a significant advantage over similar types of battery storage technologies – a very high roundtrip efficiency – in some cases discharging nearly 95% of the input electricity required to charge. This high level of efficiency has enabled further cost reductions and allowed new combinations of solar PV and wind-based grids to provide near-baseload electricity. For micro-grids, where a baseload and reliable back-up energy source has traditionally been diesel power – this presents a unique opportunity to decarbonize and clean the electric grid.

Lithium-ion batteries have limits on their discharge duration, though stacking multiple batteries with larger energy capacities are becoming a popular way to integrate batteries into island microgrids. Additionally, the rise of behind-the-meter lithium-ion storage offers a demand-side management strategy to reduce critical loads during periods.

However, lithium-ion batteries are not the only battery energy storage technology used for microgrids in islands. Vanadium-redox flow batteries have been implemented for their longer discharge duration and modularity.

With larger battery energy storage systems, microgrids can provide black-start services in the case of an outage – able to restore power back to an island without the full grid necessary to compensate.

Some islands may be able to accommodate smaller closed-loop pumped storage hydropower systems. The land-use footprint of different storage systems also influences microgrid design on islands. For instance, innovative hydropower and thermal storage may utilize <1 m^2/kW power capacity (Shan et al. 2022). Flow batteries and other lithium or sulfur-based batteries can provide energy in small areas while sparing land. Because microgrids take advantage of distributed energy resources, often times the land footprint can develop in multi-use spaces such as incorporating rooftop solar PV or building-integrated storage systems.

One critical element of microgrids that is often overlooked is the demand-side and the types of electric loads that the generators can supply power. For instance, another flexible component of microgrids being tested on islands are electric vehicles for transportation. These electric vehicles include bicycles, scooters, buses, and personal cars, which are also emerging as opportunities for microgrid development. These transportation forms when aggregated into fleets can transform into small batteries. There are charging systems that now allow for vehicle-to-grid interactions between small-scale electric buses and scooters that can also provide frequency or voltage support to microgrids in addition to meeting transportation needs. The electrification of vehicle transportation on islands – using golf carts, bikes, and scooters, is a major environmental upgrade compared to the traditional diesel use in mopeds and small trucks.

How Big Is a Microgrid?

Microgrids – having grown immensely in popularity – are not defined by their size in a standardized way. Most microgrids are serve loads on the kW scale, but MW-scale "mini-" grids are often classified under microgrids as well. Additionally, because hybrid microgrids have emerged as a more sophisticated option to scale up microgrids within larger terrestrial based electric grids, the sizing has varied. Key components include a flexible disconnecting option with a main grid – or a small grid with its own distribution feeder that typically operates at the scale of a distribution grid without bulk transmission connections to other grids. Microgrids can range from as small as one or two diesel gensets that are connected to a single load to more complex systems with multiple generators that utilize renewable and non-renewable electricity options and synchronize the frequency and voltage control in a reliable manner.

The theory behind microgrid resilience is based on expanding diverse options in a power system. Microgrids often contained meshed networks of power flowing in two directions – rather than the traditional centralized linear power flow models that dominate large-scale electric grids. Microgrids have advanced across the world to develop clever monitoring and management systems. For instance, with smart metering technology, demand-side management is possible. That means that microgrid operators can limit the load during periods where there may be solar or wind supply shortages. A prominent example of a mini-grid in Bhutan, the GridShare solution, uses a stoplight-style indicator to signal to households when cooking with rice is acceptable to avoid neighborhood brownouts (Quetchenbach et al. 2013). This is an example of peak load shifting, which can improve service quality and particularly alleviate issues on small islands where community groups may choose to cook or use large loads at the same time.

Archipelagic nations such as Indonesia and other island communities also may explore undersea connections between islands and the integration of microgrids to flexibly support remote community electricity access. Technology development driven by advances in high voltage DC lines allow for DC-DC microgrid interaction. Microgrids can be examined not only as an isolated part of the energy system, but as flexible interconnections between grids that allow for transfer of electricity. Having microgrids with black-start capabilities enables re-energizing larger grids that may be separated by water bodies.

Case of San Cristóbal Island

In the Galapagos Islands, microgrids are serving as a new opportunity to improve electricity services and reduce reliance on diesel, which is of high concern from a biodiversity and land conservation perspective. The public microgrid was developed by ELECGALAPAGOS in partnership with the Korean state electric company

KEPCO and other Korean microgrid developers. Essentially the solar and battery energy storage microgrid has a nameplate peak capacity of 1 MW with 2.2 MWh storage system. Because the total project was approximately $7 million – the system costs for an island system are high but provide environmental services in terms of reduction of diesel use and imports. As more resources are needed to accommodate tourism and electricity centers on the island, this small grid system provides a modular framework to potentially expand in the future. Furthermore, the size of the system replaces approximately 133,000 gallons of diesel per year.

Developing experience with microgrid projects is an important step toward better environmental conservation due to many of the hazards associated with diesel fuel. It also provides a test-case example for further biodiversity conservation projects in islands and parks across the world that may seek to reduce diesel consumption and reliance on fossil fuels.

There are many barriers to further implementation. Although the San Cristobal Island project is a first step, coupling the microgrid with other hybrid wind-diesel or new generation projects will be necessary to phase-out diesel on a larger scale.

Diesel is a major contributor to global greenhouse gas emissions. There are also significant concerns regarding human health impacts and biodiversity impacts. For islands surrounded by ocean waters, diesel often powers ships that bring goods to the island. Diesel also is used for power generation in absence of other resources. However, the air quality from diesel exhaust is poor. Moreover, recent studies link toxicity of soil contaminants in areas where diesel fuel is used (Hawrot-Paw et al. 2020). It may be possible to remediate soil contaminated by diesel with biological methods including specific organisms that can benefit the soil, however, this poses risks for biodiversity hotspots such as the Galapagos. Therefore, remediation of diesel contaminated soils could threaten protection of biodiversity for critical species as local policies restrict potential for introduction of non-native species.

For the Galapagos in particular, where tourism drives much of the increase for electricity demand, non-diesel powered microgrids could also serve as an opportunity to improve or expand eco-tourism. Many visitors to the Galapagos are concerned about biodiversity impacts and diesel consumption. One strategy that can be implemented by microgrid developers is to increase transparency of electricity generation options on islands – particularly islands with diesel-free microgrids. However, increased eco-tourism could lead to increased electricity demands. This could compromise conservation efforts for specific regions if not managed. At the same time, microgrids provide experiential learning environments for visitors to think more critically about the environmental impacts of humans on islands and different strategies that improve air quality for humans, wildlife, and oceans. From a biodiversity perspective, microgrids offer distributed ways to avoid large overhead transmission lines that may cut-through conservation territory or land that impacts habitats. Therefore, reduction in the need to build new transmission is another distinct advantage.

Global Microgrid Systems on Islands

Many islands have developed notable microgrid projects across the world. Table 27.1 details a few notable projects and the scale at which they have been developed. This list includes prominent island microgrid projects but is not a fully comprehensive list. However, it ranges some of the projects that have served as learning examples for other island projects around the globe.

Examples of other island-based microgrids have now accumulated decades of experience, even though power capacity remains limited compared to main centralized traditional grids. For instance, in Bonaire, the microgrid development was a direct consequence of hurricanes and wildfire that presented the impetus to rebuild the electric grid structure using microgrid. Kodiak Island microgrid in Alaska reached 99% renewable electricity integration in 2014 and is one of the larger microgrid systems to serve and island community. Diesel-based microgrids in North America such as Hartley Bay Microgrid in British Columbia serves populations accessible only by air or water. This microgrid is exclusively diesel, although smart grid technology has been implemented to try to optimize diesel dispatch and reduce diesel imports. There were plans to utilize hydropower in the microgrid system, but those plans have been stalled. Oki Island Microgrid uses hybrid battery storage systems, combining sodium-sulfur and lithium-ion battery energy storage

Table 27.1 Notable microgrid projects around the world including region and scale.

Year	Project	Continent	Scale
2001	Kythonos	Europe	kW
2003	Mawson Station	Antarctica	kW
2004	Koh Jig	Asia	kW
2004	Kodiak	North America	MW
2004	Bonaire	South America	MW
2007	Falkland Islands	South America	MW
2008	Isle of Eigg	Europe	kW
2009	Lencois	South America	kW
2009	Dongao	Asia	kW
2009	Lana'i	North America	kW
2010	King Island	Oceania	MW
2010	Marble Bar	Oceania	kW
2014	Hailuoto	Europe	kW
2015	Oki Island	Asia	MW
2015	El Hierro	Europe	MW
2015	Coral Bay	Oceania	kW
2016	Russky	Europe	kW
2016	Necker Island	South America	kW
2017	Saint Paul Island	North America	kW
2021	San Cristóbal	South America	kW
2022	Palau Archipelago	Asia	MW

technologies to increase resilience and storage capacity. These examples highlight the diversity of island-based microgrid projects and the technology composition and richness that provide examples for other islands to reduce energy costs and improve efficiency of electricity service and delivery.

Opportunities and Threats to Island Sustainability

There are multiple implications for island sustainability, health, and biodiversity. Avoiding diesel and replacement with fuel-free microgrids offers new opportunities to reduce air pollution and reduce land-use impacts on islands where land is often limited. The majority of global microgrids incorporate diesel in some way, however, the recent cost reduction of lithium-ion batteries and other battery energy storage technologies unlocks the potential for fuel-free microgrid systems, particularly on islands. One further advantage and opportunity in terms of economic development and microgrids include the opportunity for local and community ownership of the generation assets – which often can enable more local decision-making in the tariff settings and terms of use. There can be equity implications for low-income island communities, particular in the setting of electricity rates and policy measures are typically required to maintain affordability for the lowest-income user groups.

If designed with equity and communities in mind, distributed microgrids offer environmental public health, and economic benefits to island systems. Emerging storage technologies offer the technological potential and capability to phase out diesel from microgrid operations. Microgrids can build more resilience to storms and natural disasters that are more frequently occurring – particularly in islands vulnerable to hurricanes, sea level rise, deforestation, and extreme heat.

The major research areas moving forward to enable more sustainable and resilient microgrids on islands should focus on three primary areas: one is the optimal design and control for island systems that requires localized resource inventory and analysis. For instance, wind and solar generally can provide a baseline set of minimum electricity needs, but not in all cases and locations. Solar is particularly modular and scalable, yet complementary energy storage technologies are needed to ensure reliability, unless diesel is used as a backup. To ensure that the environmental and climate benefits of solar and wind electricity are maximized, microgrid operators need to regulate diesel consumption as supplemental backup electricity and utilize systems that take advantage of local resources such as the potential for small-scale hydropower generation and storage or other battery options that do not require diesel fuel. A second major area of research is the technological footprint of microgrids in terms of land area and land-use implications of increased electrification. Microgrid design and planning on islands should occur in a holistic coordinated effort with transportation plans and other building planners to take advantage of demand-response or coordinated load management opportunities. This will help lower electricity costs from the microgrid design and control and also alleviate rebound effects where microgrid growth competes with biodiversity conservation efforts for land expansion. Additionally, as a third area for researchers – microgrids can serve as synergistic technologies with biodiversity conservation – allowing for the protection of endangered species while reducing environmental impacts and matching generation sources closer to electric loads on islands. This distributed flexibility and design of microgrids enables greater uptake of rooftop solar or small-scale energy technologies that can be integrated into buildings or sites where the electric loads exist.

Conclusions

Increasing global deployment of microgrids on islands around the world can be part of a sustainability plan for islands and serve as models for microgrid development on the mainland. There may be conflicts and issues related to biodiversity conservation. However, microgrids can rapidly phase out diesel fuel as an electricity generation source, which provides air quality benefits and reduces greenhouse gas emissions. Microgrids provide resilience in the face of extreme weather events and opportunities to re-energize power grids from storm outages in a much quicker response time than traditional grids. Although battery energy storage systems present many advantages over current diesel-based systems, there should be care in terms of land suitability for microgrid development. The cost trends point toward a diesel-free microgrid future, but only if islands establish dominant energy storage technologies that can integrate with local resource designs for different island applications. Microgrids may become an increasingly larger role as part of the sustainability directive for island communities and the conservation of the ecosystems that microgrids support.

Acknowledgements Noah Kittner thanks the Center for Galapagos Studies for a seed grant and Elias Capriles for research assistance.

References

Hawrot-Paw M, Koniuszy A, Zając G, Szyszlak-Bargłowicz J (2020) Ecotoxicity of soil contaminated with diesel fuel and biodiesel. Sci Rep 10(1):1–9

Kittner N, Lill F, Kammen DM (2017) Energy storage deployment and innovation for the clean energy transition. Nat Energy 2(9):1–6

Kittner N, Schmidt O, Staffell I, Kammen DM (2020) Grid-scale energy storage. In: Technological learning in the transition to a low-carbon energy system. Academic Press, London, pp 119–143

Quetchenbach TG, Harper MJ, Robinson Iv J, Hervin KK, Chase NA, Dorji C, Jacobson AE (2013) The GridShare solution: A smart grid approach to improve service provision on a renewable energy mini-grid in Bhutan. Environ Res Lett 8(1):014018

Shan R, Reagan J, Castellanos S, Kurtz S, Kittner N (2022) Evaluating emerging long-duration energy storage technologies. Renew Sustain Energy Rev 159:112240

Noah Kittner is an Assistant Professor in the Department of Environmental Sciences and Engineering in the Gillings School of Global Public Health and the Department of City and Regional Planning at UNC-Chapel Hill. Noah studies energy systems at multiple scales, from regional and international power grids to community-owned micro-grids and household energy dynamics. Noah's work examines the relationship between energy systems, low-carbon development, and human health.

Part VIII
Island Sustainability: Paths Forward in the Galapagos & Beyond

Chapter 28
Island Digital Ecosystem Avatars (IDEA) Consortium: Infrastructure for Democratic Ecological Action

Neil Davies

Introduction: Island Earth

In 1988, Margaret Thatcher – not known as an environmental activist – implored the UN to take action on climate change, warning that "*It is life itself—human life, the innumerable species of our planet—that we wantonly destroy. It is life itself that we must battle to preserve.*" She opened her historic address on "the threat to our global environment" recounting to fellow world leaders how – to a young Charles Darwin perched on a Tahitian hillside – the South Pacific island of Moorea resembled "a framed engraving". From Darwin's contribution to coral reef science, she turned to astronomer Fred Hoyle's 1948 prediction that "once a photograph of the earth, taken from the outside is available … a new idea as powerful as any other in history will be let loose". Obscure references for such an important speech it might seem, but Thatcher was onto something: evoking the power of holistic visualization. Technology gives us new perspectives on our place in complex systems. The capacity to see the whole – perceiving the wood despite the trees – reveals interdependencies never quite grasped before. Hoyle was correct. To astronauts perched in outer space, the sight of our planet elicits an "overview effect", a quasi-spiritual awareness of the interconnection of all life and its isolation on Island Earth. Technological advances from missions like Sputnik and Apollo led to breakthroughs that eventually enabled scientists to study the Earth as an integrated whole, observing planetary-scale processes continuously in fine detail over decades. Satellite data has helped explain phenomena we experience on the planet's surface. Indeed, it sometimes

N. Davies (✉)
Gump South Pacific Research Station, University of California, Moorea, French Polynesia

Berkeley Institute for Data Science, Berkeley, CA, USA
e-mail: ndavies@berkeley.edu

S. J. Walsh et al. (eds.), *Island Ecosystems*, Social and Ecological Interactions in the Galapagos Islands, https://doi.org/10.1007/978-3-031-28089-4_28

seems easier to model processes playing out – relatively slowly – across the entire Earth than those occurring at much smaller, and noisier, scales.

Remote sensing can get you a long way. Depending on the power of your telescope, you can discern life forms, what they are doing, and what they have built. But remote sensing can only get you so far. Many of the phenomena that make the Earth unique (as far as we know) in the Universe are due to life, and to truly understand life, we need to make observations at the molecular scale too. For the time-being, many scientific observations, especially those relating to biodiversity, can only be made close-up "*in situ*". They literally require access to and physical contact with what we want to measure. For example, we cannot sequence DNA from satellites. Rather, we need to bring the molecules onboard a sensor in to read them. Most often this involves the rather messy process of physically extracting genetic material from cells and tissues. Sometimes this can be done with little impact, but all too often genetic analyses sacrifice the organism. If humanity and the rest of life on earth is to successfully navigate the next few decades, we will need to learn how to integrate *in situ* fieldwork with remotely sensed data, across all domains of scientific research, to develop much greater capacity for social-ecological foresight. We will have to model life on Earth.

Islands as Model Systems for Sustainability Science

In late 2013, following a series of conferences on Quantum Computing at the UC Gump South Pacific Research Station in French Polynesia, Matthias Troyer – a computational physicist – convened a workshop at ETH Zurich to consider the outlandish proposition of modeling an entire tropical island, from genes to satellites. At that time, large-scale modeling had become capable of measuring changes across continents and even the vast Pacific Ocean. Drawing on increasingly rich data streams, coupled with ecological understanding from experiments, it was possible to make forecasts of local-scale impacts, such as the risk of coral bleaching on a given reef. Troyer and colleagues, including ecologists, oceanographers, anthropologists, and geneticists, aimed to take this much further. Global models help understand processes like climate change and ocean acidification (OA). But how do organisms respond to these changes and feed-back on the Earth system? In OA, for example, how does the calcification process respond to lower pH in different species of coral, or among different genotypes within coral species, or among different coral microbiomes? The answers to such questions at the cellular scale affect the resilience of entire ecological communities. In other words, to really understand ecological change, we must study the Earth 'genome up' and 'planet down' (Fig. 28.1). Developing such a comprehensive view of life on Earth will require integrating diverse data, models, and understanding across vast scales (Purves et al. 2013). Needless to say, it is a massive challenge. Yet science has faced similarly overwhelming complexity before. In biomedical research, for example, great advances have been made in tackling human biology through studies of simpler,

Pacific Ocean – the world's largest biome. "Earth is a water planet: three-quarters of the surface is covered by water, and water-rich clouds fill the sky." NASA. https://eoimages.gsfc.nasa.gov/images/imagerecords/46000/46209/earth_pacific_lrg.jpg

"An Orange-fin Anemonefish (*Amphiprion chrysopterus*) and several juvenile Three-spot Dascyllus (*Dascyllus trimaculatus*) with their host sea anemone (*Heteractis magnifica*) in a lagoon of Moorea, French Polynesia." Russell J. Schmitt, Moorea Coral Reef LTER & IDEA Consortium.

Fig. 28.1 Scientific challenges: connecting large scale changes to local impact. (Adapted from the presentation "Future Pacific Ocean: Modeling the World's Largest Biome" by Nicolas Gruber & Matthias Münnich (ETH Zürich) in the "Island Avatars: Simulating Social-Ecological Systems Symposium", Berkeley Institute for Data Science, Berkeley, California. 13 April 2016)

more scientifically tractable, 'model organisms', such as the nematode worm *C. Elegans* and the fruit fly *Drosophila melanogaster*. Similarly, intensively studied islands are 'model ecosystems' for ecology (Vitousek 2002) and anthropology (Kirch 1989). In this spirit, the Zurich conference targeted the island of Moorea in French Polynesia as a model system for sustainability science (Cressey 2015).

Island Research in French Polynesia

Befitting its location in the heart of the Pacific, French Polynesia has one of the worlds' largest exclusive economic zones (EEZ) of some 5 million km^2. The country's five archipelagos and 118 islands stretch across a gradient of environmental conditions in an area the size of Europe. Most of the the country's 279,554 people live on the Windward Islands of Tahiti, Moorea-Maiao, and Tetiaroa. The cluster of islands represents a gradient in complexity for sustainability science from the small private atoll of Tetiaroa (site of an exclusive eco-resort The Brando), through Moorea (<17,000 people, 134 Km^2) to Tahiti (1045 km^2, >189,000 people). The Windward Islands also host significant research capacity with local, national (French), and international institutions, which have recently established a formal collaboration under the French Polynesia Research, Higher Education, and Innovation Consortium (RESIPOL), whose founding members include the University of French Polynesia, Institute Louis Malardé, IRD, and IFREMER on Tahiti, and the CNRS (representing its CRIOBE laboratory), and University of California Berkeley (through its Gump Station, see Fig. 28.2) on Moorea. Access to Tetiaroa is provided by Tetiaroa Society, which operates a research station on the atoll.

Moorea is a 'goldilocks' island for sustainability science. Just about the right compromise of sufficient complexity to be representative of the challenges facing coastal communities everywhere, but not too complex to be overwhelming.

Fig. 28.2 Gump Station. The University of California's Richard B. Gump South Pacific Research Station (Gump Station) on Moorea in French Polynesia supports research on land and sea spanning physical, biological, and social sciences as well as the humanities. For example, it hosts the only coral reef site in the NSF Long-Term Ecological Research (LTER) network of place-based programs collecting highest quality time-series data across different ecosystem types to understand how they respond to human activities and environmental change. Moorea is part of a growing global network of international LTER sites (Mirtl et al. 2018). The Gump Station is located on Polynesian land called Atitia and since 2002 half the property is managed by the Tahitian community-based organization Te Pu Atitia focused on traditional knowledge, culture, and educational programs. The Gump Station and Atitia Center side by side, provide a unique opportunity to explore synergies and mutualistic feedback between local traditional knowledge and global scientific understanding

Scientific progress on biophysical fronts on smaller, privately owned islands, like Tetiaroa, can be made even more rapidly, but this inevitably excludes some of the social-ecological factors that sustainability science must tackle. On the other hand, large metropolitan islands like Tahiti represent the scale of ambition for the complex places we must learn to steward effectively. The model system approach does not ignore the simpler or more complex systems; rather, it seeks to advance at multiple scales simultaneously through an intentional program of research that allocates resources where scientific progress can be made most efficiently.

Networking Island Research Stations as Innovation Hubs for Biodiversity Science

The development of model organisms for biomedical research was not accidental. They were proposed by visionary scientists like Sydney Brenner for C. elegans in 1963, who then helped build them (Brenner 2009). The approach was formalized in what one might call a systems biology roadmap (Sauer et al. 2007; Raes and Bork 2008). Inspired by this work, in the early 2000s, an international team of researchers, with support of the Gordon & Betty Moore Foundation, set out to develop Moorea as a model system for ecology. Just as Brenner and colleagues had described all the cell lineages and sequenced their worm's genome, the Moorea team proposed to sequence their island from its coral reefs to mountaintops.

Moorea Biocode

The Moorea Biocode Project (Check 2006) produced an unprecedented all-taxon biotic inventory. Applying the DNA barcoding standard first proposed by Paul Hebert (Hebert et al. 2003) the project employed an expert-driven, voucher-based methodology: collecting exemplars (individual organisms) of every species on the island, taking digital photographs, depositing specimens, subsampling tissues, extracting DNA, and sequencing at least one gene, the DNA barcode, from each species. Moorea became perhaps the best-characterized complex ecosystem in the world and served as an important use case for the development of genomic biodiversity data standards and informatics tools. For example, software developed under Moorea Biocode seeded GEOME, the Genomic Observatories Metadatabase (Deck et al. 2017, 2018), a component of the informatics stack for the international genomic observing community and contributed to the development of the internet of samples (iSamples) a national cyberinfrastructure for material samples in natural science (Davies et al. 2021).

In terms of scientific applications, studies demonstrated the value of the Biocode database as a research infrastructure for tracking species across an ecosystem, including targets that had previously been intractable for most ecological investigations: early life stages (eggs or larvae), partial tissues (e.g., legs and leaves), and homogenized mixtures, such as gut contents or environmental samples (Ransome et al. 2017; Andersen et al. 2019; Casey et al. 2019). These studies also served to confirm that the inventory was quite comprehensive, as many sequences observed in the test samples corresponded with a species in the reference database. Unidentified "dark taxa" were generally from lineages of tiny organisms that were not targeted in this phase of the project. While microbes were outside the scope of Moorea Biocode Project, microbial-host interactions were explored through preliminary studies on fungi-plant and fungi-insect associations, and through surveys of endosymbiotic bacteria across terrestrial invertebrates (Ramage et al. 2017).

The project also had impacts beyond biological science. Education and outreach components of Moorea Biocode led to a close collaboration with the Tahitian community-based organization Association Te Pu Atitia in their inventory of traditional knowledge of biodiversity on Moorea (e.g., for medicine and food). As a result, the DNA barcodes in the Biocode database serve as a potential bridge from specimens to both scientific knowledge (via the Latin name) and traditional knowledge (via the Tahitian name). Through this work with local elders (the 'Ethnocode' project), Moorea Biocode helped catalyze the Atitia Center, a cultural center located on the Gump Station property operated by Te Pu Atitia that now hosts hundreds of Polynesian school children each year. The community outreach efforts complemented work to develop Access and Benefit Sharing (ABS) policies for large biodiversity genomic studies, with education representing a significant means for sharing benefits. The Biocode ABS agreement (Davies and Hirsch 2010) is a model available to regulators and biodiversity programs worldwide.

Genomic Observatories

The next step in the systems ecology roadmap is to pivot from inventory to observatory. We know surprisingly little about how cells and organisms interact with each other and the environment to shape ecosystems. Yet, powerful biodiversity observation technologies are now more affordable than ever. They include both remote and *in situ* instruments and fall into three main categories: acoustics (e.g., hydrophones, sonar), optics (e.g., satellite imagery, digital photography), and omics (biomolecular sequencing). In addition to new sensors, rapid advances in high performance computing (e.g., machine learning) are also transforming the field of bio-observation. Individually or especially in combination, this technological triumvirate makes it possible to explore vast new realms of the living world. A focus on biodiversity observation then, is at least as much an appeal to seize low hanging fruit as it is about the importance of this dimension for sustainability science.

Just considering the molecular level (omic observations) a new age of biodiscovery is emerging. Biodiversity genomics promises to transform ecology and conservation, while providing a source of genetic parts for synthetic biology and the bio-economy. As society enters a new age of genomics (Check 2006; Field and Davies 2015), scientists have powerful new tools to address abundance, distribution, and other properties of species with the goal of developing predictive models (Casey et al. 2019). Research at island field stations will contribute to the mainstreaming of genomics and other 'omics' into sustainability science. Doing so will require global collaboration through efforts like the Genomic Observatories Network (Davies et al. 2012a, b, 2014), a collaboration of the Genomic Standards Consortium (Field et al. 2011) and the Group on Earth Observations (GEO), which demonstrated a de facto global genomic observatory through Ocean Sampling Day (Kopf et al. 2015). Further proof of concept for blending omics into sustained environmental observations has since come from programs such as Autonomous Reef Monitoring Structures (Leray and Knowlton 2015; Ransome et al. 2017; Obst et al. 2020). Island field station networks can contribute to these efforts by fostering shared capabilities for genomic sensing (Makiola et al. 2020) including the tracking of the material samples that underpin genomic research and many other domains of natural science (Davies et al. 2021). While there has been significant progress (Bork et al. 2015; Thompson et al. 2017) much remains to be done to operationalize omic observing (Buttigieg et al. 2018). The new Omic Biodiversity Observation Network (Meyer et al. 2021), stimulated by the union of the Global Omics Observatory Network (Buttigieg et al. 2019) and the Genomic Observatories Network, is one response to foster global collaboration among key networks, such as the UN Ocean Decade Program "Ocean Biomolecular Observing Network" (Leinen et al. 2022).

Island Digital Ecosystem Avatars

As important as biodiversity genomics and biocomplexity science will be in the coming decades, island research will have to go further. Addressing biodiversity and other grand challenges of environmental sustainability and justice, necessitates integration of the physical, biological and social sciences. Lessons can be learned from "precision medicine" and attempts to move away from treating disease to promoting wellness. As in ecology, the capacity to generate massive datasets using molecular technologies and wireless sensor networks is also transforming medicine (Topol 2012), leading to calls for an approach that is Personalized, Participatory, Predictive and Preventative (Hood and Flores 2012). Networks of island research centers are well placed to coordinate the application of such a "P4 approach" to sustainability through data-intensive, multi-dimensional and longitudinal studies of places (social-ecological systems). Inspired by urban data science initiatives such as ETH's Future Cities Lab in Singapore, the Moorea Island Digital Ecosystem Avatar (IDEA) workshop at ETH Zurich in 2013 laid out such a vision (see Fig. 28.3).

IDEA Consortium

The Moorea IDEA aims to enable holistic use-oriented simulations of entire social-ecological systems, starting with small oceanic islands. A roadmap was published and a consortium founded to pursue the IDEA (Davies et al. 2016). Established in

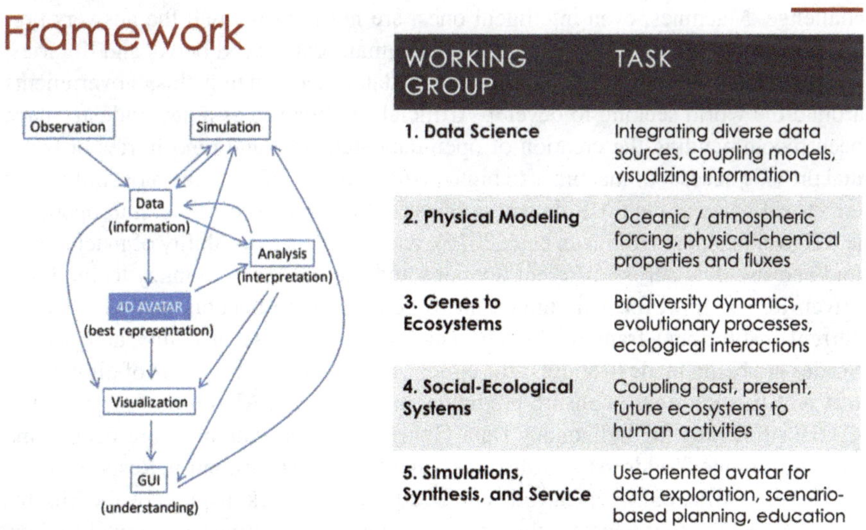

Fig. 28.3 A framework for building digital representations of complex social-ecological systems (Davies et al. 2016)

2016, the IDEA Consortium was led by an executive committee of researchers from University of California Berkeley, ETH Zurich, UC Santa Barbara, France's CNRS, and University of Oxford, and involved more than 80 scientists from 20 institutions. Its founding mission was to understand how the island would change depending on human actions, local and global. Specifically: to understand how biodiversity, ecosystem services, and society in a coupled marine-terrestrial system will co-evolve over the next several decades depending upon what actions are taken. Crucially this would involve understanding an island's history, describing its current state in detail (from genes to satellites), and then simulating different scenarios of the future. Three big questions lay behind the initiative: (1) What is the physical, biological, and social state of the island system today? (2) How did it get to this point? (3) What is its future under alternative scenarios of environmental change and human activity, including conservation efforts?

Model systems reduce the overwhelming complexity at the global scale by focusing on small local systems – microcosms of the larger problem where we can concentrate research resources and make scientific headway. It is an inherently empirical approach and goes beyond generating just big data, the key point is to build more "complete data"… ultimately data representing the entire system with an understanding of what all the components contribute to healthy functioning, and which are the most essential components to maintain resilience. The knowledge and tools gained from the island model systems can scale horizontally and help move all places – not just islands – toward greater resilience.

Work by Joachim Claudet, Lauric Thiault, and colleagues on Moorea has demonstrated the potential for integrated approaches and that they are now capable of informing policy and management (Thiault et al. 2020). But there are still many barriers to overcome. Among them, data and model curation remain a significant challenge. Machines, even intelligent ones, are not built with all the answers preloaded. They need to learn too, and just as human brains need books and libraries, machine intelligences need well-described data. Recognizing this, governments around the world seeking to develop Artificial Intelligence as a new industry have been recommending the creation of open data standards and repositories or trusts that provide access for machines to high quality datasets. One of the most influential efforts is known as the FAIR data principles for Findable, Accessible, Interoperable, and Reusable data (Wilkinson et al. 2016), which enhance the ability of machines to find and use data across different domains and sectors. As with many technology-driven trends, however, it is important to consider who benefits and whether the current institutional frameworks are also fair in the sense of equity and justice. Strides are being made to address the ethical, legal, and social aspects of digital data that will be important to ensure predictive modeling and AI is used for good. The CARE principles for Indigenous Data Governance, for example, "are people and purpose-oriented" addressing Collective benefit, Authority to control, Responsibility, and Ethics (Alliance 2019; Carroll et al. 2020). Recent work inspired by the Moorea IDEA is seeking to implement these principles in island settings as a model for how place-based research data should be managed at any site (Robinson et al. 2022).

Island Twins?

We are at an inflection point in our capacity to build digital representations of natural and human systems, and of coupled social-ecological systems. Evidence of this includes an explosion in "digital twin" (El Saddik 2018) initiatives beyond their initial use in manufacturing, such as the European Union's initiative to build a Digital Twin of Earth (Voosen 2020) and the UK's National Digital Twin program (Bolton et al. 2018). Typically, digital twins are deterministic and predictive, focusing on systems about which humans have a great deal of knowledge – often because humans built them in the first place. Like other technology-branding concepts, such as Artificial Intelligence (AI) or environmental DNA (eDNA), there is much hype and misunderstanding around what they are exactly. Indeed, the more these terms break through to the public, the more experts question whether they have any meaning at all – beyond a clever marketing campaign. The fact that they do break through, however, indicates that something significant might be happening and that the world beyond Silicon Valley should take some notice. The term twin implies an exact copy, which is clearly an impossible goal if applied to living systems. Such terminology risks disappointing at best and dangerously misleading at worst. There are relatively few studies that address the potential opportunities and risks of applying a digital twin paradigm to sustainable development goals (Tzachor et al. 2022); more will be needed. The term avatar as used by the IDEA Consortium (Davies et al. 2016) might be a useful alternative. Because social-ecological systems can be chaotic and/or the rules of the system are only partially visible, digital avatars are multiple, competing hypotheses that are all incorrect to varying degrees (none is a twin or clone of the entity being represented). The task is to weed out the avatars that are demonstrably wrong – based on observations/evidence – and build on those that cannot be ruled out (i.e., those that are the best approximations). In other words, science.

The choices we make today in designing digital avatar (or twin) programs are not solely scientific, however, and they could have profound impacts on the way decision-support infrastructure evolves. How, where, and when to implement digital twin technologies raises deep issues on the relationship between science and societal decision-making at multiple scales of governance. There is an urgent need to consider ethical, legal and social issues as well as the scientific and technological challenges. Such reflection has been taken on recently by a Swiss initiative, the Geneva Science and Diplomacy Anticipator (GESDA).

Infrastructure for Democratic Ecological Action

Among the grand challenges humanity faces, tackling biodiversity loss and increasing greenhouse gas concentrations is a prerequisite for social progress. (The reverse might also be true). In a changing environment, science provides ecological foresight, which combined with human values, guides decision-making when the future is uncertain. But science can't tell us what future to aim for. The U.N. Ocean Decade for Sustainable Development, for example, has an eloquent, inclusive, and inspiring goal: "*The science we need for the ocean we want*". If we dig into those words a little, however, nagging questions arise: What type of ocean do we want? Or even: What gives humans the right to make demands of the ocean? And perhaps even more challenging: Who is 'we' and what do we do if some of us disagree? After the techno-optimism of earlier sections, it behooves us to recall Albert Einstein's warning that "we should be on our guard not to overestimate science and scientific methods when it is a question of human problems" and we should "not assume that experts are the only ones who have a right to express themselves on questions affecting the organization of society."

In the Anthropocene, ecological questions of island sustainability are increasingly human collective action problems. How do the people of an island decide what is the best course of action to achieve their collective goals within the ecological boundaries of their island and the planet? To paraphrase the Doughnut Economics Action Lab: "*How can our [island] be a home to thriving people, in a thriving place, whilst respecting the wellbeing of all people, and the health of the whole planet?*" (Raworth 2017a; DEAL 2020). When it comes to making decisions that are in the interest of all, the cognitive diversity of the population is an empirically important resource (Landemore 2017) that is particularly well harnessed by deliberative democratic institutions (OECD 2020). Citizens assemblies, for example, bring together a random sample of the population and give them the time and information needed to address challenging questions. The digital ecosystem avatars as envisaged above, provide a framework for presenting admissible evidence for such citizen assemblies to deliberate over. The result could be an intelligent fabric of humans and machines learning together to better tend social-ecological systems. Yet, if the future is to be human-centered, public participation in science will not be enough to prevent elites dominating the new social-technological infrastructure. Superintelligence, whether wielded by humans or machines, poses well-known risks and one mitigation strategy is to ensure that systems are "*designed to be inherently uncertain about the human preferences they are required to satisfy*" (Russell 2019). Fortunately, it is impossible to be certain which policy preferences humans will prefer in the future, in part because there are an infinite range of possibilities, but also because human preferences can and do change, especially when given the opportunity for dialogue and deliberation (Fishkin 2011). This is important for Social-Ecological Foresight: First, if human preferences cannot be predicted, even by the humans concerned themselves, then future states of a social-ecological system are also impossible to predict. Second, it suggests that the purpose of digital ecosystem avatars is to support democratic deliberation by providing citizens the best-available evidence for predicting the likely consequences of their decisions (impacts on themselves, their society, and their planet) in an intelligible and transparent manner. The actual impacts are reported back through the avatars as sensor networks feed updates on the status of the social-ecological system. The iterative feedback enables society to learn and evolve towards desired future states – progressing towards the “realization of Utopias” as Oscar Wilde put it.

As decision-support tools, digital ecosystems avatars raise important questions for citizens, including: who controls the avatars (the data, code, and knowledge on which they rely), who uses avatars to make decisions, and how is this organized to ensure equity and justice? These questions have both empirical and normative dimensions and provide rich opportunities for scholarly research in political science and diplomacy. Many of the issues revolve around the concept of collective intelligence (Mulgan 2018) and how to build a better democracy (Landemore 2020). While social-ecological governance is usually territorial (place-based), there are also powerful non-territorial actors, including multinational corporations and international institutions. Networks of islands navigating to sustainable futures will share data and models pertaining to social-ecological states at multiple scales. Sharing within and between communities, including across international boundaries will raise political and diplomatic challenges including questions of “cosmopolitan democracy” and the “implementation of a multi-layered and multi-centered democratic society within, among and beyond states” (Besson 2006).

Conclusions

Navigating the Anthropocene requires much better Social-Ecological Foresight. Island observing systems will need to be better linked and harmonized with respect to the data they gather to feed modeling efforts. Models will need to be coupled with data from island observatories and connected across domains and spatial / temporal scales. In particular, mechanistic ecological, socio-economic, and social-ecological models will need to catch up with the mechanistic sophistication of physical models. As advances in science and technology transform our capacity to sense the world and to process massively diverse data streams, the collective intelligence of people and machines will expand human potential. Individuals and communities will make ever more complex decisions at multiple scales, leveraging integrated digital ecosystem avatars for environmental sustainability. Important impacts will include an enhanced appreciation for the web of life that connects our inner ecosystem (Gilbert et al. 2018) to the people and ecosystems around us, mainstreaming the concept of One Health (Coker et al. 2011; Amuasi et al. 2020) and operationalizing “Predictive, Preventive, Personalized, and Participatory” approaches to personal wellness (Hood et al. 2004) and sustainable development (Raworth 2017b). I began this chapter quoting Margaret Thatcher’s 1988 speech to the UN on the threat to our global environment. I will leave the last words to her: “We need our reason to teach us today that we are not, that we must not try to be, the lords of all we survey. We are not the lords, we are the Lord’s creatures, the trustees of this planet, charged today with preserving life itself—preserving life with all its mystery and all its wonder. May we all be equal to that task.”

Acknowledgments The creation of this chapter was partially supported by the U.S. National Science Foundation (NSF) award: "RCN: Sampling Nature: A Network to Enhance the Natural History Value Chain for Sustainability Science" (NSF 2129268) and "Internet of Samples: Toward an Interdisciplinary Cyberinfrastructure for Material Samples" (NSF 2004642). It benefited from discussion stimulated by the GESDA Summits (2021, 2022).

References

Alliance GID (2019) CARE principles for indigenous data governance. GIDA. https://www.gida-global.org/care

Amuasi JH, Lucas T, Horton R, Winkler AS (2020) Reconnecting for our future: the lancet one health commission. Lancet 395:1469–1471. https://doi.org/10.1016/S0140-6736(20)31027-8

Andersen JC, Oboyski P, Davies N et al (2019) Categorization of species as native or nonnative using DNA sequence signatures without a complete reference library. Ecol Appl 29:e01914. https://doi.org/10.1002/eap.1914

Besson S (2006) Deliberative demoi-cracy in the European Union. Towards the deterritorialization of democracy. In: Besson S, Marti JL (eds) Deliberative democracy and its discontents. Ashgate, Aldershot, pp 181–214

Bolton A, Butler L, Dabson I et al (2018) Gemini principles. Apollo – University of Cambridge Repository, Cambridge

Bork P, Bowler C, de Vargas C et al (2015) Tara Oceans studies plankton at planetary scale. Science 348:873–873. https://doi.org/10.1126/science.aac5605

Brenner S (2009) In the beginning was the worm. Genetics 182:413–415. https://doi.org/10.1534/genetics.109.104976

Buttigieg PL, Fadeev E, Bienhold C et al (2018) Marine microbes in 4D—using time series observation to assess the dynamics of the ocean microbiome and its links to ocean health. Curr Opin Microbiol 43:169–185. https://doi.org/10.1016/j.mib.2018.01.015

Buttigieg PL, Janssen F, Macklin J, Pitz K (2019) The Global Omics Observatory Network: shaping standards for long-term molecular observation. Biodiversity Information Science and Standards

Carroll SR, Garba I, Figueroa-Rodríguez OL et al (2020) The CARE principles for indigenous data governance. Data Sci J 19. https://doi.org/10.5334/dsj-2020-043

Casey JM, Meyer CP, Morat F et al (2019) Reconstructing hyperdiverse food webs: gut content metabarcoding as a tool to disentangle trophic interactions on coral reefs. Methods Ecol Evol 10:1157–1170. https://doi.org/10.1111/2041-210X.13206

Check E (2006) Treasure Island: pinning down a model ecosystem. Nature 439:378–379. https://doi.org/10.1038/439378a

Coker R, Rushton J, Mounier-Jack S et al (2011) Towards a conceptual framework to support one-health research for policy on emerging zoonoses. Lancet Infect Dis 11:326–331. https://doi.org/10.1016/S1473-3099(10)70312-1

Cressey D (2015) Tropical paradise inspires virtual ecology lab. Nature 517:255–256. https://doi.org/10.1038/517255a

Davies N, Hirsch LP (2010) Moorea biocode access and benefit sharing agreement. Zenodo. https://doi.org/10.5281/ZENODO.7150709

Davies N, Field D, The Genomic Observatories Network (2012a) A genomic network to monitor Earth. Nature 481:145–145. https://doi.org/10.1038/481145a

Davies N, Meyer C, Gilbert JA et al (2012b) A call for an international network of genomic observatories (GOs). Gigascience 1:5. https://doi.org/10.1186/2047-217X-1-5

Davies N, Field D, Amaral-Zettler L et al (2014) The founding charter of the Genomic Observatories Network. Gigascience 3:2. https://doi.org/10.1186/2047-217X-3-2

Davies N, IDEA Consortium, Field D et al (2016) Simulating social-ecological systems: the Island Digital Ecosystem Avatars (IDEA) consortium. GigaScience 5:1–4

Davies N, Deck J, Kansa EC et al (2021) Internet of Samples (iSamples): toward an interdisciplinary cyberinfrastructure for material samples. Gigascience 10. https://doi.org/10.1093/gigascience/giab028

DEAL (2020) Creating city portraits: a methodological guide from the thriving cities initiative. Doughnut Economics Action Lab, Biomimicry 3.8, C40 Cities, and Circle Economy, Oxford

Deck J, Gaither MR, Ewing R et al (2017) The Genomic Observatories Metadatabase (GeOMe): a new repository for field and sampling event metadata associated with genetic samples. PLoS Biol 15:e2002925. https://doi.org/10.1371/journal.pbio.2002925

Deck J, Gaither M, Ewing R et al (2018) Planning for field based biological sample collection: using the genomic observatories metadatabase project interface. Biodivers Inf Sci Stand; Sofia 2. https://doi.org/10.3897/biss.2.25651

El Saddik A (2018) Digital twins: the convergence of multimedia technologies. IEEE Multimedia 25:87–92. https://doi.org/10.1109/MMUL.2018.023121167

Field D, Davies N (2015) Biocode: the new age of genomics. OUP, Oxford

Field D, Amaral-Zettler L, Cochrane G et al (2011) The genomic standards consortium. PLoS Biol 9:e1001088. https://doi.org/10.1371/journal.pbio.1001088

Fishkin JS (2011) When the people speak: deliberative democracy and public consultation. Oxford University Press, Oxford

Gilbert JA, Blaser MJ, Caporaso JG et al (2018) Current understanding of the human microbiome. Nat Med 24:392–400. https://doi.org/10.1038/nm.4517

Hebert PDN, Cywinska A, Ball SL, deWaard JR (2003) Biological identifications through DNA barcodes. Proc Biol Sci 270:313–321. https://doi.org/10.1098/rspb.2002.2218

Hood L, Flores M (2012) A personal view on systems medicine and the emergence of proactive P4 medicine: predictive, preventive, personalized and participatory. N Biotechnol 29:613–624. https://doi.org/10.1016/j.nbt.2012.03.004

Hood L, Heath JR, Phelps ME, Lin B (2004) Systems biology and new technologies enable predictive and preventative medicine. Science 306:640–643. https://doi.org/10.1126/science.1104635

Kirch PV (1989) The evolution of the Polynesian chiefdoms. Cambridge University Press, Cambridge

Kopf A, Bicak M, Kottmann R et al (2015) The ocean sampling day consortium. Gigascience 4:27. https://doi.org/10.1186/s13742-015-0066-5

Landemore H (2017) Democratic reason: politics, collective intelligence, and the rule of the many. Princeton University Press, Princeton

Landemore H (2020) Open democracy: reinventing popular rule for the twenty-first century. Princeton University Press, Princeton

Leinen M, Chavez F, Meyer R et al (2022) The ocean biomolecular observing network (OBON). Mar Technol Soc J 56:106–107. https://doi.org/10.4031/MTSJ.56.3.20

Leray M, Knowlton N (2015) DNA barcoding and metabarcoding of standardized samples reveal patterns of marine benthic diversity. Proc Natl Acad Sci U S A 112:2076–2081. https://doi.org/10.1073/pnas.1424997112

Makiola A, Compson ZG, Baird DJ et al (2020) Key questions for next-generation biomonitoring. Front Environ Sci Eng China 7:962. https://doi.org/10.3389/fenvs.2019.00197

Meyer R, Buttigieg PL, Davies N (2021) Omic biodiversity observation network – Omic BON.

Mirtl M, Borer ET, Djukic I et al (2018) Genesis, goals and achievements of Long-Term Ecological Research at the global scale: a critical review of ILTER and future directions. Sci Total Environ 626:1439–1462. https://doi.org/10.1016/j.scitotenv.2017.12.001

Mulgan G (2018) Big mind: how collective intelligence can change our world. Princeton University Press, Princeton

Obst M, Exter K, Allcock AL et al (2020) A marine biodiversity observation network for genetic monitoring of hard-bottom communities (ARMS-MBON). Front Mar Sci 7. https://doi.org/10.3389/fmars.2020.572680

OECD (2020) Innovative citizen participation and new democratic institutions catching the deliberative wave: catching the deliberative wave. OECD Publishing, Paris

Purves D, Scharlemann J, Harfoot M et al (2013) Ecosystems: time to model all life on earth. Nature 493:295–297. https://doi.org/10.1038/493295a

Raes J, Bork P (2008) Molecular eco-systems biology: towards an understanding of community function. Nat Rev Microbiol 6:693–699. https://doi.org/10.1038/nrmicro1935

Ramage T, Martins-Simoes P, Mialdea G et al (2017) A DNA barcode-based survey of terrestrial arthropods in the Society Islands of French Polynesia: host diversity within the SymbioCode Project. EJT. https://doi.org/10.5852/ejt.2017.272

Ransome E, Geller JB, Timmers M et al (2017) The importance of standardization for biodiversity comparisons: a case study using autonomous reef monitoring structures (ARMS) and metabarcoding to measure cryptic diversity on Mo'orea coral reefs, French Polynesia. PLoS One 12:e0175066. https://doi.org/10.1371/journal.pone.0175066

Raworth K (2017a) Doughnut economics: seven ways to think like a 21st-century economist. Chelsea Green Publishing

Raworth K (2017b) A doughnut for the anthropocene: humanity's compass in the 21st century. Lancet Planet Health 1:e48–e49. https://doi.org/10.1016/S2542-5196(17)30028-1

Robinson E, Buys M, Chodacki J, Garza K, Monfort S, Nancarrow C, Praetzellis M, Riley B, Wimalaratne S, Davies N (2022) FAIR Island: real-world examples of place-based open science. Preprint (submitted to Gigascience). Zenodo. https://doi.org/10.5281/zenodo.7384146

Russell S (2019) Human compatible: artificial intelligence and the problem of control. Penguin Publishing Group, London

Sauer U, Heinemann M, Zamboni N (2007) Genetics. Getting closer to the whole picture. Science 316:550–551

Thiault L, Gelcich S, Marshall N et al (2020) Operationalizing vulnerability for social–ecological integration in conservation and natural resource management. Conserv Lett 13. https://doi.org/10.1111/conl.12677

Thompson LR, Sanders JG, McDonald D et al (2017) A communal catalogue reveals Earth's multiscale microbial diversity. Nature 551:457–463. https://doi.org/10.1038/nature24621

Topol E (2012) The creative destruction of medicine: how the digital revolution will create better health care. Basic Books, New York

Tzachor A, Sabri S, Richards CE et al (2022) Potential and limitations of digital twins to achieve the sustainable development goals. Nat Sustain 5:1–8. https://doi.org/10.1038/s41893-022-00923-7

Vitousek PM (2002) Oceanic islands as model systems for ecological studies. J Biogeogr 29:573–582. https://doi.org/10.1046/j.1365-2699.2002.00707.x

Voosen P (2020) Europe is building a "digital twin" of Earth to revolutionize climate forecasts. Science. https://doi.org/10.1126/science.abf0687

Wilkinson MD, Dumontier M, Aalbersberg IJJ et al (2016) The FAIR guiding principles for scientific data management and stewardship. Sci Data 3:160018. https://doi.org/10.1038/sdata.2016.18

Neil Davies graduated in Zoology from the University of Oxford and has a PhD in Genetics from University College London. He has lived in French Polynesia for over 20 years, where he is Director of the University of California's Gump South Pacific Research Station on Moorea, French Polynesia.

Chapter 29
Galapagos Genetic Barcode: A Model for Island Economic Resilience During the COVID-19 Pandemic

Jaime A. Chaves, Camille Bonneaud, Andy Russell, Carlos F. Mena, Carolina Proaño, Diego A. Ortiz, Marilyn Cruz, Alberto Velez, Jen S. Jones, Tom Chaigneau, and Diana A. Pazmino

Introduction

Nearly 200 years ago, the unique biota of Galapagos inspired amongst the greatest scientific revolutions in history – Charles Darwin's theory of evolution by natural selection. Today, this Natural World Heritage site (est. 1976) and UNESCO Biosphere Reserve (est. 1984) also inspires pioneering models of sustainability, conservation, and eco-tourism. Such models are celebrated for their long-term solutions

J. A. Chaves (✉)
Department of Biology, San Francisco State University, San Francisco, CA, USA
Department of Ornithology & Mammalogy, California Academy of Sciences, San Francisco, CA, USA
e-mail: jachaves@sfsu.edu

C. Bonneaud · A. Russell
Centre for Ecology and Conservation, University of Exeter, Exeter, UK

C. F. Mena · C. Proaño · D. A. Pazmino
College of Biological & Environmental Sciences, Universidad San Francisco de Quito, Quito, Ecuador

D. A. Ortiz
Galapagos Science Center, Universidad San Francisco de Quito USFQ, Quito, Ecuador

M. Cruz · A. Velez
Agency of Regulation and Control of the Biosecurity and Quarantine for Galapagos, Galapagos, Ecuador

J. S. Jones
Galapagos Conservation Trust, London, UK

T. Chaigneau
Environment and Sustainability Institute, University of Exeter, Exeter, UK

S. J. Walsh et al. (eds.), *Island Ecosystems*, Social and Ecological Interactions in the Galapagos Islands, https://doi.org/10.1007/978-3-031-28089-4_29

to existing tensions between the preservation of biodiversity and the social-economic well-being of local inhabitants. However, the COVID-19 pandemic has revealed their vulnerability to short-term perturbations. COVID-19 has shaken the foundations of the world as the worst pandemic in the last 40 years since HIV/AIDS outbreak in 1976. With over 601 million cases and close to 6.8 million deaths worldwide (WHO, updated February 2023), this pandemic has also destabilized the global economy. Among the most affected nations are those whose economy depends on services like tourism, mostly due to imposed travel bans. One such nation that has been impacted by this pandemic is Ecuador- in particular the Galapagos archipelago. The Galapagos Islands are a renowned destination, attracting over 275 thousand tourists from all over the world every year. Tourism on the islands is a multimillion dollar industry ($143.3 million revenue in 2006) with 78% of the islands' population employed by this sector or connected to it (Epler 2007). The hard halt of international and national flights to the islands and the complete cease of all touristic activities due to the pandemic has put these islands in a dire situation (Fig. 29.1).

Not only were the inhabitants of the Galapagos out of work, but the lack of tourism, which generates financial support for conservation spending (governmental and NGOs), was also negatively impacted. The consequence of this vulnerability is far-reaching for this community heavily reliant on tourism potentially causing damage of one of UNESCO's World Heritage Sites. Two interconnected and tangible issues were identified that required urgent attention and called for the implementation of unconventional activities outside of tourism to urgently alleviate them. First, the biodiversity from which these islands are known and upon which the Galapagos relies for its eco-tourism industry, was threatened, both from legal and illegal

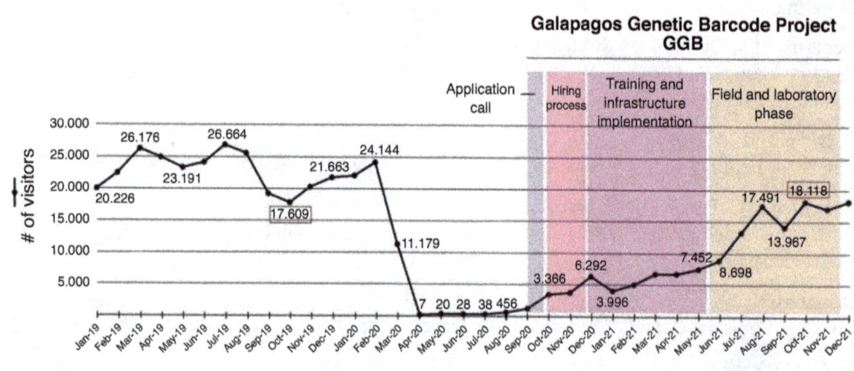

Fig. 29.1 Number of visitors to the Galapagos from January 2019 to December 2021 (source Galapagos National Park). Late February 2020 corresponds to the onset of travel bans to the islands. Colored boxes indicate the four major steps taken by the Galapagos Genetic Barcode project, in parallel with the number of tourists visiting the islands. Numbers in red boxes depict to the two time point at which similar numbers of visitors were recoreded, i.e., in pre-pandemic October 2019 and October 2021

exploitation activites (e.g., illegal exploitation of protected species). Second, naturalist guides and fishermen, who act as whistleblowers against behaviors threating the ecological integrity of the islands had lost their income. Thus, ensuring their sustained protection is important not only for the long-term economic survival of local inhabitants, but also for preserving one of the last areas of wilderness of the planet.

Here we describe an ambitious citizen science research project, the Galapagos Genetic Barcode-GGB, which we designed to catalogue the biodiversity of the Galapagos archipelago by employing some of the professions that are the hardest hit from the halt of tourism, providing them not only immediate financial relief but also, the expertise to improve their future employability and social resilience. Our vision was to (1) train and employ this critical section of the local population to genetically catalogue the biodiversity of Galapagos (i.e., genetic barcoding); (2) provide infrastructure and capacity-building in molecular techniques to be transferable to human health, ecosystem services, and conservation efforts; and (3) design a model of resilience and independence to future uncertainties on island economies and at eco-tourism-dependent regions. A key aim of this proposal was to monitor the impact of employing a subset of society to gauge the successful integration of the plan.

From a scientifific perspective, a series of barcoding projects were designed to capture diverse levels of biological organization, from bacteria found in the soil (e.g., microbiome research) to invasive insect species identification (e.g., barcoding reseach), and shark DNA collected from water samples (e.g., environmental DNA). Because barcoding all life forms is of outmost importance to science and conservation, our project aimed to establish the Galapagos as a model of future inclusion of genetic methods on sustainable development by including novel scientific approaches to tourism practices and citizen science which we hope will open a new view for educational and nature-based tourism.

Why the Galapagos Islands?

With the broad scope presented on how genetic barcoding could be used to activate population participation into research activities with immediate employment, its implementation seems a natural fit for the Galapagos islands. First, these islands experienced dire economic circumstances as result of COVID-19 pandemic and the abrupt halt in tourism (Fig. 29.2). Second, this economic stress could have triggered desperate decisions threatening the balance of its unique ecosystems (i.e., direct international flights; lift of fishing regulations, increase in illegal species trade). And third, an important part of the population was already equipped with basic scientific and field experience (naturalist guides) facilitating the kick-start of the project and allowing a potentially immediate recovery of its economic activities. Finally, and given their unparalleled biologic value, the Galapagos are an unparalleled candidate for the implementation of a large, community-based scientific project with direct

Fig. 29.2 Iconic tourist sites taken during COVID-19 pandemic in Galapagos from (**a**) Puerto Villamil on Isabela Island, August 2020, (**b**) Puerto Baquerizo Moreno on San Cristobal Island, July 2020, and (**c**) Puerto Ayora on Santa Cruz Island, August 2020

benefits to its inhabitants, to its endemic ecosystems, and to the world. Here we describe in detail some of the major aspects that prompted us to implement this project.

Genetic Research

Genetic research, based on DNA-sequencing, has been developed in the Galapagos since the 1970s and has been crucial to understand evolutionary processes and inform conservation (Hedrick 2019; Patton et al. 1975; Rassmann 1996). While the first decades of genetic work in the islands focused on genetic variation and differentiation of wild populations (Browne et al. 1997; Rick and Fobes 1975; Yang and Patton 1981) as sequencing technology developed and became more accessible, the number of scientific publications and the complexity of the studies has significantly increased (Chaves et al. 2020; Lamichhaney et al. 2015; Miller et al. 2018; Rick and Fobes 1975; Romine et al. 2022). However, despite overall advances in the molecular field, the geographical isolation of the islands remains a major challenge for genetic research. First, the logistics for sample collection and transportation can be expensive and time consuming. Therefore, biodiversity inventories and reference genetic data is still biased towards charismatic, conspicuous species and very limited, or even inexistent, for most taxa (Bernardi 2022; Bernardi et al. 2014; Wolff and Gardener 2012). Second, the technology and appropriate infrastructure, together with availability of trained personnel for sample processing is often restricted to mainland Ecuador, or in some cases, to other countries. Finally, there is limited computational capacity to handle and analyze large datasets due to connectivity restrictions that prevent access to servers and supercomputers elsewhere (Urquizo et al. 2019, 2021). Altogether, this highlights the urgent need to increase our knowledge of the natural biodiversity of the Galapagos, to support local capacity building through community' participation, and improve infrastructure to promote local processing and data analyses.

Genetic Barcoding

Genetic barcoding is a means of identifying a species using a short fraction of DNA sequence which is unique to each particular taxon (i.e., genetic fingerprint) (Pečnikar and Buzan 2014). This technology has advanced sufficiently that it can be extended to record biodiversity at the level of ecosystems from a single sample of soil or water (Pomerantz et al. 2018). From microbes to vertebrates and from marine to terrestrial environments, the Galapagos Genetic Barcode-GGB project was designed to capture the essence of an ecosystem's biodiversity through the genetic signature of its constituents (Ip et al. 2019; Roslin and Majaneva 2016). Thus, genetic barcoding was chosen to provide at least three tangible benefits. (1) Genetic barcodes allow species to be categorized based on the genetic similarity of individuals (Hebert et al. 2003), rather than on physical attributes that may or not signal species differences (Packer et al. 2009). As a result, barcodes allow not only accurate species accounts without any inherent taxonomic bias, but they can also reveal cryptic or hidden diversity (Bickford et al. 2007) and new species within taxonomic groups previously thought to be a single cohesive species (Heinrichs et al. 2011). (2) Genetic barcodes accurately describe ecosystem-level biodiversity, critical to measuring the impact of climate change and other human-driven environmental disturbance (e.g., urbanization, pollution, fishing) on species composition, abundance, and distribution. (3) Genetic barcodes can be used to quickly identify species affecting wildlife conservation and human health. For example, in preventing illegal trafficking of species, as well as to detect fraud when protected species are misidentified as commercially-approved ones (Bonaccorso et al. 2021; Johri et al. 2019), or in recognizing invasive species, which can damage and even lead to species extinction in endemic ecosystems (Chown et al. 2008; Darling and Blum 2007), and play a critical role in public and ecosystem health through the early detection of novel infectious pathogens or their insect vectors (*see next sections*). GGB uses state of the art sequencing technology to reconstruct the genetic sequence of DNA from each species, or from multiple species when extracted from one environmental sample (e.g., soil, water). The implementation of nanopore technologies (Oxford Nanopore Technologies-ONT), allows on-site sequencing, which has been a barrier in previous genetic analyses on the Galapagos.

Regardless of its applications the construction of a barcode library is needed as the source for DNA matching. This library should contain properly curated and taxonomically accurate identified samples to be analyzed (Litman et al. 2018). The comparison is done by bioinformatic algorithms able to search and match the genetic barcode from a given sample to web-based libraries around the globe resulting on an accurate identification of the species (Díaz et al. 2016; Kress 2017; Zangl et al. 2020). The information in the libraries is constantly being updated in open access online repositories. The free dissemination of data makes this platform a transparent source of external validation, allowing swift sharing of accurate information with other barcode projects worldwide and with the public in general.

Galapagos Biodiversity

Research on the Galapagos islands have, for centuries, provided information on the incredible diversity of its unique lifeforms, putting these islands in the center stage of conservation efforts. The Galapagos are home to about 500 species of native plant species with more than 180 endemic to the archipelago, 22 endemic species of reptiles, 29 endemic species of birds, and 6 endemic mammals (World Wildlife Fund n.d.). A striking contrast is the little taxonomic knowledge in the total number of insects, marine invertebrates, and fishes inhabiting this region. Barcodes could help filling this taxonomic gap by implementing a pipeline that starts with the collection of samples to the generation of DNA sequences to explore such overlooked diversity. Important deliverables are the description of new species and/or uncovering cryptic/hidden diversity within groups already known to science (Pennisi 2019). This leap into taxonomic delineations skipping the traditional taxonomic designations (mostly based on morphological traits) could accelerate the production of scientific value, and most importantly, implement immediate conservation efforts that would take several years before such discoveries. The open sharing of large amounts of newly produced data from a vast number of taxa, will be highly attractive to the scientific community around the globe. These products will incentivize the internationalization of research on the islands, often governed by non-Ecuadorian research groups. This capacity building could bring a new era of national research outburst and the development of new research projects in a truly worldwide collaboration including Galapagos locals, and thus putting these islands once again at the center stage of scientific contributions.

Invasive Species

The most serious threat to the Galapagos' unique biodiversity is the establishment of invasive species either by direct competition, hybridization, or as vectors of novel pathogens (Chaves 2018). A total of 1,476 alien species have been introduced and are now established; these introductions either intentional or unintentional are continuing in an alarming rate of 27 species per year for the past 40 years, resulting from the acceleration of tourism as well as the exponential growth of human settlements on these islands (Toral-Granda et al. 2017). Of special interest for the implementation of barcode is the identification of both marine alien species to Galapagos and their pathways of introduction. The total number of alien species recorded to date is definitely an underestimation given the little research and survey in Galapagos (Keith et al. 2016), which is further complicated by challenges in species identification. Although most efforts are directed to survey of ship hulls and ballast water as main means of introduction, visual inspection could oversee the introduction of pathogens and bacteria that could threaten human settlements and endemic species. Additionally, plastic pollution and marine debris has been identified as a novel source for marine species invasions. Roughly 25% of all plastic found at beaches in Galapagos contained at least one plant or animal adhered to its surface (Keith et al. 2018). The rapid and accurate identification of these microscopic and taxonomically challenging groups using diagnostic tools such as barcode methods in biosecurity could help determine the invasive nature of such species and if these have successfully established in new habitats (Chown et al. 2008; Nagarajan et al. 2020). These results can immediately increase biosecurity measures in conjunction with the Agency of Regulation and Control of the Biosecurity and Quarantine for Galapagos (ABG) and Port authorities to reduce the negative risks of invasions. Thus, building technological and scientific capacity in Galapagos to detect invasive species is especially important for guaranteeing biosecurity in these ports (see Madden et al. 2019).

Illegal Species Trade

The Galapagos islands are usually in the front cover of major news outlets for their uniqueness of its natural resources. Unfortunately, these islands have also been the center stage for the illegal trade for such resources. In 2017, 300 tons of illegal caught fish were seized from a Chinese vessel inside the Galapagos Marine Reserve, including 6,000 sharks of illegal precedence and belonging to several species of conservation concern (Bonaccorso et al. 2021). Internet searches for illegal trade in Galapagos returns repeated reports of wildlife-smuggling including a 2021 report of a suitcase filled with nearly two hundred baby Galapagos giant tortoises. Although not directly linked to stopping the illegal trade of species, barcode methods can serve to correctly repatriate smuggled individuals to their proper island origin (e.g., multiple tortoise species inhabiting different islands) and to accurately identify the poaching of endangered species affecting the final legal verdict (Bonaccorso et al. 2021). Natural partners in the control of illegal activities are the Galapagos National Park and the Ecuadorian Navy. The Galapagos Genetic Barcode-GGB could serve as a new avenue to connect its participants via their barcoding expertise with these agencies of control to work in conjunction in decision making-based science towards the conservation of these islands.

Galapagos Genetic Barcode (GGB) Implementation

Employment, Training, and Infrastructure

In 2020, mobility restrictions within the archipelago and between the islands and Ecuadorian mainland were in place as part of the strategy to respond to the COVID-19 pandemic. Therefore, all research activities, including those of the GGB were limited, and we required the participation of people living on each of the three main inhabited islands: Santa Cruz, San Cristobal and Isabela to achieve our sampling goals. In September 2020, an open call for applications was launched for naturalist guides, farmers, and fishermen offering job opportunities for 50 people initially. The lack of employment at the time led to a response of 446 applicants. The selection process included a three-step evaluation which consisted in: (1) determination of valid applications of permanent Galapagos residents who were unemployed; (2) evaluation of motivation, relevant experience for each position, and economic necessity due to COVID-19 impacts; (3) personality tests for top candidates on the relevant competencies identified for the four positions within the GGB: team leaders, field assistants, lab assistants, and impact evaluators. Throughout the selection process, efforts were made to ensure that gender inequalities were reduced, ensuring women candidates were available for all positions on every island, and that, they were given opportunities on all activities, including leadership. The result was 74 contracts of employment for citizen scientists from the three islands. Participants were from diverse professional backgrounds who depended either extensively or entirely on tourism for subsistence (mostly naturalist guides) with a total gender ratio nearly to 1:1 (Fig. 29.3). Salaries were standardized and were established for each position in relation to the expected time allocation and the minimum local wage. Thus, team leaders and impact evaluators, earned more as they were expected to dedicate double the time than field and laboratory assistants.

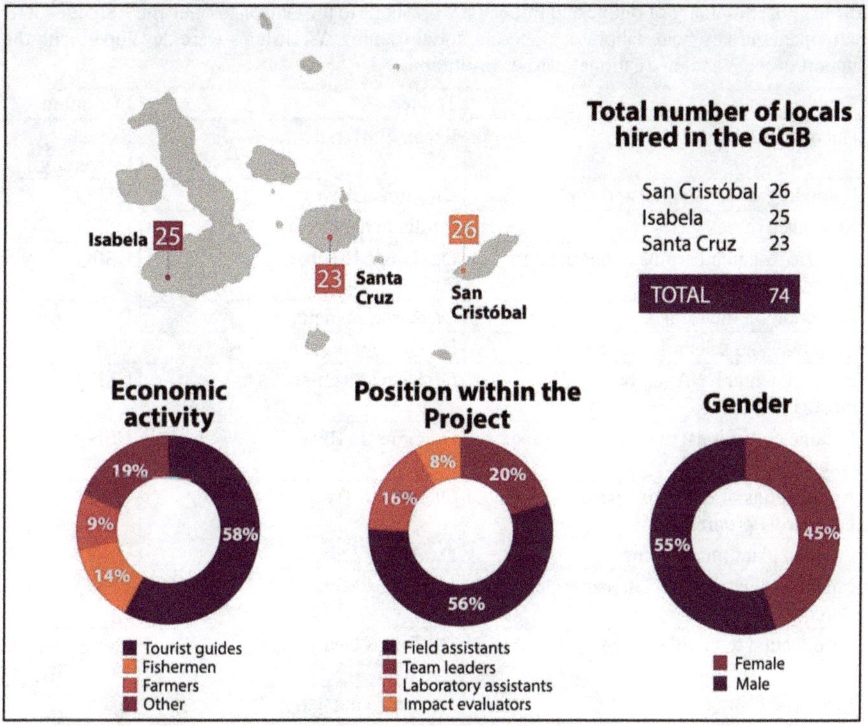

Fig. 29.3 Summary of employed local participants in the Galapagos Genetic Barcode (GGB) project, including the number of participants per island, description of background economic activity, role within the project, and gender distribution

About 370,000 US dollars (most of the project's budget) were destined to cover wages and legal responsibilities under Ecuadorian employment regulations.

All participants were trained on the different scientific methods of data collection, and a fraction of them on molecular methods to extract and sequence DNA. Guides already possessed the basic scientific and field experience, which facilitated a quick start of the project. Because the team used ground and maritime transport and food services during the fieldwork phase, the financial benefits of the project were also further spread over the community. Training was possible through a combination of in-person and video-conferencing meetings to guarantee a cohesive set of rules and norms to be applied across all participants (Table 29.1). The project was carried out simultaneously at three laboratory facilities: Galapagos Science Center-GSC in San Cristobal and Agencia de Regulación y Control de la Bioseguridad y Cuarentena para Galapagos-ABG in Santa Cruz, which were already equipped for carrying out molecular work; and a new laboratory in Puerto Villamil at the Municipal Government of Isabela was fully equipped by GGB. A team of project managers, investigators, and support staff carried out the logistics to acquire all the necessary technology, consumables and equipment from abroad, to Quito and to the islands.

Table 29.1 Summary of online and in-person workshops to the Galapagos Genetic Barcode-GGB participants in key field, laboratory, and curatorial training. Workshops were developed with the support of local and international partner institutions

Training topic	Trainer	Institution
Unconscious bias	Exeter University	Exeter University
Barcoding: use, benefits, and applications	Dr. Jaime Chaves	USFQ[a]
Relevance of databases	Emilia Peñaherrera	USFQ
Genetic research permits under the current Ecuadorian law	Dr. Selene Escobar	USFQ
Barcoding in the tropics	Dr. Aaron Pomerantz	UC Berkeley
Museums and scientific collections	Dr. Gustavo Jiménez	CDF[b]
Environmental DNA (eDNA) and metabarcoding	Dr. Diana Pazmiño	USFQ
Botanic collections: overview, relevance, taxonomy, and DNA	Dr. Gonzalo Rivas	USFQ
Applications of eDNA in species monitoring programs	Dr. Nathan Truelove	UC Berkeley
Soil and plant microbiome	Dr. Antonio León	USFQ
Impact evaluation of "Galapagos Genetic Barcode project"	Dr. Carlos Mena	USFQ
Introduction to qualitative analyses	Dr. Tomas Chaigneau	Exeter University
Industrial fishing	Dr. Ruth Thurstan	Exeter University
Using GenBank database	MSc. Diego Ortiz	USFQ
eDNA collection and processing	Dr. Diana Pazmiño	USFQ
Biosafety and good practices in the laboratory	MSc. Cristina Vintimilla/MSc. Gabriela Gavilanes	USFQ
DNA extraction and amplification	Dr. Diana Pazmiño/MSc. Alberto Vélez	USFQ and ABG[c]
Nanopore sequencing and EPI2ME analyses	MSc. Juan José Guadalupe	USFQ
Metabarcoding analyses: quality check and data filtrating	BSc. José González	USFQ

[a]USFQ: Universidad San Francisco de Quito
[b]CDF: Charles Darwin Foundation
[c]ABG: Agencia de Regulación y Control de la Bioseguridad y Cuarentena para Galapagos

Legal Framework and Key Partnerships

As barcode techniques require scientists to access species' DNA, the legal understanding of national governments' property rights over such material is mandatory. National and local agencies in charge of its control and custody were consulted on the legal implementation of the Galapagos Genetic Barcode-GGB project and granted access to such material. Three main academic institutions were involved in this endeavor: the University of Exeter-UK, Universidad San Francisco de Quito-USFQ, and the Galapagos Science Center-GSC. In addition, the implementation of the project was determined by the involvement of local and international organizations such as the Galapagos National Park, the Galapagos Conservation Trust-GCT-UK, and the Agencia de Regulación y Control de la Bioseguridad y Cuarentena para Galapagos-ABG. Training and capacitation of participants required the solid institutional support for such transfer of technology. Both USFQ and GSC provided logistical infrastructure at different levels; from on-line training, field and lab hands-on instruction, to data generation, and publication. For more than a decade, the USFQ has served as the academic body to the local community and, in particular, behind naturalist guides training through a series of classes graduating over 300 new guides in 2017. Thus, the success of this program resides in the broadly experience of its staff as well as in the trust created between these institutions, and between its professors and scientists with the community. Additionally, the on-site laboratories at the GSC provided the ideal setting for laboratory training, as well as storage of data in on-line repositories.

Achievements and Lessons Learned

A total of 172 geographic locations were visited throughout the sampling period including both terrestrial and marine sites with data collection occurring simultaneously in all three islands for biodiversity comparisons among ecosystems across these islands. Over 1,700 samples (e.g., soil, tissue) were collected (Table 29.2) and processed locally in the three laboratories for DNA extraction, amplification (i.e., polymerase chain reaction), and sequencing through barcoding techniques (e.g., Oxford Nanopore Technologies). The ambitious goal to catalogue the biodiversity of Galapagos started by identifying 355 samples of invertebrate specimens from

Table 29.2 Summary of sampling sites per islandIslands, sample type, and total processed samples

Island	Collection sites	Sample type						Sequenced samples[c]
		Soil	Water	*Rubus* - Blackberry[a]	*Scalesia* - Giant Daisy[b]	Invertebrate	Vertebrate	
Santa Cruz	42	58	30	84	196	355[d]	0	224
San Cristobal	73	152	69	96	132	85	128	171
Isabela	57	124	81	12	84	0	38	107
TOTAL	**172**	**334**	**180**	**192**	**412**	**440**	**166**	**502**

[a,b] Focal plant species: genera *Rubus* (invasive) and *Scalesia* (native)
[c] This number includes sequenced samples of each type
[d] Invertebrate samples in Santa Cruz were obtained from the entomological collection of the Agencia de Regulación y Control de la Bioseguridad y Cuarentena para Galapagos-ABG

Santa Cruz Island stored at the insect collection of the Agencia de Regulación y Control de la Bioseguridad y Cuarentena para Galapagos-ABG. Vertebrate genetic barcodes were generated from 128 samples from research projects managed by the Galapagos Science Center-GSC. Since the start of the project in 2020, the GGB has collected data from 172 sites across three islands and generated over 500 genetic analyses on site, something never before achieved to date (Table 29.2).

Overall, and despite the challenges in implementation associated with the geographic isolation of the islands (approximately 1,000 km from mainland), we have demonstrated that the Galapagos Genetic Barcode-GGB is a successful model that can provide positive outcomes both for the people involved in the project directly, and the broader local community of the islands. We identified five key components that contributed to such success: (1) *leadership* at multiple levels – from Principal Investigators to team leaders within each island, each supported a safe, conscientious work environment that resulted in motivated participants working together towards common goals; (2) *strategic planning*, which was crucial in anticipating potential logistical challenges encountered on the islands, and which allowed for flexibility during the implementation phases; (3) *collaboration* with local stakeholders (e.g., ABG and Municipal Government of Isabela) and with international partner institutions (Galapagos Conservation Trust), was essential in achieving the successful implementation of each phase of the GGB project; (4) *effective communication*, which guaranteed a good understanding of the different processes and tasks, even those that involved a technical molecular component; and (5) *transparency* throughout the duration of the project, especially in the decision making process, as it set the foundation for a solid, collaborative framework for all people and institutions involved.

Acknowledgments We want to acknowledge funding provided by the UK Research and Innovation (UKRI), the Global Challenges Research Fund (GCRF) and the Newton Fund: Agile Response call to address COVID-19 and to our partner institutions University of Exeter, Universidad San Francisco de Quito, Galapagos Science Center, Agencia de Regulación y Control de la Bioseguridad y Cuarentena para Galapagos-ABG (Agency of Regulation and Control of the Biosecurity and Quarantine for Galapagos), Galapagos Conservation Trust, and the Municipal Government of Isabela. We want to acknowledge the support of the Galapagos National Park and the Ministerio de Agua y Ambiente del Ecuador (research permit: MAATE-DBI-CM-2021-0174). We want to thank the local farmers and fishermen of the Galapagos and the naturalist guides that participated in the project. Special thanks to all the trainers and project collaborators, including Gabriela Gavilanes, Gonzalo Rivas, Antonio León, Leonardo Zurita, Pieter Van't Hof, María de Lourdes Torres, Diego Cisneros-Heredia, Selene Escobar, Cristina Vintimilla, Dario Ramírez, Emilia Peñaherrera, Gustavo Jiménez, Lorena Benitez, Patricio Vega, Ruth Thurstan, Nathan Truelove, Camila Contreras, Juan José Guadalupe, José Gonzalez, Susana Cárdenas, and Noelia Barriga. Special thanks to Aaron Pomerantz for guidance on using nanopore technologies. To all the staff at GSC and USFQ, particularly to Cristina Vintimilla, Jessenia Sotamba, Sylvia Sotamba, Ariel Pila, and Linda Chiriap.

References

Bernardi G (2022) Inter-Island local adaptation in the Galápagos archipelago: genomics of the Galápagos blue-banded goby, Lythrypnus gilberti. Coral Reefs 41(3):625–633

Bernardi G, Ramon ML, Alva-Campbell Y, McCosker JE, Bucciarelli G, Garske LE, Victor BC, Crane NL (2014) Darwin's fishes: Phylogeography of Galápagos Islands reef fishes. Bull Mar Sci 90(1):533–549

Bickford D, Lohman DJ, Sodhi NS, Ng PK, Meier R, Winker K, Ingram KK, Das I (2007) Cryptic species as a window on diversity and conservation. Trends Ecol Evol 22(3):148–155

Bonaccorso E, Ordóñez-Garza N, Pazmiño DA, Hearn A, Páez-Rosas D, Cruz S, Muñoz-Pérez JP, Espinoza E, Suárez J, Muñoz-Rosado LD (2021) International fisheries threaten globally endangered sharks in the eastern tropical Pacific Ocean: the case of the Fu yuan Yu Leng 999 reefer vessel seized within the Galápagos marine reserve. Sci Rep 11(1):1–11

Browne RA, Anderson DJ, Houser JN, Cruz F, Glasgow KJ, Hodges CN, Massey G (1997) Genetic diversity and divergence of endangered Galapagos and Hawaiian petrel populations. Condor 99(3):812–815

Chaves JA (2018) Genetic consequences of invasive species in the Galapagos Islands. In: Torres ML, Mena C (eds) Understanding invasive species in the Galapagos Islands. Springer, Cham

Chaves JA, Martinez-Torres PJ, Depino EA, Espinoza-Ulloa S, García-Loor J, Beichman AC, Stervander M (2020) Evolutionary history of the Galápagos rail revealed by ancient mitogenomes and modern samples. Diversity 12(11):425

Chown SL, Sinclair BJ, van Vuuren BJ (2008) DNA barcoding and the documentation of alien species establishment on sub-Antarctic Marion Island. Polar Biol 31(5):651–655

Darling JA, Blum MJ (2007) DNA-based methods for monitoring invasive species: a review and prospectus. Biol Invasions 9(7):751–765

Díaz J, Villanova GV, Brancolini F, Del Pazo F, Posner VM, Grimberg A, Arranz SE (2016) First DNA barcode reference library for the identification of south American freshwater fish from the lower Paraná river. PLoS One 11(7):e0157419

Epler B (2007) Tourism, the economy, population growth, and conservation in Galapagos. Charles Darwin Foundation, Galapagos

Hebert PD, Cywinska A, Ball SL, DeWaard JR (2003) Biological identifications through DNA barcodes. Proc R Soc Lond Ser B Biol Sci 270(1512):313–321

Hedrick PW (2019) Galapagos Islands endemic vertebrates: a population genetics perspective. J Hered 110(2):137–157

Heinrichs J, Kreier H, Feldberg K, Schmidt AR, Zhu R, Shaw B, Shaw AJ, Wissemann V (2011) Formalizing morphologically cryptic biological entities: new insights from DNA taxonomy, hybridization, and biogeography in the leafy liverwort Porella platyphylla (Jungermanniopsida, Porellales). Am J Bot 98(8):1252–1262

Ip YCA, Tay YC, Gan SX, Ang HP, Tun K, Chou LM, Huang D, Meier R (2019) From marine park to future genomic observatory? Enhancing marine biodiversity assessments using a biocode approach. Biodivers Data J:7

Johri S, Solanki J, Cantu VA, Fellows SR, Edwards RA, Moreno I, Vyas A, Dinsdale EA (2019) 'Genome skimming' with the MinION hand-held sequencer identifies CITES-listed shark species in India's exports market. Sci Rep 9(1):1–13

Keith I, Dawson T, Collins K, Campbell M (2016) Marine invasive species: establishing pathways, their presence and potential threats in the Galapagos marine reserve. Geogr Environ Sustain 22(4):377–385. https://doi.org/10.1071/PC15020

Keith I, Howard J, Green S, Suarez J, Vera M (2018) Plastic debris could introduce invasive species to the Galapagos marine reserve- Informe Galapagos report 2017–2018. Charles Darwin Foundation, Galapagos

Kress WJ (2017) Plant DNA barcodes: applications today and in the future. J Syst Evol 55(4):291–307

Lamichhaney S, Berglund J, Almén MS, Maqbool K, Grabherr M, Martinez-Barrio A, Promerová M, Rubin C-J, Wang C, Zamani N, Grant BR, Grant PR, Webster MT, Andersson L (2015) Evolution of Darwin's finches and their beaks revealed by genome sequencing. Nature 518(7539):371–375. https://doi.org/10.1038/nature14181

Litman J, Chittaro Y, Birrer S, Praz C, Wermeille E, Fluri M, Stalling T, Schmid S, Wyler S, Gonseth Y (2018) A DNA barcode reference library for Swiss butterflies and forester moths as a tool for species identification, systematics and conservation. PLoS One 13(12):e0208639

Madden MJ, Young RG, Brown JW, Miller SE, Frewin AJ, Hanner RH (2019) Using DNA barcoding to improve invasive pest identification at US ports-of-entry. PLoS One 14(9):e0222291

Miller JM, Quinzin MC, Edwards DL, Eaton DA, Jensen EL, Russello MA, Gibbs JP, Tapia W, Rueda D, Caccone A (2018) Genome-wide assessment of diversity and divergence among extant Galapagos giant tortoise species. J Hered 109(6):611–619

Nagarajan M, Parambath AN, Prabhu VR (2020) DNA barcoding: a potential tool for invasive species identification. In: DNA barcoding and molecular phylogeny. Springer, pp 31–43

Packer L, Gibbs J, Sheffield C, Hanner R (2009) DNA barcoding and the mediocrity of morphology. Mol Ecol Resour 9:42–50

Patton JL, Yang SY, Myers P (1975) Genetic and morphologic divergence among introduced rat populations (Rattus rattus) of the Galapagos archipelago, Ecuador. Syst Biol 24(3):296–310

Pečnikar ŽF, Buzan EV (2014) 20 years since the introduction of DNA barcoding: from theory to application. J Appl Genet 55(1):43–52

Pennisi E (2019) DNA barcodes jump-start search for new species

Pomerantz A, Peñafiel N, Arteaga A, Bustamante L, Pichardo F, Coloma LA, Barrio-Amorós CL, Salazar-Valenzuela D, Prost S (2018) Real-time DNA barcoding in a rainforest using nanopore sequencing: opportunities for rapid biodiversity assessments and local capacity building. GigaScience 7(4):giy033

Rassmann K (1996) Molecular genetics and conservation in the Galápagos

Rick CM, Fobes JF (1975) Allozymes of Galapagos tomatoes: polymorphism, geographic distribution, and affinities. Evolution 29:443–457

Romine MG, Knutie SA, Crow CM, Vaziri GJ, Chaves JA, Koop JA, Lamichhaney S (2022) The genome sequence of the avian vampire fly (Philornis downsi), an invasive nest parasite of Darwin's finches in Galápagos. G3 12(2):jkab414

Roslin T, Majaneva S (2016) The use of DNA barcodes in food web construction – terrestrial and aquatic ecologists unite! Genome 59(9):603–628

Toral-Granda MV, Causton CE, Jäger H, Trueman M, Izurieta JC, Araujo E, Cruz M, Zander KK, Izurieta A, Garnett ST (2017) Alien species pathways to the Galapagos Islands, Ecuador. PLoS One 12(9):e0184379. https://doi.org/10.1371/journal.pone.0184379

Urquizo J, Lansdale D, Singh P, Chen S, Henderson L, Pierre K, Sowe G, Cooper K, Cordova J, Martin C (2019) Improving the quality of education on the galapagos islands through a community intranet, 1–8

Urquizo J, Sánchez N, Jeong SY, Singh P, Lansdale D, Martin C (2021) Digital technology initiatives in response to covid-19 challenges in the galapagos islands, 46–51

Wolff M, Gardener M (2012) The neglected majority: biodiversity inventories as an integral part of conservation biology frank Bungartz, Frauke Ziemmeck. In: The role of science for conservation. Routledge, pp 137–160

World Wildlife Fund (n.d.) Galápagos Islands, off the coast of Ecuador. https://www.worldwildlife.org/ecoregions/nt1307

Yang SY, Patton JL (1981) Genic variability and differentiation in the Galapagos finches. Auk 98(2):230–242

Zangl L, Daill D, Schweiger S, Gassner G, Koblmüller S (2020) A reference DNA barcode library for Austrian amphibians and reptiles. PLoS One 15(3):e0229353

Jaime A. Chaves is an Ecuadorian native Evolutionary Biologist with over a decade dedicated to scientific research on the Galapagos. From finches, to Giant Daises, from microbes to fur seals, Jaime's phylogenetic approaches have shed light into the evolution of endemic species. Jaime received his undergraduate degree at the Ponticia Universidad Catolica del Ecuador, a Master's on Conservation and Ecology at San Francisco State University (SFSU), California, and his PhD at UCLA in the Department of Ecology and Evolutionary Biology. Three-year postdoctoral fellowship at UNLV and UM (Miami) led him to teach Evolution as a Professor at USFQ. He is currently an Assistant Professor at SFSU, Research Associate at the California Academy of Sciences, Research Associate at the GSC, and an Adjunct Professor at UNC.

Camille Bonneaud, Professor of Evolutionary Biology at the University of Exeter, UK. Her research mainly focuses on understanding how infectious diseases emerge and adapt to new host species, as well as how to mitigate their impacts on threatened animal populations.

Andy Russell, Professor of Animal Behaviour at the University of Exeter specializing in the evolution of cooperation in animals and humans.

Carlos F. Mena, Co-Guest Editor, Co-Director of the Galapagos Science Center, Director of the Institute of Geography, and Professor in the College of Biological and Environmental Sciences, Universidad San Francisco de Quito, Ecuador. His research assesses the importance of land use/land cover change on social and terrestrial systems in the Galapagos Islands as well as in the Ecuadorian Amazon.

Carolina Proaño, Urban Environmental Manager working as a consultant on sustainable design and urban planning. Her focus on resilient cities has motivated her search for innovative alternatives for human settlements in Ecuador that improve the connection between people, nature, and quality of life.

Diego A. Ortiz, Professor in the College of Biological and Environmental Sciences, and Researcher at the Galapagos Science Center, Universidad San Francisco de Quito, Ecuador. His research includes themes in evolutionary biology, ecology, and the conservation of vertebrates in the Galapagos Islands and Amazonia.

Marilyn Cruz, Directora Ejecutiva de la Agencia de Regulación y Control de la Bioseguridad y Cuarentena para Galápagos (ABG). Es Doctora en Medicina Veterinaria. Cuenta con amplia experiencia y entrenamiento en manejo, monitoreo y vigilancia en fauna silvestre de Galápagos (lobos marinos, tortugas, iguanas y aves en general), así como en especies de interés económico (aviar, bovina, porcina) y de compañía (canina y felina). Lidera acciones de prevención, control y respuesta rápida frente a la presencia de plagas y enfermedades que puedan afectar flora y fauna, siendo su mayor pasión la preservación de los ecosistemas insulares de Galápagos.

Alberto Vélez, Trabaja en la Agencia de Regulación y Control de la Bioseguridad y Cuarentena para Galápagos (ABG), es Responsable de la Unidad de Calidad Técnica (Laboratorios). Tiene una maestría en Biotecnología Molecular y una gran pasión por el trabajo de campo y laboratorio. Su principal actividad es el uso de las herramientas moleculares para el diagnóstico de patógenos y la identificación de organismos.

Jen Jones is the Head of Programmes at the Galapagos Conservation Trust, overseeing their programme portfolio both in Galapagos and the UK, and is an Honorary Research Fellow at the University of Exeter, specialising in the impacts of oceanic plastic pollution. Her PhD thesis focused on microplastic pollution in the Galapagos Marine Reserve.

Tom Chaigneau, Senior Lecturer at the Environment and Sustainability Institute at the University of Exeter, UK. He is an environmental social scientist who studies the relationship between the natural environment and peoples' well-being.

Diana A. Pazmiño, Professor in the College of Biological and Environmental Sciences, Universidad San Francisco de Quito, Ecuador. Her research is focused on the development and use of genetics/genomics tools for marine conservation in the Galapagos Islands. Her work beyond research includes outreach and community projects aimed to engage new generations into science.

Chapter 30
Island Innovation: Transitioning Towards a Circular Economy for Plastics in Galápagos, Ecuador

Jen S. Jones, Jess Howard, Tamara S. Galloway, Lucía Norris Crespo, and Sol Aspinosa

Introduction

Since its invention in the early twentieth century, plastic has transformed the way we live. Offering a cheap, durable material that is easy to transport, plastic has provided many benefits in areas such as the health and food industry, improving sanitation and reducing food waste (Andrady and Neal 2009). The total global production of plastic now exceeds 8000 million metric tons, of which only 9% has been recycled, with 12% incinerated and 79% discarded into landfills or lost to the natural environment (Carney Almroth and Eggert 2019; Geyer et al. 2017). The COVID-19 pandemic has exacerbated the plastic pollution problem markedly, with personal protective equipment such as masks and gloves adding a significant burden to waste management systems worldwide, also resulting in large-scale littering, evidenced by an estimated 1.56 billion face masks ending up in the ocean in 2020 (Yuan et al. 2021). The material properties that make plastic so useful also lead to long-lasting persistence in the environment when littered or leaked from waste management systems. Plastic pollution is now occurring at a monumental scale globally, found in almost all habitats investigated, from the Mariana trench in the deep ocean (Chiba et al. 2018) to the top of Mount Everest (Napper et al. 2020). This poses substantial risks to wildlife and ecosystems and also to socioeconomics and human well-being (Beaumont et al. 2019; Ryan 2015).

Pacific Plastics: Science to Solutions (https://www.pacificplasticssciencetosolutions.com/)

J. S. Jones (✉) · J. Howard · L. Norris Crespo · S. Aspinosa
Galapagos Conservation Trust, London, UK
e-mail: jen@gct.org

T. S. Galloway
University of Exeter, Exeter, Devon, UK

S. J. Walsh et al. (eds.), *Island Ecosystems*, Social and Ecological Interactions in the Galapagos Islands, https://doi.org/10.1007/978-3-031-28089-4_30

In islands such as Galápagos (Ecuador), plastic packaging is used in high quantities to wrap imported fresh foods, water and other cargo shipped from the continent in addition to the day-to-day usage of convenient single use plastics by the approx. 30,000 residents, as in cultures all around the world. However, like many other oceanic islands, the Galápagos waste management system must deal not only with its locally consumed plastics, but also incoming plastic pollution from other sources. Due to the ocean currents and the propensity of many plastic polymers to float, islands often become accumulation sites despite relatively low local consumption, a phenomenon observed across the globe (Hidalgo-Ruz et al. 2012; Lavers et al. 2019; Lavers and Bond 2017; Monteiro et al. 2018). Galápagos is no exception, with plastic accumulating on current-exposed coastlines. Key sources are postulated to be from continental South America and at-sea dumping by fisheries (Jones et al. 2021; Schofield et al. 2020; Van Sebille et al. 2019). The wildlife of Galápagos may be especially susceptible to marine pollution, as the endemic flora and fauna have evolved in relative isolation under a very specific set of environmental conditions. This could reduce their capability to deal with novel and chronic threats, such as plastic pollution (Asaad et al. 2017).

A Plastics Circular Economy

In Galápagos and around the world, it is clear that a paradigm shift is needed in our relationship with plastics. The Circular Economy concept is defined as the need for a systemic shift from the traditional linear 'cradle-to-grave' economy to a circular system that reduces waste and leakage, embracing the '4R' concept (reduce, reuse, recycle, recover). Key benefits of the Circular Economy model include economic prosperity, improvement in social equity, improvement in environmental quality and a reduction in resource use (Kirchherr et al. 2017). The circular economy is explicitly referenced in the United Nation's Sustainable Development Goals (SDGs) and is highly relevant to achieving multiple SDG targets (Schroeder et al. 2019). Not only does a plastics circular system avoid unnecessary waste and pollution, it also has the potential to significantly lower the carbon footprint of a system, particularly when the feedstock is switched from fossil fuels to biomass (Zheng and Suh 2019).

Designing out waste and pollution is a key element of the circular economy, and is increasingly urgent in Galápagos as the current system is overstretched (Torsten Hardter et al. 2010). Effective solutions require innovation and multi-disciplinary collaboration to pilot locally relevant interventions that are driven and therefore accepted by local communities. Elevating community innovations that diversify economic opportunities form a core part of the Galápagos 2030 Development Plan (Galapagos Governing Council 2021, unidosporgalapagos.com). The transition to a circular economy has the potential to boost 'green jobs', increasing economic diversification and resilience which is especially relevant in Galapagos where the pandemic has gravely impacted livelihoods. The temporary collapse of tourism due to global lockdowns caused mass unemployment creating major concern for economic resilience (Matamoros Alcivar et al. 2022).

In Ecuador, legislation is strengthening to pave the way for the introduction of a plastics circular economy. The "Organic Law on Inclusive Circular Economy" was approved in 2021, proposing to regenerate and restore ecosystems through a strategic change of production and consumption that avoids waste generation by eco-design, sustainable production and consumption, and promotes integrated and inclusive waste management. Moreover, this Law bans oxo-biodegradable plastic or any other additive turning plastic into microplastic. Plastic bags, Styrofoam, plastic packaging and single-use cutlery are still available but must include recycled material in its components or be reusable.

The Pacific Plastics: Science to Solutions Network

In 2021, the Pacific Plastics: Science to Solutions (PPSS) network was established with funding from the UK Natural Environment Research Council, co-led by the University of Exeter and Galapagos Conservation Trust (both UK based) collaborating closely with the Galápagos Science Center/ Universidad San Francisco de Quito, Conservation International Ecuador, Charles Darwin Foundation, the Galápagos National Park Directorate in Galapagos and other partners. The network is comprised of at least 15 more organisations and more than 40 researchers and practitioners in Ecuador, Perú, Chile and Europe. The mission of the PPSS network is to take a systems approach to reduce plastic waste in the Galápagos Islands and Eastern Pacific, thereby supporting marine biodiversity conservation and provision of sustainable livelihoods in the region.

The Galápagos Islands are a key focal point and developing a sustainable plastics circular economy is a major outcome the programme is aiming for. The PPSS project is exploring the opportunities and barriers to adopting a circular economy at a local level, while supporting local initiatives that are driving change. Below, we outline some of the key interventions that we are trialling to develop and support a plastics circular economy:

Shopping Malls in the Continent

One major source of plastic pollution arriving in Galápagos is from urban centres on the mainland, transported by rivers to the open ocean where currents sweep them towards the Galápagos Islands (Van Sebille et al. 2019). This demonstrates the need for initiatives that tackle plastic pollution from multiple upstream sources including diverse, and often distant locations. In mainland Ecuador shopping malls attract many visitors, and for convenience they generate large quantities of single-use plastic waste, especially in food courts. Developing solution interventions in shopping

Fig. 30.1 Reusable cups in the reusable scheme in the Mall el Jardin, Quito. (Photo: Jess Howard)

malls therefore provides an opportunity to reach a myriad of local businesses and sub-sections of the general population directly. One solution that PPSS is supporting is a **reusable crockery and cutlery initiative**, led by a local B Corp, Huella Verde. The initiative provides a cleaning service to food court businesses that adopt the reusable scheme, which has launched in some of the largest malls in Ecuador. Not only does this initiative reduce the consumption of single-use plastics (nearly 23,000 fewer single-use items used in under a month, reported from Huella Verde), but it also generates green jobs, reduces food waste, and contributes to consumer behaviour change campaigns within shopping malls that may impact pro-environmental behaviours outside of the mall context. Another important component of this project is its role in filling knowledge gaps on consumption and waste of plastics in the region, which contributes towards solution design, national policy, and influencing international instruments (Fig. 30.1).

Clean-Up and Repurposing

Preventing pollution entering the ocean from the mainland is crucial to reducing plastic pollution in Galápagos, but we are still left with the legacy problem of plastics already in the environment. To tackle this, effective monitoring and clean-up programmes are needed within the archipelago. The **Galápagos National Park Directorate** is working to remove plastic pollution from remote sites around the archipelago, preventing further breakdown of macroplastics to microplastics and reducing risk to species and habitats. However, there is an ongoing discussion about how to manage the 8–10 tonnes of plastic removed from Galápagos beaches each year. While recycling seems an obvious choice, collected plastic is often too degraded following its journey through salty surface waters, often colonised by epibiont organisms that settle on plastic surfaces, before beaching, and being fully exposed to the equatorial sun.

One potential solution emerging in Galápagos is to turn plastic cleaned from beaches into **plastic-enhanced construction materials,** whereby plastic is broken down and incorporated into concrete blocks. This project was proposed by a local resident who is passionate about transitioning to a circular economy, supported by the capacity building initiative, Co-Galapagos. The durability and thermal insulating properties of plastic make it a viable option for use in masonry blocks, pavements, asphalts and other construction applications (Pan et al. 2020). This application reduces the burden on the already strained waste management systems in Galápagos by bypassing them entirely and extending the use-life of cleaned plastic. Alongside providing a viable solution for adding value to cleaned-up plastic waste, this project also supports economic diversification in the form of green jobs (Fig. 30.2).

Fig. 30.2 Prototype plastic-enhanced concrete construction block, made from plastic collected from beaches in Galapagos. (Photo: Jess Howard)

Bioplastics

Although the local input of plastic pollution is relatively low when compared to continental and at-sea sources, there are still significant quantities of waste produced locally accumulating in urban areas and leaking into the environment. To reduce the use of destructive single-use plastics within the Archipelago, we need effective and locally relevant alternatives. Bioplastics present a promising alternative to petroleum-based plastics and can be defined as materials in which a proportion of the component monomers are derived from organic materials other than crude oil (Rosenboom et al. 2022). They can be made from a variety of source materials such as plant starches, cellulose derivatives or chitin (e.g., extracted from crustacean shells). Bioplastics have the added benefit of utilising abundant food waste, and even invasive species plant matter, that would otherwise enter the waste management system. PPSS is working with Materiom, an NGO working between Chile and the UK, to identify abundant organic waste streams in Galápagos, and to develop recipes for bioplastics using simple at-home methods that can be easily replicated. By identifying problematic local plastic items in Galápagos (such as small produce bags), bioplastic alternatives can be trialled using organic material otherwise destined for compost or landfill. This will reduce pressures on the waste management systems in Galápagos two-fold, by reducing organic waste input as well as the input of single-use plastics. The simple at-home method of producing bioplastic also opens the door for local innovations and alternative livelihoods, capitalising on existing waste streams to produce products such as souvenirs for tourists (Fig. 30.3).

Conservation starts with community, so alongside these evidence-informed practical solutions, compelling communications and awareness raising projects must also be implemented to increase knowledge of, and participation in, circular economy initiatives. There are many strong community driven outreach initiatives in existence in Galápagos trying to motivate people to adopt more sustainable practices and we aim to provide these groups with information from the PPSS network.

Fig. 30.3 Making bioplastics from agar, derived from algae (left), an array of bioplastics made from coffee waste, fish gelatine, tea waste and algae (right). (Photos: Jess Howard)

Pathways to the Future

As with many issues that require a major shift at a systems level, both top-down and bottom-up strategies are required, enabling political, educational, and technical innovations to be delivered with an integrated approach (Ford et al. 2022). Increasing consumer awareness, encouraging innovation in industry, and providing government incentives for the lifestyle and industrial changes needed are all key strategies to achieving a circular economy (Yuan et al. 2021).

Embracing a circular economy for plastics in Galapagos and mainland Ecuador would significantly reduce waste accumulating on shorelines throughout the archipelago, although due to the high incidence of international input (i.e., from fisheries and other countries in the Americas), a global effort is required. Ecuador is one of the countries leading the way with calls for a Global Treaty on plastics, resulting in negotiations that will continue into 2024. Ecuador's commitment to the circular economy is strong, with enthusiastic political support for the concept. Supporting communities and businesses to transition towards a circular economy for plastics will require multidisciplinary collaboration from a range of sectors and sciences, and grassroots action supported by a global community. This needs to be backed up by local, national, and global policies that prioritise communities, wildlife, and livelihoods, particularly supporting vulnerable populations to adapt.

Plastic pollution and climate change are inherently linked, both triggered by the same root causes of mass consumption of finite, polluting resources (Ford et al. 2022). Adopting a circular economy for plastics is especially important in island

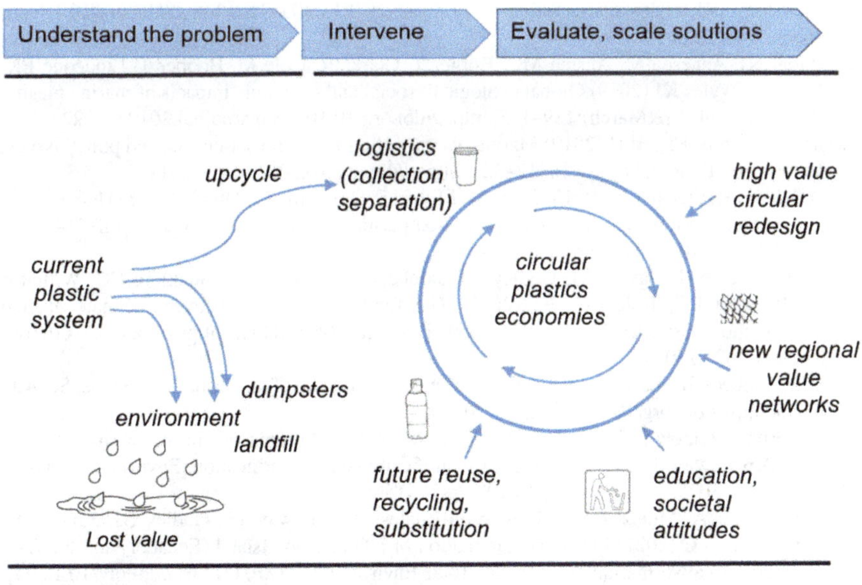

Fig. 30.4 A systematic overview of a circular economy for plastics

systems, where ecological and economic resilience is naturally lower, and where the impacts of global catastrophes such as plastic pollution and climate change are most keenly felt. Galapagos is often held up as a 'living laboratory', both for its importance in the study of evolution and adaptation, and for inspiring Darwin to produce his theories on natural selection. However, Galapagos also has lessons to teach us about the balance of conservation with community well-being and socioeconomics. The prestigious status of Galapagos allows us to pilot innovative solutions in an isolated, semi-closed system with international support. From the solutions trialled in Galapagos, we can provide a blueprint for the transition towards a circular economy for plastics that can be adapted to other islands, cities, countries, and regions across the world (Fig. 30.4).

Acknowledgments Funding was provided by the Global Challenges Research Fund (NE/V005448/1) as part of the Pacific Plastics: Science to Solutions programme. The authors are grateful for the support provided by Universidad San Francisco de Quito colleagues in developing the Pacific Plastics: Science to Solutions network across Ecuador, Peru, and Chile, particularly Juan Pablo Muñoz Pérez, Susana Cardenas, Alex Hearn, Daniela Alarcón, Judith Denkinger and Carlos Mena. The support of all project teams mentioned in this article is truly valued, to guide research outcomes into practical on-the-ground impact.

References

Andrady AL, Neal MA (2009) Applications and societal benefits of plastics. Philos Trans R Soc B Biol Sci 364(1526):1977–1984. https://doi.org/10.1098/rstb.2008.0304

Asaad I, Lundquist CJ, Erdmann MV, Costello MJ (2017) Ecological criteria to identify areas for biodiversity conservation. Biol Conserv 213:309–316. https://doi.org/10.1016/j.biocon.2016.10.007

Beaumont NJ, Aanesen M, Austen MC, Börger T, Clark JR, Cole M, Hooper T, Lindeque PK, Pascoe C, Wyles KJ (2019) Global ecological, social and economic impacts of marine plastic. Mar Pollut Bull 142(March):189–195. https://doi.org/10.1016/j.marpolbul.2019.03.022

Carney Almroth B, Eggert H (2019) Marine plastic pollution: sources, impacts, and policy issues. Rev Environ Econ Policy 13(2):317–326. https://doi.org/10.1093/reep/rez012

Chiba S, Saito H, Fletcher R, Yogi T, Kayo M, Miyagi S, Ogido M, Fujikura K (2018) Human footprint in the abyss: 30 year records of deep-sea plastic debris. Mar Policy 96(April):204–212. https://doi.org/10.1016/j.marpol.2018.03.022

Ford HV, Jones NH, Davies AJ, Godley BJ, Jambeck JR, Napper IE, Suckling CC, Williams GJ, Woodall LC, Koldewey HJ (2022) The fundamental links between climate change and marine plastic pollution. Sci Total Environ 806:150392. https://doi.org/10.1016/j.scitotenv.2021.150392

Geyer R, Jambeck JR, Law KL (2017) Production, uses, and fate of all plastics ever made. Sci Adv 3(7):5. https://doi.org/10.1126/sciadv.1700782

Hidalgo-Ruz V, Gutow L, Thompson RC, Thiel M (2012) Microplastics in the marine environment: A review of the methods used for identification and quantification. Environ Sci Technol 46(6):3060–3075. https://doi.org/10.1021/es2031505

Jones JS, Porter A, Muñoz-Pérez JP, Alarcón-Ruales D, Galloway TS, Godley BJ, Santillo D, Vagg J, Lewis C (2021) Plastic contamination of a Galapagos Island (Ecuador) and the relative risks to native marine species. Sci Total Environ 789:147704. https://doi.org/10.1016/j.scitotenv.2021.147704

Kirchherr J, Reike D, Hekkert M (2017) Conceptualizing the circular economy: an analysis of 114 definitions. Resour Conserv Recycl 127(April):221–232. https://doi.org/10.1016/j.resconrec.2017.09.005
Lavers JL, Bond AL (2017) Exceptional and rapid accumulation of anthropogenic debris on one of the world's most remote and pristine islands. Proc Natl Acad Sci 114(23):6052–6055. https://doi.org/10.1073/pnas.1619818114
Lavers JL, Dicks L, Dicks MR, Finger A (2019) Significant plastic accumulation on the Cocos (Keeling) Islands, Australia. Sci Rep 9(1):1–9. https://doi.org/10.1038/s41598-019-43375-4
Matamoros Alcivar E, de Llano García D, Vaca Chanatasig C (2022) Review of impacts on sustainable development in Ecuador of the COVID-19 pandemic and lockdown measures. MDPI 15(38):10. https://doi.org/10.3390/environsciproc2022015038
Monteiro RCP, Ivar do Sul JA, Costa MF (2018) Plastic pollution in islands of the Atlantic Ocean. Environ Pollut 238:103–110. https://doi.org/10.1016/j.envpol.2018.01.096
Napper IE, Davies BFR, Clifford H, Elvin S, Koldewey HJ, Mayewski PA, Miner KR, Potocki M, Elmore AC, Gajurel AP, Thompson RC (2020) Reaching New Heights in plastic pollution—preliminary findings of microplastics on Mount Everest. One Earth 3(5):621–630. https://doi.org/10.1016/j.oneear.2020.10.020
Pan D, Su F, Liu C, Guo Z (2020) Research progress for plastic waste management and manufacture of value-added products. Adv Compos Hybrid Mater 3(4):443–461. https://doi.org/10.1007/s42114-020-00190-0
Rosenboom JG, Langer R, Traverso G (2022) Bioplastics for a circular economy. Nat Rev Mater 7(2):117–137. https://doi.org/10.1038/s41578-021-00407-8
Ryan PG (2015) A brief history of marine litter research. In: Bergmann M, Gutow L, Klages M (eds) Marine anthropogenic litter. Springer, pp 1–25. https://doi.org/10.1007/978-3-319-16510-3_1
Schofield J, Wyles KJ, Doherty S, Donnelly A, Jones J, Porter A (2020) Object narratives as a methodology for mitigating marine plastic pollution: multidisciplinary investigations in Galápagos. Antiquity 94(373):228–244. https://doi.org/10.15184/aqy.2019.232
Schroeder P, Anggraeni K, Weber U (2019) The relevance of circular economy practices to the sustainable development goals. J Ind Ecol 23(1):77–95. https://doi.org/10.1111/jiec.12732
Torsten Hardter U, Larrea Ona I, Butt K, Chitwood J (2010) Waste management blueprint for the Galápagos Islands. In WWF & Toyota https://c402277.ssl.cf1.rackcdn.com/publications/329/files/original/Waste_Management_Blueprint_for_the_Galápagos_Islands_%28PDF__2.01_MB%29_2010.pdf?1345732665
Van Sebille E, Delandmeter P, Schofield J, Hardesty D, Jones J, Donnelly A (2019) Basin-scale sources and pathways of microplastic that ends up in the Galápagos Archipelago. Ocean Science Discuss:1–15. https://doi.org/10.5194/os-2019-37
Yuan X, Wang X, Sarkar B, Ok YS (2021) The COVID-19 pandemic necessitates a shift to a plastic circular economy. Nat Rev Earth Environ 2(10):659–660. https://doi.org/10.1038/s43017-021-00223-2
Zheng J, Suh S (2019) Strategies to reduce the global carbon footprint of plastics. Nat Clim Chang 9(5):374–378. https://doi.org/10.1038/s41558-019-0459-z

Jen S. Jones is the Head of Programmes at the Galapagos Conservation Trust, overseeing their programme portfolio both in Galapagos and the UK, and is an Honorary Research Fellow at the University of Exeter, specialising in the impacts of oceanic plastic pollution. Her PhD thesis focused on microplastic pollution in the Galapagos Marine Reserve.

Jess Howard is a marine biologist and the Galapagos Conservation Trust's Project Manager for the Pacific Plastics: Science to Solutions network working in the Eastern Pacific region. Jess specialises in science communications and network management.

Tamara S. Galloway is a Professor of EcoToxicology at the University of Exeter and is a leading expert in the impacts of microplastics and the potential of circular economy solutions.

Lucía Norris Crespo is the Galapagos Conservation Trust's Programmes & Public Policy Manager who specialises in facilitating linkages between community wisdom and policy.

Sol Aspinosa is a social scientist specialising in the engagement of civil society with environmental programmes. She is one of Galapagos Conservation Trust's consultants for the Pacific Plastics: Science to Solutions network.

Chapter 31
From Building Resilience to Adaptive Transformation: Exploring the Rationale for Inclusive Governance in Galapagos

Maria Soledad Garcia Ferrari, Amelia A. Bain, and Stephanie Crane De Narváez

Introduction

This chapter presents a reflection on research carried out between 2020 and 2022 in collaboration with the Government Council of Galapagos (*Consejo de Gobierno del Régimen Especial de Galápagos*), the British-Ecuadorian Chamber of Commerce (*Câmara Ecuatoriano Britânica*), academics from San Francisco de Quito University (*Universidad San Francisco de Quito*), and local institutions and community groups in the Ecuadorian province of Galapagos. The aim of this research was to identify drivers at the local level to enable sustainable and inclusive development, including a sustainable energy transition, and a transformation in resource management in Galapagos, rooted in local communities' needs and livelihoods (Garcia Ferrari et al. 2021a). Through interviews and focus groups (Garcia Ferrari et al. 2021a, b), the research has engaged local stakeholders to identify pathways towards more balanced and inclusive governance that promote the active engagement and empowerment of communities to build resilience. A key finding is that conservation goals in Galapagos can be strengthened by ensuring that policy and actions are grounded in integrated and inclusive governance frameworks that seek shared responsibility in managing resources within a complex socio-ecosystem (Garcia Ferrari et al. 2021a).

Here, we further explore the basis for inclusive decision-making in the province based on a series of interviews focused on investigating collective actions in response to the COVID-19 crisis, and focus groups aimed at building understanding

M. S. Garcia Ferrari (✉) · S. Crane De Narváez
School of Architecture and Landscape Architecture, University of Edinburgh, Edinburgh, UK
e-mail: s.garcia@ed.ac.uk

A. A. Bain
School of Architecture and Landscape Architecture & School of Geosciences, University of Edinburgh, Edinburgh, UK

S. J. Walsh et al. (eds.), *Island Ecosystems*, Social and Ecological Interactions in the Galapagos Islands, https://doi.org/10.1007/978-3-031-28089-4_31

of local perceptions of resilience. This work has contributed to the development of the 2030 Galapagos Islands Strategic Plan (Galapagos Government Council 2021), which is structured according to the following five pillars: governance, community, environment, habitat, and economy. Within the design of this policy, our research initially aimed to identify pathways towards achieving zero fossil fuel use in the province by 2040, which represents a flagship climate change mitigation policy for Ecuador. For such a transition to be effective and sustainable in a province with ca. 30,000 residents and over 275,000 tourists per year (prior to the COVID-19 pandemic), our research explored opportunities to develop a governance framework that enables solutions to be co-produced directly with Galapagos communities, to build resilience through ensuring real engagement with needs, priorities, and knowledge at the local level. This approach requires establishing inclusive spaces to develop solutions that involve a diverse range of stakeholders throughout the design and implementation of actions, while legitimizing and valuing the different types of knowledge they bring to the discussion (Smith et al. 2020; Garcia Ferrari et al. 2021c). To this aim, our research has explored an alternative framework for resource management in Galapagos developed through intersecting the key pillars of the water-energy-food (WEF) nexus and adaptive co-management (ACM) approaches (Garcia Ferrari et al. 2021a). This interconnected framework has contributed to the identification of governance mechanisms that could contribute to building resilience in the province, by opening up a space for deliberation and conflict resolution between stakeholders in the management of critical resources.

The Galapagos economy has been severely impacted by the crisis brought on by the COVID-19 pandemic which began in 2020, due to a heavy reliance on tourism, with jobs both directly in and indirectly dependent on tourism severely affected (Burbano et al. 2022). During the crisis, our research explored how local communities, NGOs, and government institutions identified new ways to diversify livelihoods, with a greater focus on harnessing local skills and products. The pandemic has therefore brought new opportunities to rebuild the local economy towards achieving greater sustainability (e.g., reducing imports, and the associated threat of invasive species) and resilience (i.e., to shocks such as economic crises and climate change). However, the crisis has also brought demands from local residents for economic development, for example, pressure to open more direct air traffic routes to boost tourism, and resume long-line fishing in the protected marine reserve to support fishers, bringing threats to conservation. This illustrates the important interlinkages between Galapagos communities, their wellbeing and livelihoods, the protection of the environment, and resilience to crises.

Our work so far has provided principles for an inclusive transformation of resource governance in Galapagos, based on data collection in relation to energy and resources (Bain et al. 2020), and mapping these findings with government policy and community actions (Garcia Ferrari et al. 2021b). More broadly, through our ongoing research on the co-production of risk management strategies and adaptation to climate change in various countries of Latin America (Ecuador, Colombia, Mexico, and Guatemala), we have explored alternative governance approaches that build trust between diverse stakeholders through a 'dialogue of knowledges', leading to socio-technical and technological solutions that promote co-responsibility and community agency (Garcia Ferrari et al. 2021a, c, Montejano-Castillo et al. 2022; Smith et al. 2020, 2021). In this chapter, we first present a theoretical framework defining our approach to building resilience through co-production, which underpins our argument for the development of inclusive decision-making processes around natural resources, and progressive governance mechanisms more widely, considering the key aspects of scale, knowledge and power (Brugnach et al. 2017). We then outline the methodology used in this research and discuss our findings. Finally, we present a set of reflections on the development of a governance framework in Galapagos that integrates diverse knowledges and rebalances power through the co-production of actions and interventions at different geographical scales, to support the system transformation necessary to build long-term resilience in the province.

Theoretical Framework

Building Resilience

The concept of resilience typically considers the assets and attributes of a society that allow it to return to its regular functioning when a shock occurs, or to maintain its functioning in the face of chronic stressors (CARE 2014). Whereas chronic stressors might include small-scale disasters, such as small-scale losses due to repeated landslides in the rainy season, shocks may include pandemics, economic crises, or high-impact environmental hazards. Resilience also encompasses the ways in which a socio-environmental system can be adjusted (i.e., through changes to governance or infrastructure for example) to ensure fewer losses. For a community, resilience represents the ability to use its resources and capacities to absorb, resist and recover from the effects of such shocks or stressors (Manyena 2006).

However, the concept of resilience extends further than reducing existing risks and preventing the creation of new ones, to improving "*a system's performance in the face of multiple hazards rather than preventing or reducing the loss of assets caused by specific events*" (ARUP 2014, p. 11). Such a 'system' could, for example, represent the Galapagos archipelago, including its environment, local people and visitors, with the opportunity to challenge and alter its functioning to adapt to risks and reduce their impact. In the context of climate change, Pelling (2011) defines three levels of adaptation to risks, including: (1) resilience, in which the functional integrity of a system is maintained, and existing practices are improved; (2) transition, in which incremental changes in governance are made by altering rules and decision-making; and (3) transformation, which involves a regime change in which the underlying values of the system are questioned, addressing the root causes of vulnerability and deprivation.

Drawing on Pelling's (2011) approach, our work in Galapagos is exploring pathways to move beyond achieving resilience, towards a transformation of the socio-ecosystem to face existing and emerging risks. Building this transformation requires solutions that are designed and implemented based on the knowledge of different stakeholder groups (e.g., Aguilar-Barajas et al. 2019). Adopting a complex systems perspective that acknowledges the synergies and conflicts between different sectors and stakeholders, the process of building resilience can be viewed as a social and environmental capacity-building process, through acquiring knowledge, learning to adapt, and organising around specific goals (Borquez et al. 2017).

In the context of Galapagos, an insular province with strong conservation goals and limited resources, building resilience is best approached through a socio-ecosystem lens: in other words, an ecosystem in which humans play a fundamental role that should not be viewed as separate from the conservation of the natural environment (e.g., Tapia et al. 2009; Consejo de Gobierno del Régimen Especial de Galápagos 2016; Rousseaud et al. 2017; Espin et al. 2019). In this view, sustainable development for the people of Galapagos is tightly connected to the conservation of the natural environment (e.g., Burbano and Meredith 2021), and appropriate decision-making and governance can allow progress in both of these areas in conjunction (Garcia Ferrari et al. 2021a). To begin to address this topic, our research has begun to explore policies, mechanisms and institutional structures that might allow conservation approaches in synergy with sustainable development, promoting risk reduction, economic diversification and satisfying, decent livelihoods. Based on our broader research in Latin America, these elements are typically linked with actions at a range of geographical scales and require the participation of a diverse group of stakeholders. Local communities therefore represent fundamental participants in the identification, development, and implementation of actions towards more sustainable development, resilience, and adaptive transformation. Our approach considers that appropriate decision-making and policies in Galapagos can and should be determined through the involvement of all relevant stakeholders, including for example fishers, farmers, people working in the tourism industry, the national park, and local and provincial government, local NGOs and conservation organisations.

Co-production of Resilience

As detailed above, solutions and strategies that build resilience to current and future environmental and economic shocks require the active engagement of the relevant groups of actors that will implement them (Ingram 2013; Brugnach et al. 2017). Collaborative governance structures and decision-making processes are therefore required (Bouwen and Taillieu 2004; Gray 2007; Huxham et al. 2000) to facilitate and support the equitable and effective implementation of actions and policies (Forsyth 2010; Brugnach et al. 2017). Collaboration in collective decision-making through multi-scale dynamics and negotiations, together with knowledge- and power-sharing structures, have been defined as key elements for building resilience and fostering transformative change (e.g., Brugnach et al. 2017; Smith et al. 2021; Garcia Ferrari et al. 2021a).

Within this conceptual framework, co-production is defined as a collaborative approach in which various stakeholders participate in and benefit from the delivery of a service or a policy. The concept of co-production arose in the context of the implementation or improvement of urban services, such as water supply or sewerage, typically in communities where these services are inadequate. However, our work has explored co-production in the context of developing strategies for risk and resource management in growing urban areas (Smith et al. 2020, 2021; Garcia Ferrari et al. 2021a, c). In this context, co-production represents an inclusive approach for identifying management strategies (e.g., in relation to resources, risk, environmental conservation, urban development policies, etc.) in which those active within different sectors of a socio-ecosystem collaborate, negotiate, and reach consensus in decision-making. This process is built around a 'dialogue of knowledges', bringing together local, traditional, technical, and academic knowledge to identify solutions.

Whereas scientific or technical solutions identified in isolation have been criticised for "failing to inform environmental management and decision making in an effective way" due to separating "science, policy, and society in ways that inhibit […] problem solving (Gibbons et al. 1994; Jasanoff et al. 1998; Latour 1998; Jasanoff 2009, cited in Djenontin and Meadow 2018, p. 885), a marked increase in interest in co-production approaches to knowledge production has been noted within climate sciences, climate change adaptation, and more broadly within environmental management and governance (Lemos and Morehouse 2005; Visbeck 2008; Ziervogel et al. 2016; Wamsler 2017, cited in Djenontin and Meadow 2018). Furthermore, Brugnach and Ozerol (2019), argue that disciplinary knowledge and science alone do not provide a comprehensive understanding of the complex, interdependent and multi-faceted nature of resource-related issues, because scientific and technical knowledge do not fully capture and account for what happens on the ground, and are often mediated by power dynamics. In contrast, the interaction of diverse knowledges provides the basis for creating more comprehensive and inclusive understandings of resource-related issues, as well as improving science-policy interactions that facilitate democratic resource governance (Brugnach and Ozerol 2019).

In this sense, co-production represents an innovative and inclusive approach to governance as it recognises the knowledge and expertise of the multiple stakeholders involved in the process. In essence, co-production moves beyond consultative participatory processes to actively involve stakeholders in designing appropriate solutions, integrating various types of knowledge and avoiding an exclusive preference or reliance on scientific or technical understandings, to the detriment of other knowledge types (Howarth et al. 2018, p. 78). Co-production aspires to empower stakeholders by valuing their participation and acknowledging their various contributions towards identifying and implementing solutions for the improvement of their community (Ostrom 1996). Valuing this social capital offers the necessary social viewpoint and relationships that bring communities together and build capacity and resilience (Randolph 2011).

The concepts of co-production, in particular, and citizen participation more generally, are therefore especially valuable for building resilience: the understanding and evaluation of risks, such as climate change-related risks, and their impacts can be enriched by leveraging information and knowledge provided by those who are directly exposed and affected at a local level, allowing for more informed decision-making (Bennett et al. 2014). Participation can be seen either as an 'institutional arrangement' that is part of decision-making processes in governance, or as a 'right' in collective actions of claiming citizenship (Kalandides 2018). Participative processes should allow the practice of social learning and enable the exchange of knowledge and information between the actors involved, regardless of their expertise (Milupi et al. 2017).

Engaging actors in the co-production and negotiation of solutions and strategies has been effective in building resilience and empowering communities to take a central role in managing risks (Armitage et al. 2011). By linking co-production processes to the concept of building resilience and 'building back better' following a shock (such as the COVID-19 crisis), rather than simply reverting to existing practices that may include unsustainable or unfair practices to continue, it is possible to imagine alternative governance structures or policies that support a more equitable, inclusive and sustainable society. Co-production mechanisms, in which all stakeholders collaborate, have the potential to address these deeper changes, enabling the Galapagos socio-ecosystem to manage risk and resources in a way that moves beyond building resilience, towards transformation.

Innovative Governance Mechanisms for Resilience

As outlined in the previous section, there is a considerable body of work demonstrating that sustainable development can only be achieved by considering the complexities and interconnections between societies and their environment. The interdependence between these systems generates non-linear relations and unstable conditions, characterised by periods of change, adaptation, and transformation (Boyd and Folke 2011). Effective governance for building resilience therefore requires an approach that focuses on the social challenges of the system, as well as on the environmental hazards and stressors, and establishes a framework that enables policies and actions to be negotiated and agreed upon. Furthermore, flexibility in governance structures is key for compatibility with an ecosystem-based management approach (Folke et al. 2005), allowing adaptive decision-making.

Examples of innovation have been identified in policy development and in collective actions at the community level based on specific practices and activities, however, long term support for these top-down and bottom-up initiatives to grow and interconnect has been less evident in the literature. Understanding the capacity of socio-ecological systems to adapt to environmental change can help to develop and agree interconnected actions at different scales that can protect natural resources and at the same time ensure human wellbeing and sustainability. Governance frameworks that enable this balance must consider institutional and community dynamics and power relations at a range of geographical scales. These interconnected frameworks will include knowledge co-production, negotiation, and agreement across geographical and institutional scales, to allow the development of solutions to complex problems. Addressing local resilience within the approach of co-production therefore requires considering three essential elements (Brugnach et al. 2017):

i. Scale: Understanding actions at the local scale as well as at the wider institutional and geographical scale is essential to achieve long term sustainability, linking bottom-up actions with government-led initiatives (Maskrey 1989; Petcou and Petrescu 2015; Stevenson and Petrescu 2016).
ii. Knowledge: the integration of multifaceted knowledge, built collaboratively between actors is essential when building resilience (Hallegatte et al. 2018; Aguilar-Barajas et al. 2019).
iii. Power: Complex power relationships emerge when building resilience across diverse social, economic, and political dimensions. Vulnerable groups must be empowered to influence decision-making at the local level, using community knowledge to respond to risks, assuming an active and participatory role in risk assessment, mitigation planning, capacity building, and monitoring (Pandey and Okazaki 2005).

The Galapagos archipelago presents a range of challenges at the local level arising from the interaction of social and environmental systems, coupled with external pressures linked with an economic over-reliance on the high tourism demand. Gonzalez et al. (2008) suggested that achieving a sustainable model for Galapagos will require the province to "modify traditional practices to produce a more adaptive resilience-based, co-management model, adopt a more comprehensive approach to territorial planning, strengthen participative approaches and institutional networks, and promote transdisciplinary research at the frontiers of social and biophysical sciences" (de Haan et al. 2019). In line with these findings, our research has identified that the role of communities in resource management (e.g., water, energy, food) is critical for sustainable development in Galapagos, in particular, for designing future blueprints for economic and energy transitions based on local needs and priorities (Garcia Ferrari et al. 2021a). Through intersecting the key pillars of adaptive co-management (ACM) and water-energy-food (WEF) nexus approaches to resource management, we proposed a new framework for decision-making and governance in Galapagos. ACM seeks to empower local stakeholders through processes of experimentation, monitoring, deliberation and responsive resource management, in articulation with governmental agencies, educational institutions and NGOs at multiple territorial scales (Hasselman 2017). ACM also aims to harness socio-ecological knowledge and experience, and engage diverse and multi-scale interest groups, ranging from local communities, municipalities, regional and national institutions and international-level organisations (Folke et al. 2002). The WEF nexus approach promotes the adoption of a complex systems view to maximise synergies and optimise trade-offs (e.g., Bleischwitz et al. 2018; Simpson and Jewitt 2019), requires a focus on sustainable livelihoods within WEF resources (e.g., Biggs et al. 2015), seeks to maintain the environmental integrity of ecosystems while facilitating equitable access to resources (e.g., Leese and Meisch, 2015; Simpson and Jewitt 2019), and advocates for multi-stakeholder involvement in deliberative scenario planning (e.g., Howarth and Monasterolo 2017; Wicaksono et al. 2017). Our proposed ACM-WEF framework represents a possible innovative approach to decision-making around resources at the local level in Galapagos, which could be integrated with other framings, for example at the scale of each inhabited island and the scale of the province. Building on this work, the following sections further explore pathways towards innovative governance mechanisms in Galapagos, through building understanding of local perceptions and experiences in relation to resilience.

Methodology

The research discussed below is based on four main components. First, between December 2019 and June 2020, a literature review was carried out by researchers at the University of Edinburgh in relation to the social, environmental, and energy context of the Galapagos Islands (Bain et al. 2020). This review allowed the identification of research questions in relation to sustainability, development, and conservation in the province. In a second stage, nine semi-structured interviews were completed online in June–July 2020 with representatives of the provincial government, local NGOs, and local businesses (Garcia Ferrari et al. 2021a). These interviews aimed to build understanding of challenging issues related with natural resources management, community participation, and economic development. In several cases, interviewees had held positions in different types of organizations (for example, government, NGOs, tourism) and were able to provide an integrated viewpoint across complex interconnected issues. These interviews led to the proposed ACM-WEF framework for natural resource management described above. In a third research stage, a further 15 semi-structured interviews were conducted in May–June 2020 with representatives from local government, local NGOs, professionals, and farmers, to gather data relating to collective actions that arose in response to the COVID-19 pandemic in Galapagos. Participants were identified using a snowball sampling approach through an online review of social media, focusing on individuals and organizations engaged in collective actions responding to the economic impacts of the crisis.

Finally, three focus groups were conducted online in June 2021 to build understanding of resilience to shocks and stresses on a local level (Garcia Ferrari et al. 2021b). These focus groups included one session with provincial and local government representatives (7 participants), one with NGOs (7 participants), and one with community members (7 participants), including farmers, a youth council member, an independent community researcher, and local community group leaders. In these focus groups, participants were asked to share their perspectives on resilience, local initiatives, and potential drivers for scaling up actions from an individual to a collective level. In addition, researchers from the University of Edinburgh shared the findings from the previous phases of the research.

The analysis presented below is grounded in the need identified through the initial set of interviews for more inclusive and equitable decision-making, to support sustainable development and build resilience in Galapagos. For this analysis, the findings from the second set of interviews (collective actions in response to COVID-19) and the focus groups (perspectives on resilience) were mapped with the key aspects relating to inclusive governance mechanisms identified in the theoretical framework above, i.e., scale, knowledge, and power. The mapping results were then reconciled through discussion and debate between the authors to reach consensus, and the results are reported below.

Discussion

Scale

The need for resilience in Galapagos to both local and global events was highlighted in both the institutional and NGO focus groups. In particular, the COVID-19 pandemic clearly emphasised the need for local resilience to global shocks due to the significant impact in Galapagos of the economic crisis. Strengthening local food chains was considered a key aspect of local resilience, guaranteeing product quality, reducing food loss and waste, and strengthening the agricultural sector. A resilient food system was noted to be a cornerstone of strengthening regional autonomy.

Furthermore, an important point raised in the focus group with representatives from NGOs was the need to consider the meaning of resilience on multiple scales. For example, building resilience on the island of Santa Cruz, with a population of ca. 16,000 (according to the 2015 census, INEC, 2015) and through which most tourists arrive in the archipelago, will be different from building resilience in Isabela, with a population of ca. 2500 and a smaller tourist sector (INEC 2015). Participants emphasised the need to develop a vision and 'road map' for each island as well as the province as a whole to recognize differences in vulnerability and adaptation capacity. In parallel, it was also considered valuable for each sector (such as, tourism, fisheries, agriculture) to assess its own vulnerabilities and resilience, as well as their dependency on other sectors. For example, the fisheries and agricultural sectors were impacted by the collapse of the tourism industry during the COVID-19 crisis, which reduced demand. However, these sectors displayed a high adaptive capacity, which contributed to resilience in the province.

Reflections in the focus group with community members indicated the importance of systemic change. For example, concerns were raised that upscaling regenerative agriculture would require committed changes in buying habits in the tourism industry to ensure a sustainable market to support local production. Systemic change was also brought up in connection with upscaling local initiatives, for similar reasons, and these changes were seen as requiring support from both institutions and the private sector.

In addition, several local initiatives were noted as having increased, or currently contributing to increase, resilience in Galapagos. For example, the '*Canastas ACA*' initiative provided products from regenerative agriculture to ensure that Galapagueños had sufficient food during the pandemic, via a collaborative network of actors. An interviewee who worked in hospitality prior to the crisis also highlighted the value of this initiative in creating a weekly market where he can sell his specialist products. The '*Yo Solo Vendo lo que Produzco*' ('I Only Sell What I Produce') project used online platforms and tools to develop innovative marketing strategies and new marketing channels (e.g., a mobile application, website and a 'farmer's store') to directly link consumers and the tourism industry with local producers, reducing the costs generated by intermediaries. These initiatives were recognized as having improved food security during the COVID-19 crisis. However, an interviewee who supported an initiative to distribute food baskets also highlighted the challenge of upscaling and ensuring continuity of some small-scale, community-driven initiatives.

Knowledge

On a broad level, participants in the institutional focus group highlighted the need to develop a strategic framework and vision for the composition of public and private administrative structures. This was linked with strengthening communication channels to more effectively share available knowledge during crises. Inter-institutional knowledge-sharing and collaboration at the provincial level, along with better integration with national institutions, was also seen as key to address complex challenges and build resilience. This was echoed in our interviews with provincial and local government representatives, and a medical professional, who highlighted the importance of this institutional coordination during the COVID-19 crisis, including collaboration with academic institutions. Institutional actors in the focus group also viewed the training of new leaders as key to guaranteeing the consistency and sustainability of policies, plans and programmes. For example, an ability to collaborate and to work for the common good were seen as important qualities of future leaders, and these skills were also highlighted as valuable during the COVID-19 crisis in our interviews.

Furthermore, participants in the community focus group voiced that there is a need for sustainability policies to be better informed by local experiences on the ground, integrated between sectors, and aimed at the long term. They also highlighted the importance of conserving the natural environment to build resilience against food insecurity, to better manage energy and to fight corruption. Community members connected these needs with the co-development of visions and plans for the future of Galapagos (i.e., co-develop "*imaginarios*").

Institutional focus group participants also raised the importance of handling information and knowledge appropriately to avoid disinformation, curb biases, and increase understanding of particular issues, such as the risks to Galapagos from climate change. In terms of sharing and disseminating knowledge, participants in the NGO focus groups highlighted the "multiplying effect" of individuals who are active within several different organizations and noted the importance of the participation of young people connected to youth groups. The importance of involving younger generations in building resilience to challenges such as climate change was also voiced in our interviews. Collaborative initiatives were highlighted by institutional representatives as examples of local actions that build a sense of 'community' and develop a sense of responsibility to conserve the shared environment. For example, the '*Galapagos Infinito*' initiative, led by the Naveducando Foundation and the Galapagos Government Council, allows children and young people from Galapagos to connect with and learn more about their natural heritage. Several interviewees belonging to a collective against violence towards women ('*Colectivo MAGMA*') also highlighted the role of civil society organisations in bringing issues into the public space for debate, increasing visibility, and promoting cultural and political change.

Considering the lessons from the COVID-19 crisis, participants in the institutional focus group noted that the global economic downturn associated with the pandemic revealed the need to support the diversification of livelihoods through training and developing technical skills (e.g., for fisheries, agriculture, and tourism sectors). This was supported by comments in the focus group with community members, who highlighted challenges in accessing information and training when outside the formal structures of academia or larger, well-funded NGOs. This lack of knowledge and experience was also cited as a barrier to obtaining funding for grassroots initiatives, due to competition with larger organisations. Co-authorship of research between academics and local people was highlighted as a desirable mechanism to improve knowledge accessibility.

Conversely, some local initiatives, such as the *Huertos Tranquilos* ('Tranquil Gardens') collective, are disseminating knowledge on growing vegetables and empowering families to engage in this activity together, through providing knowledge, materials, games, and ongoing support. An interviewee who has engaged with this initiative indicated that these efforts are increasing food security and contributing to a sense of autonomy. Another interviewee who was involved in several initiatives related to food solidarity during the COVID-19 crisis highlighted the importance of sharing knowledge and skills between community members for strengthening local level initiatives. For example, local people with technology skills assisted in the distribution of food baskets, through improving and systematising communication. This interviewee also noted the value of these local initiatives for raising awareness of the diversity of possible livelihoods, beyond tourism.

An important element that was raised in the focus groups and interviews was the re-emergence of traditional practices during the COVID-19 crisis, based on inter-generational knowledge exchange. For example, participants noted that the re-emergence of bartering (*trueque*, for example, vegetables for fish), which used to be widely practiced, increased resilience during the economic crisis. There was also a recognition of the need to retain the knowledge and skills developed during the pandemic, especially given the marked preference of those who diversified their activities to return to the tourism industry once tourism resumed. One of the NGO representatives indicated that they were actively documenting the local changes that took place in response to the crisis to prevent a loss of resilience to future events. One of our interviewees engaged in regenerative farming also indicated that she expected the high demand for her products during the crisis to continue beyond the pandemic, as local people gained a greater appreciation of local products. This greater appreciation for local produce as a result of the crisis, and solidarity actions that were taken, was echoed in several interviews and the focus groups. The knowledge, skills, and experiences gained during the pandemic were therefore considered to represent a contribution towards future resilience, through the people who lived through it and adapted their behavior and attitudes. Furthermore, the pandemic revealed important vulnerabilities that were previously unrecognized. For example, certain groups, such as those working in the tourism industry and naturalist guides, were highly vulnerable to the collapse of the tourism sector, whereas these groups were not previously understood as vulnerable.

Power

Based on their experiences during the COVID-19 crisis, participants in the institutional focus group voiced the need for more collaborative dynamics to build resilience, including within and across government institutions and civil organizations, as well at the provincial level and across sectors, to allow greater focus on common goals. A representative of the collective against violence towards women noted in an interview that cooperation and negotiation between civil society and government institutions is, in their view, key to driving change. This participant also indicated that they expect the relationship between these actors to evolve over the long term, from resistance to more productive ways of working together. This interviewee also noted that the dynamic they seek is not related with replacing the responsibility of the state, but with increasing the agency of civil society to raise awareness of policy gaps or failures and achieve social change. Similarly, interviewees from the provincial government and an NGO criticised the notion of overly paternalistic government institutions, suggesting that this disempowers civil society.

Institutional representatives also highlighted the importance of planning, to define priorities and strategies for resilience in the face of external factors and risks. Representatives from NGOs indicated that collaboration and associativity increased resilience during the pandemic, however, trust issues limited some people's willingness to work with government institutions. Conversely, participants in the NGO focus group noted resistance in some institutions to the participation of civil society in influencing policies, revealing a need to strengthen institutional capacities for supporting participative processes. In this respect, an interviewee from the collective against violence towards women noted the potential 'gateway' effect of initially working more closely with more willing institutions. Participants in the community focus group also highlighted a need for local knowledge to feed into policies, to avoid short-term solutions that do not solve the issues experienced at the local level.

Furthermore, community members noted that sustainable policies should aim to build trust between different actors, ensuring that the community has ownership over the policies, otherwise they risk failure. In the view of these participants, policies must also be developed on a local level, as national laws sometimes clash with important priorities in Galapagos. In addition, community members noted that the inclusion of different actors in policy development may help in balancing different interests and priorities.

To build resilience, NGO representatives highlighted the need to valorize community-oriented actions that were taken during the COVID-19 crisis, to recognize the value in these efforts and encourage continuity and replication. Community members also raised the importance of developing adequate social policies to avoid overworking or marginalising certain groups in the process of building resilience. For example, the caring responsibilities of women increased during the pandemic, which, when coupled with new agricultural work, led to significant overworking. This point was echoed in interviews with members of the collective against violence towards women, who noted that stronger social policies are needed, whereas policies relating to conservation have typically been prioritised.

Reflections on Building Resilience and Transformation in Galapagos

The analysis above demonstrates that the COVID-19 crisis brought a new recognition in Galapagos that the province needs to build resilience to global shocks as well as local events. Strengthening food security is a crucial element of this resilience and was a key focus during the economic crisis. Collaboration between government institutions at all levels, as well as with civil society, was highlighted as a priority. Better integration of institutions, plans and programs, rooted in local community needs and priorities, is seen as essential. We found that there is a strong recognition of the need for systemic change in Galapagos, to support sustainable development, and encourage economic diversification with decent and secure livelihoods. These systemic changes must be accompanied by adequate social policies, considering the diverse groups that exist in the province and their distinct needs. Knowledge exchange and accessibility are also considered vital to empower valuable local initiatives that strengthen resilience and cooperation. Despite a widespread return to economic activity associated with tourism, we found a desire to valorise and document community-focused actions and activities that were taken during the pandemic, to ensure the survival of the knowledge and skills developed or rediscovered during the crisis. In this sense, the economic crisis brought on by the COVID-19 pandemic represented an opportunity for learning about vulnerabilities, adaptive capacities, and experimenting with alternative realities.

A key question that came to light during our research was how to conserve the gains in resilience achieved through systemic changes during the pandemic. In other words, how can the adaptive capacity demonstrated in the crisis lead to transformation? Galapagos has seen various examples of participative processes, including in the development of the economic reactivation plan following the COVID-19 crisis, and the 2030 Development Plan. However, such short-term consultation efforts are unlikely to put underlying systems under scrutiny, nor alter power relations (Hordijk et al. 2014), and are therefore likely to lead to maintaining the status quo and building resilience through improving existing practices (Pelling 2011). Such an approach does not tackle the root causes of vulnerability and will not achieve a deeper level of adaptation to local and global risks. However, incremental changes in decision-making, evidencing a process of transition, "can be fostered through 'learning-by-doing', with experimentation and constant monitoring as practised in adaptive management." (Hordijk et al. 2014). Furthermore, opportunities for transformation in Galapagos, which require radical changes in decision-making structures and power relations (Pelling 2011) may open up if opportunities for deliberation and social learning are created (Hordijk et al. 2014). There exists a precedent for such innovative decision-making structures in Galapagos, in the form of the now-defunct Participatory Management Board of the Galapagos Marine Reserve (Garcia Ferrari et al. 2021a, and references therein). This experience represented a world-leading example of co-management of a shared resource. However, this co-management governance structure has now been superseded by a more exclusionary decision-making process that has removed power from certain groups, such as local fishers (Burbano et al. 2020).

The experiences and learning gained in Galapagos from the COVID-19 crisis should therefore be embedded within innovative decision-making structures, to allow a deeper level of adaptation to take place. Our research shows that inclusion is a strong element in adaptation, allowing the knowledge existing at the local level to feed into the development of policies adapted for each island, through creating opportunities for civil society to gain power in decision-making structures. Building trust and encouraging cooperation is also a significant goal documented in our interviews and focus groups. Furthermore, increasing food security requires a careful consideration of the interconnections between the food sector and other resource sectors, such as energy and water, when building new forms of production at the local level (e.g., Garcia Ferrari et al. 2021a). Inclusion, building trust, adopting a complex systems view, and encouraging experimentation are key elements of the ACM-WEF framework for resource management proposed in our previous work (Garcia Ferrari et al. 2021a), and co-production approaches to governance more generally.

Beyond building resilience in Galapagos, there is therefore a strong case for adopting innovative and inclusive approaches to decision-making, to achieve a deeper level of adaptation that significantly reduces vulnerability. Key aims of opening up spaces for co-production include restructuring relations between civil society and the state, empowering communities to contest power and negotiate around their needs, and unlocking economic and political resources that allow communities to reduce their vulnerability (Garcia Ferrari et al. 2021c, and references therein). Actions rooted in co-production enable the integration of different types of knowledge on a level platform, creating spaces for negotiation and agreement, thereby re-balancing asymmetric power dynamics. These actions also promote the long-term sustainability of adaptation measures by ensuring community acceptability and trust (Garcia Ferrari et al. 2021c, and references therein). Through deliberative scenario planning, co-production may also offer an adaptive form of governance, appropriate and necessary in the face of evolving socio-environmental crises, such as the climate emergency. The scale of Galapagos, and the existence of four inhabited, interconnected islands with distinct vulnerabilities, make the province an especially interesting living laboratory for experimentation with such progressive approaches to governance. Such innovative and effective governance frameworks that champion inclusion in decision-making must articulate and mediate multi-scale and multi-stakeholder processes relating to resource management, planning and economic development. Indeed, moving beyond participation in governance will allow individuals to claim citizenship. This acknowledges a social and political struggle between the community and the state, which in the case of Galapagos, is critical to balance conservation goals and socioeconomic development in the face of current and forthcoming challenges. Decision-making structures aimed at developing and sharing knowledge and strengthening community agency may contribute to greater autonomy at the local level. Such an approach may increase communities' capacity to respond to growing environmental challenges, and can build new forms of democracy, equity and social justice.

Based on the research described in this paper, we argue that Galapagos would benefit from a new phase of experimentation with inclusive governance structures, beginning with the co-production of visions and policies for the future of each island, and the province as a whole, with close attention to interconnections with governance at the national (e.g., policy of zero fossil fuel use in Galapagos by 2040) and international (e.g., UN Sustainable Development Goals and Sendai Framework for Disaster Risk Reduction) levels. As the province recovers from the COVID-19 crisis, and the climate emergency intensifies, we suggest that it is vital to open spaces for co-developing policies relating to water, energy, and food, through balancing local community knowledge, institutional knowledge, and scientific/technical knowledge. A shift in the balance of power in Galapagos through co-production may help resolve the struggle that local communities experience when accessing funding and other specialist resources to support grassroots initiatives, and the institutional desire to avoid overly paternalistic governance.

Acknowledgements We thank Lucía Norris, Maritina Iliadi, and Katherina Kaesehage for their help in conducting the interviews and focus groups and producing the community toolkit 'Tools and Perspectives for Sustainability, Innovation and Resilience in Galapagos'. We thank Maritina

Iliadi and Salam Mare'e for their ongoing collaboration in research on innovative governance mechanisms for risk reduction. We also wish to thank all the participants who gave their time to participate in this research.

References

Aguilar-Barajas I, Sisto NP, Ramirez AI, Magaña-Rueda V (2019) Building urban resilience and knowledge co-production in the face of weather hazards: flash floods in the monterrey metropolitan area (Mexico). Environ Sci Policy 99(August 2018):37–47. https://doi.org/10.1016/j.envsci.2019.05.021

Armitage D, Berkes F, Dale A, Kocho-Schellenberg E, Patton E (2011) Co-management and the co-production of knowledge: learning to adapt in Canada's Arctic. Glob Environ Chang 21(3):995–1004

ARUP (2014) City resilience index. Available online: https://www.arup.com/perspectives/publications/research/section/city-resilience-index

Bain AA, Maximov Gajardo S, Bravo Vargas R, Crane De Narváez S, Garcia Ferrari S (2020, August) Social, environmental, and energy context of the Galapagos Islands. Available online: https://e0b4b077-81a3-4e4b-84ba-9dea58ffefea.filesusr.com/ugd/649d7b_429e4235c86a44f58adcfcfcec6bbccd.pdf

Bennett JE, Blangiardo M, Fecht D, Elliott P, y Ezzati, M. (2014) Vulnerability to the mortality effects of warm temperature in the districts of England and Wales. Nature. Climate Change 4(4):269–273. https://doi.org/10.1038/nclimate2123

Biggs EM, Bruce E, Boruff B, Duncan JM, Horsley J, Pauli N et al (2015) Sustainable development and the water–energy–food nexus: a perspective on livelihoods. Environ Sci Pol 54:389–397. https://doi.org/10.1016/j.envsci.2015.08.002

Bleischwitz R, Spataru C, VanDeveer SD, Obersteiner M, van der Voet E, Johnson C et al (2018) Resource nexus perspectives towards the United Nations sustainable development goals. Nat Sustain 1:737–743. https://doi.org/10.1038/s41893-018-0173-2

Borquez R, Aldunce P, Adler C (2017) Resilience to climate change: from theory to practise through co-production of knowledge in Chile. Sustain Sci 12(1):163–176. https://doi.org/10.1007/s11625-016-0400-6

Bouwen R, Taillieu T (2004) Multi-party collaboration as social learning for interdependence: developing relational knowing for sustainable natural resource management. J Community Appl Soc Psychol 14:137–153

Boyd E, Folke C (eds) (2011) Adapting institutions: governance, complexity and social-ecological resilience. Cambridge University Press

Brugnach M, Özerol G (2019) Knowledge co-production and transdisciplinarity: opening Pandora's box. Water 11(10):1997

Brugnach M, Craps M, Dewulf ARPJ (2017) Including indigenous peoples in climate change mitigation: addressing issues of scale, knowledge and power. Clim Chang 140(1):19–32

Burbano DV, Meredith TC (2021) Effects of tourism growth in a UNESCO world heritage site: resource-based livelihood diversification in the Galapagos Islands, Ecuador. J Sustain Tour 29(8):1270–1289

Burbano DV, Meredith TC, Mulrennan ME (2020) Exclusionary decision-making processes in marine governance: the rezoning plan for the protected areas of the 'iconic'Galapagos Islands, Ecuador. Ocean Coastal Manage 185:105066

Burbano DV, Valdivieso JC, Izurieta JC, Meredith TC, Ferri DQ (2022) "Rethink and reset" tourism in the Galapagos Islands: Stakeholders' views on the sustainability of tourism development. Ann Tour Res Emp Insight 3(2):100057

CARE (2014) Participatory monitoring, evaluation, reflection and learning for community-based adaptation: PMERL A revised manual for local practitioners. CARE International UK, London. Available online: https://insights.careinternational.org.uk/media/k2/attachments/CARE_PMERL_revised_manual.pdf

Consejo de Gobierno del Régimen Especial de Galapagos (2016) Plan de Desarrollo Sustentable y Ordenamiento Territorial del Régimen Especial de Galápagos – Plan Galápagos. Available online at: https://www.gobiernogalapagos.gob.ec/wp-content/uploads/downloads/2017/04/Plan-Galapagos-2015-2020_12.pdf

de Haan FJ, Quiroga Ferri D, Walsh SJ, Bettencourt LMA (2019) Scales and transformative change: transitions in the Galapagos. In: Kvan, T., Karakiewicz, J. (eds) Urban Galapagos. Social and Ecological Interactions in the Galapagos Islands. Springer, Cham, https://doi.org/10.1007/978-3-319-99534-2_4

Djenontin INS, Meadow AM (2018) The art of co-production of knowledge in environmental sciences and management: lessons from international practice. Environ Manag 61(6):885–903

Espin PA, Mena CF, Pizzitutti F (2019) A model-based approach 1672 to study the tourism sustainability in an island environment: the case 1673 of Galapagos Islands. In: Kvan T, Karakiewicz J (eds) Urban Galapagos: transition to sustainability in complex adaptive systems. Springer, Cham, pp 97–113. https://doi.org/10.1007/978-3-319-99534-2_7

Folke C, Carpenter S, Elmqvist T, Gunderson L, Holling CS, Walker B (2002) Resilience and sustainable development: building adaptive capacity in a world of transformations. Ambio 31:437–440. https://doi.org/10.1579/0044-7447-31.5.437

Folke C, Hahn T, Olsson P, Norberg J (2005) Adaptive governance of social-ecological systems. Annu Rev Environ Resour 30:441–473

Forsyth T (2010) Panacea or paradox? Cross-sector partnerships, climate change, and development. Wiley Interdiscip Rev Clim Chang 1(5):683–696

Galapagos Government Council (2021) Galápagos 2030 – Galápagos Islands Strategic Plan. Available online: https://unidosporgalapagos.files.wordpress.com/2021/05/galacc81pagos-2030-strategic-plan.pdf

Garcia Ferrari S, Bain AA, Crane De Narváez S (2021a) Drivers, opportunities, and challenges for integrated resource co-management and sustainable development in Galapagos. Front Sustain Cities 3:666559

Garcia Ferrari S, Bain AA, Crane De Narváez S, Kaesehage K, Iliadi M, Norris L (2021b) Tools and perspectives for sustainability, innovation and resilience in Galapagos. Available online: https://www.forgalapagos.com/_files/ugd/e71f5c_12376895662b474c9fe5804f402fa005.pdf

Garcia Ferrari S, Kaesehage K, De Narvaez SC, Bain AA (2021c) Adaptation strategies for people: mitigating climate-change-related risks in low-income and informal urban communities through co-production. J British Acad 9(s9):7–37

Gibbons M, Limoges C, Nowotny H, Schwartzman S, Scott P, Trow M (1994) The new production of knowledge: the dynamics of science and research in contemporary societies. Sage, London, p 179

Gonzalez JA, Montes C, Rodriguez J, Tapia W (2008) Rethinking the Galapagos Islands as a complex social-ecological system: implications for conservation and management. Ecol Soc 13(2):13

Gray B (2007) The process of partnership construction: anticipating obstacles and enhancing the likelihood of successful partnerships for sustainable development. In: Glasbergen P, Biermann F, Mol APJ (eds) Partnerships, governance and sustainable development. Edward Elgar, Cheltenham

Hallegatte S, Rentschler J, Walsh B (2018) Building back better – achieving resilience through stronger, faster, and more inclusive post-disaster reconstruction 47. https://doi.org/10.1523/JNEUROSCI.2676-06.2007

Hasselman L (2017) Adaptive management; adaptive co-management; adaptive governance: what's the difference? Austr J Environ Manage 24:31–46. https://doi.org/10.1080/14486563.2016.1251857

Hordijk M, Sara LM, Sutherland C (2014) Resilience, transition or transformation? A comparative analysis of changing water governance systems in four southern cities. Environ Urban 26(1):130–146

Howarth C, Monasterolo I (2017) Opportunities for knowledge co-production across the energy-food-water nexus: making interdisciplinary approaches work for better climate decision making. Environ Sci Pol 75:103–110. https://doi.org/10.1016/j.envsci.2017.05.019

Howarth C, Morse-Jones S, Brooks K, Kythreotis AP (2018) Co-producing UK climate change adaptation policy: an analysis of the 2012 and 2017 UK climate change risk assessments. Environ Sci Pol 89:412–420

Huxham C, Vangen S, Eden C (2000) The challenge of collaborative governance. Public Manag Rev 2:337–358. https://doi.org/10.1080/14719030000000021

INEC (2015) Principales resultados Censo de Población y Vivienda Galápagos 2015. Available online: https://www.ecuadorencifras.gob.ec/documentos/web-inec/Poblacion_y_Demografia/CPV_Galapagos_2015/Presentacion_CPVG15.pdf

Ingram H (2013) No universal remedies: design for contexts. Water Int 38(1):6–11

Jasanoff S (2009) The fifth branch: science advisers as policymakers. Harvard University Press, Cambridge

Jasanoff S, Wynne B, Buttel F, Charvolin F, Edwards P, Elzinga A, Miller C (1998) Science and decisionmaking. In: Human choice and climate change, vol 1: the societal framework, pp 1–87

Kalandides A (2018) Citizen participation: towards a framework for policy assessment. Journal of Place Management and Development, 11(2). https://doi.org/10.1108/JPMD-02-2018-0017

Latour B (1998) From the world of science to the world of research? Science 280(5361):208–209

Leese M, Meisch S (2015) Securitising sustainability? Questioning the 'water, energy and food-security nexus'. Water Alternat 8:695–709

Lemos MC, Morehouse BJ (2005) The co-production of science and policy in integrated climate assessments. Glob Environ Change 15(1):57–68

Manyena SB (2006) The concept of resilience revisited. Disasters 30:434–450. https://doi.org/10.1111/j.0361-3666.2006.00331.x

Maskrey A (1989) Disaster mitigation: a community based approach. Oxfam International

Milupi ID, Somers MJ, Ferguson JWH (2017) A review of community-based natural resource management

Montejano-Castillo M, Moreno-Villanueva M, Cabrera-Sanchez BN, Reséndiz-Vázquez A, Makita M, Smith H, Garcia-Ferrari S (2022) Exploring digital co-management of flood risk reduction in Mexico City, Mexico: conditions and potentials. WIT Trans Ecol Environ 258:15–26

Ostrom E (1996) Crossing the great divide: coproduction, synergy, and development. World Dev 24(6):1073–1087

Pandey B, Okazaki K (2005) Community based disaster management: empowering communities to cope with disaster risks. United Nations Centre for Regional Development, Japan

Pelling M (2011) Adaptation to climate change: from resilience to transformation. Routledge, London, New York

Petcou C, Petrescu DM (2015) R-URBAN or how to co-produce a resilient city. Ephemera: Theory Polit Organ 15(1):249–262

Randolph J (2011) Creating the climate change resilient community. In: Goldstein BE (ed) Collaborative resilience: moving through crisis to opportunity. MIT Press, pp 127–148

Rousseaud A, Cruz E, Naula E, Ramos A, Granda M, Calvopifta M (2017) Plan Galapagos: an instrument for the holistic sustainable development of the province. In Galapagos Report 2015–2016. GNPD; GCREG; CDF; GC, Puerto Ayora, 13–19

Simpson GB, Jewitt GP (2019) The development of the water-energy-food nexus as a framework for achieving resource security: a review. Front Environ Sci 7:8. https://doi.org/10.3389/fenvs.2019.00008

Smith H, Coupé F, Garcia Ferrari S, Rivera H, Mera WEC (2020) Toward negotiated mitigation of landslide risks in informal settlements: reflections from a pilot experience in Medellín, Colombia. Ecol Soc 25(1):19. https://doi.org/10.5751/ES-11337-250119

Smith H, Garcia Ferrari S, Medero M, Rivera H, Coupé F, Mejía Escalante M, Castro Mera W, Montoya Correa C, Abiko A, Marinho F (2021) Exploring appropriate sociotechnical arrangements for the co-production of landslide risk management strategies in informal Neighbourhoods in Colombia and Brazil. Int J Urban Sustain Dev. https://doi.org/10.1080/19463138.2021.1872082

Stevenson F, Petrescu D (2016) Co-producing neighbourhood resilience. Build Res Inf 44(7):695–702

Tapia W, Ospina P, Quiroga D, Reck G, Gonzales JA, Montes C (2009) Toward a shared vision of Galapagos: the archipelago as a socioecological system. In Galapagos Report 2007–2008. GNPS; GCREG; CDF; GC, Puerto Ayora, 11–15

Visbeck M (2008) From climate assessment to climate services. Nat Geosci 1(1):2–3

Wamsler C (2017) Stakeholder involvement in strategic adaptation planning: transdisciplinarity and co-production at stake? Environ Sci Pol 75:148–157

Wicaksono A, Jeong G, Kang D (2017) Water, energy, and food nexus: review of global implementation and simulation model development. Water Policy 19:440–462. https://doi.org/10.2166/wp.2017.214

Ziervogel G, Archer van Garderen E, Price P (2016) Strengthening the knowledge–policy interface through co-production of a climate adaptation plan: leveraging opportunities in Bergrivier municipality, South Africa. *Environ Urban* 28(2):455–474

Maria Soledad Garcia Ferrari is a Senior Lecturer in the School of Architecture and Landscape Architecture, University of Edinburgh, UK. Her research focuses on current processes of urban development and regeneration in Latin America and Europe.

Amelia A. Bain is an interdisciplinary researcher affiliated with the School of Architecture and Landscape Architecture and the School of Geosciences at the University of Edinburgh, UK. Her work relates to physical processes related to volcanic hazards, as well as community-based risk and resource management, and integrated risk management.

Stephanie Crane De Narváez is a Research Fellow in the Edinburgh College of Art at the University of Edinburgh. She has research experience in sustainable urban development with a particular interest in community-based bottom-up processes, participation and social equality towards vulnerability reduction and resilience building in Latin America.

Chapter 32
The Extinction Market: Reflections on the Possible Future of the Illegal Galapagos Wildlife Trade

Evelyn Vega Barrera, Diego Quiroga Ferri, and Carlos F. Mena

Introduction

The unprecedented loss of biodiversity (Gibson et al. 2011) is the tangible result of a combination of animal trade, habitat loss, and the climate crisis (Milligan et al. 2009; Bakare et al. 2020; Raftowicz 2021) that is occurring in every region of the planet. Animal specimens from all the families are bought and sold as pets and trophies, meat, used as clothing, medicines, or for religious or cultural purposes (Li et al. 2020). Each year, billions of wild plants and animals are traded to meet a rapidly expanding global demand (Scheffers et al. 2019) and US$7 billion to $23 billion is gained annually from this illegal trade, making it one of the world's largest illegitimate businesses (Wyler and Sheikh 2008; Lehmacher 2016; Nellemann et al. 2016).

The global prohibition of the illegal wildlife trade has not prevented its growth in the last decades. The opportunity for large profits and low risk of detection fuels the frequent participation of organized crime in the wildlife black market. It is a complex problem that has proven difficult to restrain for countries with more financial resources as well as for those with insufficient economic means to address

E. Vega Barrera (✉)
Criminal Procedural Law, Universidad Complutense de Madrid, Madrid, Spain

ESCOEX (Social and Legal Sciences), Universidad del Atlántico Medio, Gran Canaria, Spain
e-mail: evelyn.vega@pdi.atlanticomedio.es

D. Quiroga Ferri
College of Social Sciences and Humanities, Universidad San Francisco de Quito, Quito, Ecuador

C. F. Mena
College of Biological & Environmental Studies, Universidad San Francisco de Quito, Quito, Ecuador

S. J. Walsh et al. (eds.), *Island Ecosystems*, Social and Ecological Interactions in the Galapagos Islands, https://doi.org/10.1007/978-3-031-28089-4_32

environmental challenges, and it is often aggravated by a lack of political will and effective public policies.

The illegal wildlife trade extends to every corner of the globe, including the iconic Galapagos Islands. Current governmental and individual actions to tackle these crimes fall short of providing real protection for the endemic and unique native species that exist in very low numbers on these islands.

History of Species Extraction

The accidental, or unexpected, voyage of discovery of the Galapagos Archipelago by the Spaniard Fray Tomás de Berlanga, Bishop of Panama, who was traveling to Peru to stop the fighting between the different conquistadors in 1535, was the event that began the social history of the islands.[1]

The Galapagos Islands were annexed by Ecuador in 1832 during Juan José Flores's government and named the Archipelago of Colón (Villacrés 1960; Ayala Mora 1995). The initial 1936 Galapagos National Park declaration was ratified in 1959[2] and extended to 97% of the land area due to its high ecological value, making it the first natural protected area in the country. The Galapagos Marine Reserve was later established through the Organic Law of Special Regime[3] in 1998 with an area of 133,000 km^2 that included the inland waters of the archipelago (50,100 km^2) and all those contained within 40 nautical miles measured from the outer coastal tip of the archipelago.[4]

At the beginning, the islands were frequented by pirates, whalers, and occasional fishermen (Ambrose Cowley 1699; Quiroga 2009; Hennessy 2019a). They searched refuge along their routes, where they stocked up on fresh water, huge land tortoises (*Chelonoidis sp.*) that could be easily hunted, dragged aboard and stored upside

[1] In the natural history of the Galapagos Islands, it is the endemism of the animals that inhabit it that has given it worldwide recognition. During the first period of human settlement on the islands, these animals were not considered in the same way as they are today without great scientific or tourist value. The meat of the Galapagos tortoise was once eaten with great relish, and the local animals were considered 'as much animals' as the introduced ones (CDF and WWF 2002; Ahassi 2007; Hennessy 2019b).

[2] Acuerdo Ministerial n° 31 (R.O. n° 189, May 14, 1936) y Decreto Ejecutivo n° 17, July 4, 1959 (R.O. n° 873, July 20, 1959), to commemorate the first centenary of the publication of the book "The Origin of Species" by Charles Darwin.

[3] Organic Law of the Special Regime for the Conservation and Sustainable Development of Galapagos (LOREG). R.O. n° 278, March 18, 1998.

[4] The islands were declared a UNESCO World Heritage Site in 1978 and to recognize the enormous ecological, cultural, and economic value of the GMR, it was included in 2001. In recognition of their importance, the islands were also designated by UNESCO as a Biosphere Reserve in 1984 and Ramsar Site in 2002, Whale Sanctuary in 1990, Sensitive Marine Area in 2005 by resolution of the IMO, and in 2016 the area between Darwin and Wolf Islands was established as a Marine Sanctuary, where fishing and any other extractive activities are completely prohibited.

down, where they lived for months with nothing to eat or drink, serving as a meat reserve (Hughes 1999), or fur seals (*Arctocephalus galapogoensis*) to extract their meat, skins, and oil (MAE and PNG 2014).

Charles Darwin's famous visit in 1835 on the Beagle and his prolonged 5 weeks exploring the Galapagos, led to the islands' global recognition as a site of scientific interest owing to the peculiarities of their unique populations (Hughes 1999).[5] Since then, the importance of endemic and native species, and the ecosystems to which they belong, have been acknowledged worldwide, as evident in the many declarations that they should be protected.

In the nineteenth century, colonists from Ecuador settled in the islands, first in Floreana and later in San Cristobal and Isabela. These colonists established haciendas and plantations and brought people from the mainland; many people were runaways that had problems with the law. The main industries were sugar cane, orchilla, and cattle. In the twentieth century other native and endemic species were exploited legally and illegally such as sharks, groupers, sea cucumbers, tuna, tortoises, iguanas, turtles, and trees.

There is a long history of extraction of endemic species in Galapagos. The pirates, whalers and the Ecuadorian colonists extracted tortoises, sea lions, whales, and other native and endemic species. Besides this, scientists also collected and took hundreds of thousands of specimens to museums, private and public collections. For example, the California Academy of Science (CAS) expedition of 1905–1906 removed more than 75,000 specimens from the islands, including 400 tortoises.[6] Wealthy patrons in North America and in Europe, such as Walter Rothchild, financed many of these expeditions with the main purpose of extracting live and dead animals.

The codfish fisheries legally extracted hundreds of tons of endemic and native demersal fish that were taken to the mainland. As the result of recent changes of the economy of the Galapagos, such as the collapse of the sea cucumber and lobster fisheries, there is now more interest in the illegal export of animals such as tortoises, sea horses, and land iguanas. The trade of illegal species has also become more attractive for the local population since the collapse of tourism, which had been the main source of income in Galapagos, due to the start of the COVID-19 pandemic in 2020.

[5] Darwin's most important contribution to the explanation of evolution was the scientific theory that populations evolve over the course of generations through a process known as natural selection (Darwin 1845, 1859).

[6] Rollo Beck made five voyages to the Galapagos, eventually leading the 1905–1906 California Academy of Sciences expedition. He would go on to become one of the most skilled ornithological collectors of his generation (Van Denburgh 1907, 1914; Pitelka 1986; James 2017).

Current Situation

Although it is possible that introduced species now constitute the greatest threat to much of the flora and fauna of the Galapagos, illegal and legal extraction of terrestrial and marine fauna remains nonetheless very significant as well. There are also several reports of local people extracting or killing and eating endemic animals illegally, including tortoises, sea lions, and marine invertebrates.

A recent source of illegal extraction of animal parts has to do with shark finning. Foreign vessels come every year to fish around Galapagos, they stay in international waters just outside a maritime border around the islands, being a major threat to highly migratory endangered species, such as sharks (Hearn and Bucaram 2017). In 2020, about 243 Chinese flag vessels (including companies with suspected records of IUU fishing[7]) were present and the same situation occurred on August 2021, with another 290 Chinese vessels southwest of the Ecuadorian Insular EEZ.

In 2017, Fu Yuan Yu Leng 999 vessel, was found within Galapagos Marine Reserve (RMG) with 300 tons of whitefish and 6623 sharks, included hammerhead, silky, bigeye thresher, pelagic thresher and mako sharks (Bonaccorso et al. 2021).[8] The expert opinion of a marine biologist, as part of a judicial process, valued the damages at US$36 billion. Twenty crew members were convicted to 1–3 years in prison, plus individual fines according to their degree of participation, the vessel was confiscated, and an order was issued to pay US$ 6,137,753.42 as reparation for the damages caused.

In this context, it is important to highlight that the market for shark fins is extensive and they are highly sought in parts of Asia, where shark fin soup can cost between US$ 100 and US$ 200, Hong Kong paying the highest price for fins at US$30/kg. The global trade in shark and ray meat is worth US$ 2.6 billion, exceeding the value of shark fins, about US$1.5 billion (Niedermüller et al. 2021).

There is also a legal and regulated as well as illegal fishing of sea cucumbers (*isostichopus fuscus*), and there are cases of possession of sea cucumbers in the process of being exported to markets in China and Hong Kong. In 2015, 10,852 dry-salted sea cucumbers were found ready to be shipped by air cargo, containing species in danger of extinction and at that time, the fisheries of these species were banned. In Galapagos, they were valued between US$40 and US$70 per kilo and can reach a price between US$78,000 and US$132,000 in the international market. In 2017, in a similar case, people were captured trying to smuggle 1934 sea cucumbers (1724 *horrens* and 210 *fuscus*) and criminal charges were made.

In 2015, a Mexican citizen was arrested in Santa Cruz, with 9 hatchlings of marine iguanas (*amblyrhynchus cristatus*) and 2 juvenile land iguanas (*conolophus subcristatus*) that he was trying to take out of the country. He was convicted,

---

[7] According to research by C4ADS, a data analysis NGO.

[8] 12 species listed as vulnerable or most at risk by the IUCN and 8 of these species are protected by CITES.

imprisoned for 2 years, and fined US$ 20,000 as reparation for the damage caused to biodiversity and native species, in favor of the Galapagos National Park.

As mentioned above, the iconic giant tortoises of the Galapagos have been extracted from the islands since the eighteenth century. In the last 4 years, there have been several cases that deserve special mention. In 2018, 123 tortoise hatchlings, bred in captivity at a breeding center on Isabela, disappeared. The expert opinion presented as part of the legal process highlighted the relevance of forensic DNA evidence to combat the illegal wildlife trade (Smart et al. 2021). Among the evidence in the case, the genetic analysis of the turtles conducted by Yale University (USA), determined that 16 belonged to the Isabela breeding center and 8 to the San Cristóbal breeding center. However, there is still no criminal conviction as the case is still in the trial phase. The expert evidence constitutes an indispensable proof and *sine qua non* condition to determine the existence of the crime against the Galapagos fauna, although one must recognize the difficulties in the elaboration of an expert opinion in environmental crimes, the analysis of the damages produced and their valuation (Vega Barrera 2020).

In March 2021, 185 baby chelonians were found packed with plastic inside a suitcase at the Baltra Airport, the goal of the smugglers having been to fly them to Guayaquil. A giant Galapagos tortoise could be sold for up to US$60,000 on the international black market. According to Pacífico Libre, a young tortoise can cost between US$5000 and US$7000 and adults can cost up to US$60,000 each. The seizure in this case, would reach a total of US$1,295,000. For this crime against wild flora and fauna, one person was convicted, according to Criminal Code Article 247 to 3 years in prison and must pay US$639,100 as full reparation for the damages caused. He was released after serving 8 months in prison because of the low risk of absconding and the problem of overcrowding and violence in Ecuadorian prisons in 2021.

On June 25, 2022, the Ecuadorian Navy found protected species coming from the Archipelago, in a tourist boat, that was being towed from Floreana Island, by a barge, to Guayaquil. Six people were arrested, and 84 giant tortoises and 5 yellow land iguanas were seized. The case is under evaluation according to the competent authorities. The species were transported in 10 jute bags, five golden iguanas, one in each bag and in the rest of the bags, tortoises of different sizes were found. The Ecuadorian Navy in a press release announced that "Transnational organizations dedicated to committing environmental crimes were involved." The GNPD confirmed that the numbers of tortoises recorded in the monitoring of the breeding centers of the three islands are correct and there are no tortoises missing, so it is presumed that the species found on the boat were taken from their natural habitat.[9]

Legalization of certain illegal activities could be argued as an alternative, for example, in the case of drugs, but it will not work in the case of the illicit trade of animals. There are three different economic considerations that would be influenced by legalizing animal trade: "some demand will be satisfied, a potential increase in

[9] Press release from the Ministry of Environment, Water and Ecological Transition, June 25, 2022.

the consumer base, and reduced cost of smuggling." The illegal wildlife market is more complicated compared to the black market for drugs (Nefedova 2019), in part, because the availability of rare and often fragile animals is limited and in many cases can lead to the extinction of the species.

However, what could work in some cases is the creation of captive breeding centers to sell animals. In the case of rhinoceroses in Africa, the creation of ranches has been proposed so they can be grown for the horn market. Captive breeding centers already exist in the case of birds, some reptiles, amphibians and with certain marine species such as seahorses. If this can be done affordably, such centers could present an alternative to the continuance of the illegal market. Furthermore, the money collected from the selling of animals, or their parts, raised in captivity can be used to protect those species in the wild.

Future Prospects in Galapagos

As in many other geographic areas rich in biodiversity and endemism, there is a strong connection between illegal wildlife trafficking and different types of organized crime (Maher and Sollund 2016). Moreover, in Latin America the connection between wildlife trafficking and drugs trade is diverse, strong, and complex (Van Uhm et al. 2021). Assuming that the Galapagos is one steppingstone within one of the routes of the drug trade from mainland South America to Central America (Santos 2016), the main concern is how the drug trade's logistical and funding networks will drive illegal wildlife trade in Galapagos. So far, there are no systematic studies that shed light onto the synergies between the two illegal activities at the local level.

The escalation of wildlife trafficking is largely influenced by market forces, offenders being motivated by the potential of large economic gain, and the expansion of consumer markets (Maher and Sollund 2016). In Galapagos, these driving forces, confounded with the porous controls of law enforcement, especially in the formal and informal ports, will drive illegal wildlife trade, in particular, reptiles and fisheries. Law enforcement, including that of the Galapagos National Park and the Ecuadorian Armed Forces, must be prepared to face an increase in cases linked to aforementioned factors.

In South America, NGOs have been an integral part of nature conservation, including within the fight against illegal wildlife trade. In Galapagos, except for illegal fishing, conservation NGOs, in contrast, have treated the problem as an isolated issue. The role of NGOs in the fight against wildlife trafficking is crucial in Galapagos, not only to serve as a part of practical enforcement schemes, but also acting as policy advisors, long term monitors, field researchers, expert witnesses at court, casework managers, etc. (Nurse 2016).

The recent case of Galapagos land tortoise hunting on Isabela Island could indicate a strong link between the "fortress" model of park conservation, the lack of the spillover of benefits from species conservation to local people, and the cultural "load" and the lack of education for local inhabitants about the potential benefits of species conservation. It should be point out that the cases about tortoise hunting are under a *previous investigation* stage by the Office of the Prosecutor and has not been judicialized at court to date. It is also important to underline that every person, related or not with the investigation, has the right to the presumption of innocence in any circumstances, until a competent judge, within due process, declares the criminal liability and the conviction.

As a final remark, engaging the local community of Galapagos – including fishers, farmers, and other groups that do not benefit directly from ecotourism – will be key in finding the solutions in the uphill battle against illegal wildlife trafficking. Community-based approaches will not be sufficient, but it is a necessary step to achieve a stronger national coordination of actions between law enforcement, administrative and judicial authorities, conservation NGOs, science sector, and the local community.

References

Ahassi C (2007) Lo galapagueño, los galapagueños: proceso de construcción de identidades en las Islas Galápagos. Revista de Antropología Experimental 7:169–176

Ambrose Cowley W (1699) Cowley's voyage round the globe. In: Hacke W (ed) William Hacke's collection of original voyages. James Knapton, London

Ayala Mora E (1995) Nueva Historia del Ecuador. Documentos de la Historia del Ecuador. Corporación Editora Nacional

Bakare AG et al (2020) Impact of climate change on sustainable livestock production and existence of wildlife and marine species in the South Pacific Island countries: a review. Int J Biometeorol 64(8):1409–1421

Bonaccorso E et al (2021) International fisheries threaten globally endangered sharks in the eastern tropical Pacific Ocean: the case of the Fu Yuan Yu Leng 999 reefer vessel seized within the Galápagos marine reserve. Sci Rep 11. https://doi.org/10.1038/s41598-021-94126-3

CDF and WWF (2002) A biodiversity vision for the Galapagos Islands. Puerto Ayora

Darwin C (1845) Journal of researches into the natural history and geology of the countries visited during the voyage of H.M.S. beagle round the world, under the command of Capt. Fitz Roy, R.A, 2nd edn. John Murray, London

Darwin C (1859) On the origin of species by means of natural selection, or the preservation of favoured races in the struggle for life. John Murray, London

Gibson L et al (2011) Primary forests are irreplaceable for sustaining tropical biodiversity. Nature 478(7369):378–381. https://doi.org/10.1038/nature10425

Hearn AR, Bucaram SJ (2017) Ecuador's sharks face threats from within. Science 358(6366):1009

Hennessy E (2019a) In Darwin's footsteps. In which the tortoises are soup, and souvenirs. In: On the backs of tortoises: Darwin, the Galapagos, and the fate of an evolutionary Eden. Yale University Press, pp 17–42

Hennessy E (2019b) Restoring evolution. In: On the backs of tortoises: Darwin, the Galapagos, and the fate of an evolutionary Eden. Yale University Press, pp 147–179

Hughes JD (1999) Darwin in the Galápagos. Capital Nat Social 10(2):107–114. https://doi.org/10.1080/10455759909358860

James MJ (2017) Collecting evolution: the Galapagos expedition that vindicated Darwin. Oxford University Press, Oxford

Lehmacher W (2016) Wildlife crime: a $23 billion trade that's destroying our planet. In World economic forum
Li J et al (2020) Game consumption and the 2019 novel coronavirus. Lancet Infect Dis 20(3):275–276. https://doi.org/10.1016/S1473-3099(20)30063-3
MAE and PNG (2014) Plan de Manejo de las Áreas Protegidas de Galápagos para el Buen Vivir. Puerto Ayora
Maher J, Sollund R (2016) Wildlife trafficking: harms and victimization. In: Sollund R, Stefes CH, Germani AR (eds) Fighting environmental crime in Europe and beyond: the role of the EU and its member states. Palgrave Macmillan, London, pp 99–128
Milligan SR, Holt WV, Lloyd R (2009) Impacts of climate change and environmental factors on reproduction and development in wildlife. Philos Trans R Soc B Biol Sci 364(1534):3313–3319
Nefedova Y (2019) Illicit market of animals. In: Morrone M et al (eds) Perspectives on black markets, vol 3. Indiana University
Nellemann C et al (2016) The rise of environmental crime – a growing threat to natural resources peace, development and security. UNEP, INTERPOL, Norwegian Center for Global Analyses
Niedermüller S et al (2021) The shark and ray meat network. A deep dive into a global affair. WWF
Nurse A (2016) A global movement: NGOs and the policing of international wildlife trafficking. J Traff Organ Crime Securit 2(1):50–61
Pitelka FA (1986) Rollo Beck: old-school collector, member of an endangered species. American Birds 40(3):385–387
Quiroga D (2009) Crafting nature: the Galapagos and the making and unmaking of a 'natural laboratory. J Polit Ecol 16(1):123–140. https://doi.org/10.2458/v16i1.21695
Raftowicz M (2021) The climate crisis as a product of globalization. In SHS web of conferences. EDP Sciences, p 6029. doi: https://doi.org/10.1051/shsconf/20219206029
Santos CP (2016) Disrupting cocaine trafficking networks: interdicting a combined social-functional network model. Naval Postgraduate School Monterey CA Monterey
Scheffers BR et al (2019) Global wildlife trade across the tree of life. Science 366(6461):71–76
Smart U, Cihlar JC, Budowle B (2021) International wildlife trafficking: a perspective on the challenges and potential forensic genetics solutions. Forensic Sci Int Genet 54:102551
Van Denburgh J (1907) Expedition of the California Academy of Sciences to the Galápagos Islands, 1905–1906. I. Preliminary descriptions of four new races of gigantic land tortoises from the Galapagos Islands. In Proceedings of the California Academy of Sciences 4th series, pp 1–6
Van Denburgh J (1914) Expedition of the California Academy of Sciences to the Galapagos Islands, 1905–1906. X. The Gigantic Land Tortoises of the Galapagos Archipelago. In Proceedings of the California Academy of Sciences 4th series, vol. 2, pt. 1. The Academy, San Francisco, pp 203–374
Van Uhm D, South N, Wyatt T (2021) Connections between trades and trafficking in wildlife and drugs. Trends Organ Crime 24(4):425–446
Vega Barrera E (2020) La prueba pericial en delitos medioambientales en lugares de especial protección: las Islas Galápagos. In: Picó i Junoy J, Andino López JA, Cerrato Guri E (eds) La prueba pericial a examen. Propuestas de "lege ferenda.". JM Bosch, Barcelona, pp 677–688
Villacrés MJW (1960) El Archipiélago de Colón (Galápagos), Centro Científico Internacional. Rev Geogr 26(52):63–76
Wyler LS, Sheikh PA (2008) International illegal trade in wildlife: threats and US policy. In: Congressional research service. Library of Congress

Evelyn Vega Barrera Lawyer, earning her law degree from Complutense University of Madrid, Master in Private Law from Madrid Bar Association. Currently, she is a PhD candidate within the Faculty of Law at Complutense University of Madrid and Professor in ESCOEX, Universidad del Atlántico Medio, Gran Canaria, Spain. She has done research stays at the Galapagos Science Center and at the Max Planck Institute for Foreign and International Criminal Law in Freiburg, Germany. Evelyn studies the problems in investigating and prosecuting environmental crimes, focusing on the Galapagos Islands as a place of particular global interest.

Diego Quiroga Ferri Diego has led the research and international programs at USFQ for over 10 years. His vision has led USFQ to stand out in the rankings as the No. 1 University in Internationalization in Latin America (Times Higher Education). His most recent position was Dean of Research and International Affairs as well as Co-Director of the Galapagos Campus in San Cristóbal, Galapagos. Diego obtained his Ph.D. in Anthropology of University of Illinois at Urbana, Champaign. His research focus is Urban Ecology, Political Ecology and Medical Anthropology in the Galapagos, Amazonian region and the Andes. Diego has been working constantly in biodiversity and vulnerability research as well as traditional medicine. Diego believes that the role of universities is to be a part of the innovation ecosystem and that it should be tied to the community needs to develop public policy and support the private sector. His work philosophy is based on Liberal Arts, which promotes interdisciplinary groups of work where each person contributes with his area of expertise to go forward in research.

Carlos F. Mena, Co-Guest Editor, Co-Director of the Galapagos Science Center, Director of the Institute of Geography, and Professor in the College of Biological and Environmental Sciences, Universidad San Francisco de Quito, Ecuador. His research assesses the importance of land use/land cover change on social and terrestrial systems in the Galapagos Islands as well as in the Ecuadorian Amazon.

Index

S. J. Walsh et al. (eds.), *Island Ecosystems*, Social and Ecological Interactions in the Galapagos Islands, https://doi.org/10.1007/978-3-031-28089-4

CPSIA information can be obtained
at www.ICGtesting.com
Printed in the USA
LVHW051316140523
746955LV00001B/1

9 783031 280887